Lecture Notes in Computer Science   10502

Commenced Publication in 1973
Founding and Former Series Editors:
Gerhard Goos, Juris Hartmanis, and Jan van Leeuwen

More information about this series at http://www.springer.com/series/7412

Walter G. Kropatsch · Nicole M. Artner
Ines Janusch (Eds.)

Discrete Geometry for Computer Imagery

20th IAPR International Conference, DGCI 2017
Vienna, Austria, September 19–21, 2017
Proceedings

 Springer

Editors
Walter G. Kropatsch
TU Wien
Vienna
Austria

Ines Janusch
TU Wien
Vienna
Austria

Nicole M. Artner
TU Wien
Vienna
Austria

ISSN 0302-9743   ISSN 1611-3349   (electronic)
Lecture Notes in Computer Science
ISBN 978-3-319-66271-8   ISBN 978-3-319-66272-5   (eBook)
DOI 10.1007/978-3-319-66272-5

Library of Congress Control Number: 2017950082

LNCS Sublibrary: SL6 – Image Processing, Computer Vision, Pattern Recognition, and Graphics

Printed on acid-free paper

This Springer imprint is published by Springer Nature
The registered company is Springer International Publishing AG
The registered company address is: Gewerbestrasse 11, 6330 Cham, Switzerland

Preface

Vienna was the location of the 20th international conference on Discrete Geometry for Computer Imagery (DGCI). It followed the tradition of previous events, which have been held alternately in France and outside France at an approximate interval of 18 months. DGCI 2017 attracted a number of research contributions from academic and research institutions in our field. A total of 36 papers were submitted from all over the world. Papers were first assigned to one of the main topics of the conference, each one led by two or three members of the Program Committee (PC). Then, the members of the PC selected at least three reviewers for each paper. After all the reviews had been collected by the PC members, they generated a meta-review based on which the authors could explain potential misunderstandings and propose some corrections for the final version. Following this "rebuttal" phase, 28 papers were accepted by the program chairs and their authors were invited to present their work at DGCI 2017.

The program consisted of 18 oral and 10 poster presentations. In addition, three internationally well-known speakers were invited for keynote lectures: Helmut Pottmann (TU Wien, Austria), Michael Wilkinson (Univ. of Groningen, The Netherlands), and Eric Andres (Univ. Poitiers, France). We are happy that all invited speakers were willing to also contribute a written article to the proceedings. These proceedings also follow the tradition of DGCI and appear in the LNCS series of Springer, a publisher celebrating its 175th birthday this year.

We are also very thankful to Springer for sponsoring a best student paper award for DGCI 2017. Furthermore, DGCI 2017 received the sponsorship of the International Association of Pattern Recognition (IAPR). It is planned to produce a special issue of the Journal of Mathematical Imaging and Vision from extended versions of excellent contributions presented at DGCI 2017.

We would like to thank all contributors, the invited speakers, all PC members and reviewers, the members of the Steering Committees, and all those who made this conference happen. Furthermore, we are grateful to our financial support institutions. Last but not least, we thank all the participants and hope that everyone found great interest in the DGCI 2017 program and also had a very good time in our city of Vienna.

July 2017

Walter G. Kropatsch
Nicole M. Artner
Ines Janusch

Preface

Organization

Program Chairs

Walter G. Kropatsch	TU Wien, Austria
Nicole M. Artner	TU Wien, Austria
Ines Janusch	TU Wien, Austria
David Coeurjolly	LIRIS, France

Program Committee

Joost K. Batenburg	CWI, Amsterdam, The Netherlands
Isabelle Bloch	LTCI, Télécom ParisTech, Paris, France
Srecko Brlek	Université du Québec à Montréal, Canada
Sara Brunetti	University of Siena, Italy
Michel Couprie	LIGM, University of Paris-Est, France
Guillaume Damiand	LIRIS, CNRS, Lyon, France
Yan Gerard	ISIT, Auvergne University, France
Rocio Gonzalez-Diaz	University of Seville, Spain
Yukiko Kenmochi	LIGM, University of Paris-Est, France
Bertrand Kerautret	Loria, University of Lorraine, France
Christer Kiselman	Uppsala University, Sweden
Jacques-Olivier Lachaud	LAMA, University Savoie Mont Blanc, France
Pawel Pilarczyk	IST Austria
Xavier Provençal	LAMA, University Savoie Mont Blanc, France
Gabriella Sanniti di Baja	ICAR, CNR, Italy
Robin Strand	Centre for Image Analysis, Uppsala, Sweden
Imants Svalbe	Monash University, Melbourne, Australia

Reviewers

Andreas Alpers	Arindam Biswas	Fabien Feschet
Frosini Andrea	Gunilla Borgefors	Largeteau-Skapin Gaëlle
Vialard Anne	Nicolas Boutry	Yan Gerard
Teo Asplund	Shekhar Chandra	Aldo Gonzalez-Lorenzo
Peter Balazs	Isabelle	Jeanpierre Guedon
Fabien Baldacci	Debled-Rennesson	Lajos Hajdu
Reneta Barneva	Eric Domenjoud	Cris L. Luengo Hendriks
Etienne Baudrier	Paolo Dulio	Damien Jamet
Partha Bhowmick	Henri-Alex Esbelin	Maria Jose Jimenez
Silvia Biasotti	Jean-Marie Favreau	Andrew Kingston

Ullrich Köthe
Jean-Philippe Labbé
Claudia Landi
Pascal Lienhardt
Joakim Lindblad
Filip Malmberg
Loïc Mazo
Christian Mercat
Benoît Naegel
Benedek Nagy
Laurent Najman
Phuc Ngo
Nicolas Normand

Laszlo Nyul
Silvia Pagani
Nicolas Passat
Samuel Peltier
Christophe Picouleau
Daniel Prusa
Eric Remy
Ana Romero
Tristan Roussillon
Apurba Sarkar
Isabelle Sivignon
Natasa Sladoje
Robin Strand

Akihiro Sugimoto
Hugues Talbot
Oriol Ramos Terrades
Edouard Thiel
Darren Thompson
Laure Tougne
Antoine Vacavant
Jonathan Weber
Günter M. Ziegler
Rita Zrour
Henri der Sarkissian

Contents

Discrete Modelling and Visualization

Morphological Analysis

Discrete Shape Representation, Recognition and Analysis

Discrete and Combinatorial Topology

Discrete Models and Tools

Models for Discrete Geometry

Invited Talks

Freeform Architecture and Discrete Differential Geometry

Helmut Pottmann[1]([⊠]) and Johannes Wallner[2]

[1] Center for Geometry and Computational Design, TU Wien,
Wiedner Hauptstraße 8–10, 1040 Wien, Austria
pottmann@geometrie.tuwien.ac.at
[2] Institut für Geometrie, TU Graz, Kopernikusgasse 24, 8010 Graz, Austria

Abstract. Freeform structures play an important role within contemporary architecture. While there is a wealth of excellent tools for the digital design of free-form geometry, the actual fabrication on the architectural scale is a big challenge. Key issues in this context are free-form surfaces composed of panels which can be manufactured at reasonable cost, and the geometry and statics of the support structure. The present article is an extended abstract of a talk on the close relation between geometric computing for free-form architecture and discrete differential geometry. It addresses topics such as skins from planar, in particular quadrilateral panels, geometry and statics of supporting structures, structures in force equilibrium.

1 Introduction

The mathematical and computational challenges posed by free-form shapes in architecture are twofold. One is *rationalization* which means approximating a given design surface by a collection of smaller parts which can be individually manufactured and put together. There is a great variety of constraints imposed on the individual parts, most having to do with manufacturing. The second challenge is *design* of free forms. The goal here is to develop tools which allow the user to interactively design free forms, such that key aspects of statics and fabrication are taken into account directly in the design phase. Meanwhile there is a wealth of results on these topics, and we want to point to the survey article [13].

2 Freeform Skins from Planar Panels and Associated Support Structures

Steel-glass constructions usually require a decomposition of freeform skins into *flat* panels, which leads us to the question of rationalization with polyhedral surfaces, and designing with polyhedral surfaces. The combinatorics of meshes plays an important role here: It is very easy to represent a given shape by a triangle mesh, and in fact the majority of freeform skins which exist are based on triangle meshes. However there are drawbacks: On average 6 edges meet in a vertex,

© Springer International Publishing AG 2017
W.G. Kropatsch et al. (Eds.): DGCI 2017, LNCS 10502, pp. 3–8, 2017.
DOI: 10.1007/978-3-319-66272-5_1

so structures based on triangle meshes have complicated nodes. Further, they tend to be heavier than structures based on quad meshes. This has led to a new line of research into PQ (planar quad) meshes, which are meshes whose faces are planar quadrilaterals. Here an important link to discrete differential geometry is established: Combinatorially regular quad meshes decompose into two families of mesh polylines which can be seen as discrete versions of the parameter lines on a smooth surface. A quad mesh is then interpreted as a discrete version of a parametric surface. Properties of quad meshes relevant to architecture turn out to be equivalent to properties relevant in discrete differential geometry (in particular, the integrable systems viewpoint of discrete differential geometry, see [3]).

This connection between smooth surfaces and discrete surfaces is very important in investigating the degrees of freedom available for rationalization and design: E.g. a PQ mesh constitutes a discrete version of a so-called conjugate network of curves [9]. Meshes where the edge polylines appear smooth will need to approximate a conjugate network of curves. The conjugate networks are known and in theory there are many, but we nevertheless can draw the conclusion that in connection with practical considerations (e.g. angles between edges) there might be little flexibility or even no satisfactory network at all which serves as guidance for a PQ mesh (see Fig. 1).

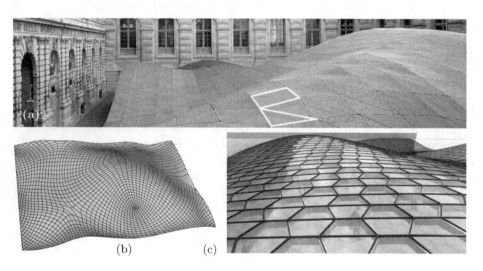

(a)

(b)   (c)

Fig. 1. Differential geometry informing rationalization. (a) The *Cour Visconti* roof in the Louvre (image courtesy Waagner-Biro Stahlbau). It was intended to be built in a lightweight way, possibly as a quad mesh. (b) A rationalization of this surface as a quad mesh with planar faces and smooth edge polylines must follow a conjugate network of curves, but these networks have unacceptable singularities. (c) If zigzags are allowed, rationalization as a PQ mesh with regular combinatorics is possible. For the actual roof, however, a different solution with both triangular and quadrilateral faces was found.

An important special case are *nearly rectangular panels*. Aside from aesthetics there are fundamental geometric reasons for constraining a PQ mesh to some form of orthogonality between edges. As it turns out, such meshes are discrete versions of principal curve networks, and the known nice behaviour of the surface normals along the curves of such a network translates to good properties of the *support structure* associated with the quad mesh, see Fig. 2. One is able to design so-called torsion-free nodes [9,11]. Research in this direction also led to progress in discrete differential geometry itself, in particular a new curvature theory for discrete surfaces [4]. Direct design of torsion-free support structures with quad combinatorics is related to special parametrizations of *congruences*. This word refers to a 2-dimensional system of straight lines and constitutes a classical topic of differential geometry. Its discrete incarnation turned out to be quite useful and has been explored systematically. We used it in connection with shading and guiding light by reflection, see [17] and Fig. 3.

Fig. 2. Torsion-free support structures. The Chadstone shopping mall in Melbourne features a steel-glass roof in the shape of a planar quad mesh. The member corresponding to an edge is aligned along the *support plane* (yellow) of that edge, and the intersection of members in a node is defined by the *node axis* (red) where support planes meet. This behaviour of node axes is analogous to the behaviour of surface normals along principal curvature lines (original photo: T. Burgess, imageplay).

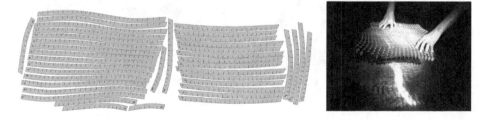

Fig. 3. Torsion-free support structures for shading and lighting. By cutting out and assembling the strips shown above one creates a torsion-free support structure capable of reflecting light into prescribed patterns. This arrangement of planes and lines discretizes the notion of *torsal parametrization of a line congruence*. The strips correspond to the two families of developable surfaces which make up the congruence (the system of normals of a surface along principal curvature lines is a special instance of this).

The previous paragraphs did not give an exhaustive list of the correspondences between discrete surfaces and smooth parametric surfaces which have already been used in the context of freeform architecture. In particular we did not mention *semidiscrete* surfaces relevant to structures with bent glass [12].

3   Structures in Static Equilibrium

Statics obviously is of paramount importance in architecture and building construction. It is therefore important that aspects of statics play a role already in the first stages of design. It is a long-term goal to create design tools which incorporate constraints relating to geometry, fabrication and statics while being still fast enough to allow interactive modeling. We are currently far from this goal, but partial results have been achieved. We start our summary by mentioning the *thrust network method* [1,2]: Maxwell's ideas on graphical statics are the basis of a method to treat systems of equilibrium forces which act in surface-like geometries. By separating vertical and horizontal components one is led to a discrete *Airy potential* polyhedron, which is a finite element discretization of the Airy stress potential well known in 2D elasticity theory. Compressive stresses are characterized by convexity of the stress potential.

A particularly nice application of this method is self-supporting masonry which is stable even without mortar, see Fig. 4. It is possible to interpret forces resp. stresses in differential-geometric terms, and we refer to [14,16] for this "geometrization" of the force balance condition, and for a treatment of the so-called isotropic differential geometry which occurs here. The direct interactive design of meshes (in particular polyhedral surfaces) with additional force balance conditions is a special case of constrained geometric modeling, see [15].

Recently we have worked on material-minimizing structures, see Fig. 4. This optimization problem was originally proposed in a groundbreaking paper by

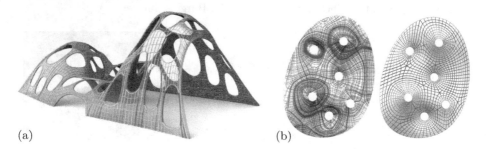

(a)   (b)

Fig. 4. Self-supporting and weight-optimal structures. (a) This masonry vault with holes contains a network of compressive forces which is in equilibrium with the deadload, implying the remarkable fact of stability of the structure when built of bricks even without mortar. Interactive design of such self-supporting surfaces is possible [15,16]. (b) The search for quad meshes with planar faces and minimal weight in the sense of M.G.M. Michell's *limit of economy* is converted into computing a variant of principal curves, by a suitable differential-geometric interpretation [8].

M.G.M. Michell [10] and is meanwhile formulated in modern language [18]. Our work, like others mentioned in this paper, is based on a differential-geometric interpretation of the subject of interest which, in this case, is the volume of members of a structure based on a mesh, and also the forces acting in these members. For example, 2-dimensional optimal trusses are characterized by Airy potential surfaces of minimal total curvature in the sense of isotropic geometry. This topic and its extension to shells is treated by [8] (Fig. 4).

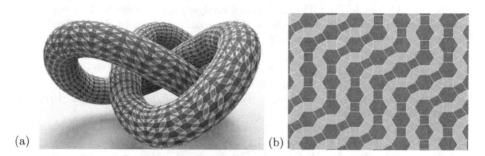

(a)   (b)

Fig. 5. Nonstandard notions of fairness. Here a given smooth surface is approximated by a polyhedral surface (a) of prescribed local combinatorics (b). The concept of fairness employed in the computation is based on existence of local approximate symmetries.

4   On Fairness, the Importance of Regularizers and Structures Beyond Discrete Differential Geometry

In all examples mentioned above, *fairness* plays an important role in identifying those discrete structures which meaningfully correspond to smooth objects. On a technical level, fairness functionals are used as regularizers in optimization and in iterative constraint solvers. There are, however, many different ways to express fairness computationally. The standard quadratic fairness energies composed of iterated differences might not be appropriate for meshes like the one shown by Fig. 1c. The zigzag polylines might be fully intentional, but they cause high (bad) values of such a fairness energy. Recently, alternative approaches to fairness have been successfully employed in creating polyhedral patterns [6]. They are based on existence of *local approximate symmetries*. An interpretation in terms of standard concepts of discrete differential geometry is still open. A difficult topic in general are fairness functionals of high nonlinearity, e.g. those involving kink angles. A fairness measure based on angles only [7] has led to a new concept of *smoothness of discrete surfaces* [5]. Recently we have investigated a functional defined as the sum of edge lengths times absolute value of kink angles. Its "isotropic" version surprisingly turns up in connection with material minimization (see previous paragraph). The shape of minimizers is a topic of current research; we conjecture that at least in negatively curved areas they are principal meshes.

Acknowledgments. This research has been supported by the Austrian Science Fund (FWF) under grant no. I-2978, within the framework of the SFB-Transregio Programme *Discretization in Geometry and Dynamics.*

References

1. Adriaenssens, S., Block, P., Veenendaal, D., Williams, C. (eds.): Shell Structures for Architecture. Taylor & Francis, Routledge (2014)
2. Block, P., Ochsendorf, J.: Thrust network analysis: a new methodology for three-dimensional equilibrium. J. Int. Assoc. Shell Spatial Struct. **48**, 167–173 (2007)
3. Bobenko, A., Suris, Y.: Discrete differential geometry: Integrable structure. American Math. Soc. (2009)
4. Bobenko, A., Pottmann, H., Wallner, J.: A curvature theory for discrete surfaces based on mesh parallelity. Math. Annalen **348**, 1–24 (2010)
5. Günther, F., Jiang, C., Pottmann, H.: Smooth polyhedral surfaces. Preprint (arXiv:1703.05318)
6. Jiang, C., Tang, C., Vaxman, A., Wonka, P., Pottmann, H.: Polyhedral patterns. ACM Trans. Graph. **34**(6), article 172 (2015)
7. Jiang, C., Günther, F., Wallner, J., Pottmann, H.: Measuring and controlling fairness of triangulations. In: Advances in Architectural Geometry 2016, VDF Hochschulverlag, ETH Zürich, 2016, pp. 24–39 (2016)
8. Kilian, M., Pellis, D., Wallner, J., Pottmann, H.: Material-minimizing forms and structures (2017, submitted for publication)
9. Liu, Y., Pottmann, H., Wallner, J., Yang, Y.-L., Wang, W.: Geometric modeling with conical meshes and developable surfaces. ACM Trans. Graph. **25**(3), 681–689 (2006)
10. Michell, A.G.M.: The limit of economy of material in frame-structures. Phil. Mag., Ser. VI **8**, 589–597 (1904)
11. Pottmann, H., Liu, Y., Wallner, J., Bobenko, A., Wang, W.: Geometry of multilayer freeform structures for architecture. ACM Trans. Graph. **26**(3), article 65 (2007)
12. Pottmann, H., Schiftner, A., Bo, P., Schmiedhofer, H., Wang, W., Baldassini, N., Wallner, J.: Freeform surfaces from single curved panels. ACM Trans. Graph. **27**(3), article 76 (2008)
13. Pottmann, H., Eigensatz, M., Vaxman, A., Wallner, J.: Architectural geometry. Comput. Graph. **47**, 145–164 (2015)
14. Strubecker, K.: Airy'sche Spannungsfunktion und isotrope Differentialgeometrie. Math. Zeitschrift **78**, 189–198 (1962)
15. Tang, C., Sun, X., Gomes, A., Wallner, J., Pottmann, H.: Form-finding with polyhedral meshes made simple. ACM Trans. Graph. **33**(4), article 70 (2014)
16. Vouga, E., Höbinger, M., Wallner, J., Pottmann, H.: Design of self-supporting surfaces. ACM Trans. Graph. **31**(4), article 87 (2012)
17. Wang, J., Jiang, C., Bompas, P., Wallner, J., Pottmann, H.: Discrete line congruences for shading and lighting. Comput. Graph. Forum **32**, 53–62 (2013)
18. Whittle, P.: Networks - Optimisation and Evolution. Cambridge University Press, Cambridge (2007)

A Guided Tour of Connective Morphology: Concepts, Algorithms, and Applications

Michael H.F. Wilkinson[✉]

Johann Bernoulli Institute for Mathematics and Computer Science,
University of Groningen, P.O. Box 407, 9700 AK Groningen, The Netherlands
m.h.f.wilkinson@rug.nl

Abstract. Connective morphology has been an active area of research for more than two decades. Based on an abstract notion of connectivity, it allows development of perceptual grouping of pixels using different connectivity classes. Images are processed based on these perceptual groups, rather than some rigid neighbourhood imposed upon the image in the form of a fixed structuring element. The progress in this field has been threefold: (i) development of a mathematical framework; (ii) development of fast algorithms, and (iii) application of the methodology in very diverse fields. In this talk I will review these developments, and describe relationships to other image-adaptive methods. I will also discuss the opportunities for use in multi-scale analysis and inclusion of machine learning within connected filters.

Keywords: Connectivity · Connected filters · Mathematical morphology · Algorithms

1 Introduction

Connected filters, and more generally connective morphology [1,2] are relative newcomers in the field of mathematical morphology. Rather than being based on adjunctions of erosions and dilations by structuring elements, they are based on a generalised notion of connectivity, and connectivity openings form the building blocks of operators. The key property of these image operators is that they do not consider the image as a collection of pixels, but more as a collection of connected structures, referred to as (quasi) flat-zones or connected components. They work by merging these connected structures and assigning new grey-levels or colours to them, based on certain criteria. This property means that they are strictly edge preserving: they cannot introduce new edges in an image. Their use ranges from simple image filtering, through detection and enhancement of specific structures to hierarchical segmentation.

As so often happens, the development of these filters originally stemmed from practical needs, and only later a rigorous mathematical framework was set up. Likewise, the initial algorithms were comparatively inefficient, but as the potential power of these filters became apparent, a great deal of effort was

© Springer International Publishing AG 2017
W.G. Kropatsch et al. (Eds.): DGCI 2017, LNCS 10502, pp. 9–18, 2017.
DOI: 10.1007/978-3-319-66272-5_2

made into development of efficient algorithms. In this review I will first describe the history of these image operators, introducing the most important types of connected operators. After that, I will present the lattice-theoretical framework of connectivity classes, and several generalisations that have been proposed. This is followed by a discussion of the most important algorithms, both sequential and parallel, for efficient computation of these operators. Finally, I will discuss future perspectives.

2   A Brief History

The very first connected filter was published in the thesis of Klein [3], in which he presented reconstruction operators. The basic reconstruction operator is the reconstruction f from marker g. In the binary case, all connected components of f which have a non-empty intersection with g re preserved, all others are removed. The *opening-by-reconstruction* is obtained by choosing

$$g = f \ominus B \tag{1}$$

with B some structuring element. This preserves all connected components of f into which some translate of B fits entirely. If B is a ball of radius r, this preserves only those connected components with an erosion width larger than $2r$. An example is shown in Fig. 1. Quite clearly, the opening by reconstruction preserves shapes exactly, whereas the regular, structural opening does not.

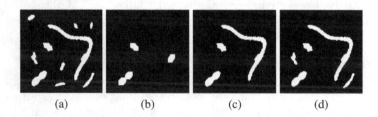

    (a)   (b)   (c)   (d)

Fig. 1. A comparison of openings: (a) original image of bacteria; (c) structural opening by 11×11 square; (b) reconstruction of (a) by (b); (d) moment-of-inertia opening with $\lambda = 11^4/4$.

Area openings were proposed in the early 1990s [4,5] as filters which remove connected components with an area smaller than some threshold λ from images. After this, Breen and Jones [6] proposed *attribute openings* as a generalisation. In this case some increasing measure $\mu(C)$ of each connected component is computed, and compared to some threshold. An example is shown in Fig. 1(d), where the measure is the trace of the moment of inertia, and the threshold is set to that of an 11×11 square. The latter is clearly more selective for elongated features.

The above filters are anti-extensive, i.e. the remove bright, or foreground structures. Extensive filters which remove dark, or background structures can be defined by duality, i.e., by inverting the image, applying the opening (or thinning) and inverting the result.

The grey-scale case can be obtained by threshold superposition [7]. The easiest way to visualise this process is through the concept variously known as component trees [8,9], max-trees and min-trees [10], and opening trees [11]. An example of a max-tree of a simple grey scale image is shown in Fig. 2. It can be constructed by thresholding the grey level image at all threshold levels, and computing the connected components, called *peak components*, of every threshold set. It is easy to see that each peak component at grey level h is nested within a single peak component at a lower level. Therefore, the peak components can be organized into a tree structure called the max-tree, because the leaves are the local maxima. A min-tree is is simply a max-tree of the inverted image, and component tree may be either.

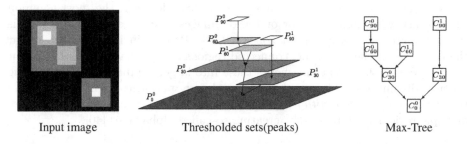

| Input image | Thresholded sets(peaks) | Max-Tree |

Fig. 2. A simple grey-scale image, its peak components and max-tree

Once a max-tree has been computed filtering can be done by applying some decision function which determines which nodes of the tree to keep, and which to remove. In the simplest case, some increasing attribute, such as area, is computed for each node, and this is compared to some threshold. In this case we end up pruning those branches of the tree which contain nodes that are too small to meet the criterion. All pixels in the removed nodes are assigned the grey levels of the nearest surviving ancestor. In more complicated cases, the attributes are non-increasing, and removed and retained nodes along the root path may alternate. In this we need to make more complicated filtering rules to assign new grey levels [10,12]. Although we add complexity, it does allow implementation of scale invariant shape filters [12,13], which allow filtering of features such as blood vessels based on shape, rather than size [14,15]. Non-increasing filtering also takes place in so-called vector-attribute filters [16,17]. In this case each node of the tree contains a feature vector describing size, shape, or grey-level statistics. In the simplest case, we can then retain or remove nodes based on how closely they resemble some prototype vector, using distance measures like euclidean or Mahalanobis distances. Machine learning and clustering can also be used to distinguish different classes of nodes [18]. An interesting development is that of building a component tree of a component tree, and steering the filtering process from that second tree [19].

Several self-dual filters have been developed: levelings [20,21] as self-dual versions of the reconstruction operators, and level-line tree [22–24] also called tree-of-shapes [25] methods in the case of attribute filters, which can be seen as combining a max-tree and min-tree in such a way that the extrema are the leaves.

All these tree structures allow more than just filtering. As these trees are multi-scale representations of the image, all kinds of multi-scale analysis is very readily performed using these trees, including computation of pattern spectra [26–29] and morphological profiles [30,31]. There are also important relations between the various partition trees and hierarchical segmentation methods [32–34].

Component trees and their relatives require a total order of the pixel values in order to work, so adaptations are need for work in colour, vector and tensor images. Approaches include total pre-orders in component trees [35], merging of trees of shapes from different colour bands [36], and area scale spaces for colour images using local extremal values [37,38]. However, the most common way to deal with colour, vector or tensor images in connective morphology is through hierarchies of quasi flat zones [32,39–41]. In principle we start out at the flat-zones of the image, i.e. connected regions of maximal extent of constant colour. We then hierarchically merge these into larger components in increasing order of dissimilarity. The simplest approach only performs binary merges, and thus creates a *binary partition trees* [39]. Alternatively, we can allow mergers of multiple regions simultaneously, creating so-called alpha trees [40].

3   Connectivity Theory

In the discussion above, we implicitly assumed the usual definition of a connected component, as a maximal connected subset of some set X. We did not, however specify what we actually mean by "connected". Serra [42,43] introduced the general notion of *connectivity classes*. These are essentially families of all elements of the lattice under consideration that are connected. In the case of binary images that means the family of all connected sets. In all cases, a connectivity class $\mathcal{C}$ must have the following three properties

1. $\mathbf{0} \in \mathcal{C}$
2. $\mathcal{C}$ is sup-generating
3. For any set $\{C_i\} \subset \mathcal{C}$, with i from some index set, $\bigwedge_i C_i \neq \mathbf{0} \quad \Rightarrow \quad \bigvee_i C_i \in \mathcal{C}$

This means that the least element of the lattice ($\emptyset$ in the set theoretical case) is connected, that any element of the lattice can be generated by taking the supremum of elements of $\mathcal{C}$, and that if the infimum of elements of $\mathcal{C}$ is not $\mathbf{0}$, their supremum is connected. In the set theoretical case this means that if the intersection of any number of connected sets is not empty, their union is connected.

It is easily verified that the standard notions of (path) connectivity in images adhere to these three axioms. Many others have been proposed in a framework often called *second-generation connectivity*, where the standard connectivity is modified either by applying operators like dilation, opening or closings to the image [42,44,45], or by using some second image in so-called mask-based second-generation connectivity [46]. These methods allow objects broken up by noise to be considered a single, connected entity, or objects linked by single pixels to be considered separate. This is done by considering whether pixels are connected at a given grey level in a modified image, rather than in the original. A further extension also assigns grey levels to the edges between pixels, allowing further freedom in deciding what is connected [47].

3.1   Beyond Connectivity

In [42] suggested that there are generalisations to connectivity that have been explored further by other researchers. In particular, the notion of *hyperconnectivity* has received much attention [48–52]. Initially, Serra changed the third axiom of connectivity classes, from merely requiring a non-empty intersection for a group of connected sets to have a connected union, to some stronger requirement called the overlap criterion. This might be that the intersection contains a ball of a certain diameter, leading to *viscous hyperconnections* [53,54] It has been shown that this can lead to a family of filters that bridges the gap between connected filters and structural morphological filters [53]. It has since been shown that the overlap criterion is not required to define these hyperconnections, but that other axiomatics can be used [48,52].

Other extensions of connectivity are *partial connections* [34], and *attribute-space connections* [55].

4   Algorithms

In support of all the theoretical developments, algorithms for fast computation of connected operators have been an important area of research [5,9,10,40,56,57]. For a recent review see [58]. Many of the algorithms centre around component trees (min-tree, max-tree, tree-of-shapes) on the one hand, and partition trees on the other (binary partition trees and alpha trees, mainly). In the former category, there were two different approaches, i.e. building the tree from root to leaf, by recursive or non-recursive flood-filling [10,59,60], and the reverse approach, starting from the leaves joining up branches, and ultimately aggregating them into a single tree. These latter fall into two categories: (i) those using a priority queue [8], derived from the area opening algorithm of Vincent [5], and (ii) using Tarjan's union-find [9,61,62] based on the area opening algorithm in [57]. All algorithms have their pros and cons as reviewed in [58].

An important aspect of efficient computation on modern compute architectures is the ability to use parallellism, available either through multi-core processors, or even the graphics processing unit (GPU). Connected filters are particularly troublesome for parallel implementation since the processing order of most algorithms is very much data driven, and the usual approach of dividing the data over different cores (with some degree of overlap) is hampered by the fact that we do not a priori know how much overlap is needed, because we do not know the location or borders of connected components. This problem was addressed in [63], in which it was shown that max-trees or min-trees of disjoint image regions could be merged into a complete tree, complete with the correct attribute information. The cost of merging is proportional to the number of grey levels, worst case, so the algorithm does not work efficiently beyond 16 bits per pixel. A solution has recently been presented by Moschini et al. [64], in which first a root-to-leaf flood-filling algorithm algorithm is used on a uniformly quantised approximation image, in which the number of grey levels equals the number of processors, after which a variant of Berger's leaf-to-root algorithm is used to refine the tree. These algorithms have been used in various multi-scale analysis tasks in remote sensing [27,65] and astronomical imaging [28,66].

A limitation of both the above parallel algorithms is that they only work on shared-memory parallel computers. In practice this means that images up to a few gigapixel can be processed. This is because the entire max-tree must be stored in memory. In a recent development, a distributed memory algorithm suitable for use on clusters has been developed [67]. This does not build a single max-tree, but rather a forest of trees, each for one tile of the image, which can be stored on one node of the cluster. After building these trees, only the subtree connected to the tile boundaries are exchanged and merged hierarchically to correct the topology. Once these corrections have been made, filtering the individual trees yields the same result as filtering the entire max-tree. The results show a speed up of up to 103 on 128 cores, in the case of a 153 Gpixel image.

The above algorithms focus on regular path connectivity, but they have been extended to the case of second-generation connectivity, without loss of efficiency [46,68], and even some hyperconnections [54,69].

Other important algorithms concern building the binary partition tree [39], the sequential alpha tree algorithm [40], and a parallel algorithm for alpha trees [70], based on the same merge strategy as [63].

5   Conclusions and Perspectives

Connected filters are versatile tools for image filtering, feature extraction, object recognition, and segmentation. An attractive property of these methods is that they allow modelling of perceptual grouping through changing the notion of connectivity, and simultaneously allow modelling of object properties through the choice of attributes used. Finally, it is relatively straightforward to introduce machine learning into the framework, in the form of a decision function used to accept or reject connected components.

There are still many issues to be addressed. Algorithmically, we still not have a solution that would allow us to use the power of GPUs on these methods. Likewise, the new distributed algorithm for attribute filters only works up to 16-bits per pixel images, because it uses the same merging algorithm as in [63]. The same holds for [70]. Therefore, extreme-dynamic range versions of these algorithms are needed, capable of handling floating point values.

Integration of machine learning is also an area of research that is still in its infancy, and more work is needed here. This is not restricted to the decision function needed in attribute filters, but also in e.g. computation of connectivity masks, where currently only ad hoc approaches are used in determining the best connectivity mask to achieve a particular task. Whether it is possible to learn the best mask is an open question.

References

1. Salembier, P., Wilkinson, M.H.F.: Connected operators: a review of region-based morphological image processing techniques. IEEE Signal Process. Mag. **26**(6), 136–157 (2009)
2. Salembier, P., Serra, J.: Flat zones filtering, connected operators, and filters by reconstruction. IEEE Trans. Image Proc. **4**, 1153–1160 (1995)
3. Klein, J.C.: Conception et réalisation d'une unité logique pour l'analyse quantitative d'images. PhD thesis, Nancy University, France (1976)
4. Cheng, F., Venetsanopoulos, A.N.: An adaptive morphological filter for image processing. IEEE Trans. Image Proc. **1**, 533–539 (1992)
5. Vincent, L.: Morphological area openings and closings for grey-scale images. In: O, Y.L., Toet, A., Foster, D., Heijmans, H.J.A.M., Meer, P. (eds.) Shape in Picture. NATO ASI Series (Series F: Computer and Systems Sciences), vol. 126, pp. 197–208. Springer, Heidelberg (1993)
6. Breen, E.J., Jones, R.: Attribute openings, thinnings and granulometries. Comp. Vis. Image Understand. **64**(3), 377–389 (1996)
7. Maragos, P., Ziff, R.D.: Threshold decomposition in morphological image analysis. IEEE Trans. Pattern Anal. Mach. Intell. **12**(5), 498–504 (1990)
8. Jones, R.: Connected filtering and segmentation using component trees. Comp. Vis. Image Understand. **75**, 215–228 (1999)
9. Najman, L., Couprie, M.: Building the component tree in quasi-linear time. IEEE Trans. Image Proc. **15**, 3531–3539 (2006)
10. Salembier, P., Oliveras, A., Garrido, L.: Anti-extensive connected operators for image and sequence processing. IEEE Trans. Image Proc. **7**, 555–570 (1998)
11. Vincent, L.: Granulometries and opening trees. Fundamenta Informaticae **41**, 57–90 (2000)
12. Urbach, E.R., Roerdink, J.B.T.M., Wilkinson, M.H.F.: Connected shape-size pattern spectra for rotation and scale-invariant classification of gray-scale images. IEEE Trans. Pattern Anal. Mach. Intell. **29**, 272–285 (2007)
13. Urbach, E.R., Wilkinson, M.H.F.: Shape-only granulometries and grey-scale shape filters. In: Proceeding of International Symposium Mathematical Morphology (ISMM) 2002, pp. 305–314 (2002)
14. Wilkinson, M.H.F., Westenberg, M.A.: Shape preserving filament enhancement filtering. In: Niessen, W.J., Viergever, M.A. (eds.) MICCAI 2001. LNCS, vol. 2208, pp. 770–777. Springer, Heidelberg (2001). doi:10.1007/3-540-45468-3_92
15. Westenberg, M.A., Roerdink, J.B.T.M., Wilkinson, M.H.F.: Volumetric attribute filtering and interactive visualization using the max-tree representation. IEEE Trans. Image Proc. **16**, 2943–2952 (2007)
16. Urbach, E.R., Boersma, N.J., Wilkinson, M.H.F.: Vector-attribute filters. In: Mathematical Morphology: 40 Years On, Proceedings of International Symposium Mathematical Morphology (ISMM) 2005, Paris, 18–20 April 2005 95–104

17. Naegel, B., Passat, N., Boch, N., Kocher, M.: Segmentation using vector-attribute filters: methodology and application to dermatological imaging. In: Proceeding International Symposium on Mathematical Morphology (ISMM) 2007, pp. 239–250 (2007)
18. Kiwanuka, F., Wilkinson, M.: Cluster based vector attribute filtering. Math. Morphol. Theory Appl. **1**(1), 116–135 (2016)
19. Xu, Y., Carlinet, E., Géraud, T., Najman, L.: Hierarchical segmentation using tree-based shape space. IEEE Trans. Pattern Anal. Mach. Intell. **39**(3), 457–469 (2016)
20. Meyer, F.: From connected operators to levelings. In: Fourth International Symposium on Mathematical Morphology, ISMM 1998, pp. 191–198. Kluwer, Amsterdam, The Netherlands (1998)
21. Meyer, F.: Levelings, image simplification filters for segmentation. J. Math. Imag. Vis. **20**(1–2), 59–72 (2004)
22. Monasse, P., Guichard, F.: Fast computation of a contrast invariant image representation. IEEE Trans. Image Proc. **9**, 860–872 (2000)
23. Monasse, P., Guichard, F.: Scale-space from a level lines tree. J. Vis. Commun. Image Repres. **11**, 224–236 (2000)
24. Caselles, V., Monasse, P.: Grain filters. J. Math. Imag. Vis. **17**, 249–270 (2002)
25. Géraud, T., Carlinet, E., Crozet, S., Najman, L.: A Quasi-linear algorithm to compute the tree of shapes of nD images. In: Hendriks, C.L.L., Borgefors, G., Strand, R. (eds.) ISMM 2013. LNCS, vol. 7883, pp. 98–110. Springer, Heidelberg (2013). doi:10.1007/978-3-642-38294-9_9
26. Urbach, E.R., Roerdink, J., Wilkinson, M.H.F.: Connected rotation-invariant size-shape granulometries. In: Proceeding 17th International Conference on Pattern Recognition, vol. 1, pp. 688–691 (2004)
27. Wilkinson, M.H.F., Moschini, U., Ouzounis, G.K., Pesaresi, M.: Concurrent computation of connected pattern spectra for very large image information mining. In: Proceeding of ESA-EUSC-JRC 8th Conference on Image Information Mining, Oberpfaffenhofen, Germany, pp. 21–25 (2012)
28. Moschini, U., Teeninga, P., Trager, S.C., Wilkinson, M.H.F.: Parallel 2D local pattern spectra of invariant moments for galaxy classification. In: Azzopardi, G., Petkov, N. (eds.) CAIP 2015. LNCS, vol. 9257, pp. 121–133. Springer, Cham (2015). doi:10.1007/978-3-319-23117-4_11
29. Bosilj, P., Wilkinson, M.H.F., Kijak, E., Lefèvre, S.: Local 2D pattern spectra as connected region descriptors. In: Benediktsson, J.A., Chanussot, J., Najman, L., Talbot, H. (eds.) ISMM 2015. LNCS, vol. 9082, pp. 182–193. Springer, Cham (2015). doi:10.1007/978-3-319-18720-4_16
30. Pesaresi, M., Benediktsson, J.: A new approach for the morphological segmentation of high-resolution satellite imagery. IEEE Trans. Geosci. Remote Sens. **39**(2), 309–320 (2001)
31. Benediktsson, J., Palmason, J., Sveinsson, J.: Classification of hyperspectral data from urban areas based on extended morphological profiles. IEEE Trans. Geosci. Remote Sens. **43**(3), 480–491 (2005)
32. Soille, P.: Constrained connectivity and connected filters. IEEE Trans. Pattern Anal. Mach. Intell. **30**(7), 1132–1145 (2008)
33. Najman, L.: On the equivalence between hierarchical segmentations and ultrametric watersheds. J. Math. Imag. Vis. **40**(3), 231–247 (2011)
34. Ronse, C.: Partial partitions, partial connections and connective segmentation. J. Math. Imag. Vis. **32**(2), 97–125 (2008)

35. Naegel, B., Passat, N.: Component-trees and multi-value images: a comparative study. In: Wilkinson, M.H.F., Roerdink, J.B.T.M. (eds.) ISMM 2009. LNCS, vol. 5720, pp. 261–271. Springer, Heidelberg (2009). doi:10.1007/978-3-642-03613-2_24
36. Carlinet, E., Géraud, T.: MToS: a tree of shapes for multivariate images. IEEE Trans. Image Process. **24**(12), 5330–5342 (2015)
37. Evans, A.N.: Color area morphology scale-spaces. Adv Imaging Electr. Phys. **160**, 35–74 (2010)
38. Gimenez, D., Evans, A.N.: An evaluation of area morphology scale-spaces for colour images. Comp. Vis. Image Understand. **110**, 32–42 (2008)
39. Salembier, P., Garrido, L.: Binary partition tree as an efficient representation for image processing, segmentation and information retrieval. IEEE Trans. Image Proc. **9**(4), 561–576 (2000)
40. Ouzounis, G.K., Soille, P.: The Alpha-Tree algorithm. Publications Office of the European Union, December 2012
41. Aptoula, E., Pham, M.-T., Lefèvre, S.: Quasi-flat zones for angular data simplification. In: Angulo, J., Velasco-Forero, S., Meyer, F. (eds.) ISMM 2017. LNCS, vol. 10225, pp. 342–354. Springer, Cham (2017). doi:10.1007/978-3-319-57240-6_28
42. Serra, J.: Connectivity on complete lattices. J. Math. Imag. Vis. **9**(3), 231–251 (1998)
43. Serra, J.: Connections for sets and functions. Fundam. Inf. **41**(1–2), 147–186 (2000)
44. Braga-Neto, U., Goutsias, J.: A theoretical tour of connectivity in image processing and analysis. J. Math. Imag. Vis. **19**, 5–31 (2003)
45. Braga-Neto, U., Goutsias, J.: Object-based image analysis using multiscale connectivity. IEEE Trans. Pattern Anal. Mach. Intell. **27**(6), 892–907 (2005)
46. Ouzounis, G.K., Wilkinson, M.H.F.: Mask-based second generation connectivity and attribute filters. IEEE Trans. Pattern Anal. Mach. Intell. **29**, 990–1004 (2007)
47. Oosterbroek, J., Wilkinson, M.H.F.: Mask-edge connectivity: Theory, computation, and application to historical document analysis. In: Proceeding 21st International Conference on Pattern Recognition, pp. 3112–3115 (2012)
48. Wilkinson, M.H.F.: An axiomatic approach to hyperconnectivity. In: Wilkinson, M.H.F., Roerdink, J.B.T.M. (eds.) ISMM 2009. LNCS, vol. 5720, pp. 35–46. Springer, Heidelberg (2009). doi:10.1007/978-3-642-03613-2_4
49. Wilkinson, M.H.F.: Hyperconnections and openings on complete lattices. In: Soille, P., Pesaresi, M., Ouzounis, G.K. (eds.) ISMM 2011. LNCS, vol. 6671, pp. 73–84. Springer, Heidelberg (2011). doi:10.1007/978-3-642-21569-8_7
50. Perret, B., Lefevre, S., Collet, C., Slezak, E.: Hyperconnections and hierarchical representations for grayscale and multiband image processing. IEEE Trans. Image Process. **21**(1), 14–27 (2012)
51. Perret, B., Lefèvre, S., Collet, C.: Toward a new axiomatic for hyper-connections. In: Soille, P., Pesaresi, M., Ouzounis, G.K. (eds.) ISMM 2011. LNCS, vol. 6671, pp. 85–95. Springer, Heidelberg (2011). doi:10.1007/978-3-642-21569-8_8
52. Perret, B.: Inf-structuring functions: a unifying theory of connections and connected operators. J. Math. Imaging Vis. **51**(1), 171–194 (2015)
53. Wilkinson, M.H.F.: Hyperconnectivity, attribute-space connectivity and path openings: theoretical relationships. In: Wilkinson, M.H.F., Roerdink, J.B.T.M. (eds.) ISMM 2009. LNCS, vol. 5720, pp. 47–58. Springer, Heidelberg (2009). doi:10.1007/978-3-642-03613-2_5
54. Moschini, U., Wilkinson, M.H.F.: Viscous-hyperconnected attribute filters: a first algorithm. In: Benediktsson, J.A., Chanussot, J., Najman, L., Talbot, H. (eds.) ISMM 2015. LNCS, vol. 9082, pp. 669–680. Springer, Cham (2015). doi:10.1007/978-3-319-18720-4_56

55. Wilkinson, M.H.F.: Attribute-space connectivity and connected filters. Image Vis. Comput. **25**, 426–435 (2007)
56. Vincent, L.: Morphological grayscale reconstruction in image analysis: application and efficient algorithm. IEEE Trans. Image Proc. **2**, 176–201 (1993)
57. Meijster, A., Wilkinson, M.H.F.: A comparison of algorithms for connected set openings and closings. IEEE Trans. Pattern Anal. Mach. Intell. **24**(4), 484–494 (2002)
58. Carlinet, E., Géraud, T.: A comparative review of component tree computation algorithms. IEEE Trans. Image Proc. **23**(9), 3885–3895 (2014)
59. Hesselink, W.H.: Salembier's Min-tree algorithm turned into breadth first search. Inf. Process. Lett. **88**(5), 225–229 (2003)
60. Wilkinson, M.H.F.: A fast component-tree algorithm for high dynamic-range images and second generation connectivity. In: Proceeding International Conference on Image Processing 2011, pp. 1041–1044 (2011)
61. Tarjan, R.E.: Efficiency of a good but not linear set union algorithm. J. ACM **22**, 215–225 (1975)
62. Berger, C., Geraud, T., Levillain, R., Widynski, N., Baillard, A., Bertin, E.: Effective component tree computation with application to pattern recognition in astronomical imaging. In: Proceeding of International Conference Image Processing 2007, San Antonio, Texas, USA, 16–19 September 2007, vol. IV, pp. 41–44 (2007)
63. Wilkinson, M.H.F., Gao, H., Hesselink, W.H., Jonker, J.E., Meijster, A.: Concurrent computation of attribute filters using shared memory parallel machines. IEEE Trans. Pattern Anal. Mach. Intell. **30**(10), 1800–1813 (2008)
64. Moschini, U., Meijster, A., Wilkinson, M.H.F.: A hybrid shared-memory parallel max-tree algorithm for extreme dynamic-range images. IEEE Trans. Pattern Anal. Mach, Intell. (2017, in press)
65. Wilkinson, M.H.F., Pesaresi, M., Ouzounis, G.K.: An efficient parallel algorithm for multi-scale analysis of connected components in gigapixel images. ISPRS Int. J. Geo-Inf. **5**(3), 22 (2016)
66. Moschini, U., Teeninga, P., Wilkinson, M.H.F., Giese, N., Punzo, D., Van der Hulst, J.M., Trager, S.C.: Towards better segmentation of large floating point 3d astronomical data sets: first results. In: Proceedings of the 2014 conference on Big Data from Space BiDS14, pp. 232–235. Publications Office of the European Union (2014)
67. Kazemier, J.J., Ouzounis, G.K., Wilkinson, M.H.F.: Connected morphological attribute filters on distributed memory parallel machines. In: Angulo, J., Velasco-Forero, S., Meyer, F. (eds.) ISMM 2017. LNCS, vol. 10225, pp. 357–368. Springer, Cham (2017). doi:10.1007/978-3-319-57240-6_29
68. Ouzounis, G.K., Wilkinson, M.H.F.: A parallel dual-input max-tree algorithm for shared memory machines. In: Proceeding of International Symposium Mathematical Morphology (ISMM) 2007, pp. 449–460 (2007)
69. Ouzounis, G.K., Wilkinson, M.H.F.: Hyperconnected attribute filters based on k-flat zones. IEEE Trans. Pattern Anal. Mach. Intell. **33**(2), 224–239 (2011)
70. Havel, J., Merciol, F., Lefèvre, S.: Efficient tree construction for multiscale image representation and processing. J. Real-Time Image Process. 1–18 (2016)

Collaborating with an Artist in Digital Geometry

Eric Andres[(⊠)], Gaelle Largeteau-Skapin, and Aurélie Mourier

Laboratory XLIM, Team ASALI, UMR CNRS 6712, University of Poitiers,
BP 30179, 86962 Futuroscope Chasseneuil Cedex, France
{eric.andres, gaelle.largeteau.skapin}@univ-poitiers.fr
contact@aureliemourier.net

Abstract. In this invited paper, we are going to present an ongoing collaboration with a local artist, Aurélie Mourier. The artist works with voxel shapes and this led our digital geometry team to develop new shape modeling tools and explore a particular class of unfolding problems.

1 Introduction and Context of the Artistic Work

In this paper, we are going to present the results of an ongoing collaboration of our Digital Geometry Research team with a local graphical artist, Aurélie Mourier, that happens to be working with cubes. Let us present the motivations behind her work and then we'll present some examples of our research that were partly or completely driven by the artists demands and questions.

An artist typically tries to understand the world around him by proposing his own reproduction and/or by dissecting specific elements of it in order to propose an original point of view. The artist will embody his unique and personal point of view in an art work so his experience of the world can be shared with the public. Art is not meant to be didactic but singular. The inspiration for his work may come from many different sources and points of view including scientific ones. There are many similarities in the approach used by artists and scientists in particular in the attempts to propose an abstract representation of the world. The artist will use his art form while the scientist will use mathematics as common shared language.

In A. Mourier's case, she focuses on shapes. Those shapes can be extracted from reality or invented. In order to study only the shape of things, she willingly discards parameters such as color, texture and size: a planet, a ball and a marble have all the same shape. It is interesting here to make a parallel with the ideas behind the invention of topology. In her case however, the geometric shape (although in a digital abstract form) still plays an important role. She proceeds by injecting the object she wants to reproduce into a 3D cubic grid similar to what happens when one *pixelizes* a shape in 2D. A resolution was chosen: the grid size of $25 \times 25 \times 25$. This sizes was chosen arbitrarily so that the shapes are just big enough to be recognizable and allow some expressibility. A shape is *well formed* for A. Mourier if it is in one piece, with each cube touching another

© Springer International Publishing AG 2017
W.G. Kropatsch et al. (Eds.): DGCI 2017, LNCS 10502, pp. 19–30, 2017.
DOI: 10.1007/978-3-319-66272-5_3

one by face and touches at least two opposite sides of the grid. She orders the shapes according to their volume (their number of voxels). She did not know, in the beginning of her artistic work, that her process was called digitization and her *one piece, face-connected* constraint translates in topology as *a single 6-connected component.*

Her inspiration came from the observation that when we look at the world through computer screens, our world view is formed of digital images with a finite fixed resolution. There exists only finite, although very big, number of possible images. If we could create a catalog of all the possible images, it would, for instance, include all the images of all the people in the world, dead, alive and yet to be born. Let us note here that our biological eye has also a finite number of cones and that our brain has a limit on the number of different colors it is able to distinguish. A. Mourier works on a shape repository but more importantly, on how a shape can be represented or coded and how such shapes can then lead to sculptures that reinstate a size and a material with physical properties. This idea of repository is inspired by the novel written in 1941 by J.-L. Borges, "La biblioteca de Babel" [6]. The story of Borges describes a library with all the possible books of 410 pages, each made of 40 lines of about 80 characters each in a alphabet composed of 22 regular letters, and the characters space, coma and point (check out https://libraryofbabel.info/ for a virtual example of this library). Of course, just as for the Library of Babel, and even though well formed objects represent only a fraction of the $2^{25^3} \approx 3.9 \times 10^{4703}$ possible digital objects that can be represented in her grid, it gives a setting for shape exploration.

Before we met, A. Mourier used to model her voxel objects with generic modeling software. She then printed and cut the result by hand to obtain either the unfolded pattern of the shape or its slices. An example of a stereo microscope shape is presented Fig. 1. The number in the name of the shape corresponds to its number of voxels (797) and the order of the object among all the shapes having the same number of voxels. Once a shape has been defined, A. Mourier is interested in all the ways such a shape can be represented: as a set of voxels in 3D, as a set of 2D slices, unfolded as a net, etc.

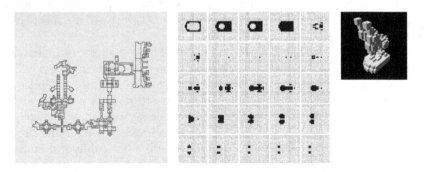

Fig. 1. A. Mourier stereo microscope shape 00797.001: the unfolded net, the slices in the z axis and the 3D numeric object.

2   Creating Digital Shapes for an Artist

As an artist working on her shape depository, A. Mourier used to design her digital objects by hand. Coming from a completely different background, terms like *digitization*, *6-connectivity* or *digital geometry* were unknown to her, on the other hand she had manipulated such shapes and representation forms for a couple of years and often she has a better intuition than we could possibly have on how to do certain operations. For her, it was interesting to put scientific words on notions she was manipulating intuitively. For us, it was interesting to discover new classes of problems with specific applicative constraints.

The starting point of our collaboration was through a student project that created a simple voxel modeling software that lets you freely create voxel objects. This tool can be found online at http://www.aureliemourier. net/logiciel/25aucube.html. However, with this tool, shapes still had to be constructed by hand. We proposed to design algorithms to help create specific classes of voxel objects. For this, several constraints had to be met so that our client, A. Mourier, could use these objects: one of the main constraints was to incorporate a manual design possibility so that the artist may express herself. As a first proposal, we developed a online software to generate digital surfaces of revolution based on a hand-drawn generatrix and hand-drawn curves of revolution. This work will be presented in Sect. 2.1. We then proposed a method for creating tubes where the 3D curve and the section of the tube can be designed freely. This will be presented in Sect. 2.2. More recently, we looked into the problem of unfolding the surface of a voxel object. This was motivated by the nets created, by the artist, by hand. This will be presented in Subsect. 2.3.

2.1   Digital Surface of Revolution with Hand-Drawn Generatrix and Curve of Revolution

Working with a visual art artist, our goal was to propose flexible, intuitive to use tools for designing 3D voxel surfaces. For this we considered 2D (hand-) drawings which are a natural shape representation form for most artists. A recent paper proposed a digitization method for surfaces of revolution [3] based on a very simple and straightforward method for digitizing almost any implicit nD surfaces [16]. It is based on a morphological type digitization method called *flake digitization*. The flake digitization allows to define digital surfaces with a controlled topology (control on tunnels of the digital surface). The paper [16] can be seen as an extension to all tunnel connectivities and dimensions of the paper of S. Laine [14].

A surface of revolution is defined by two 2D curves: the curve of revolution and the generatrix. We propose three ways of defining a curve of revolution and three ways of defining a generatrix in order to allow maximum flexibility. A curve of revolution can be defined as an implicit curve (that separates space into positive and negative valued regions), as a closed hand-drawn curve or as binary pixel image that serves as a look-up matrix. The curve of revolution is not limited to the traditional circle and not limited to a unique connected component, which offers a great liberty in designing complex shapes. Contrary to the curve of revolution, the generatrix is not necessarily a closed curve. Three similar ways of defining generatrix curves are proposed: as an explicit function, as hand-drawn curves or as contour curves extracted from a binary image. This leads to nine different ways of defining digital surfaces of revolution. Two of those methods have already been published by Andres et al. respectively in [3,4].

Let us detail a little bit the overall method. A curve of revolution can be basically defined in any way as long as we are able, slice by slice, to define regions where with a interior and an exterior. The curve(s) of revolution are then defined as the boundaries between those regions. For the digitization, we determine if a digital point belongs to the digital surface of revolution by considering the vertices of the three dimensional structuring element (k-Flakes [16]) centered on this point and computing their position relatively to the curve of revolution. If some of the vertices are inside and some outside then the corresponding voxel is cut by the surface and therefore belongs to the Digital Surface of Revolution. When the curve of revolution is implicitly defined the vertex localisation is straightforward. When the curve of revolution is given as a hand-drawn curve, which is more natural for an artist, we record the sequence of Euclidean points, while the artist draws the curve of revolution. We ensure that this curve is a closed one (to define one or several interior(s) and exterior(s)) by adding the first point at the end of the list. This list of point is then treated as a closed polygon and the critical localisation information can be obtained using a *Point In Polygon* (PIP) algorithm [12].

The other curve, the generatrix g, typically plays the role of an homothetic factor for the curve of revolution. A slice $z = z_0$ of the surface of revolution is the curve of revolution scaled by a factor $g(z_0)$ for a generatrix that is defined by an explicit function $y = g(z)$. When the generatrix is drawn and represented by a list of points, there are two main issues: firstly, there can be more than one value $g(z)$ per z and secondly, the curve may not be closed. To handle the multiplicity of $g(z)$ values, the point sequence is divided into strictly monotonic (in z, increasing or decreasing) or horizontal subsequences. The end point of one subsequence is duplicated as the starting point of the next one. Each subsequence can be digitized completely independently because the flake digitization method we use is a morphological type digitization (i.e. consistant with the union operator). One last problem had to be addressed: the generatrix may be an open curve and therefore it may have extremities. This case needs to be handled specifically because the digitization of the surface of revolution supposes that we are able, for all points, to compute its localisation relatively to the surface. At the generatrix extremities, some of the vertices of the voxels' flake are outside the domain of definition of the generatrix. There can be, in this case, a defined localisation for some vertices and not for others. In this case, we consider only the parts of the flake that are inside the domain and therefore we take, as substitute for the undefined vertices, the endpoints of a *cropped flake* line segments. See [4] for more details. See http://imgur.com/a/eDFbY for some examples of swept digital tubes and digital surfaces of revolution. The method is not limited to curves of revolution of dimension two. You can see an example at http://imgur.com/a/eDFbY of a four-dimensional torus defined by a 3D surface of revolution (a sphere) and a 2D generatrix (a circle) (Fig. 2).

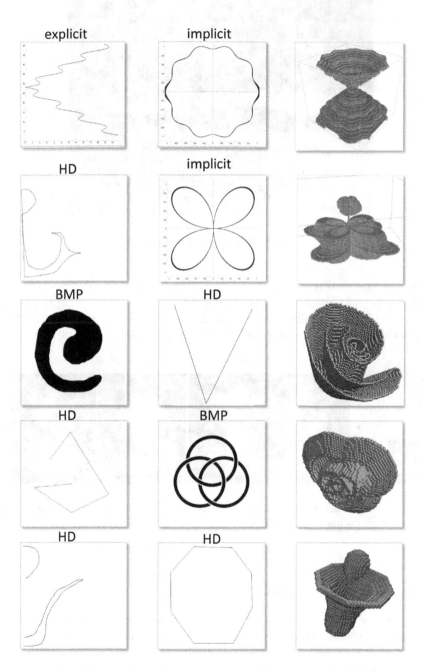

Fig. 2. Examples of 3D surfaces of revolution (HD: hand-drawn, BMP: bitmap image).

Fig. 3. Spinning top sculptor

The software that implements these methods is available at http://xlim-sic. labo.univ-poitiers.fr/demonstrateurs/DSoR_Generator/. This software has been used by A.Mourier to build chess pieces that have been 3D printed (See Fig. 4 for an example), and a sculptor representing a spinning top (Fig. 3).

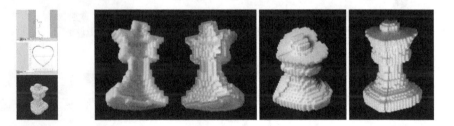

Fig. 4. The 3D printed chess white queen. This piece was built using an implicit curve of revolution and a hand-drawn generatrix.

2.2 Swept Tubular Surfaces

As a complement to surfaces of revolution, we designed a tool that allows to create *Swept Tubular surfaces*. Such tubes can also be defined as digital implicit surfaces [16]. These digital surfaces have applications well beyond our visual art interest: modeling of body parts [13], medicine [7], etc.

Formally, a swept tube is defined by a 3D curve, the *spine curve*, and a 2D closed curve, the *profile*. The *profile* (sometimes called *cross-section* in the literature) is swept out in a plane normal to the 3D *spine curve* (or *trajectory*). Using a mapping of the profile on the spine together with the implicit surface digitization method [16], it is actually quite easy to build digitized swept tubes. In Fig. 5, one can see three interlaced digital swept tubes generated by a parametric 3D spine curve and a bitmap image of three disks that served as profile (please check http://imgur.com/a/eDFbY for an animated version). For the mapping, we chose the Frenet-Serret formulas [10] which describes the motion of a particle along a 3D continuous curve (any other mapping intended to define a moving frame like the *Darboux frame* for instance could also be used). The Serret-Frenet frame is defined by the tangent, the normal and the binormal unit vectors in any point of the 3D curve.

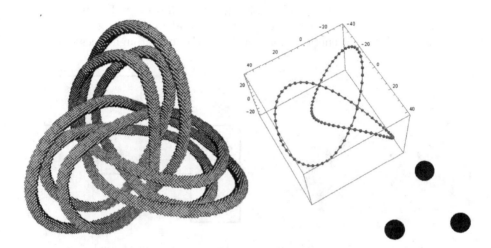

Fig. 5. An example of a swept tube defined by a parametric closed 3D spine curve (on the upper right side) and a bitmap image of three disks as profile (lower right side)

For the animations that we created or that the artist created, the idea was simply to rotate the profile (see http://imgur.com/a/eDFbY for some examples).

2.3 Unfolding Voxel Surfaces

A central question in graphical arts has always been the representation of 3D socalled *reality* in 2D. For a long time Hierarchical representation (where important personae tended to be painted in big and in the middle of a painting while others where painted in small and in the periphery) or perspective representation was the norm. Let us not that for hierarchical representation, although less realistic, conteined additional information with an added, otherwise abstract, social and political dimension. *Cubists*, at the beginning of the twentieth century, proposed a new form of representation that tried to incorporate multiple view points in an abstract recomposition. One of the motivations was to represent elements that would otherwise be hidden. The abtract nature of such recomposition make the paintings sometimes difficult to understand while at the same time they potentially represent a more complete representation of the reality than a more classical representation with only one point of view. That was the starting point of A. Mouriers' interest in nets and the problem of unfolding voxel surfaces. A net shows the complete 3D object in 2D with all the complete surface visible although the net makes it difficult to imagine what the corresponding 3D object looks like. Representing a complete 3D *reality* comes with a price.

The unfolding problem is an old problem already discussed by A. Durer [11]. Since then, unfolding problems have been extensively studied with a wide range of applications ranging from industrial manufacturing, storage problems, to texture mapping, etc. The unfolding problem can be stated as follows: *can the surface of a 3D closed object be unfolded flat to a single component without overlap?* [5]. There are two ways of unfolding: edge-following unfolding and general unfoldings. In edge-unfoldings, one can only make cuts along the edges of the polyhedra while for general unfoldings, one can cut through faces (Fig. 6).

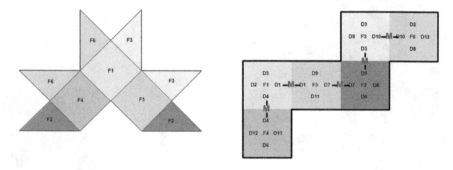

Fig. 6. General unfolding net and edge unfolding net of a cube.

2.4   State of the Art

It has been shown that a non convex polyhedra can not always be edge unfolded [5] and that a convex polyhedra has always a general unfolding [1]. To our best knowledge, it is not known if a convex polyhedra always has an edge unfolding or if a non convex polyhedra always has a general unfolding. There is however a subclass of unfolding problems where does not deal with an arbitrary polyhedra but with what is called, orthogonal polyhedra. An Orthogonal polyhedron is a polyhedron whose faces meet only with a 0 angle (both faces are coplanar), a $\pi/2$ angle (both faces have a so-called *valley* fold) or a $3\pi/2$ angle (both faces have a so-called *mountain* fold). The terms valley and mountain fold comes originally from the origami community. In 2007, it has been shown that arbitrary genus 0 orthogonal polyhedra always have a general unfolding [9]. More recently, in 2016, an algorithm for genus 2 orthogonal orthogonal polyhedra has been proposed [8] as well as a one layer genus g method [15]. A particular case of general unfoldings that are considered are so called grid-unfoldings where cuts across orthogonal faces are only allowed along the edges of a subgrid that may be arbitrarily small in some cases.

2.5   Our Unfolding Problem

Our work on these problems have been driven by the questions and needs as presented to us by A. Mourier. An *art-and-science* project funded by the regional direction of Art and Culture (DRAC Nouvelle Aquitaine) was proposed with the question of peeling (unfolding) genus 0 egg shaped objects (not necessarily convex). For this, we proposed, firstly, a new way of generating digital objects based on Focus points which was accepted for DGCI 2017 [2]. Once we had a way of generating such objects, we were asked if it was possible to generate nets randomly. The problem, for A. Mourier, being the question of having different nets that represent the same final 3D object.

In our case, we consider a 6-connected voxel object whose surface is divided into square voxel faces. We are therefore looking into an edge unfolding problem of orthogonal polyhedra that bears some resemblance with the general grid-unfolding of orthogonal polyhedra class of problems. The key difference here is that the voxel faces can not be subdivided. The classical examples counterexamples, to prove that edge-unfolding for orthogonal polyhedra is not always possible, have nets in our case. The question of the existence of an edge-unfolding solution for all such voxel surface object remains open. And if such a solution always exists, what algorithm could be proposed in the general case ?

For the problem on hand, with the demand of the artist to not propose a deterministic solution, we developed an algorithm based on the following basis:

1. we start with an empty 2D grid in which the voxel faces will be set in order to define a net.
2. First we extract a list of the surface faces of the 6-connected voxel object.
3. for each face we numerotate the edges. An edge is of course shared by two and only two faces.
4. We start with a random face and put it in the center of our grid.
5. We compute all the faces that can be put down next as neighbors for the already settled faces. The key point is that you can put down a new face next to a face in the grid if the edge corresponds. Of course it may happen that a grid square next to a settled face can not be filled because the face that corresponds to this edge has already been put down elsewhere in the grid for another edge.
6. we choose randomly, among those faces, the next face that can be put down as long as not all the faces have been put down or that we are not blocked.
7. if we are not blocked or haven't finished we go to step 5.

This method of building a net does not always provide a solution. It can be blocked simply because all the (four) grid places where we could put down a face are already occupied by other faces.

There are however several ways to improve the convergence: If we are blocked and have several faces left, we can try to correct our net. For the faces that we could not put down, we can determine where we could have put them. If the location where we could have put down the face is an isolated face or a face in a cycle such that it is not the place where the cycle is linked to the other faces of the net, then we can remove the isolated or cycle face and put down our blocked face. We have now swapped one face for another. The idea is that, may be, this new face will find a free grid spot to put it down. The reason why we can only swap with isolated or non splitting cycle faces is that otherwise the face represents the root of a tree of faces that would be disconnected from the net. Since the aim is to create a net in one piece, that can not be allowed. In some cases, a face can not be placed at any spot where we would have isolated faces. We have then the choice, either to abandon and start over or to extract all the faces of the smallest sub-tree and start over with trying to place all those new faces.

A last method we have implemented to get out of a blocking situation, without starting all over, is to remove all the isolated and non splitting cycle faces from the blocked net and start over at step 6.

2.6   Results

To our surprise, the algorithm we have developed works surprisingly well with what we thought would be topologically complicated voxel objects. Figure 7 shows the net for a voxel object formed of a 5^3 voxel cube with 4 traversing tunnels on each side. There are 270 faces and the algorithm takes only a couple of seconds to find a net for such an object with very often only a couple of attempts. The algorithm has also found a solution for the much more complex digital object shown below in the figure. We didn't include here the image of the net because it is simply too big to distinguish any details. It can take usually up to a couple of hours to find a solution for such an object. There are 1350 faces

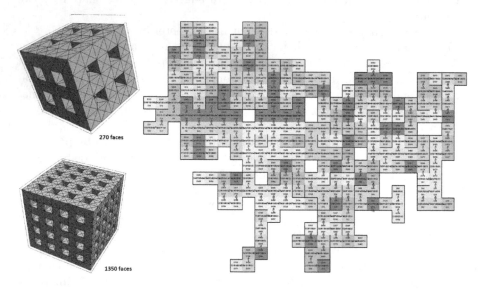

Fig. 7. Voxel cube with 4 holes by face and one of the resulting nets. Below Cube with 16 holes where a solution was found as well.

in this case. The other surprise was that the method works not very well with topologically simple objects. The algorithm struggles with simple objects such as digital spheres of relatively small radii. It is however not completely surprising, all things considered. Our net generator is based on random choices which tends to produce relatively compact nets. The algorithm will typically not stretch the net in one direction to provide more space to put down faces.

3 Conclusion and Perspectives

In this paper we have presented some of the scientific work that has resulted from the collaboration between a scientific team working in digital geometry and a visual art artist. As an artist, A. Mourier explores shapes in a voxel form. Several modeling tools for digital objects have been developed with a focus on giving the artist the possibility to express herself graphically. This has led to develop new methods for generating digital surfaces of revolution and digital tubular swept objects. A more recent collaboration has focused on the unfolding problem for voxel surface objects. A new algorithm for generating nets has been proposed. There are several open questions that have been raised in these different works: we have developed a method where the generatrix may be an open curve. What about open curves of revolution? The same question can be asked for digital swept tubes. Can they be defined with open profile curves? Lastly, we have considered a particular type of unfolding problem: orthogonal edge based grid unfolding. Is there always a solution for this particular case? Right now we have developed a random search algorithm to generate nets that are different each time. This method may need several attempts before it proposes a result. Is there a way to create deterministic algorithms for the general case of convex, non-convex voxel surface unfoldings?

References

1. Agarwal, P.K., Aronov, B., O'Rourke, J., Schevon, C.A.: Star unfolding of a polytope with applications. In: Gilbert, J.R., Karlsson, R. (eds.) SWAT 1990. LNCS, vol. 447, pp. 251–263. Springer, Heidelberg (1990). doi:10.1007/3-540-52846-6_94
2. Andres, E., Biswas, R., Bhowmick, P.: Digital primitives defined by weighted focal set. In: Kropatsch, W.G., Artner, N.M., Janusch, I. (eds.) DGCI 2017. LNCS, vol. 10502, pp. 388–398. Springer, Heidelberg (2017)
3. Andres, E., Largeteau-Skapin, G.: Digital surfaces of revolution made simple. In: Normand, N., Guédon, J., Autrusseau, F. (eds.) DGCI 2016. LNCS, vol. 9647, pp. 244–255. Springer, Cham (2016). doi:10.1007/978-3-319-32360-2_19
4. Andres, E., Richaume, L., Largeteau-Skapin, G.: Digital surface of revolution with hand-drawn generatrix. J. Math. Imaging Vis. **59**(1), 40–51 (2017)
5. Biedl, T., Demaine, E., Demaine, M., Lubiw, A., Overmars, M., O'Rourke, J., Robbins, S., Whitesides, S.: Unfolding some classes of orthogonal polyhedra. In: Proceedings of the 10th Canada Conference on Computing and Geometry, pp. 70–71 (1998)

6. Borges, J.L.: La Biblioteca de Babel: prólogos. Obras de Borges, Emecé Editores (2000)

7. Bornik, A., Reitinger, B., Beichel, R.: Simplex-mesh based surface reconstruction and representation of tubular structures. In: Meinzer, H.P., Handels, H., Horsch, A., Tolxdorff, T. (eds.) Bildverarbeitung für die Medizin 2005. Informatik aktuell, pp. 143–147. Springer, Heidelberg (2005)

8. Damian, M., Demaine, E., Flatland, R., O'Rourke, J.: Unfolding genus-2 orthogonal polyhedra with linear refinement. CoRR, abs/1611.00106 (2016)

9. Damian, M., Flatland, R., O'Rourke, J.: Epsilon-unfolding orthogonal polyhedra. Graphs Comb. **23**, 179–194 (2007)

10. Chand De, U.: Differential Geometry of Curves and Surfaces in E3. Anshan (2007)

11. Dürer, A.: Unterweysung der Messung mit der Zirkel un Richtscheyt in Linien Ebnen uhnd Gantzen Corporen (1525)

12. Michael Galetzka and Patrick O Glauner. A correct even-odd algorithm for the point-in-polygon (pip) problem for complex polygons. arXiv preprint arXiv:1207.3502 (2012)

13. Hyun, D.E., Yoon, S.H., Kim, M.S., Juttler, B.: Modeling and deformation of arms and legs based on ellipsoidal sweeping. In: 11th Pacific Conference on Computer Graphics and Applications, Proceedings, pp. 204–212. IEEE (2003)

14. Laine, S.: A topological approach to voxelization. Comput. Graph. Forum **32**(4), 77–86 (2013)

15. Liou, M.-H., Poon, S.-H., Wei, Y.-J.: On edge-unfolding one-layer lattice polyhedra with cubic holes. In: Cai, Z., Zelikovsky, A., Bourgeois, A. (eds.) COCOON 2014. LNCS, vol. 8591, pp. 251–262. Springer, Cham (2014). doi:10.1007/978-3-319-08783-2_22

16. Toutant, J.-L., Andres, E., Largeteau-Skapin, G., Zrour, R.: Implicit digital surfaces in arbitrary dimensions. In: Barcucci, E., Frosini, A., Rinaldi, S. (eds.) DGCI 2014. LNCS, vol. 8668, pp. 332–343. Springer, Cham (2014). doi:10.1007/978-3-319-09955-2_28

Geometric Transforms

Honeycomb Geometry: Rigid Motions on the Hexagonal Grid

Kacper Pluta[1(✉)], Pascal Romon[2], Yukiko Kenmochi[3], and Nicolas Passat[4]

[1] LIGM (UMR 8049), LAMA (UMR 8050), UPEM, UPEC, CNRS,
Université Paris-Est, Marne-la-Vallée, France
kacper.pluta@univ-paris-est.fr
[2] LAMA (UMR 8050), UPEM, UPEC, CNRS, Université Paris-Est,
Marne-la-Vallée, France
pascal.romon@u-pem.fr
[3] LIGM (UMR 8049), UPEM, CNRS, ESIEE Paris, ENPC, Université Paris-Est,
Marne-la-Vallée, France
yukiko.kenmochi@esiee.fr
[4] CReSTIC, Université de Reims Champagne-Ardenne, Reims, France
nicolas.passat@univ-reims.fr

Abstract. Euclidean rotations in $\mathbb{R}^2$ are bijective and isometric maps, but they generally lose these properties when digitized in discrete spaces. In particular, the topological and geometric defects of digitized rigid motions on the square grid have been studied. This problem is related to the incompatibility between the square grid and rotations; in general, one has to accept either relatively high loss of information or non-exactness of the applied digitized rigid motion. Motivated by these facts, we study digitized rigid motions on the hexagonal grid. We establish a framework for studying digitized rigid motions in the hexagonal grid—previously proposed for the square grid and known as neighborhood motion maps. This allows us to study non-injective digitized rigid motions on the hexagonal grid and to compare the loss of information between digitized rigid motions defined on the two grids.

1 Introduction

Rigid motions on $\mathbb{R}^2$ are fundamental transformations in 2D image processing. They are isometric and bijective. Nevertheless, digitized rigid motions are, in general, non-bijective. Despite many efforts during the last twenty years, the topological and geometric defects of digitized rigid motions on the square grid are still not fully understood. According to Nouvel and Rémila, this problem is not related to a limitation of arithmetic precision, but arises as a deep incompatibility between the square grid and rotations [1].

Pioneering contributions to two-dimensional digital geometry in the square grid were made in 1970s (see [2,3] for a survey). The main reason for taking this Cartesian approach was that objects defined on the square grid are easy to address in the computer memory, as well as image acquisition devices, whose

W.G. Kropatsch et al. (Eds.): DGCI 2017, LNCS 10502, pp. 33–45, 2017.
DOI: 10.1007/978-3-319-66272-5_4

sensors are generally organized on square grids. Moreover, the square grid has a direct extension to higher dimensions. However, such a Cartesian framework suffers from fundamental topological problems and therefore one has to choose different connectivity relations for objects of interest and their complements [2], or use a cell-complex framework [4], for example.

On the contrary, the regular hexagonal grid (called hereafter the hexagonal grid) suffers less from these problems since it possesses the following properties: *equidistant neighbors* – each hexagon has six equidistant neighbors; *uniform connectivity* – there is only one type of connectivity [5,6]. However, operations such as a sampling of continuous signals with the hexagonal grids are often computationally more expensive than ones defined on the square grid. Indeed, a digitization operator on the square grid can be defined by a rounding function, while there is no such counterpart on the hexagonal grid. Nevertheless, a practically useful digitization operator on a hexagonal grid has been proposed by Her [7].

Motivated by aforementioned issues, we study the differences between digitized rigid motions defined on both types of grids. To understand these differences, we establish a framework for studying digitized rigid motions on the hexagonal grid at a local scale. Our approach is an extension to the hexagonal grid of the former study of digitized rigid motions of the square grid, which is based on neighborhood motion maps [8]. This enables us to study the structure of groups induced by images of digitized rigid motions. It also allows us to focus on the issue of the information preservation under digitized rigid motions defined on the two grids. We provide a comparison of the information loss between the hexagonal and square grids. In addition, we present a complete list of neighborhood motion maps for the 6-neighborhood in Appendix.

2   Digitizing on the Hexagonal Grid

2.1   Hexagonal Grid

A hexagonal grid $\mathcal{H}$ is a grid formed by a tessellation of $\mathbb{R}^2$ by regular hexagons of side length $1/\sqrt{3}$, as illustrated in Fig. 1. Let us consider the center points of these hexagons, i.e. the points of the underlying lattice $\Lambda = \mathbb{Z}\boldsymbol{\epsilon}_1 \oplus \mathbb{Z}\boldsymbol{\epsilon}_2$ (see Fig. 1(a)) where $\boldsymbol{\epsilon}_1 = (1,0)^t$ and $\boldsymbol{\epsilon}_2 = \left(-\frac{1}{2}, \frac{\sqrt{3}}{2}\right)^t$. We can then regard $\mathcal{H}$ as the boundary of the Voronoi tessellation of $\mathbb{R}^2$ associated with the lattice Λ.

2.2   Digitized Rigid Motions

Rigid motions on $\mathbb{R}^2$ are isometric maps defined as

$$\left| \begin{array}{l} \mathcal{U} : \mathbb{R}^2 \to \mathbb{R}^2 \\ \quad \mathbf{x} \; \mapsto \; \mathbf{R}\mathbf{x} + \mathbf{t} \end{array} \right. \tag{1}$$

where $\mathbf{t} \in \mathbb{R}^2$ is a translation vector and $\mathbf{R}$ is a rotation matrix.

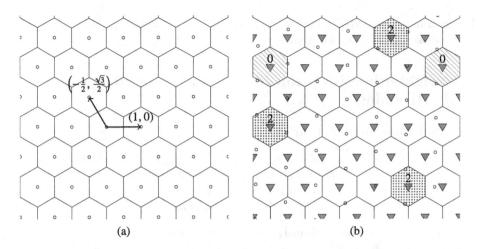

(a)   (b)

Fig. 1. A visualization of a part of the hexagonal grid (a) The arrows represent the base of the underlying lattice Λ with the white dots representing its elements. (b) Examples of three different point statuses induced by a rigid motion $\mathcal{U}$ on the hexagonal grid: digitization cells corresponding to 0- and 2-points are marked by green lined, red dotted patterns and labeled, with their status numbers, respectively. The white dots indicate the positions of the images of the points of the initial set Λ under $\mathcal{U}$

According to Eq. (1), we generally have $\mathcal{U}(\Lambda) \nsubseteq \Lambda$. As a consequence, in order to define digitized rigid motions as maps from Λ to Λ, we commonly apply rigid motions on Λ as a part of $\mathbb{R}^2$ and then combine the results with a digitization operator. To define such a digitization operator on Λ, let us first define a *digitization cell* of $\kappa \in \Lambda$

$$\mathcal{C}(\kappa) = \left\{ \mathbf{x} \in \mathbb{R}^2 \mid \forall \beta \in B, (\, \|\mathbf{x} - \kappa\| \leq \|\mathbf{x} - \kappa + \beta\|) \wedge (\|\mathbf{x} - \kappa\| < \|\mathbf{x} - \kappa - \beta\|) \right\},$$

where $B = \{\epsilon_1, \epsilon_2, \epsilon_1 + \epsilon_2\} \subset \Lambda$. The *digitization operator* is then defined as a function $\mathcal{D} : \mathbb{R}^2 \to \Lambda$ such that $\forall \mathbf{x} \in \mathbb{R}^2, \exists ! \mathcal{D}(\mathbf{x}) \in \Lambda$ and $\mathbf{x} \in \mathcal{C}(\mathcal{D}(\mathbf{x}))$. We finally define digitized rigid motions as $U = \mathcal{D} \circ \mathcal{U}_{|\Lambda}$. Note that this definition of the digitization operator is rather theoretical but not computationally relevant. For readers who are interested in implementing digitization operators in the hexagonal grid, we advise the method proposed by Her [7].

Due to the behavior of $\mathcal{D}$, that maps $\mathbb{R}^2$ onto Λ, digitized rigid motions are, most of the time, non-bijective. This leads us to define a notion of point status with respect to digitized rigid motions, as in the case of the square grid [9].

Definition 1. *Let $\lambda \in \Lambda$. The set of preimages of $\lambda \in \Lambda$ with respect to U is defined as $S_U(\lambda) = \{\kappa \in \Lambda \mid U(\kappa) = \lambda\}$, and λ is referred to as an s-point where $s = |S_U(\lambda)|$ and is called the* status *of λ.*

Remark 2. For any $\lambda \in \Lambda$, $|S_U(\lambda)| \in \{0, 1, 2\}$. From the digitization cell geometry, we have that $|S_U(\lambda)| = 2$ only when two preimages $\kappa, \sigma \in S_U(\lambda)$ satisfy $\|\kappa - \sigma\|_2 = 1$.

The non-injective and non-surjective behaviors of digitized rigid motions result in the existence of 2- and 0-points, respectively (see Fig. 1(b)).

3 Neighborhood Motion Maps and Remainder Range Partitioning

3.1 Neighborhood Motion Maps

In $\mathbb{R}^2$, an intuitive way to define the neighborhood of a point $\mathbf{x}$ is to consider the set of points that lie within a ball of a given radius centered at $\mathbf{x}$. This metric definition actually remains valid in Λ and it allows us to retrieve the classical notion of neighborhood based on adjacency relations.

Definition 3 (Neighbourhood). *The* neighborhood *of $\kappa \in \Lambda$ (of a squared radius $r \in \mathbb{R}_+$), denoted $\mathcal{N}_r(\kappa)$, is defined as $\mathcal{N}_r(\kappa) = \{\kappa + \delta \in \Lambda \mid \|\delta\|_2^2 \leq r\}$.*

Remark 4. $\mathcal{N}_1$ corresponds to the 6-neighborhood which is the smallest, non-trivial and isotropic neighborhood of $\kappa \in \Lambda$.

As stated above, the non-injective and/or non-surjective behavior of a digitized rigid motion may result in the existence of 2- and/or 0-points. In other words, given a point $\kappa \in \Lambda$, the image of its neighborhood $\mathcal{N}_r(\kappa)$ may be distributed in a non-homogeneous fashion within the neighborhood of the image $\mathcal{N}_{r'}(U(\kappa)), r \leq r'$, of κ with respect to the digitized rigid motion U.

In order to track these local alterations of the neighborhood of $\kappa \in \Lambda$, we consider the notion of a *neighborhood motion map* on Λ defined as a set of vectors, each representing the rigid motion of a neighbor.

Definition 5 (Neighborhood motion map). *The* neighborhood motion map *of $\kappa \in \Lambda$ with respect to a digitized rigid motion U and $r \in \mathbb{R}_+$ is the function defined as*

$$\left| \begin{array}{rcl} \mathcal{G}_r^U(\kappa) : \mathcal{N}_r(\mathbf{0}) & \longrightarrow & \mathcal{N}_{r'}(\mathbf{0}) \\ \delta & \longmapsto & U(\kappa + \delta) - U(\kappa) \end{array} \right.$$

where $r' \geq r$.

In other words, $\mathcal{G}_r^U(\kappa)$ associates to each relative position of a point $\kappa + \delta$ in the neighborhood of κ, the relative position of the image $U(\kappa + \delta)$ in the neighborhood of $U(\kappa)$.

Remark 6. The maximal squared radius r' of the new neighborhood $U(\mathcal{N}_r(\kappa))$ is slightly larger than r, due to digitization effect. In particular, we have $r' = 4$ for $r = 1$.

For a better understanding of $\mathcal{G}_r^U(\kappa)$, we will consider a visual representation of the $\mathcal{G}_r^U(\kappa)$ functions as label maps. A first—reference—map $\mathcal{L}_r$ associates a specific color label to each point δ of $\mathcal{N}_r(\mathbf{0})$ for a given squared radius r (see

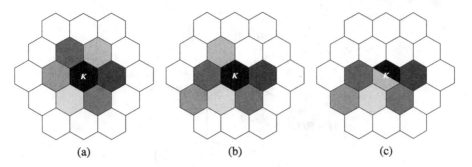

(a)   (b)   (c)

Fig. 2. (a) The reference label map $\mathcal{L}_1$. (b–c) Examples of label maps $\mathcal{L}_1^U(\kappa)$. (b) Each point contains at most one label: the rigid motion $U_{|\mathcal{N}_1(\kappa)}$ is then locally injective. (c) One point contains two labels: $U_{|\mathcal{N}_1(\kappa)}$ is then non-injective

Fig. 2(a) for the map $\mathcal{L}_1$). A second map $\mathcal{L}_r^U(\kappa)$—associated to $\mathcal{G}_r^U(\kappa)$, i.e. to a point κ and a digitized rigid motion U—associates to each point σ of $\mathcal{N}_{r'}(\mathbf{0})$ the labels of all the points $\delta \in \mathcal{N}_r(\mathbf{0})$ such that $U(\kappa + \delta) - U(\kappa) = \sigma$. Each σ may contain $0, 1$ or 2 labels, due to the various possible statuses of points of Λ under digitized rigid motions (see Remark 2). Figures 2(b–c) provide some examples.

Note that a similar idea was previously proposed to study local alterations of the neighborhood $\mathcal{N}_1$ under 2D digitized rotations [1], and local alterations of $\mathcal{N}_r$ under 2D digitized rigid motions [8] defined on the square grid.

3.2   Remainder Range Partitioning

Digitized rigid motions $U = \mathcal{D} \circ \mathcal{U}_{|\Lambda}$ are piecewise constant, which is a consequence of the nature of $\mathcal{D}$. In other words, the neighborhood motion map $\mathcal{G}_r^U(\kappa)$ evolves non-continuously according to the parameters of $\mathcal{U}$ that underlies U. Our purpose is now to express how $\mathcal{G}_r^U(\kappa)$ evolves.

Let us consider a point $\kappa + \delta \in \Lambda$ in the neighborhood $\mathcal{N}_r(\kappa)$ of κ. From Formula (1), we have

$$\mathcal{U}(\kappa + \delta) = \mathbf{R}\delta + \mathcal{U}(\kappa). \tag{2}$$

We know that $\mathcal{U}(\kappa)$ lies in a digitization cell $\mathcal{C}(U(\kappa))$ centered at $U(\kappa)$, which implies that there exists a value $\rho(\kappa) = \mathcal{U}(\kappa) - U(\kappa) \in \mathcal{C}(\mathbf{0})$.

Definition 7. *The coordinates of $\rho(\kappa)$, called the* remainder *of κ under $\mathcal{U}$, are the fractional parts of the coordinates of $\mathcal{U}(\kappa)$ and ρ is called the* remainder map *under $\mathcal{U}$.*

As $\rho(\kappa) \in \mathcal{C}(\mathbf{0})$, this range $\mathcal{C}(\mathbf{0})$ is called the *remainder range*. Using ρ, we can rewrite Eq. (2) as $\mathcal{U}(\kappa + \delta) = \mathbf{R}\delta + \rho(\kappa) + U(\kappa)$.

Without loss of generality, we can consider that $U(\kappa)$ is the origin of a local coordinate frame of the image space, i.e. $\mathcal{U}(\kappa) \in \mathcal{C}(\mathbf{0})$. In such local coordinate frame, the former equation rewrites as $\mathcal{U}(\kappa + \delta) = \mathbf{R}\delta + \rho(\kappa)$. Still under this assumption, studying the non-continuous evolution of the neighborhood motion

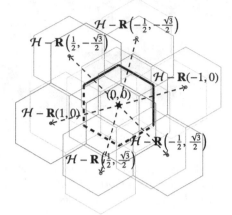

Fig. 3. The remainder range $\mathcal{C}(\mathbf{0})$ intersected with the translated hexagonal grids $\mathcal{H} - \mathbf{R}\boldsymbol{\delta}, \boldsymbol{\delta} \in \mathcal{N}_1(\mathbf{0})$ for rotation angle $\theta = \frac{\pi}{12}$. Each hexagonal grid $\mathcal{H} - \mathbf{R}\boldsymbol{\delta}$ is colored with respect to each $\boldsymbol{\delta} \in \mathcal{N}_1(\mathbf{0})$ in the reference label map $\mathcal{L}_1$ (see Fig. 2)

map $\mathcal{G}_r^U(\boldsymbol{\kappa})$ is equivalent to studying the behavior of $U(\boldsymbol{\kappa} + \boldsymbol{\delta}) = \mathcal{D} \circ \mathcal{U}(\boldsymbol{\kappa} + \boldsymbol{\delta})$ for $\boldsymbol{\delta} \in \mathcal{N}_r(\mathbf{0})$ and $\boldsymbol{\kappa} \in \Lambda$, with respect to the rotation angle θ defining $\mathbf{R}$ and the translation embedded in $\rho(\boldsymbol{\kappa}) = (x, y) \in \mathcal{C}(\mathbf{0})$, that deterministically depends on (x, y, θ). The discontinuities of $U(\boldsymbol{\kappa} + \boldsymbol{\delta})$ occur when $\mathcal{U}(\boldsymbol{\kappa} + \boldsymbol{\delta})$ is on the boundary of a digitization cell. These critical cases related to $\mathcal{U}(\boldsymbol{\kappa} + \boldsymbol{\delta})$ can be observed via the relative positions of $\rho(\boldsymbol{\kappa})$, which are formulated by the translated hexagonal grid $\mathcal{H} - \mathbf{R}\boldsymbol{\delta}$, more precisely $\mathcal{C}(\mathbf{0}) \cap (\mathcal{H} - \mathbf{R}\boldsymbol{\delta})$. Let us consider $\mathcal{H} - \mathbf{R}\boldsymbol{\delta}$ for all $\boldsymbol{\delta} \in \mathcal{N}_r(\mathbf{0})$ in $\mathcal{C}(\mathbf{0})$, namely $\mathscr{H} = \bigcup_{\boldsymbol{\delta} \in \mathcal{N}_r(\mathbf{0})} (\mathcal{H} - \mathbf{R}\boldsymbol{\delta})$. Then $\mathcal{C}(\mathbf{0}) \cap \mathscr{H}$ subdivides the remainder range into regions, as illustrated in Figs. 3 and 4 for $r = 1$, called *frames*. Compared to the remainder range of the square grid [8], the geometry of the frames is relatively complex (see Fig. 4). From the definition, we have the following proposition.

Proposition 8. *For any* $\boldsymbol{\lambda}, \boldsymbol{\kappa} \in \Lambda$, $\mathcal{G}_r^U(\boldsymbol{\lambda}) = \mathcal{G}_r^U(\boldsymbol{\kappa})$ *iff* $\rho(\boldsymbol{\lambda})$ *and* $\rho(\boldsymbol{\kappa})$ *are in the same frame.*

3.3   Generating Neighborhood Motion Maps for $r = 1$

From the above discussion, it is plain that a partition of the remainder range given by $\mathcal{C}(\mathbf{0}) \cap \mathscr{H}$ depends on the rotation angle θ. In order to detect the set of all neighborhood motion maps for $r = 1$, i.e. equivalence classes of rigid motions $U_{|\mathcal{N}_1(\boldsymbol{\kappa})}$, we need to consider critical angles that lead to topological changes of $\mathcal{C}(\mathbf{0}) \cap \mathscr{H}$. Indeed, from Proposition 8, we know that this is equivalent to computing all different frames in the remainder range. Such changes occur when at least one frame has a null area, i.e. when at least two parallel line segments which bound a frame have their intersection equal to a line segment.

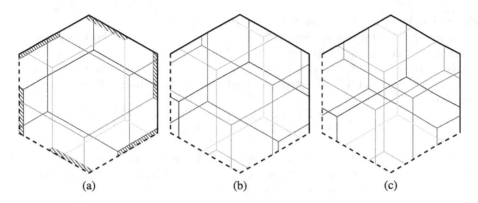

Fig. 4. Three different partitions of the remainder range $\mathcal{C}(\mathbf{0})$ depending on rotation angles. (a) $\theta_1 = \pi/12 \in (\alpha_0, \alpha_1)$, together with non-injective zones marked by a lined pattern. (b) $\theta_2 = 357/1000 \in (\alpha_1, \alpha_2)$ and (c) $\theta_3 = 19/40 \in (\alpha_2, \alpha_3)$. The color of each $\mathcal{H} - \mathbf{R}\boldsymbol{\delta}$ corresponds to that of each neighbor $\boldsymbol{\delta}$ in the label map $\mathcal{L}_1$

To illustrate this issue, let us consider the minimal distance among the distances between all pairs of the parallel line segments. Thanks to rotational symmetries by an angle of $\frac{\pi}{6}$, and based on the above discussion, we can restrict, without loss of generality, the parameter space of (x, y, θ) to $\mathcal{C}(\mathbf{0}) \times \left[0, \frac{\pi}{6}\right]$. Then, we found that there exist two critical angles between 0 and $\frac{\pi}{6}$, denoted by α_1 and α_2, with $0 < \alpha_1 < \alpha_2 < \frac{\pi}{6}$. Note that the angles 0 and $\frac{\pi}{6}$ are also critical, and denoted α_0 and α_3, respectively. Figure 4 presents three different partitions of the remainder range for angles $\theta_1 \in (\alpha_0, \alpha_1)$, $\theta_2 \in (\alpha_1, \alpha_2)$ and $\theta_3 \in (\alpha_2, \alpha_3)$, respectively.

Based on this knowledge on critical angles, we can observe the set of all distinct neighborhood motion maps $\mathbb{M}_r = \bigcup_{\mathcal{U} \in \mathbb{U}} \bigcup_{\boldsymbol{\kappa} \in \Lambda} \{\mathcal{G}_r^U(\boldsymbol{\kappa})\}$ where $\mathbb{U}$ is the set of all rigid motions $\mathcal{U}$ defined by the restricted parameter space $\mathcal{C}(\mathbf{0}) \times \left[0, \frac{\pi}{6}\right]$. The cardinality of $\mathbb{M}_1$ is equal to 67. It should be noticed that $\left| \bigcup_{\boldsymbol{\kappa} \in \Lambda} \{\mathcal{G}_1^U(\boldsymbol{\kappa})\} \right|$ is constant: 49 for any $\mathcal{U}$, except for $\theta = \alpha_i, i \in \{0, 1, 2, 3\}$. Indeed, we have $\left| \bigcup_{\boldsymbol{\kappa} \in \Lambda} \{\mathcal{G}_1^U(\boldsymbol{\kappa})\} \right| = 1$ for $\theta = \alpha_0$, 43 for $\theta = \alpha_1$, 37 for $\theta = \alpha_2$ and 30 for $\theta = \alpha_3$. Such elements of the set $\mathbb{M}_1$ are presented in Appendix.

4   Eisenstein Rational Rotations

In this section, we study the images of the remainder map ρ with respect to the parameters of the underlying rigid motion. In the case of the square grid, it is known that only for rotations with rational cosine and sine, i.e. rotations given by *Pythagorean primitive triples*, the remainder map does have a finite number of images; furthermore these images form a group [1,8,10]. When, on the contrary, cosine or/and sine are irrational, the images form an infinite and dense set: any ball of non-zero radius in the remainder range intersects images of the remainder map. We will show that, in the hexagonal grid case, a similar result is obtained for rotations corresponding to the counterparts of the Pythagorean primitive triples in the hexagonal lattice. These are called *Eisenstein primitive triples*, namely triples $(a, b, c) \in \mathbb{Z}^3$ such that $0 < a < c < b$, $\gcd(a, b, c) = 1$ and $a^2 - ab + b^2 = c^2$ [11,12]. We shall say that the rotation matrix $\mathbf{R}$ is *Eisenstein rational* if

$$\mathbf{R} = \begin{bmatrix} \frac{2a-b}{2c} & -\frac{\sqrt{3}b}{2c} \\ \frac{\sqrt{3}b}{2c} & \frac{2a-b}{2c} \end{bmatrix} = \frac{a}{c}\begin{bmatrix} 1 & 0 \\ 0 & 1 \end{bmatrix} + \frac{b}{c}\begin{bmatrix} -\frac{1}{2} & -\frac{\sqrt{3}}{2} \\ \frac{\sqrt{3}}{2} & \frac{1}{2} \end{bmatrix}$$

where (a, b, c) is an Eisenstein primitive triple. We must note that *any* rotation matrix can be written in this form, with a, b, c real numbers, not Eisenstein triples (and not even integers in general).

To begin, we focus on the rotation part of a rigid motion and define a group $\mathcal{G} = \mathbb{Z}\mathbf{R}\epsilon_1 \oplus \mathbb{Z}\mathbf{R}\epsilon_2 \oplus \mathbb{Z}\epsilon_1 \oplus \mathbb{Z}\epsilon_2$ and its translation $\mathcal{G}' = \mathcal{G} + \mathbf{t}$. We state the following result.

Proposition 9. *The group $\mathcal{G}$ is a rank two lattice if and only if the rotation matrix $\mathbf{R}$ is Eisenstein rational.*

Proof. First, we note that the density properties of the underlying group is not affected by affine transformations, i.e. a lattice (resp. dense group) is transformed into another lattice (resp. dense group). Here we consider $\mathbf{X} = [\epsilon_1 \mid \epsilon_2]^{-1}$ so that $\{\mathbf{X}\kappa \mid \kappa \in \Lambda\} = \mathbb{Z}^2$. Then we obtain

$$\check{\mathbf{R}} = \mathbf{X}\mathbf{R}\mathbf{X}^{-1} = \mathbf{X}\begin{bmatrix} \frac{2a-b}{2c} & -\frac{\sqrt{3}b}{2c} \\ \frac{\sqrt{3}b}{2c} & \frac{2a-b}{2c} \end{bmatrix}\mathbf{X}^{-1} = \begin{bmatrix} \frac{a}{c} & -\frac{b}{c} \\ \frac{b}{c} & \frac{a-b}{c} \end{bmatrix}, \tag{3}$$

and study $\check{\mathcal{G}} = \check{\mathbf{R}}\mathbb{Z}\left(\begin{smallmatrix}1\\0\end{smallmatrix}\right) \oplus \check{\mathbf{R}}\mathbb{Z}\left(\begin{smallmatrix}0\\1\end{smallmatrix}\right) \oplus \mathbb{Z}\left(\begin{smallmatrix}1\\0\end{smallmatrix}\right) \oplus \mathbb{Z}\left(\begin{smallmatrix}0\\1\end{smallmatrix}\right)$, instead of $\mathcal{G}$.

The generators of $\check{\mathcal{G}}$ are given by the columns of the rational matrix $\mathbf{B} = [\check{\mathbf{R}} \mid \mathbf{I}_2]$ where $\mathbf{I}_2$ stands for the 2×2 identity matrix. As $\mathbf{B}$ is a rational, full row rank matrix, it can be brought to its Hermite normal form $\mathbf{H} = [\mathbf{T} \mid \mathbf{0}_{2,2}]$. The problem of computing the Hermite normal form $\mathbf{H}$ of the rational matrix $\mathbf{B}$ reduces to that of computing the Hermite normal form of an integer matrix: $c \in \mathbb{Z}$ is the least common multiple of all the denominators of $\mathbf{B}$; compute the Hermite normal form $\mathbf{H}'$ for the integer matrix $c\mathbf{B}$; finally, the Hermite normal form $\mathbf{H}$ of $\mathbf{B}$ is obtained by $c^{-1}\mathbf{H}'$. The columns of $\mathbf{H}$ are the minimal generators of $\check{\mathcal{G}}$. Since the rank of $\mathbf{B}$ is equal to 2, $\mathbf{H}$ gives a base (σ, ϕ), so that $\check{\mathcal{G}} = \mathbb{Z}\sigma \oplus \mathbb{Z}\phi$. As $\mathbf{H}'$ gives an integer base, $c\check{\mathcal{G}}$ is an integer lattice. Finally, $\mathcal{G} = \mathbb{Z}\mathbf{X}^{-1}\sigma \oplus \mathbb{Z}\mathbf{X}^{-1}\phi$.

Conversely, let us prove that $\mathcal{G}$ is dense if (a, b, c) is not an Eisenstein primitive triple (up to scaling). Again, we consider instead $\check{\mathbf{R}} = [\mathbf{b}_1 \mid \mathbf{b}_2]$, and prove that for any $\varepsilon > 0$ there exists $\mathbf{e}, \mathbf{e}' \in \check{\mathcal{G}}$, linearly independent, such that $\|\mathbf{e}\| < \varepsilon$, $\|\mathbf{e}'\| < \varepsilon$. Let $\{.\}$ stand for the fractional part function. We study the images of $\{\mathbb{Z}\mathbf{b}_1\} \in [-1/2, 1/2)^2$, where $\mathbf{b}_1 = \left(\begin{smallmatrix} a/c \\ b/c \end{smallmatrix}\right)$ denotes the first column of $\check{\mathbf{R}}$. If $\mathbf{b}_1$ contains irrational elements, the $\{\mathbb{Z}\mathbf{b}_1\}$ contains infinitely many distinct points. By compactness of $[-1/2, 1/2]^2$, we can extract a subsequence $(\{n_k\mathbf{b}_1\})_{k\in\mathbb{N}}$, converging to some point in $[-1/2, 1/2]^2$. Thus $\{(n_{k+1}-n_k)\mathbf{b}_1\} = \{n_{k+1}\mathbf{b}_1\} - \{n_k\mathbf{b}_1\}$ converges to $(0, 0)$. In particular, we can find integers m, p, q, where $m = n_{k+1} - n_k$

for k large enough, such that $\boldsymbol{e} = m\boldsymbol{b}_1 + (p,q) \in \check{\mathcal{G}}$ has norm smaller than $\frac{\varepsilon}{3}$. Note now that the second column of $\check{\mathbf{R}}$ satisfies $\boldsymbol{b}_2 = \left[\begin{smallmatrix} 0 & -1 \\ 1 & -1 \end{smallmatrix}\right] \boldsymbol{b}_1$. Then we claim that $\boldsymbol{e}' = \left[\begin{smallmatrix} 0 & -1 \\ 1 & -1 \end{smallmatrix}\right] \boldsymbol{e} = m\boldsymbol{b}_2 + (-q, p-q)$ also lies in $\check{\mathcal{G}}$, has norm less than $3\|\boldsymbol{e}\|$ (hence less than ε) and is linearly independent from $\boldsymbol{e}$ (the matrix $\check{\mathbf{R}}$ has no eigenvectors).

Consequently, $\boldsymbol{b}_1$ has rational coefficients, and we may take a, b, c integers with gcd equal to 1. Since $\cos\theta = \frac{2a-b}{c}$ and $\sin\theta = \frac{\sqrt{3}b}{2c}$, we conclude that these form an Eisenstein primitive triple. $\qquad\square$

In the next section, we focus on a subgroup of $\mathcal{G}$ obtained from the intersection with the remainder range, i.e. $\bar{\mathcal{G}} = \mathcal{G}/\Lambda$ and its translation by $\mathbf{t}$ (modulo Λ).

Corollary 10. *If* $\cos\theta = \frac{2a-b}{2c}$ *and* $\sin\theta = \frac{\sqrt{3}b}{2c}$ *where* (a,b,c) *is a primitive Eisenstein triple, the group* $\bar{\mathcal{G}} = \mathcal{G}/\Lambda$ *is cyclic and* $|\bar{\mathcal{G}}| = c$.

Proof. Up to the affine transformation of (3), we may consider the quotient group $\check{\mathcal{G}} = \check{\mathcal{G}}/\mathbb{Z}^2$ and we note that for a primitive Eisenstein triple (a,b,c), any two integers in the triple are coprime [12, p. 12]. Let us first give a characterization of $\check{\mathcal{G}}$. From the proof of Proposition 9, we know that any element $\mathbf{x} \in \check{\mathcal{G}}$ is a rational vector of the form $(\frac{q_1}{c}, \frac{q_2}{c})$. By definition, $\{\mathbf{x}\} \in \check{\mathcal{G}}$ iff there exist $n, m \in \mathbb{Z}$ such that $\{x\} = \{n\boldsymbol{b}_1 + m\boldsymbol{b}_2\}$, i.e. there exist integers n, m, u, v such that

$$\begin{cases} \frac{q_1}{c} + u = n\frac{a}{c} - m\frac{b}{c}, \\ \frac{q_2}{c} + v = n\frac{b}{c} + m\frac{a-b}{c} \end{cases} \text{ or, equivalently, } \begin{cases} q_1 + uc = an - bm, \\ q_2 + vc = bn + am - bm. \end{cases}$$

A linear combination of both lines yields directly $bq_1 - aq_2 = c(-cm - bu + av)$, hence $bq_1 - aq_2 \equiv 0 \pmod{c}$ is a necessary condition. It is also sufficient. Indeed, let us suppose that $bq_1 - aq_2 = kc$, $k \in \mathbb{Z}$.

Then, since $\gcd(a,b) = 1$, we know that the solutions to this Diophantine equation are of the form $(q_1, q_2) = \ell(a, b) + kc(\beta, -\alpha)$, where $\alpha a + \beta b = 1$ (Bézout identity) and $\ell \in \mathbb{Z}$. Consequently, $(\frac{q_1}{c}, \frac{q_2}{c}) = \ell\boldsymbol{b}_1 + (k\beta, -k\alpha)$ lies in $\check{\mathcal{G}}$.

Moreover, we deduce that $\check{\mathcal{G}}$ is cyclic with generator $\boldsymbol{b}_1$. Finally, $\{\ell\boldsymbol{b}_1\} = (0,0)$ implies $\ell a = uc$ and $\ell b = vc$ for some integers u, v. Applying the Gauss' lemma to coprimes a, c, we see that ℓ needs to be a multiple of c. Therefore $|\check{\mathcal{G}}| = c$. $\square$

5   Non-injective Digitized Rigid Motions

5.1   Non-injective Frames of the Remainder Range

In Remark 2, we observed that some frames of the remainder range correspond to neighborhood motion maps which exhibit non-injectivity of the corresponding digitized rigid motion. The set of all the neighborhood motion maps $\mathbb{M}_1$, presented in Appendix, allows us to identify such non-injective zones i.e., unions of frames of the remainder range. For instance, the frames related to the neighborhood motion maps of the axial coordinates; $(-2, 4), (-1, 4)$ and $(0, 3)$ (see Fig. 6(top)), constitute a case of such a non-injective zone for rotation angles in

(α_0, α_1). Based on this observation, we can characterize the non-injectivity of a digitized rigid motion by the presence of $\rho(\kappa)$ in these specific zones (illustrated in Fig. 4(a)).

Conjecture 11. Let C_6 stand for the 6-fold discrete rotation symmetry group. Given $\mathcal{U} \in \mathbb{U}$ and $\kappa \in \Lambda$, $U(\kappa)$, has two preimages κ and $\kappa + \delta$ where $\delta \in \mathcal{N}_1(\kappa)$ iff $\rho(\kappa)$ is in one of the zones $c_k(\mathcal{F}), c_k \in C_6$, where $\mathcal{F}$ is the parallelogram region whose vertices are: $\left(\cos\theta - 1, \frac{\cos\theta}{\sqrt{3}} \right)$, $\left(0, \frac{1}{\sqrt{3}} \right)$, $\left(\frac{1}{2} \left(2 - \cos\theta - \sqrt{3}\sin\theta \right), \frac{1}{6} \left(\sqrt{3}\cos\theta + 3\sin\theta \right) \right)$, $\left(\frac{1}{2} \left(\cos\theta - \sqrt{3}\sin\theta \right), \frac{1}{6} \left(3\sin(\theta) + 3\sqrt{3}\cos(\theta) - 2\sqrt{3} \right) \right)$.

5.2   Non-injective Digitized Rigid Motions in Square and Hexagonal Grids

In this section, we compare the loss of information induced by rigid motions on the square and the hexagonal grids. Indeed, we aim to determine on which type of grid digitized rigid motions preserve more information.

In accordance with the discussion in Sect. 4 and the similar discussion for the square grid in [1,13], the density of images of the remainder map in the non-injective zones is related to the cardinality and the structure of $\bar{\mathcal{G}}$. On the one hand, when $\bar{\mathcal{G}}$ is dense, it is considered as the ratio between the area of the union of all the non-injective zones $\bigcup\limits_{k=1,...,6} c_k(\mathcal{F})$, and the area of the remainder range [13]. On the other hand, when $\bar{\mathcal{G}}$ forms a lattice, it is estimated as $\frac{|\bar{\mathcal{G}} \cap (\bigcup c_k(\mathcal{F}))|}{c}$, $k = 1, \ldots, 6$, where c is an element of a primitive Eisenstein (resp. Pythagorean) triple, i.e. $|\bar{\mathcal{G}}|$ [13].

Nevertheless, to facilitate our study, we will consider the former ratio between the areas as an approximation of the information loss measure. Indeed, the rota-

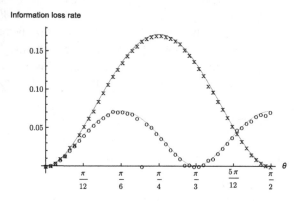

Fig. 5. Comparison of the loss of information induced by digitized rigid motions on the hexagonal and the square grids. The red and blue curves correspond to the ratios between the areas of non-injective zones and the remainder range for the square and the hexagonal grids, respectively. The "x" (resp. "o") markers correspond to the values of the information loss rate $1 - \frac{|U(S_Z)|}{|S_Z|}$ (resp. $1 - \frac{|U(S_L)|}{|S_L|}$)

tions induced by primitive Eisenstein (resp. Pythagorean) triples are dense; one can always find a relatively near rotation angle such that c is relatively high. The area-ratio density measure can be then seen as a limit for the cardinality based ratio. Figure 5 presents the analytical curves of the area ratios for the hexagonal and the square grids with respect to rotation angles.

In order to validate this approximation, we also measure the loss of points for different sampled rotations and the following finite sets $S_L = \Lambda \cap [-100, 100]^2$ (resp. $S_Z = \mathbb{Z}^2 \cap [-100, 100]^2$) in the hexagonal (resp. square) grid. In this setting, we use $1 - \frac{|U(S_L)|}{|S_L|}$ (resp. $1 - \frac{|U(S_Z)|}{|S_Z|}$) as the measure. They are plotted as well in Fig. 5. We note that the experimental results follow the area-ratio measure that provides consequently a good approximation[1]. The obtained results allow us to conclude that digitized rigid motions on the hexagonal grid preserve more information than their counterparts defined on the square grid.

6   Conclusion

In this article, we have extended the framework of neighborhood motion maps to rigid motions defined on the hexagonal grid—previously proposed for digitized rigid motions defined on $\mathbb{Z}^2$ [8, 10]. Then, we studied the density of images of the remainder map and the structure of the induced groups. Finally, we have shown that the loss of information induced by digitization of rigid motions on the hexagonal grid is relatively low, in comparison with those on the square edge.

Our main perspective is to use the proposed framework for studying bijective digitized rigid motions and geometric and topological alterations induced by digitized rigid motions on the hexagonal grid.

Following a paradigm of reproducible research, the source code of the tool designed for studying neighborhood motion maps on the hexagonal grids is provided at the following URL: https://github.com/copyme/NeighborhoodMotionMapsTools

Appendix: Neighborhood motion maps for $\mathcal{G}_1^U$ and their graph

In Sect. 3.3, we observed that there exist three topologically different partitions of the remainder range. They depend on rotation angles: $\theta_1 \in (\alpha_1, \alpha_2), \theta_2 \in (\alpha_2, \alpha_3)$ and $\theta_3 \in (\alpha_2, \alpha_3)$, as illustrated in Fig. 4. From Proposition 8, we also know that each frame corresponds to a different neighborhood motion map. Figure 6 illustrates all the neighborhood motion maps of $\mathbb{M}_1$, with the different rotation angles: $\theta_1 \in (\alpha_1, \alpha_2), \theta_2 \in (\alpha_2, \alpha_3)$ and $\theta_3 \in (\alpha_2, \alpha_3)$.

The dual graph of the remainder range partitioning, in Fig. 4, is considered. In this graph $G_1 = \{\mathbb{M}_1, E\}$, each node is represented by a neighborhood motion map (or a frame), while each edge between two nodes corresponds to a line segment shared by frames in the remainder range. Moreover, each edge is

---

[1] Similar attempt was made in [14], but with a different approach of measuring.

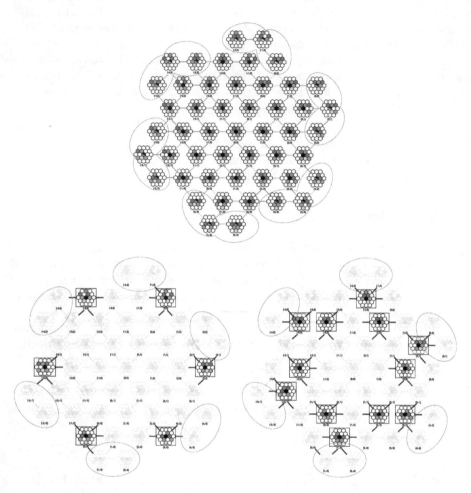

Fig. 6. All the neighborhood motion maps $\bigcup_{\kappa \in \Lambda} \left\{ \mathcal{G}_r^U(\kappa) \right\}$ for any angle in (α_0, α_1) (top), (α_1, α_2) (bottom, left), and (α_2, α_3) (bottom, right), visualized by the label map $\mathcal{L}_1^U$ (see Fig. 2). Neighborhood motions maps are considered as graph vertices and linked by edges with respect to the adjacency relations of respective frames, i.e. each edge in the graph represents a side shared by two frames in the remainder range. The neighborhood motion maps which correspond to non-injective zones are surrounded by pink ellipses. The elements which are changed with respect to the rotation angles in the bottom figures are surrounded by black squares, while those which are not changed are faded. A high resolution version of the figure can be found via the URL: https://doi.org/10. 5281/zenodo.820911

labeled with the color of the corresponding line segment (see Fig. 3). We note that an edge color denotes the color of a point in the transition between the two corresponding neighboring motion maps. For instance, if there exists a red horizontal edge between two nodes, then we observe a transition of the red neighbor ($\delta = (1,0)$) between two neighborhood motion maps connected by this edge. We invite the reader to consider Fig. 6(top) and neighborhood motion maps of the indices $(-1, 2)$ and $(0, 2)$.

Note that, neighborhood motion maps in Fig. 6 are arranged with respect to the hexagonal lattice and each can be identified thanks to its axial coordinates [5]. We also observe that, in such an arrangement, neighborhood motion maps are symmetric with respect to the origin—the frame of index $(0, 0)$. For example, the neighborhood motion map of index $(-4, 3)$ is symmetric to that of the index $(4, -3)$ (see Fig. 6).

References

1. Nouvel, B., Rémila, E.: Configurations induced by discrete rotations: periodicity and quasi-periodicity properties. Discrete Appl. Math. **147**, 325–343 (2005)
2. Kong, T., Rosenfeld, A.: Digital topology: introduction and survey. Comput. Vis. Graph. Image Process. **48**, 357–393 (1989)
3. Klette, R., Rosenfeld, A.: Digital Geometry: Geometric Methods for Digital Picture Analysis. Elsevier, Amsterdam (2004)
4. Kovalevsky, V.A.: Finite topology as applied to image analysis. Comput. Vis. Graph. Image Process. **46**, 141–161 (1989)
5. Middleton, L., Sivaswamy, J.: Hexagonal Image Processing: A Practical Approach. Advances in Pattern Recognition. Springer, Berlin (2005)
6. Serra, J.: Image Analysis and Mathematical Morphology. Academic Press, London (1982)
7. Her, I.: Geometric transformations on the hexagonal grid. IEEE Trans. Image Process. **4**, 1213–1222 (1995)
8. Pluta, K., Romon, P., Kenmochi, Y., Passat, N.: Bijective digitized rigid motions on subsets of the plane. J. Math. Imaging Vis. **59**, 84–105 (2017)
9. Ngo, P., Kenmochi, Y., Passat, N., Talbot, H.: Topology-preserving conditions for 2D digital images under rigid transformations. J. Math. Imaging Vis. **49**, 418–433 (2014)
10. Nouvel, B., Rémila, É.: On colorations induced by discrete rotations. In: Nyström, I., Sanniti di Baja, G., Svensson, S. (eds.) DGCI 2003. LNCS, vol. 2886, pp. 174–183. Springer, Heidelberg (2003). doi:10.1007/978-3-540-39966-7_16
11. Gilder, J.: Integer-sided triangles with an angle of 60°. Math. Gaz. **66**, 261–266 (1982)
12. Gordon, R.A.: Properties of Eisenstein triples. Math. Mag. **85**, 12–25 (2012)
13. Berthé, V., Nouvel, B.: Discrete rotations and symbolic dynamics. Theoret. Comput. Sci. **380**, 276–285 (2007)
14. Thibault, Y.: Rotations in 2D and 3D discrete spaces. Ph.D. thesis, Université Paris-Est (2010)

Large Families of "Grey" Arrays with Perfect Auto-Correlation and Optimal Cross-Correlation

Imants Svalbe[1], Matthew Ceko[1($\boxtimes$)], and Andrew Tirkel[2]

[1] School of Physics and Astronomy, Monash University, Melbourne, Australia
{imants.svalbe,matthew.ceko}@monash.edu
[2] Scientific Technology, Melbourne, Australia
atirkel@bigpond.net.au

Abstract. Digital watermarking applications have a voracious demand for large sets of distinct 2D arrays of variable size that possess both strong auto-correlation and weak cross-correlation. We use the discrete Finite Radon Transform to construct "perfect" $p \times p$ arrays, for p any prime. Here the array elements are comprised of the integers $\{0, \pm 1, +2\}$. Each array exhibits perfect periodic auto-correlation, having peak correlation value p^2, with all off-peak values being exactly zero. Each array, by design, contains just $3(p-1)/2$ zero elements, the minimum number possible when using this "grey" alphabet. The grey alphabet and the low number of zero elements maximises the efficiency with which these perfect arrays can be embedded into discrete data. The most useful aspect of this work is that large families of such arrays can be constructed. Here the family size, M, is given by $M = p^2 - 1$. Each of the $M(M-1)/2$ intra-family periodic cross-correlations is guaranteed to have one of the three lowest possible merit factors for arrays with this alphabet. The merit factors here are given by $v^2/(p^2 - v^2)$, for $v = 2$, 3 and 4. Whilst the strength of the auto-correlation rises with array size p as p^2, the strength of the many (order p^4) cross-correlations between all M family members falls as $1/p^2$.

Keywords: Perfect arrays · Low cross-correlation arrays · Discrete projection · Finite Radon Transform · Watermarking

1 Introduction

We are motivated by the many watermarking applications, like [2,7,8], for which one needs large families of arrays that have both low off-peak auto-correlation and cross-correlation. For example, to provide watermark tags for each frame of a 5 min YouTube 120 fps video requires about 36,000 arrays. If all of these tags are unique and have a low cross-correlation, it is possible to easily isolate and verify any individual frame within a 5 min sequence. A family comprised of 39,600 perfect arrays, each of size 199×199, would suffice for such an application.

© Springer International Publishing AG 2017
W.G. Kropatsch et al. (Eds.): DGCI 2017, LNCS 10502, pp. 46–56, 2017.
DOI: 10.1007/978-3-319-66272-5_5

The cross-correlation between functions f and g is given by

$$C_{fg}(s) = f \otimes g = \sum f(s) \cdot g(s - r) \tag{1}$$

where r, a shift variable, is taken over all coordinates of g, and s covers the domain of f. Auto-correlation corresponds to the case where $f = g$. Perfect arrays have periodic auto-correlation with constant off-peak values. For a $p \times p$ array, the peak is p^2 with zero elsewhere (or $p^2 - 1$ peak and -1 elsewhere), and cross-correlation between all family members have only $\pm p$ values.

Previous work [11] used the Finite Radon Transform (FRT) to construct $p \times p$ pseudo-noise arrays in families of size $M = p$ (where p is a $4N - 1$ prime) that had optimal periodic auto-correlation and cross-correlation, that meet the Welch correlation bounds [12]. These families of (Legendre) arrays have an alphabet of a single zero element with the remainder being equal numbers of ± 1 elements. "Grey" versions of these array families were also constructed that have integer alphabets (with integer values ranging between $\pm\sqrt{p}$). Recovery of a "grey" array, A, embedded in "grey" data, B, can be advantageous, as $A \otimes (A + B) \approx 2A \otimes A$ if we choose to embed A in those parts of B where $A \approx B$.

Subsequent work [10] extended the size of these array families (M) to multiples of p, typically $M \approx 3p$. This was done by blending the original array family with distinct arrays either derived from the original array auto-correlations, or with new arrays, also built using the FRT, but with their families generated using different (but equivalent) Hadamard matrices. The only concession made when extending the family size beyond p is that the strength of each cross-correlation now lies in a range of statistically predictable values, at or just above the lowest possible levels.

Further extension of the size of a family of arrays well beyond p is difficult as it is hard to constrain the range of cross-correlation values. The rapidity of this rise is, in part, due to the depth of the array alphabet. A binary array (or any array with mostly ± 1 values) has only so many combinations that can simultaneously sustain high auto- and low cross-correlation. The "grey" versions of the $p \times p$ Legendre arrays constructed in [11] can support a much larger and more diverse range of well-correlated structures. The combinatorial diversity of grey perfect arrays also makes them significantly more secure and resistant to hacking.

However, the balance theorem ensures that the sum of the array values dictates the sum over all correlation values [3]. This ensures that alphabets spanning a wide range of greys also require a rapid increase in the number of zero elements in those perfect arrays (see Sect. 5). The number of zero elements in a perfect-correlation array increases with the square of the values of the non-zero elements. Arrays containing a large number of zero elements have reduced operational efficiency, as the zero terms "change nothing" when embedded into any local data.

For these reasons, we have investigated construction of perfect $p \times p$ arrays with a restricted grey alphabet of just $\{0, \pm 1, +2\}$. We introduce a minimal number, $(p - 1)/2$, of elements having value $+2$, thus requiring $3(p - 1)/2$ zero terms in each array. The balance of these array values (always a clear majority) are either ± 1. The presence of a relatively few extra zeroes reduces the efficiency of these arrays, by $\mathcal{O}(1/p)$, but this becomes less significant for large p. Very large numbers of such arrays can be made, where each array contains a fixed proportion of each grey element. We can then select large families of arrays, where the intra-family array cross-correlations are restricted to the lowest possible levels.

Section 2 reviews the important link between the correlations of arrays and the correlations between projection of those arrays. This link permits the construction of 2D perfect arrays from 1D perfect projections. Section 3 reviews the Finite Radon Transform, a discrete projection scheme whose inverse back-projection permits exact reconstruction of any $p \times p$ set from $p + 1$ discrete projections, for p prime. Section 4 reviews the use of affine transforms as a means to produce many distinct variants of a perfect array that retain the original array and correlation properties. Section 5 introduces a method to construct perfect arrays with a fixed "grey" alphabet and tightly bounded cross-correlation values. It then shows how to assemble a large family of such arrays. Section 6 presents some results for example array families. Ways to improve this technique and future work are highlighted in Sect. 7.

2   Projection Preserves Moments and Correlations

The central slice theorem [1] states that projected views of a distribution preserve the Fourier transform of the distribution. This is the main result underlying image reconstruction methods for computed tomography. As a corollary of the central slice theorem, moments and correlations of a distribution are also preserved under projection. This means, for example, that the auto-correlation of the projected view of some object is equal to the same projected view of the full object auto-correlation [4,6]. The projections of any distribution inherit that distribution's correlation properties.

We use this result in reverse to construct arrays with any desired correlation properties. For example, we assemble a set of 1D projections, each having perfect auto-correlation. We then reconstruct from that set a 2D object that inherits their perfect auto-correlation [10,11].

The cross-correlation C_{fg} between function f and g is defined in Eq. (1). Correlations are termed periodic where the sum of the products of overlapped functions is taken over cyclic boundary conditions and termed aperiodic when zeroes extend the function boundaries. Auto-correlation is the case where $f = g$. The correlation results as presented here are for periodic arrays. In practice, the aperiodic correlation results are more relevant. However, there the boundary conditions are usually not zero, but depend on the values of the data in which the arrays are embedded.

3   Using 1D FRT Projections to Build 2D Arrays

We exploit the correlation-preserving property of projections as given in Sect. 2. The discrete Finite Radon Transform [5] is used to provide a unique and exact reconstruction of any 2D $p \times p$ object from its $p + 1$ 1D projected views. Here p must be prime to ensure the (cyclically wrapped) projections are uncoupled, as each projection fully tiles a $p \times p$ array, exactly once, at all positions, in a distinct pattern.

Projection $R(t, m)$ of an image $I(x, y)$ starts from translate t, $0 \leq t < p$. Usually t is defined as one of the pixels along the top row of a $p \times p$ image. Each 1D projection is comprised of p parallel rays, where each ray sums p pixel values in $I(x, y)$ that are located at p steps beginning from t, each step being m pixels across and one pixel down, wrapping periodically around the ray as required, where $0 \leq m \leq p$. Projections 0 and p are column and row sums of $I(x, y)$, respectively.

$$R(t, m) = \sum I(\langle t + my \rangle_p, y) \qquad (2)$$

where $\langle j \rangle_p$ means j modulus p. Back-projecting each of the $p + 1$ 1D projections across a zeroed $p \times p$ array at the complemented angles ($m' = p - m$) and normalising the result recovers, exactly, the 2D data that was projected.

4 Affine Transforms Preserve Correlations

Just as discrete projection preserves correlation, so too does affine transformation. For example, a 2D affine transformation (reversibly) maps each pixel (x, y) of a prime $p \times p$ 2D image to a new location (x', y'). Under matrix multiplication in homogeneous coordinates (modulus p):

$$\begin{bmatrix} x' \\ y' \\ 1 \end{bmatrix} = \begin{bmatrix} a & b & e \\ c & d & f \\ 0 & 0 & 1 \end{bmatrix} \begin{bmatrix} x \\ y \\ 1 \end{bmatrix}$$

where the values, $0 \leq a, b, c, d, e, f < p$, are arbitrary integer transform coefficients, provided only that the upper matrix $[a\ b;\ c\ d]$ has non-zero determinant (modulus p).

The coefficients e and f serve as a discrete translation vector; hence we always set these to 0, as simple translations of an array exhibit the same (periodic) correlations. When $[a\ b;\ c\ d] = [j\ -i;\ i\ j]$, the affine transform rotates the array by the discrete angle $i{:}j$, when $[a\ b;\ c\ d] = [j\ i;\ i\ j]$, the affine transform skews the array by vector $i{:}j$.

With 4 arbitrary affine transform coefficients, a single 13×13 perfect array A thus has 12^4 distinct affine variations, each of which has perfect correlation, since the original A is perfect. The very many cross-correlations between these arrays will vary from being optimally low through to many cases where the transformed array is a cyclic shift of A (equivalent under periodic correlations). For watermarking applications, sign changed, reflected or transposed arrays should be avoided.

For $p \times p$ arrays, we know in advance the exact set of angles $i{:}j$ that correspond to the complete set of $p + 1$ discrete projections of the FRT for a $p \times p$ array [9]. If we avoid the simple axial rotations 1:0 (90°) and 0:1 (0°), we can, without redundancy, rotate each original array A by affine coefficients $i{:}j$ to obtain up to $p - 3$ distinct copies A' of each A, whilst preserving the original correlation properties. Affine skews for angles $\pm 1{:}1$ are skipped, because they yield a zero determinant. Similarly, for all $p = 4n + 1$ primes, there is one set of degenerate rotations $\pm i{:}j$ (and $\pm j{:}i$) that are skipped, for the case $i^2 + j^2 = p^2$, which also has zero determinant [9].

5   Construction of a Family of Perfect Arrays with Alphabet $\{0, \pm 1, +2\}$

In this section, we detail the construction of families of perfect $p \times p$ arrays using the alphabet $\{-1, 0, +1, +2\}$. The FRT is employed to construct perfect arrays using distinct cyclic shifts. Affine rotations can then be used to extend the size of the array family.

5.1   Array Construction

Discrete 1D "delta" functions (or unit impulses) of length p, for example [1 0 0 0 0 0 0], can be used to create the $p + 1$ FRT projections (as done in [11]). A delta function has a perfect 1D auto-correlation, hence so too will any 2D array reconstructed from these 1D delta projections by applying the FRT inverse transform.

We want to minimise the number of zeroes in the 2D array reconstructed from these projections. This requires that the 1D rays back-projected from each view angle (m_1) must intersect with the rays from other angles (m_2) at as many distinct array positions as possible. This condition can be achieved by judicious adjustment of the (circular) phase shift of each delta function, for example from [1 0 0 0 0 0 0] to something like [0 0 0 1 0 0 0].

In the FRT, each ray (m, t) is back-projected [5] as the line that passes through the image points (x, y), where

$$x = -my + t. \tag{3}$$

We want to ensure that the delta impulse from projection m_1 intersects with the delta impulse from projection m_2 at a distinct point (x_{12}, y_{12}) for each pair m_1, m_2. We assign

$$t = 1/m \tag{4}$$

for $1 \leq m \leq p - 1$, and substitute into (3) so the rays for projections m_1 and m_2 have

$$x_{12} = -m_1 y_{12} + \frac{1}{m_1} = -m_2 y_{12} + \frac{1}{m_2}$$

$$y_{12}(m_2 - m_1) = \frac{1}{m_2} - \frac{1}{m_1} = \frac{-(m_2 - m_1)}{m_1 m_2}$$

which intersects at $(x_{12}, y_{12}) = \left(\frac{1}{m_1} + \frac{1}{m_2}, \frac{-1}{m_1 m_2}\right)$. Alternatively, we can assign

$$t = m^2 \tag{5}$$

for $1 \le m \le p - 1$, and substitute into (3) so the rays m_1 and m_2 have

$$x_{12} = -m_1 y_{12} + m_1^2 = -m_2 y_{12} + m_2^2$$
$$y_{12}(m_2 - m_1) = m_2^2 - m_1^2 = (m_2 - m_1)(m_2 + m_1)$$
$$y = m_1 + m_2 \text{ and } x = -m_1 m_2$$

which intersects at $(x_{12}, y_{12}) = (-m_1 m_2, m_1 + m_2)$.

Assigning projection m to have cyclic shift $t = 1/m$ or $t = m^2$ imposes a strong symmetry on the FRT matrix, as each projection m has a negative counterpart $m' = p - m = -m$, and then $m_2 = m_2'$, and $1/m = -1/m'$. The "near-orthogonality" of these shift assignments is evident in the "pseudo-Hadamard" constructed from the $p \times p$ matrix product of the shifted delta functions of the FRT, B, with the shifted impulses of its transpose, shown as $B * B^T$, in Fig. 1 (a) for $t = 1/m$ and (b) for $t = m^2$. We use the term near-orthogonal to reference the fact that there are few non-zero elements that lie off the diagonal of $B * B^T$.

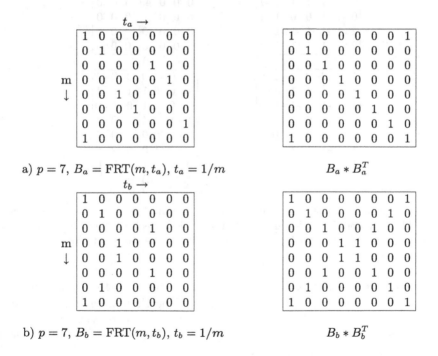

a) $p = 7$, $B_a = \mathrm{FRT}(m, t_a)$, $t_a = 1/m$ $B_a * B_a^T$

b) $p = 7$, $B_b = \mathrm{FRT}(m, t_b)$, $t_b = 1/m$ $B_b * B_b^T$

Fig. 1. FRT projection matrices for a 2D array built from 1D phase shifted delta functions, for $p = 7$, (a) $t = 1/m$, (b) $t = m^2$. The pseudo-orthogonality of these phase shifts is shown on the right via the matrix product of their 2D FRT arrays

Typical arrays reconstructed from FRT's that are built using delta functions where the phase-shift t for projection m are given by $t = 1/m$ are shown in Fig. 2a and for $t = m^2$ in Fig. 2b. Note that these perfect $p \times p$ arrays all have sum $= p$. All arrays made this way will have the same histograms: for the 49 elements in each 7×7 array, 18 elements have value -1, 9 are zero, 19 are $+1$ and 3 are $+2$ elements, giving sum $= 7$.

We extend the size of the array families by computing FRT matrices with shifts, t, that are linear multiples of $(4, 5)$, modulus p, which then undergo many affine rotations and skews. Combining these operations can produce some duplicated arrays. Although every distinct FRT set corresponds to a unique array, some scaled mapping of the FRT variables m and t can be degenerate. For example, the FRT of the transpose of a $p \times p$ array, $B^T(m, t)$, can be obtained from shuffling the FRT of the original array, $B(m, t)$, by $m' = 1/m$ and $t' = -t/m$ (as the transpose maps projection angle $x{:}y$ to $y{:}x$). The shifts of Eq. (4) are thus very close to a transpose operation. Similar structural overlaps in reconstructed arrays can result from axial rotations or symmetric reflections.

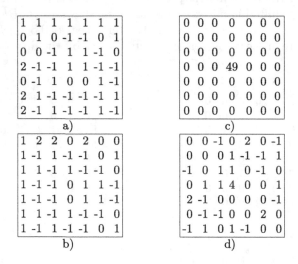

Fig. 2. 7×7 arrays reconstructed from (a) FRT B_a from Fig. 1 (a). (b) B_b from Fig. 1 (b). (c) The auto-correlation for arrays (a) and (b) is perfect. (d) The cross-correlation between (a) and (b) has type L_2. This example shows the strongest cross we accept.

We assign frequencies k, l, m and n to the occurrence of grey elements -1, 0, $+1$ and $+2$ respectively in any $p \times p$ array. The FRT translates, t, arrange the $(p-1)$ projections to have distinct intersections as pairs, yielding $(p-1)/2$ elements with value $+2$ in the final array, thus fixing $n = (p-1)/2$. The sum over all p^2 elements of a $p \times p$ array is then

$$-1 \cdot k + 0 \cdot l + 1 \cdot m + 2 \cdot (p-1)/2 = p. \tag{6}$$

The sum of the array auto-correlation values, by the balance theorem [3], means

$$(-1)^2 \cdot k + (0)^2 \cdot l + (+1)^2 \cdot m + (+2)^2 \cdot (p-1)/2 = p^2. \qquad (7)$$

From (6), $m - k = 1$ and from (7), $m + k = p^2 - 2p + 2$, giving $m = (p-1)^2/2$, $k = m + 1$ and $l = 3(p-1)/2$. Any $p \times p$ array (with p prime) made using the FRT with these values of t will have a fixed histogram for its element values, -1 through $+2$, as $[(p-1)^2/2, \ 3(p-1)/2, \ (p-1)^2/2 + 1, \ (p-1)/2]$.

The fixed histogram of element values permits quantification of the merit factors for the periodic cross-correlation of these arrays. The merit factor (MF) is defined as the square of the peak correlation value divided by the sum of all $p^2 - 1$ off-peak values. Perfect arrays are, by definition, spectrally flat. All cross-correlations between pairs of spectrally flat arrays are also spectrally flat by the convolution theorem, and hence those cross-correlations are also perfect arrays themselves.

The lowest possible maximum value of any cross-correlation is $\pm 1 \cdot 2 = \pm 2$ (it cannot be ± 1, as one of the 2's will line up at least once with the majority of ± 1 terms). The sum of all array terms squared is always p^2, hence L_0, the lowest possible MF value, is given by $L_0 = 2^2/(p^2 - 2^2) = 4/(p^2 - 4)$.

The next level possible cross level, L_1, corresponds to a maximum cross-correlation sum of 3, giving $L_1 = 3^2/(p^2 - 3^2)$. The next possible level has cross-correlation value $= 4$, thus $L_2 = 16/(p^2 - 16)$. The next level $L_3 = (p-1)^2/(2p - 1)$, corresponds to arrays where all the ones line up, and finally $L_4 = p^2/0 = \infty$, when the two arrays are identical (and the cross becomes a perfect auto-correlation).

Here L_0, L_1, L_2, L_3 and L_4 are the only possible periodic cross-correlation values between these arrays when built using symmetric 1D projections, for any array size p. For $p = 7$, L_3 corresponds to a peak cross value of $p - 1 = 6$. A correlation peak value of 5 is not possible for a cross between these arrays. However we can construct different arrays (that require an asymmetric set of cyclic shifts for different projections m in the FRT) to give an alphabet $\{0, \pm 1, 2, 5\}$. This array, for $p = 7$, contains 27 zeroes, compared to just 9 zeroes for the arrays with alphabet $\{0, \pm 1, 2\}$. These "asymmetrically made" grey arrays also, of course, still retain perfect auto-correlations.

Note the merit factors L_0, L_1 and L_2 are all < 1 for any prime $p > 5$, while L_3 denotes a strong cross-correlation with MF of order p/2 and L_4 means the two arrays are a perfect match. When selecting arrays to build an extended family, we restrict the choice of arrays to be only those that yield cross-correlations of L_0, L_1 or L_2, preferably choosing the sets of arrays that have a larger fraction of crosses being either L_0 or L_1.

5.2   Building Array Families

To construct a family of arrays, a set A_1 of $p - 1$ seed arrays is made using the FRT with delta functions as 1D projections. Each array is made using (4) and a distinct cyclic shift $t = \alpha/m$, for $1 \le \alpha \le p - 1$. A second set A_2 of $p - 1$ seed arrays is made using (5) and cyclic shifts $t = \alpha m^2$, again for $1 \le \alpha \le p-1$. The translates chosen for the remaining FRT projections, for $m = 0$ and p in set A_1, can be fixed independently of the assigned shifts necessary for the $(p-1)$ paired intersecting rays. If the $m = 0$ and p rays are all set to $t = 0$, the arrays A_2 are the transpose of the arrays in A_1, listed in reverse order. A free choice is possible in the assignments of parameter t for the perpendicular rays $m = 0$ and $m = p$ when constructing the arrays for set A_1. This permits several distinct perfect A_1 families to be constructed whilst using the same pairing rules that fix the array histogram.

The sets A_1 and A_2 are pooled to form A_P. The seed family A_P is then affine rotated by a selection of the FRT projection angles $i{:}j$ for that prime p to produce array family A_R. The seed family A_P is skewed by selected valid skew vectors to form array family A_S. The two families A_R and A_S are then pooled to form a (large) family A_T. Any duplicate copies of arrays formed by matched affine transformations are removed. Each array in A_T has perfect periodic auto-correlation. All intra-family correlations for A_T are checked and those pairings that produce correlation values L_3 and L_4 are discarded from A_T to produce the final optimal family A.

This method of pooling rotated and skewed arrays ensures all valid possible variants of A_1 and A_2 are produced. The final thinned set A then is always comprised of $p^2 - 1$ arrays. There are also many different ways to thin and discard pairs of arrays from A_T to select the final family A. We select arrays with the most favourable distribution of cross-correlation values L_0, L_1 and L_2 that best suit a given application.

6   Results

Table 1 presents example results for families of $p \times p$ arrays where $p = 7$, 23 and 43 with family sizes 48, 528 and 1848, respectively. In each case, all auto-correlations within each family are perfect, having peak values of 49, 529 and 1849 respectively. The merit factor (MF) for any intra-family periodic cross-correlations is either L_0, L_1 or L_2, having the values as listed. The relative frequencies of the L_0, L_1 or L_2 occurrences are given to highlight the distribution of cross-correlation values between all family members.

The family of arrays for $p = 23$ is generated as follows: each of the initial seed sets A_1 and A_2 contains $p - 1 = 22$ perfect arrays. Each of these 23×23 arrays contains 529 elements; 242 elements being -1, 33 zeroes, 243 $+1$ elements and 11 $+2$ values. Their periodic auto-correlation peak $= 529$, with all off-peak entries $= 0$. Each of the 231 cross-correlations between the 22 arrays in A_1 (and also between all A_2 members) has the minimal MF $L_0 = 0.00762$.

Set $A_P = (A_1 + A_2)$ has 44 distinct arrays. Affine rotation of the set A_P, by 11 FRT angles, produces 484 more perfect arrays, 396 of which are distinct (duplicate free). When A_P is skewed by 10 FRT angles, another 440 perfect arrays are produced, of which 308 are duplicate-free. The resulting $396 + 308 + 44 = 748$, 23×23 arrays are pooled as set A_T. A_T is checked for cross-correlations,

Table 1. Cross-correlation values and their relative frequencies for two sets.

48 7×7 perfect arrays			
MF	$L_0 = 0.08889$	$L_1 = 0.22500$	$L_2 = 0.48485$
Set 1	0.44681	0.44681	0.10638
Set 2	0.36171	0.57447	0.06383
528 23×23 perfect arrays			
MF	$L_0 = 0.00762$	$L_1 = 0.01731$	$L_2 = 0.03119$
Set 1	0.03985	0.84788	0.11227
Set 2	0.03985	0.83428	0.12582
1848 43×43 perfect arrays			
MF	$L_0 = 0.00227$	$L_1 = 0.00489$	$L_2 = 0.00873$
Set 1	0.02220	0.83881	0.15216
Set 2	0.02220	0.83452	0.14328

from which 220 arrays, whose crosses give an MF > 1, are discarded, leaving the final family A of $748 - 220 = 528$ perfect arrays. Two different families of 528 arrays were selected from A_T, each has a slightly different distribution of cross-correlations over the same low cross-correlation merit factors.

Table 1 shows that the MF for the bulk of the intra-family cross-correlation values are mainly of type L_1, but that selecting arrays in different ways can, at least marginally, alter the proportion of L_0, L_1 and L_2 results. However, the $p = 7$ case shows that it is possible to select different families of arrays from A_T that favour the production of more L_0 crosses and that minimise the number of L_2 crosses.

7   Conclusions and Further Work

The families of $p \times p$ arrays constructed here have perfect auto-correlation with guaranteed low cross-correlation values between all family members. These arrays have a restricted grey alphabet (elements of $\{0, \pm 1, +2\}$) with a small number, $3(p-1)/2$, of zero elements. This makes them highly efficient and secure when embedded as watermarks. The fixed frequency of array element values permits selection of $p^2 - 1$ family members that have cross-correlation values with just three of the lowest possible merit factors, MF $= v^2/(p^2 - v^2)$, for $v = 2, 3$ and 4.

The size of A_T needs to be kept small to avoid excessive computation to do the cross-correlation checks required to remove arrays that have MF > 1. If A_T has size M, $M(M - 1)/2$ crosses of size $p \times p$ need to be checked. However, pooling A_T to be too small may also restrict the range of distinct or partially overlapped solution sets A that can be selected from A_T. It may be possible to define a set of affine rotations and skews to directly produce a final set A without computing the larger intermediate set A_T.

At present, we select the final A arrays by deleting of one of each pair of arrays in A_T that has MF > 1. The order of these deletions can be done in several different ways. This selection and deletion process could be organised more strategically.

The FRT can be adapted to reconstruct higher dimensional arrays from 1D projections. The technique developed here can then be extended to produce families of nD arrays.

We have yet to examine perfect arrays over other, larger alphabets, even in 2D, to see if the merit factors of their cross-correlations can also be fixed algebraically.

Acknowledgments. The School of Physics and Astronomy at Monash University, Australia, has supported and provided funds for this work. M.C. has the support of the Australian government's Research Training Program (RTP) and the J.L. William scholarship from the School of Physics and Astronomy at Monash University.

References

1. Bracewell, R.: Strip integration in radio astronomy. Aust. J. Phys. **9**(2), 198–217 (1956)
2. Chauhan, S., Rizvi, S.: A survey: Digital audio watermarking techniques and applications. In: 2013 4th International Conference on Computer and Communication Technology (ICCCT), pp. 185–192. IEEE (2013)
3. Golomb, S., Gong, G.: Signal Design for Good Correlation: For Wireless Communication, Cryptography, and Radar. Cambridge University Press, Cambridge (2005)
4. Guédon, J.: The Mojette Transform: Theory and Applications. ISTE John Wiley & Sons, London (2009)
5. Matúš, F., Flusser, J.: Image representation via a finite Radon transform. IEEE Trans. Pattern Anal. Mach. Intell. **15**(10), 996–1006 (1993)
6. Phillipé, O.: Image representation for joint source channel coding for QoS networks (Ph.D. Thesis). University of Nantes (1998)
7. Potdar, V., Han, S., Chang, E.: A survey of digital image watermarking techniques. In: 2005 3rd IEEE International Conference on Industrial Informatics, INDIN 2005, pp. 709–716. IEEE (2005)
8. Sethuraman, P., Srinivasan, R.: Survey of digital video watermarking techniques and its applications. Eng. Sci. **1**(1), 22–27 (2016)
9. Svalbe, I.: Sampling properties of the discrete Radon transform. Discrete App. Math. **139**(1), 265–281 (2004)
10. Svalbe, I., Tirkel, A.: Extended families of 2D arrays with near optimal auto and low cross-correlation. EURASIP J. Adv. Sig. Process. **2017**(1), 18 (2017)
11. Tirkel, A., Cavy, B., Svalbe, I.: Families of multi-dimensional arrays with optimal correlations between all members. Electron. Lett. **51**(15), 1167–1168 (2015)
12. Welch, L.: Lower bounds on the maximum cross correlation of signals. IEEE Trans. Inf. Theory **20**(3), 397–399 (1974)

The Minimum Barrier Distance: A Summary of Recent Advances

Robin Strand[1]([✉]), Krzysztof Chris Ciesielski[2,3], Filip Malmberg[1], and Punam K. Saha[4]

[1] Centre for Image Analysis, Uppsala University, Uppsala, Sweden
robin@cb.uu.se
[2] Department of Mathematics, West Virginia University, Morgantown, USA
[3] Department of Radiology, MIPG, University of Pennsylvania, Philadelphia, USA
[4] Department of Electrical and Computer Engineering and Department of Radiology, University of Iowa, Iowa City, USA

Abstract. In this paper we present an overview and summary of recent results of the *minimum barrier distance* (MBD), a distance operator that is a promising tool in several image processing applications. The theory constitutes of the continuous MBD in $\mathbb{R}^n$, its discrete formulation in $\mathbb{Z}^n$ (in two different natural formulations), and of the discussion of convergence of discrete MBDs to their continuous counterpart. We describe two algorithms that compute MBD, one very fast but returning only approximate MBD, the other a bit slower, but returning the exact MBD. Finally, some image processing applications of MBD are presented and the directions of potential future research in this area are indicated.

1 Introduction

Distance functions and their transforms (DTs, where each pixel is assigned the distance to the closest seed pixel) have been used extensively in image processing applications. Since Rosenfeld and Pfaltz defined distance transforms on the binary images in the 1960's [16,17], a huge number of distance transform methods have been developed in theoretical and application setting. The classical, binary DTs are useful, for example, for measurement and description of binary objects. Distance functions and transforms that are defined as minimal cost paths in general images include geodesic distance [9], fuzzy distances [19], and the minimax cost function; they have been used, for example, in segmentation and saliency detection [18]. The most common algorithms for computing distance transforms are: (i) raster scanning methods, where distance values are propagated by sequentially scanning the image in a pre-defined order [1,5]; (ii) wave-front propagation methods, where distance values are propagated from low distance value points (the object border) to the points with higher distance values, using data structure containing previously visited points and their corresponding distance values, until all points have been visited [15,21]; and (iii) separable algorithms, where one-dimensional subsets of the image are scanned separately, until all principal directions have been scanned [4,20].

© Springer International Publishing AG 2017
W.G. Kropatsch et al. (Eds.): DGCI 2017, LNCS 10502, pp. 57–68, 2017.
DOI: 10.1007/978-3-319-66272-5_6

Recently, we introduced in [22] the *minimal barrier distance* (MBD) function, based on the minimal length of the interval of intensity values along a path between two points, see Fig. 1. The MBD differs from traditional distance functions in a number of aspects. For example, the length of a path may remain constant during its growth until a new stronger barrier is met on the path. This subtle shift in the notion of path length allows the new distance function to capture separation between two points in a "connectivity"-sense. Our experiments have shown that the MBD is stable to noise, seed point position [3,10,22,23]. The MBD has many interesting theoretical properties, and has been shown to be a potentially useful tool in image processing [3,7,10,22,25].

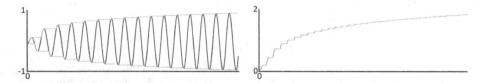

Fig. 1. One-dimensional example of the minimum barrier. Left: The graph of a function (blue) and, for each x on the horizontal axis, the maximum and minimum value attained in $[0, x]$ (red). Right: The pointwise difference between the maximum and minimum values – the minimum barrier. The minimum barrier distance between two points in an image is defined by a path with minimum barrier between the points. (Color figure online)

This paper collects and compares published results on the MBD and presents them in a unified framework. In Sect. 2, the MBD is defined on $\mathbb{R}^n$ with the two equivalent formulations ρ (Eq. 2) and φ (Eq. 3). In Sect. 3, their discrete counterparts $\widehat{\rho}$ (Eq. 4) and $\widehat{\varphi}$ (Eq. 5) are defined in $\mathbb{Z}^n$. Section 4 gives results on convergence between the different versions of MBD. The different MBD versions are related as follows:

$$\rho \xleftarrow{\quad\text{Thm. 1}\quad}_{=} \varphi$$

$$\begin{array}{cc} \text{Eq. 4} \Big\downarrow & \Big\downarrow \text{Eq. 5} \\ \text{Thm. 2} & \text{Thm. 2} \end{array}$$

$$\widehat{\rho} \xleftarrow{\quad}_{\approx}\xrightarrow{\quad} \widehat{\varphi}$$
$$\text{Cor. 1}$$

The arrows indicate conceptual relations. The top row shows the continuous formulations and the bottom row the two different discrete versions.

Section 5 describes the algorithms for computation of DTs. The MBD based on $\widehat{\rho}$ is, similar to [8], not smooth in the sense of [2,6]. As a consequence, exact DTs for $\widehat{\rho}$ cannot be computed as efficiently as for $\widehat{\varphi}$, but both approximate algorithms and efficient algorithms for exact computation have been developed.

These are given in Sect. 5.1. Section 5.2 shows that DTs for $\hat{\varphi}$ can be efficiently computed by standard wave-front propagation techniques. Applications in image processing are presented in Sect. 6.

2 The Minimum Barrier Distance in $\mathbb{R}^n$

We consider bounded maps $f\colon D \to \mathbb{R}$ and their graphs $A = \{(x, f(x)) : x \in D\}$. When $f\colon D \to [0,1]$ the set A can be seen as a fuzzy subset of D with f being its membership function.

A *path from p to q* (where $p, q \in D$ and $D \subset \mathbb{R}^n$) is any continuous function $\pi\colon [0,1] \to D$ with $p = \pi(0)$ and $q = \pi(1)$. The symbol $\Pi_{p,q}$ (without subscripts, when p and q are clear from the context) is used to denote the family of all such paths. We consider functions $f\colon D \to \mathbb{R}$ such that

- f is bounded;
- f is continuous;
- $D \subset \mathbb{R}^n$ is path connected, that is, for every $p, q \in D$ there exists a path $\pi\colon [0,1] \to D$ from p to q.

The *barrier* of a path $\pi\colon [0,1] \to D$ is the number

$$\tau(\pi) = \max_t f(\pi(t)) - \min_t f(\pi(t)) = \max_{t_0, t_1}\big(f(\pi(t_1)) - f(\pi(t_0))\big). \qquad (1)$$

The maxima and minima in Eq. 1 are attained due to the Extreme Value Theorem. The *minimum barrier distance* between $p, q \in D$ is the number

$$\rho(p, q) = \inf_{\pi \in \Pi_{p,q}} \tau(\pi). \qquad (2)$$

2.1 Metricity

Definition 1. *A function $d\colon D \times D \to [0, \infty)$ is a metric on a set D provided, for every $x, y, z \in D$,*

(i) $d(x, x) = 0$ *(identity)*;
(ii) $d(x, y) > 0$ *for all $x \neq y$ (positivity)*;
(iii) $d(x, y) = d(y, x)$ *(symmetry)*;
(iv) $d(x, z) \leq d(x, y) + d(y, z)$ *(triangle inequality)*.

It is easy to see that metricity property (ii) does not hold for the minimum barrier distance (take a constant function f for example). Metricity properties (i), (iii), and (iv) are obeyed, and the minimum barrier distance is therefore a *pseudo-metric* [22].

2.2   Alternative Formulation

Now, the mapping $\varphi \colon D \times D \to [0, \infty)$ is defined by two separate paths, via the formula

$$\varphi(p, q) = \inf_{\pi_1 \in \Pi_{p,q}} \max_t f(\pi_1(t)) - \sup_{\pi_0 \in \Pi_{p,q}} \min_t f(\pi_0(t)). \qquad (3)$$

In Theorem 1 below, we see that the mappings φ and ρ are identical under mild assumptions on the set D.

Recall, that a set $D \subset \mathbb{R}^n$ is *simply connected*, provided it is path connected and for all $p, q \in D$ the paths $\pi_0, \pi_1 \in \Pi_{p,q}$ are homotopic, that is, there exists a continuous function $h \colon [0, 1]^2 \to D$, known as a *homotopy* between π_0 and π_1, such that $h(\cdot, 0) = \pi_0(\cdot)$, $h(\cdot, 1) = \pi_1(\cdot)$, and the maps $h(0, \cdot)$, $h(1, \cdot)$ are constant. Intuitively, the homotopy condition means that D has no holes.

Theorem 1 ([22]). *If $D \subset \mathbb{R}^n$ is simply connected, then the mappings ρ and φ are equal, that is, $\rho(p, q) = \varphi(p, q)$ for all $p, q \in D$.*

3   The Minimum Barrier Distance in $\mathbb{Z}^n$

In the digital setting, we consider the (bounded) functions $\widehat{f} \colon \widehat{D} \to \mathbb{R}$, where the digital scene $\widehat{D}$ is a finite subset of a digital space $\langle \phi \mathbb{Z}^n, \alpha \rangle$, where ϕ is a positive number and α is an adjacency relation such that two points in $\phi \mathbb{Z}^n$ are α-adjacent provided that no coordinate differs by more than ϕ and that the points differ in exactly one coordinate. Note that this is equivalent to the standard 6-adjacency [11] in a 3D digital space.

A digital path in a subset $\widehat{D}$ of the space $\langle \phi \mathbb{Z}^n, \alpha \rangle$ is any ordered sequence $\widehat{\pi} = \langle \widehat{\pi}(0), \widehat{\pi}(1), \ldots, \widehat{\pi}(k) \rangle$ of points in $\widehat{D}$ such that $\widehat{\pi}(i)$ and $\widehat{\pi}(i - 1)$ are α-adjacent for all $i \in \{1, 2, \ldots, k\}$. If $\widehat{\pi}(0) = p$ and $\widehat{\pi}(k) = q$, we say that the path $\widehat{\pi}$ is from p to q. For a fixed set $\widehat{D}$, a family of all paths in $\widehat{D}$ from p to q is denoted by $\widehat{\Pi}_{p,q}$ (with the subscripts omitted when p and q are clear from the context). Note that the digital paths are denoted by $\widehat{\pi}$, while the paths in the continuous space $\mathbb{R}^n$ are denoted by π.

In what follows, we assume that the digital scenes $\widehat{D}$ are of the rectangular form $\widehat{D}_\phi = D \cap \phi \mathbb{Z}^n$, where $D = \{x \in \mathbb{R}^n \colon L_i \leq x(i) \leq U_i\}$ for some real numbers L_i, U_i such that $L_i < U_i$ for all i. In particular, any two points in $\widehat{D}$ are connected by a path.

In view of Theorem 1, there are two natural ways of defining the discrete minimum barrier distance for $\widehat{f} \colon \widehat{D} \to \mathbb{R}$, the discretization of the formula for $\rho(p, q)$ and of that for $\varphi(p, q)$:

– **Discretization I**

$$\widehat{\rho}(p,q) = \min_{\widehat{\pi} \in \widehat{\Pi}_{p,q}} \left(\max_i \left[\widehat{f}(\widehat{\pi}(i)) \right] - \min_j \left[\widehat{f}(\widehat{\pi}(j)) \right] \right), \qquad (4)$$

– **Discretization II**

$$\widehat{\varphi}(p,q) = \min_{\widehat{\pi_1} \in \widehat{\Pi}_{p,q}} \max_i \left[\widehat{f}(\widehat{\pi}_1(i)) \right] - \max_{\widehat{\pi_0} \in \widehat{\Pi}_{p,q}} \min_j \left[\widehat{f}(\widehat{\pi}_0(j)) \right]. \qquad (5)$$

We know from [22] that each of the functions $\widehat{\rho}$ and $\widehat{\varphi}$ is a pseudo-metric on $\widehat{D}$ and that $\widehat{\varphi}(p,q) \leq \widehat{\rho}(p,q)$ for all $p, q \in \widehat{D}$.

4   Convergence Properties

Next we will see, in Theorem 2, that if $\widehat{f}: \widehat{D}_\phi \to \mathbb{R}$ is a discretization of a continuous function f defined on a rectangular region D, then, for a sufficiently small ϕ, the numbers $\widehat{\varphi}(p,q)$ and $\widehat{\rho}(p,q)$ well approximate $\varphi(p,q) = \rho(p,q)$.

Theorem 2 (Theorem 2 in [22]). *Let D be a rectangular region in $\mathbb{R}^n$ and $f: D \to \mathbb{R}$ be continuous. Let $\widehat{\rho}$ and $\widehat{\varphi}$ be the discrete minimum barrier distance functions for the sampling $\widehat{f}$ of f on $\widehat{D}_\phi$, that is, with $\widehat{f}(p) = f(p)$ for all $p \in \widehat{D}_\phi$. Then, for every $\varepsilon > 0$ there exists a $\phi_0 > 0$ such that for every $\phi \in (0, \phi_0]$*

$$|\widehat{\rho}(p,q) - \rho(p,q)| < \varepsilon \text{ and } |\widehat{\varphi}(p,q) - \varphi(p,q)| < \varepsilon \text{ for all } p, q \in \widehat{D}_\phi.$$

More precisely, this holds for any $\phi_0 > 0$ such that $|f(x) - f(y)| < \varepsilon/4$ for any $x, y \in D$ with $\|x - y\| \leq \phi_0 \sqrt{n}/2$.

Since, by Theorem 2, the values $\widehat{\rho}(p,q)$ and $\widehat{\varphi}(p,q)$ converge, as $\phi \to 0$, to $\rho(p,q) = \varphi(p,q)$, we obtain the following corollary.

Corollary 1 (Corollary 1 in [22] and Theorem 1 in [3]). *Let $\widehat{D}_\phi$, $\widehat{f}$, $\widehat{\rho}$ and $\widehat{\varphi}$ be as in Theorem 2. Then*

$$\max_{p,q \in \widehat{D}_\phi} |\widehat{\rho}(p,q) - \widehat{\varphi}(p,q)| \to 0 \text{ as } \phi \to 0.$$

More precisely, let $\varepsilon = \max \left\{ |\widehat{f}(x) - \widehat{f}(y)| : x, y \in \widehat{D}_\phi \text{ are } (3^n - 1)\text{-adjacent} \right\}$, with $\widehat{f}$ as in Theorem 2. Then

$$0 \leq \widehat{\rho}(p,q) - \widehat{\varphi}(p,q) \leq 2\varepsilon \text{ for all } p, q \in \widehat{D}_\phi.$$

5   Discrete Distance Transform Computation

Efficient distance transform computation is crucial for most applications of distance transforms. As described in the introduction, many different computation approaches have been proposed, including raster scanning and wave-front propagation. In image segmentation, it is natural to compute DTs from seed points in the background and in the object and then assign each point to a seed from which it has the minimal distance. In practice, this is often computed efficiently by propagating different labels from object and background seed points together with the distance values. In this way, the points get labeled during the DT computation, resulting in efficient computation of the labeling/segmentation. As described above, the MBD originally is formulated in the continuous space and offers two natural discretizations. This leads to two different problems to solve when developing methods for computing the DT, one for each discretization. Computing the DT of Discretization I turns out to be fairly easy, whereas computing it for Discretization II is an intricate problem.

5.1   Distance Transform Computation of Discretization I, $\widehat{\rho}$

Approximate Computation. By propagating the minimum barrier, i.e., the minimal interval of minimum and maximum value by a wave-front propagation approach using auxiliary data structures that hold the minimal and maximal attained values, an approximate distance transform can be computed [3,22]. We call the algorithm the *Dijkstra approximation algorithm* (Algorithm 1 in [3]). The minimum barrier distance based on $\widehat{\rho}$ is not smooth in the sense of [2,6] and, as a result, the obtained distance transforms is not error-free [3,14,22,25] with this approach. However, Zhang et al. [25] gave an error bound and also showed that for a very restricted class of 2D images, the approach gives exact minimum barrier distance transforms, see Theorem 3 which is here adapted to the Dijkstra approximation algorithm.

Theorem 3. [25] *Let $\widehat{f}$ be an integer-valued 2D image ($n = 2$) on a rectangular domain $\widehat{D}$ and let*

$$\widehat{f}' = \left\lfloor \frac{\widehat{f}}{\epsilon} \right\rfloor \epsilon,$$

where $\epsilon = \max\left\{ |\widehat{f}(x) - \widehat{f}(y)| : x, y \in \widehat{D} \text{ are } (3^n - 1)\text{-adjacent} \right\}$. If the set of seed points is an α-connected set, then the absolute error in the distance map obtained by the Dijkstra approximation algorithm on $\widehat{f}'$ is strictly less than ϵ.

The following Corollary holds since if $\epsilon = 1$ in Theorem 3, then the maximum absolute error is integer valued and strictly less than 1, that is, equal 0.

Corollary 2. *Let $\widehat{f}$ be an integer-valued 2D image ($n = 2$) on a rectangular domain $\widehat{D}$. If $\max\left\{|\widehat{f}(x) - \widehat{f}(y)| : x, y \in \widehat{D} \text{ are } (3^n - 1)\text{-adjacent}\right\} = 1$ (i.e., if ϵ in Theorem 3 is 1) and if the set of seed points is an α-connected set, then the absolute error in the distance map obtained by the Dijkstra approximation algorithm on $\widehat{f}$ is error-free.*

Exact Distance Transform Computation. Computing the exact discrete MBD *efficiently* for general images is far from trivial, but it is fairly easy to check if there exists a path between two points in a digital space within a given interval: threshold the image at the lower and upper limits of the interval and check if the two points are connected in the so-obtained connected regions. This approach gives a simple, but computationally very inefficient way to compute the minimum barrier distance transform: for a given seed point, compute for each interval the set of points it is connected to after the thresholding procedure explained above. The distance between the seed point and another given point is the minimum of all such intervals, for which they are connected.

A slightly more efficient approach is to loop over all possible lower bounds of the interval and compute a minimax transform from a given seed point. The minimum barrier distance is then obtained by a min-operator of the so-obtained distance maps.

This idea can be further optimized by sorting the priority queue in an efficient way. By popping points from the queue based on the upper limit of the attained barrier and propagating and pushing points to the queue based on the lower limit of the barrier while storing the minimum barrier attained at each point, it has been proved that an error-free algorithm can be obtained [3].

5.2 Distance Transform Computation of Discretization II, $\widehat{\varphi}$

The transform $\widehat{\varphi}(p, \cdot)$ can be efficiently computed since the path cost functions $\max_i\left[\widehat{f}(\widehat{\pi}_1(i))\right]$ and $\min_j\left[\widehat{f}(\widehat{\pi}_0(j))\right]$ are smooth in the sense of [2,6] and can therefore be computed by Dijkstra's algorithm, where wave-front propagation is used to compute lowest cost paths by local propagation [3,22].

6 The Minimum Barrier Distance in Image Processing Applications

Since color images are used in most applications, an important extension of the MBD is the *vectorial* MBD, where vectorial, i.e., multi-band images, are used as input. Different versions of the vectorial MBD were developed by Kårsnäs et al. [10] and applied to interactive color image segmentation, where images are segmented by manually placed seed points in the image background and in the object together with a label propagation approach to segment color images. Instead of considering the one-dimensional interval, Kårsnäs et al. consider the diameter and volume of the bounding box as well as the diameter of the (hyper-)volume of a convex hull in feature space (e.g. the RGB color space) as the basis for the path cost. This shift in the MBD cost function definition leads to additional problems in how the exact distance transform is computed. However, in [10], an approximate method based on Dijkstra-like wave-front propagation is used to compute the vectorial MBD.

The vectorial MBD on color images is used as a pre-processing step by Grand-Brochier et al. [7] in a comparison of different pre-processing step methods for image segmentation. They assume that the object is centered within the image and compute the DT from a single seed point in the center of the image. The pixels with low distance values are then assumed to be object pixels.

A slightly different, and more successful, approach based on a real-time implementation of the MBD by raster-scanning until convergence is presented by Zhang et al. [25]. They detect salient objects in images as follows. As initialization, the image border pixels are set to seed points. Using the assumption that the object does not touch the image border, the pixels with high distance values are those that belong to the object. They use a raster-scanning approach, where each row is scanned, first from upper left to lower right and the from lower right to upper left, until convergence, which is very well suited for parallel implementations. By a GPU implementation based on the raster-scanning technique, real-time performance, 80 MBD DTs per second, is achieved.

6.1   Example Applications of Different Versions of the Minimum Barrier Distance

In this section, different MBD DTs with different sets of seed points are illustrated by DTs of a single image from the MSRA database ([13]). In Fig. 2, all border pixels of the image are set to seed points (c.f. [25]) and in Fig. 3, a single seed point is centered in the image (as used in [7]).

In Fig. 2 and Fig. 3, some of the methods described in this paper are illustrated by applying them to a color image and its gray-scale version. The examples show:

- how restrictive the conditions in Corollary 2 are in order to guarantee that the obtained MBD DT is error-free,
- the exact MBD DT of Discretization I, $\widehat{\rho}$ (Sect. 5.1),
- the Dijkstra approximation of Discretization I, $\widehat{\rho}$ (Sect. 5.1),
- the exact MBD DT of Discretization II, $\widehat{\varphi}$ (Sect. 5.2),
- the color/vectorial MBD DT using the L_1-diameter of the bounding box in the RGB and Lab color spaces, see [10] for details (Sect. 6).

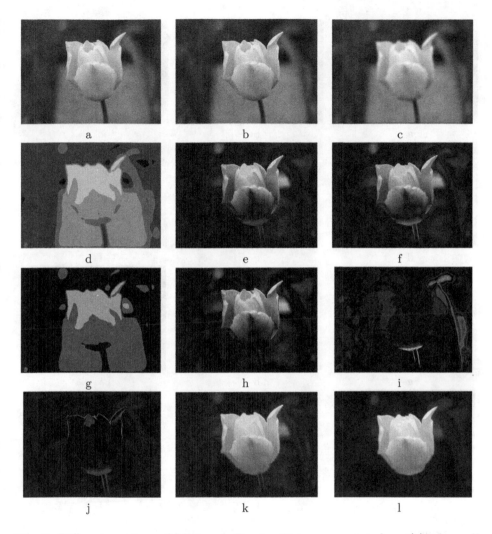

Fig. 2. Different versions of Minimum Barrier Distance computed on (a) when *all border pixels are set to seed points.* a: Original color image; b: Gray scale image; c: b smoothed; d: c quantized such that the condition in Corollary 2 holds; e: exact MBD DT, Discretization I, of b; f: Dijkstra approximation of MBD, Discretization I, of b; g: MBD, Discretization I of d (exact MBD computed by Dijkstra approximation); h: MBD DT, Discretization II, of b; i: Absolute pointwise difference between e and f (gray scale between 0 (no difference) and 28 (maximum difference)) j: Absolute pointwise difference between e and h (gray scale between 0 (no difference) and 99 (maximum difference)) k: Color MBD of a in the RGB color space. l: Color MBD of a in the Lab color space. (Color figure online)

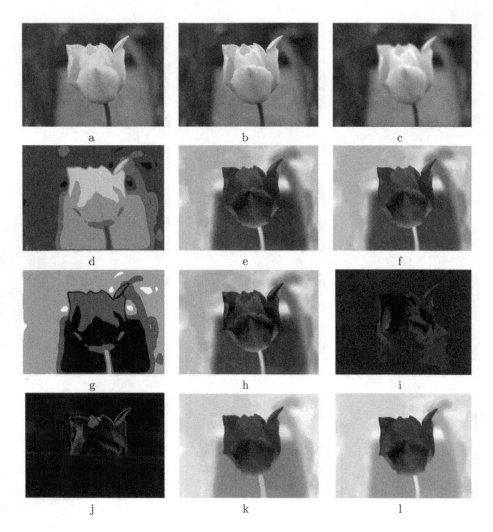

Fig. 3. Different versions of Minimum Barrier Distance computed on (a) when *only the center pixel is set to seed point.* a: Original color image; b: Gray scale image; c: b smoothed; d: c quantized such that the condition in Corollary 2 holds; e: exact MBD DT, Discretization I, of b; f: Dijkstra approximation of MBD, Discretization I, of b; g: MBD, Discretization I of d (exact MBD computed by Dijkstra approximation); h: MBD DT, Discretization II, of b; i: Absolute pointwise difference between e and f (gray scale between 0 (no difference) and 38 (maximum difference)) j: Absolute pointwise difference between e and h (gray scale between 0 (no difference) and 62 (maximum difference)) k: Color MBD of a in the RGB color space. l: Color MBD of a in the Lab color space. (Color figure online)

7   Discussion

The basic idea behind the MBD is very easy to explain and MBD is straight-forward to define, also for color images. Still, the theory of MBD is surprisingly intricate and holds many interesting and to some extent surprising results, such as the equivalence between the two formulations ρ and φ. Efficient algorithms for DT computation, together with the algorithm for exact computation, makes MBD easy to apply in real-life applications. Even though the methods presented and illustrated here are rather simple applications of MBD, it is a promising tool for image processing such as segmentation and saliency detection. A more complex method based on a minimum spanning tree representation together with the MBD for saliency detection was recently presented in [24].

In $\mathbb{R}^n$, we assume that we have continuous images $f: D \to \mathbb{R}$, which we in practice do not have. However, most image acquisition methods induce smooth-ing of the image scene by a point spread function, leading to continuous f. See for example [12] for a discussion. A smooth f can also be found by interpolation of a digital function (often, just an image intensity function).

Open problems include properties and efficient algorithms for DT computa-tion of the vectorial MBD, using other feature spaces than the RGB color space, error bound for the Dijkstra approximation method in arbitrary dimensions, and parallel implementation of the algorithm for exact DT computation.

References

1. Borgefors, G.: Distance transformations in digital images. Comput. Vision Graph. Image Process. **34**, 344–371 (1986)
2. Ciesielski, K.C., Falcão, A.X., Miranda, P.A.V.: Path-value functions for which Dijkstra's algorithm returns optimal mapping (2017, submitted)
3. Ciesielski, K.C., Strand, R., Malmberg, F., Saha, P.K.: Efficient algorithm for find-ing the exact minimum barrier distance. Comput. Vis. Image Underst. **123**, 53–64 (2014)
4. Coeurjolly, D., Vacavant, A.: Separable distance transformation and its applica-tions. In: Brimkov, V., Barneva, R. (eds.) Digital Geometry Algorithms. LNCVB, vol. 2, pp. 189–214. Springer, Dordrecht (2012)
5. Danielsson, P.E.: Euclidean distance mapping. Comput. Graph. Image Process. **14**, 227–248 (1980)
6. Falcão, A.X., Stolfi, J., de Alencar Lotufo, R.: The image foresting transform: the-ory, algorithms, and applications. IEEE Trans. Pattern Anal. Mach. Intell. **26**(1), 19–29 (2004)
7. Grand-Brochier, M., Vacavant, A., Strand, R., Cerutti, G., Tougne, L.: About the impact of pre-processing tools on segmentation methods applied for tree leaves extraction. In: 2014 International Conference on Computer Vision Theory and Applications (VISAPP), vol. 1, pp. 507–514. IEEE (2014)
8. Holuša, M., Sojka, E.: A k-max geodesic distance and its application in image segmentation. In: Azzopardi, G., Petkov, N. (eds.) CAIP 2015. LNCS, vol. 9256, pp. 618–629. Springer, Cham (2015). doi:10.1007/978-3-319-23192-1_52
9. Ikonen, L., Toivanen, P.: Shortest routes on varying height surfaces using grey-level distance transforms. Imag. Vis. Comp. **23**(2), 133–141 (2005)

10. Kårsnäs, A., Strand, R., Saha, P.K.: The vectorial minimum barrier distance. In: 2012 21st International Conference on Pattern Recognition (ICPR), pp. 792–795. IEEE (2012)
11. Kong, T.Y., Rosenfeld, A.: Digital topology: introduction and survey. Comput. Vis. Graph. Image Process. **48**(3), 357–393 (1989)
12. Köthe, U.: What can we learn from discrete images about the continuous world? In: Coeurjolly, D., Sivignon, I., Tougne, L., Dupont, F. (eds.) DGCI 2008. LNCS, vol. 4992, pp. 4–19. Springer, Heidelberg (2008). doi:10.1007/978-3-540-79126-3_2
13. Liu, T., Sun, J., Zheng, N.N., Tang, X., Shum, H.Y.: Learning to detect a salient object. In: 2007 IEEE Conference on Computer Vision and Pattern Recognition, pp. 1–8 (2007)
14. Mansilla, L.A., Miranda, P.A.V., Cappabianco, F.A.: Image segmentation by image foresting transform with non-smooth connectivity functions. In: 2013 XXVI Conference on Graphics, Patterns and Images, pp. 147–154. IEEE (2013)
15. Piper, J., Granum, E.: Computing distance transformations in convex and non-convex domains. Pattern Recogn. **20**(6), 599–615 (1987)
16. Rosenfeld, A., Pfaltz, J.L.: Distance functions on digital pictures. Pattern Recogn. **1**, 33–61 (1968)
17. Rosenfeld, A., Pfaltz, J.L.: Sequential operations in digital picture processing. J. ACM **13**(4), 471–494 (1966)
18. Saha, P.K., Strand, R., Borgefors, G.: Digital topology and geometry in medical imaging: a survey. IEEE Trans. Med. Imaging **34**(9), 1940–1964 (2015)
19. Saha, P.K., Wehrli, F.W., Gomberg, B.R.: Fuzzy distance transform: theory, algorithms, and applications. Comput. Vis. Image Underst. **86**, 171–190 (2002)
20. Saito, T., Toriwaki, J.I.: New algorithms for Euclidean distance transformation of an n-dimensional digitized picture with applications. Pattern Recogn. **27**(11), 1551–1565 (1994)
21. Sethian, J.A.: A fast marching level set method for monotonically advancing fronts. In: Proceedings of National Academy of Sciences, vol. 93, pp. 1591–1595 (1996)
22. Strand, R., Ciesielski, K.C., Malmberg, F., Saha, P.K.: The minimum barrier distance. Comput. Vis. Image Underst. **117**(4), 429–437 (2013). Special Issue on Discrete Geometry for Computer Imagery
23. Strand, R., Malmberg, F., Saha, P.K., Linnér, E.: The minimum barrier distance – stability to seed point position. In: Barcucci, E., Frosini, A., Rinaldi, S. (eds.) DGCI 2014. LNCS, vol. 8668, pp. 111–121. Springer, Cham (2014). doi:10.1007/978-3-319-09955-2_10
24. Wei-Chih, T., Shengfeng He, Q.Y., Chien, S.Y.: Real-time salient object detection with a minimum spanning tree. In: IEEE Conference on Computer Vision and Pattern Recognition (CVPR 2016), pp. 2334–2342 (2016)
25. Zhang, J., Sclaroff, S., Lin, Z., Shen, X., Price, B., Mech, R.: Minimum barrier salient object detection at 80 fps. In: Proceedings of the IEEE International Conference on Computer Vision, pp. 1404–1412 (2015)

Convexity-Preserving Rigid Motions
of 2D Digital Objects

Phuc Ngo[1]([✉]), Yukiko Kenmochi[2], Isabelle Debled-Rennesson[1],
and Nicolas Passat[3]

[1] Université de Lorraine, LORIA, UMR, 7503, Villers-lès-Nancy, France
hoai-diem-phuc.ngo@loria.fr
[2] Université Paris-Est, LIGM, CNRS, Paris, France
[3] Université de Reims Champagne-Ardenne, CReSTIC,
Reims, France

Abstract. Rigid motions on $\mathbb{R}^2$ are isometric and thus preserve the geometry and topology of objects. However, this important property is generally lost when considering digital objects defined on $\mathbb{Z}^2$, due to the digitization process from $\mathbb{R}^2$ to $\mathbb{Z}^2$. In this article, we focus on the convexity property of digital objects, and propose an approach for rigid motions of digital objects which preserves this convexity. The method is extended to non-convex objects, based on the concavity tree representation.

Keywords: Digital rigid motion · Digital convexity · Half-plane representation · Concavity tree · Quasi-regularity

1 Introduction

Rigid motions (*i.e.,* transformations based on translations and rotations) in $\mathbb{R}^n$ are well-known topology- and geometry-preserving operations. They are frequently used in image processing and image analysis. Due to digitization effects, these important properties are generally lost when considering digital images defined on $\mathbb{Z}^n$, as illustrated in Fig. 1.

A part of these problems, topological preservation, was studied for 2D images under rigid transformations [19]. In a similar context, we investigate the geometrical issue, in particular convexity, in this article.

Convexity plays an important role in geometry. This notion is well studied and understood in the continuous space. A convex object in $\mathbb{R}^n$ is defined as a region such that the segment joining any two points within the region is also within the region. Nevertheless, this notion is not adapted to $\mathbb{Z}^n$, due to the digitization process required to convert continuous objects to digital objects, as illustrated in Fig. 2.

In the literature, several definitions have been proposed for convexity defined on $\mathbb{Z}^n$, namely *digital convexity*, such as MP-convexity [18], S-convexity [22], D-convexity [12], H-convexity [12]. The latter will be used in this article due

© Springer International Publishing AG 2017
W.G. Kropatsch et al. (Eds.): DGCI 2017, LNCS 10502, pp. 69–81, 2017.
DOI: 10.1007/978-3-319-66272-5_7

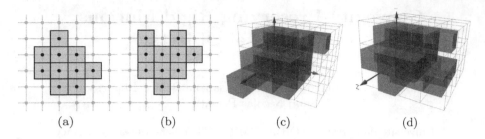

Fig. 1. Convexity alterations under rotation: (a) a convex digital object in 2D, and (b) its non-convex transformed image by a rotation; (c) a convex digital object in 3D, and (d) its non-convex transformed image by a rotation.

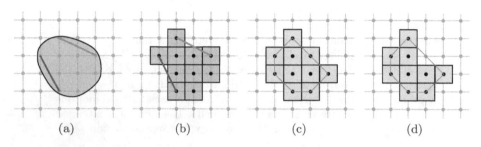

Fig. 2. (a–b) Convexity defined on $\mathbb{R}^n$ is not applicable to $\mathbb{Z}^n$ due to the digitization. (a) A convex continuous object defined on $\mathbb{R}^2$ and the grid $\mathbb{Z}^2$, (b) its digital image represented by pixels. The segments in red and blue belong fully to (a) but not to (b) H-convex (c) and non H-convex (d) objects. Red polygons are their convex hulls. (Color figure online)

to its simplicity and usefulness. Roughly speaking, a digital object $\mathsf{X} \subset \mathbb{Z}^2$ is *H-convex* if its convex hull does contain exactly the points of X (see Fig. 2(c–d)).

In this article, we propose a method allowing us to preserve the H-convexity of digital objects under any rigid motions. Such a method relies on a representation of an H-convex digital object by half-planes. The rigid motion is then applied on such half-plane representation and followed by a digitization. Furthermore, we investigate the condition under which the H-convexity of digital objects is preserved by arbitrary rigid motions.

The paper is organized as follows. After recalling in Sect. 2 basic notions about rigid motions on digital objects, Sect. 3 presents a representation of digital object, well-adapted to H-convexity recognition. A convexity-preserving method for digital objects under rigid motions is then proposed in Sect. 4. Section 5 investigates the problem of rigid motions of non-convex digital objects and opens perspectives summarized in the conclusion.

2   Digital Objects and Rigid Motions on $\mathbb{Z}^2$

2.1   Objects and Rigid Motions on $\mathbb{R}^2$

Let us consider an object X in the Euclidean plane $\mathbb{R}^2$ as a closed connected finite subset of $\mathbb{R}^2$. Rigid motions on $\mathbb{R}^2$ are defined by a mapping:

$$\left| \begin{array}{l} \mathfrak{T} : \mathbb{R}^2 \to \mathbb{R}^2 \\ \quad x \ \mapsto Rx + t \end{array} \right. \tag{1}$$

where R is a rotation matrix and $t \in \mathbb{R}^2$ is a translation vector. Such bijective transformation $\mathfrak{T}$ is isometric and orientation-preserving. Thus, $\mathfrak{T}(X)$ has the same shape (*i.e.*, the same geometry and topology) as X.

2.2   Digitization of Objects and Topology Preservation

As the digitization process may cause topological alterations, conditions for guaranteeing the topology of shape boundaries have been studied [16, 20].

Definition 1. *An object* $X \subset \mathbb{R}^2$ *is r-regular if for each boundary point of X, there exist two tangent open balls of radius r, lying entirely in X and its complement* $\overline{X}$, *respectively.*

This notion, based on classical concepts of differential geometry, establishes a topological link between a continuous shape and its digital counterpart as follows.

Proposition 1 ([20]). *An r-regular object* $X \subset \mathbb{R}^2$ *has the same topological structure as its digitized version* $X \cap \mathbb{Z}^2$ *if* $r \geq \frac{\sqrt{2}}{2}$.

Note that the digitization defined by the intersection of a continuous object X and $\mathbb{Z}^2$ is called Gauss digitization [15]. It was shown that the digitization process of an r-regular object yields a well-composed shape [16], whose definition relies on the following concepts of digital topology (see e.g., [15]). Given a point $p \in \mathbb{Z}^2$, the k-neighborhood of p is defined by $\mathcal{N}_k(p) = \{q \in \mathbb{Z}^2 : \|p - q\|_\ell \leq 1\}$ for $k = 4$ (resp. 8) where $\ell = 1$ (resp. ∞). We say that a point q is k-adjacent to p if $q \in \mathcal{N}_k(p) \setminus \{p\}$. From the reflexive-transitive closure of this k-adjacency relation on a finite subset $S \subset \mathbb{Z}^2$, we derive the k-connectivity relation on S. The k-connectivity relation on S is an equivalence relation. If there is exactly one equivalence class for this relation, namely S, then we say that S is k-connected.

Definition 2 ([16]). *A finite subset* $S \subset \mathbb{Z}^2$ *is well-composed if each 8-connected component of* S *and of its complement* $\overline{S}$ *is also 4-connected.*

This definition implies that the boundary[1] of S is a set of 1-manifolds whenever S is well-composed. As stated above, there exists a strong link between r-regularity and well-composedness.

---

[1] Here, the boundary of S is associated to the continuous boundary induced by the pixels of S, *i.e.*, the union of pixel edges shared by pixels of S and $\overline{S}$.

Proposition 2 ([16]). *If an object* $X \subset \mathbb{R}^2$ *is* r-*regular with* $r \geq \frac{\sqrt{2}}{2}$, *then* $X \cap \mathbb{Z}^2$ *is well-composed.*

We define a digital object X as a connected finite subset of $\mathbb{Z}^2$. In the context of topological coherence of shape boundary, we hereafter assume that X is well-composed, and thus 4-connected.

2.3 Problem of Point-by-Point Rigid Motions on $\mathbb{Z}^2$

If we simply apply a rigid motion $\mathfrak{T}$ of Eq. (1) to every point in $\mathbb{Z}^2$, we generally have $\mathfrak{T}(\mathbb{Z}^2) \not\subset \mathbb{Z}^2$. In order to get the points back on $\mathbb{Z}^2$, we then need a digitization operator $\mathfrak{D} : \mathbb{R}^2 \to \mathbb{Z}^2$, commonly defined as a standard rounding function. A discrete analogue of $\mathfrak{T}$ is then obtained by $\mathcal{T}_{point} = \mathfrak{D} \circ \mathfrak{T}_{|\mathbb{Z}^2}$, so that the discrete analogue of the point-by-point rigid motion of a digital object X on $\mathbb{Z}^2$ is given by $\mathcal{T}_{point}(X)$.

Figure 3 illustrates some examples of $\mathcal{T}_{point}$ of digital lines with different thicknesses. These examples show that the topology and geometry of digital objects are not always preserved, even when the initial shapes are very simple. If a digital line is sufficiently thick, then it preserves topology, but not always geometry. This led us to consider a digital counterpart of regularity condition for guaranteeing topology during point-by-point rigid motions on $\mathbb{Z}^2$ [19]. However, finding rigid motions on $\mathbb{Z}^2$ preserving geometry is still an open problem. In this article, we focus on one important geometrical property, namely convexity. We present a new method for rigid motions on $\mathbb{Z}^2$ that preserves convexity.

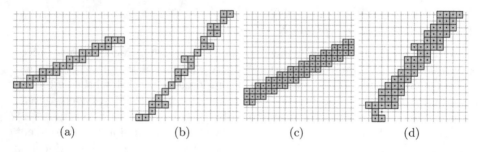

| (a) | (b) | (c) | (d) |

Fig. 3. Well-composed digital lines (a, c) with different thicknesses, which remain well-composed (d) or not (b) after a point-by-point digitized rigid motion $\mathcal{T}_{point}$. In both cases, the convexity of the digital lines is lost by $\mathcal{T}_{point}$.

3 Digital Convexity

In $\mathbb{R}^2$, an object X is said to be convex if and only if for any pair of points $x, y \in X$, every point on the straight line segment joining x and y, defined by $[x, y] = \{\lambda x + (1 - \lambda)y \mid 0 \leq \lambda \leq 1\}$, is also within X. This continuous notion, however, cannot be directly applied to digital objects X in $\mathbb{Z}^2$ since for $p, q \in X$, we generally have $[p, q] \not\subset \mathbb{Z}^2$. In order to tackle this issue, several extensions of the notion of convexity from $\mathbb{R}^2$ to $\mathbb{Z}^2$ have been proposed.

3.1 Definitions

First of all, we introduce the definition of Minsky and Papert [18], called MP-convexity, which is a straightforward extension of the continuous notion.

Definition 3. *A digital object* X *is MP-convex if for any pair of points* $p, q \in X$, $\forall r \in [p, q] \cap \mathbb{Z}^2$, $r \in X$.

Sklansky proposed a different definition based on digitization process [22], called S-convexity.

Definition 4. *A digital object* X *is S-convex if there exists a convex object* $X \subset \mathbb{R}^2$ *such that* $X = X \cap \mathbb{Z}^2$.

Later, Kim gave a geometrical definition that is based on convex hull [12], named *H-convexity*. The convex hull of X is defined by

$$\mathcal{C}onv(X) = \left\{ x \in \mathbb{R}^2 \; \middle| \; x = \sum_{i=1}^{|X|} \lambda_i p_i \; \wedge \; \sum_{i=1}^{|X|} \lambda_i = 1 \; \wedge \; \lambda_i \geq 0 \; \wedge \; p_i \in X \right\}.$$

Definition 5. *A digital object* X *(connected, by definition) is H-convex if* $X = \mathcal{C}onv(X) \cap \mathbb{Z}^2$.

Kim also showed the equivalence between MP-convexity and H-convexity for 4-connected digital objects [12, Theorem 5]. Similar results under the assumption of 8-connectivity can be found in [10] via the chord property, which relates the MP- and H-convexities to another digital convexity notion based on digital lines, called D-convexity [13]. On the other hand, discussing the relation between S-convexity and H-convexity requires some conditions. An element p of a digital object X is said to be isolated if $|\mathcal{N}(p) \cap X| \leq 1$. Under the condition that X has no isolated element, it was proven that X is H-convex if and only if X is S-convex [12, Theorem 4]. A detailed description of various notions of digital convexity can be found in [5, Chap. 9]. In this article, we adopt the H-convexity notion, as it allows us to propose a method for rigid motions of digital objects which preserves H-convexity thanks to the half-plane representation (see Sect. 4).

It should be also mentioned that a similar property to the intersection property known from ordinary convex sets, is preserved; a proof based on an approach of the chord property can be found in [5, Corollary 3.5.1].

Property 1. *Let* X *and* Y *be two digital objects. If* X *and* Y *are H-convex and* $X \cap Y$ *is connected, then* $X \cap Y$ *is H-convex.*

Let us remark that convexity does not imply connectivity in $\mathbb{Z}^2$ by contrast to $\mathbb{R}^2$. This is a reason why an additional connectivity condition is reasonable in the discrete setting.

3.2 Digital Half-Plane Representation

If a digital object X consists of more than two elements, then the convex hull $\mathcal{C}onv(X)^2$ is a convex polygon whose vertices are some points of X. As these vertices are grid points, $\mathcal{C}onv(X)$ is thus represented by the union of closed half-planes with integer coefficients such that

$$\mathcal{C}onv(X) = \bigcap_{H \in \mathcal{R}(\mathcal{C}onv(X))} H,$$

where $\mathcal{R}(P)$ is the minimal set of closed half-planes that constitute a convex polygon P and each H is a closed half-plane in the following form:

$$H = \{(x,y) \in \mathbb{R}^2 \mid ax + by + c \le 0,\ a,b,c \in \mathbb{Z},\ \gcd(a,b) = 1\}. \qquad (2)$$

Note that the integer coefficients of H are uniquely obtained by the pairs of the corresponding consecutive vertices of $\mathcal{C}onv(X)$, denoted by $u, v \in \mathbb{Z}^2$, which are in the clockwise order, such that $(a,b) = \frac{1}{\gcd(w_x, w_y)}(-w_y, w_x)$, and $c = (a,b) \cdot u$ where $(w_x, w_y) = v - u \in \mathbb{Z}^2$.

Therefore, from Definition 5, if X is H-convex, then we have

$$X = \left(\bigcap_{H \in \mathcal{R}(\mathcal{C}onv(X))} H \right) \cap \mathbb{Z}^2 = \bigcap_{H \in \mathcal{R}(\mathcal{C}onv(X))} \left(H \cap \mathbb{Z}^2 \right) \qquad (3)$$

where each $H \cap \mathbb{Z}^2$ is called a *digital half-plane*. It is obvious that any digital half-plane is H-convex. An example of an H-convex digital object with its half-plane representation is illustrated in Fig. 4.

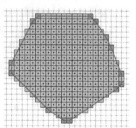 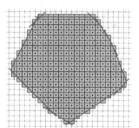

Fig. 4. Example of the digital half-plane representation of a digital object. Left: Digital object and its convex hull (convex hull vertices are in red). Right: Digital half-planes extracted from this convex hull. (Color figure online)

3.3 Verification and Recognition Algorithms

To determine if a digital object is H-convex, several discrete methods were proposed with different approaches. In particular, they are based on the observation of the 8-connected curve corresponding to the inner contour of digital object.

---

[2] In this paper, we consider only X such that the area of $\mathcal{C}onv(X)$ is not null.

In [6], a linear algorithm determines if a given polyomino is convex and, in this case, it returns its convex-hull. It relies on the incremental digital straight line recognition algorithm [7], and uses the geometrical properties of leaning points of maximal discrete straight line segments on the contour. The algorithm scans the contour curve and decomposes it into discrete segments whose extremities must be leaning points. The tangential cover of the curve [11] can be used to obtain this decomposition. Then, the half-planes corresponding to Eq. (3) are directly deduced from the characteristics of discrete segments obtained in the curve decomposition. Similar approaches [8,21] were used to decompose a discrete shape contour into a faithful polygonal representation respecting convex and concave parts.

On the other hand, a combinatorial approach presented in [4] uses tools of combinatorics on words to study contour words: the linear Lyndon factorization algorithm [9] and the Christoffel words. A linear time algorithm verifies convexity of polyominoes and can also compute the convex hull of a digital object. It is presented as a discrete version of the classical Melkman algorithm [17]. This latter can also be used to compute the convex hull of a digital object, so that the half-planes of $\mathcal{R}(\mathcal{C}onv(\mathsf{X}))$ of a digital object X are then deduced from the consecutive vertices (see Eq. (2)). Note that $\mathcal{R}(P)$ is the minimal set of half-planes whose support lines are the edges of a convex polygon P. Then, the H-convexity is tested with elementary operations (see Eq. (3)).

4   Convexity-Preserving Rigid Motions of Digital Objects

In order to preserve the H-convexity of a given H-convex digital object X, we will not apply rigid motions to each grid point of X, as discussed in Sect. 2.3, but to each half-plane H of $\mathcal{R}(\mathcal{C}onv(\mathsf{X}))$ according to Eq. (3). We first explain how to perform a rigid motion of a closed integer half-plane. Then, we propose a method for rigid motions of the whole H-convex digital object.

4.1   Rational Rigid Motions of a Digital Half-Plane

A notion of rigid motion of a digital half-plane $\mathsf{H} \cap \mathbb{Z}^2$ is given by:

$$\mathcal{T}_{\mathcal{C}onv}(\mathsf{H} \cap \mathbb{Z}^2) := \mathfrak{T}(\mathsf{H}) \cap \mathbb{Z}^2 \qquad (4)$$

where $\mathfrak{T}(\mathsf{H})$ is defined analytically as follows.

Digital objects and digital half-planes involve exact computations with integers. Thus, we assume hereafter that all the parameters R and t of rigid motions $\mathfrak{T}$ are rational. More precisely, we consider $R = \frac{1}{r}\begin{pmatrix} p & -q \\ q & p \end{pmatrix}$ where $p, q, r \in \mathbb{Z}$ constitute a Pythagorean triple, i.e., $p^2 + q^2 = r^2$, $r \neq 0$, and $t = (t_x, t_y) \in \mathbb{Q}^2$. This assumption is reasonable, as we can always find rational parameter values sufficiently close to any real values (see [2] for finding such a Pythagorean triple).

An integer half-plane H defined by Eq. (2) is transformed by such $\mathfrak{T}$ to the rational half-plane:

$$\mathfrak{T}(H) = \{(x, y) \in \mathbb{R}^2 \mid \alpha x + \beta y + \gamma \leq 0, \ \alpha, \beta, \gamma \in \mathbb{Q}\}, \tag{5}$$

whose coefficients α, β, γ are given by $\begin{pmatrix} \alpha \\ \beta \end{pmatrix} = R \begin{pmatrix} a \\ b \end{pmatrix}$ and $\gamma = c + \alpha t_x + \beta t_y$.

Note that any rational half-plane can be easily rewritten as an integer half-plane in the form of Eq. (2).

4.2   Rigid Motions of H-convex Digital Objects

In the previous subsection, we showed how to transform a digital half-plane with Eq. (4) via Eq. (5). Since an H-convex digital object X is represented by a finite set of digital half-planes H, as shown in Eq. (3), we can define a rigid motion of X on $\mathbb{Z}^2$ using its associated digital half-planes such that

$$\mathcal{T}_{Conv}(X) := \mathfrak{T}\left(\bigcap_{H \in \mathcal{R}(Conv(X))} H \right) \cap \mathbb{Z}^2 = \left(\bigcap_{H \in \mathcal{R}(Conv(X))} \mathfrak{T}(H) \right) \cap \mathbb{Z}^2. \tag{6}$$

For each $H \in \mathcal{R}(Conv(X))$, we obtain $\mathfrak{T}(H)$ by Eq. (5), and then re-digitize the transformed convex polygon $P = \bigcap_{H \in \mathcal{R}(Conv(X))} \mathfrak{T}(H)$.

We now show that $\mathcal{T}_{Conv}(X)$ is H-convex under a condition on the transformed convex polygon P, called *quasi-r-regularity*. Thanks to Property 1, what we need here is to characterize convex polygons P whose re-discretization $P \cap \mathbb{Z}^2$ preserves the 4-connectivity since any digital half-plane is H-convex.

Property 2. *Let P be a (closed) convex polygon in $\mathbb{R}^2$. If P includes a (closed) ball of diameter $\sqrt{2}$, then $P \cap \mathbb{Z}^2$ is not empty.*

This property is trivial due to the fact that we work on the grid $\mathbb{Z}^2$ of size 1. From a morphological point of view, it can be reformulated as follows.

Property 3. *If the erosion of P by a ball $B_{\frac{\sqrt{2}}{2}}$ of radius $\frac{\sqrt{2}}{2}$ is not empty, namely, $P \ominus B_{\frac{\sqrt{2}}{2}} \neq \emptyset$, then $P \cap \mathbb{Z}^2$ is not empty.*

Indeed, if $P \ominus B_{\frac{\sqrt{2}}{2}} \neq \emptyset$, then the opening $(P \ominus B_{\frac{\sqrt{2}}{2}}) \oplus B_{\frac{\sqrt{2}}{2}}$ is the $(\frac{\sqrt{2}}{2})$-regular part of P, noted $\mathcal{R}eg(P)$. Its digitization with grid interval 1 is guaranteed to have the same topological structure of P (see Proposition 1). In this context, the regular part of P is defined by

$$\mathcal{R}eg(P) = (P \ominus B_{\frac{\sqrt{2}}{2}}) \oplus B_{\frac{\sqrt{2}}{2}},$$

and $P \setminus \mathcal{R}eg(P)$ is called *non-regular part*. If, for each boundary half-plane of $H \in \mathcal{R}(P)$, there exists a ball $B_{\frac{\sqrt{2}}{2}} \subset P$ being tangent to H, then we can define

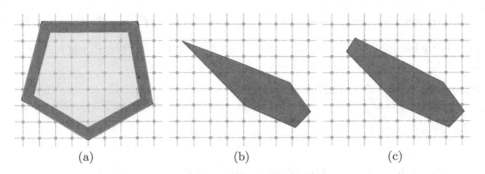

Fig. 5. Convex polygons P, which are quasi-$(\frac{\sqrt{2}}{2})$-regular (a) and not (b-c): (b) there is a vertex of P with angle $< \frac{\pi}{2}$; (c) there is an half-plane $H \in \mathcal{R}(P)$ where no ball $B_{\frac{\sqrt{2}}{2}} \subset P$ exists such that H is tangent to $B_{\frac{\sqrt{2}}{2}}$. In the figures, P are in blue, $B_{\frac{\sqrt{2}}{2}}$ are in red, $P \ominus B_{\frac{\sqrt{2}}{2}}$ is in yellow, $\mathcal{R}eg(P)$ are in pink, and $Cor(P)$ are in green. (Color figure online)

the *corner parts* of P as $P \setminus \mathcal{R}eg(P)$. More precisely, each connected component of $P \setminus \mathcal{R}eg(P)$ corresponds to a corner part of P. In other words, the set of the corner parts is defined by $Cor(P) = \mathcal{C}(P \setminus \mathcal{R}eg(P))$ where $\mathcal{C}$ is the (continuous) connected component function (see Fig. 5).

Property 4 *Let us consider a corner part $A \in Cor(P)$. If the angle $\theta(A)$ at the corresponding vertex of P verifies $\theta(A) \geq \frac{\pi}{2}$ and $A \cap \mathbb{Z}^2$ is not empty, then any point $\mathsf{p} \in A \cap \mathbb{Z}^2$ has at least one 4-adjacent point in $\mathcal{R}eg(P)$.*

Proof. As the Euclidean distance from p to $\mathcal{R}eg(P)$ cannot be higher than $\frac{\sqrt{2}}{2}$, thus inferior to 1, the circle of radius 1 whose center is p always has a non-empty intersection with $\mathcal{R}eg(P)$. Due to the angle condition of $\theta(A)$, this intersection forms a circular arc whose central angle is not less than $\frac{\pi}{2}$. The 4-neighbours of p are on the circle of radius 1 and at least one of them appears on this circular-arc intersection with $\mathcal{R}eg(P)$. Therefore, there always exists at least one of the 4-neighbours of p in $\mathcal{R}eg(P)$.

Let us now introduce the notion of *quasi-regularity*.

Definition 6. *A convex polygon P is quasi-r-regular if $\forall H \in \mathcal{R}(P)$, $\exists B_r \subset P$ such that H is tangent to B_r, and if $\forall A \in Cor(P)$, $\theta(A) \geq \frac{\pi}{2}$.*

From Property 4, we then obtain the following lemma.

Lemma 1. *Let X be an H-convex digital object. If $Conv(\mathsf{X})$ is quasi-$(\frac{\sqrt{2}}{2})$-regular, then $\mathcal{T}_{Conv}(\mathsf{X})$ is 4-connected.*

Finally, we obtain the main proposition from this lemma and Property 1.

Proposition 3. *Let X be an H-convex digital object. If $Conv(\mathsf{X})$ is quasi-$(\frac{\sqrt{2}}{2})$-regular, then $\mathcal{T}_{Conv}(\mathsf{X})$ is H-convex.*

Moreover, the transformed convex polygon $P = \bigcap\limits_{H \in \mathcal{R}(Conv(X))} \mathfrak{T}(H)$ generally does not correspond to the convex hull P' of $\mathcal{T}_{Conv}(X)$, i.e., $P' = Conv(P \cap \mathbb{Z}^2)$. However, we always have the following inclusion relation.

Property 5. *Let* X *be an H-convex digital object. Let us consider the two convex polygons such that* $P = \bigcap\limits_{H \in \mathcal{R}(Conv(X))} \mathfrak{T}(H)$ *and* $P' = Conv(\mathcal{T}_{Conv}(X))$. *Then, we always have* $P' \subseteq P$.

5   Rigid Motions of a Non-convex Digital Object

In the previous section, we showed how to carry out rigid motions of H-convex digital objects with preservation of their H-convexity under the quasi-$(\frac{\sqrt{2}}{2})$-regular condition. In this section, we first present a hierarchical representation of a non-convex digital objects based on convex hulls, called a concavity tree [23]. Then, we show how to perform rigid motions of a non-convex digital object via this concavity-tree representation.

5.1   Concavity Tree of a Digital Object

A concavity tree, initially introduced in [23], is a hierarchical representation of the convex and concave parts of the contour of a digital object, with different levels of detail. This tree structure has been used, *e.g.*, for minimum-perimeter polygon computation [14,23], hierarchical shape analysis [3], polygonal approximation [1]. The root of the tree corresponds to a given digital object. Then, each node corresponds to a concavity part of its parent. Note that each concavity part X' of a digital object X is obtained as a connected component of the subtraction of X from the digitized convex hull $Conv(X) \cap \mathbb{Z}^2$. This is written by $X' \in \mathfrak{C}((Conv(X) \cap \mathbb{Z}^2) \setminus X)$ where $\mathfrak{C}(S)$ denotes the set of all connected components of a finite set $S \in \mathbb{Z}^2$. In other words, we have

$$X = \left(Conv(X) \cap \mathbb{Z}^2\right) \setminus \left(\bigcup_{X' \in \mathfrak{C}((Conv(X) \cap \mathbb{Z}^2) \setminus X)} X'\right), \tag{7}$$

where each concavity part X' is recursively replaced by the subtraction of the concavity parts of X' from $Conv(X') \cap \mathbb{Z}^2$ until no concavity part is found. An illustration of concavity tree is given in Fig. 6.

5.2   Digital Object Rigid Motions Using Concavity Tree

As observed in Eq. (7), a digital object X can be decomposed into the set of digitized convex hulls of hierarchical concavity parts, $Conv(X) \cap \mathbb{Z}^2$, $Conv(X') \cap \mathbb{Z}^2$, ..., via the subtraction operations. As we are rather interested in such a hierarchical decomposition by digitized convex polygons, we consider a concavity tree such that each node corresponds to the half-plane representation of the

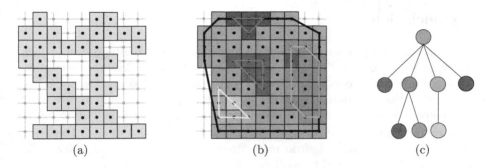

Fig. 6. (a) A digital object. (b) The convex and concave parts of (a) with their convex hulls. (c) The concavity tree of (a) where each colored node corresponds to a concavity part of its parent and the root (in orange) corresponds to (a). (Color figure online)

convex hull of a concavity part Y, $\mathcal{C}onv(Y) = \bigcap_{H \in \mathcal{R}(\mathcal{C}onv(Y))} H$, instead of Y itself.
Once the hierarchical object decomposition is obtained, we simply apply $\mathcal{T}_{Conv}$ to each digitized convex polygon $\mathcal{C}onv(Y) \cap \mathbb{Z}^2$, as shown in the previous section, and then carry out the set subtraction operations guided by the tree.

Figure 7 illustrates comparisons between applications of point-by-point transformation $\mathcal{T}_{point}$ and convexity-preserving transformation $\mathcal{T}_{Conv}$ using concavity tree to non-convex digital objects.

From Proposition 3, we can preserve the H-convexity of each digitized convex polygon in the tree if the convex polygon is quasi-$(\frac{\sqrt{2}}{2})$-regular. Practically, such hypothesis is not satisfied in general (see Fig. 6(b)). However up-sampling strategies on the digital objects can allow us to generate convex polygons which guarantee the quasi-$(\frac{\sqrt{2}}{2})$-regularity. This subject is one of our perspectives.

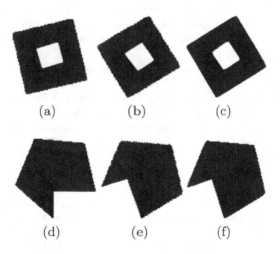

Fig. 7. Original digital objects (a, d) with their point-by-point digital rigid motions (b, e) and their convex-preserving digital rigid motions using concavity tree (d, f).

6   Conclusion

In this article, we proposed a method for rigid motions of digital objects which preserves the H-convexity. The method is based on the half-plane representation of H-convex digital objects. In order to guarantee the H-convexity, we introduced the notion of quasi-r-regularity for convex polygons, and showed that the H-convexity is preserved under rigid motions if the convex hull of an initial H-convex digital object is quasi-($\frac{\sqrt{2}}{2}$)-regular. The necessity of such condition is caused by the fact that the convexity does not imply the connectivity in $\mathbb{Z}^2$ contrary to $\mathbb{R}^2$. As we need to re-discretize the transformed convex polygon at the end, it is natural to have a similar notion of the r-regularity of continuous objects, which guarantees the topology after its digitization. The method was also extended to non-convex digital objects using the hierarchical object representation.

The proposed method works only with digital objects satisfying the quasi-($\frac{\sqrt{2}}{2}$)-regular condition on the convex hull. In practice, this condition is difficult to obtain, and this would limit the direct use of the proposed method. However, an up-sampling approach would provide a promising strategy to guarantee the quasi-($\frac{\sqrt{2}}{2}$)-regularity of the convex hull of any digital object. It is also observed that the quasi-($\frac{\sqrt{2}}{2}$)-regularity is sufficient but not necessary for the H-convexity preservation. It would be one of our perspectives to find such a sufficient and necessary condition. Another perspective would be to extend the method into higher dimensions.

References

1. Aguilera-Aguilera, E., Carmona-Poyato, A., Madrid-Cuevas, F., Medina-Carnicer, R.: The computation of polygonal approximations for 2D contours based on a concavity tree. J. Vis. Commun. Image Represent. **25**(8), 1905–1917 (2014)
2. Anglin, W.S.: Using Pythagorean triangles to approximate angles. Am. Math. Monthly **95**(6), 540–541 (1988)
3. Borgefors, G., di Baja, G.S.: Analyzing nonconvex 2D and 3D patterns. Comput. Vis. Image Underst. **63**(1), 145–157 (1996)
4. Brlek, S., Lachaud, J., Provençal, X., Reutenauer, C.: Lyndon + Christoffel = digitally convex. Pattern Recogn. **42**(10), 2239–2246 (2009)
5. Cristescu, G., Lupsa, L.: Non-Connected Convexities and Applications. Kluwer Academic Publishers, Dordrecht (2002)
6. Debled-Rennesson, I., Rémy, J.L., Rouyer-Degli, J.: Detection of the discrete convexity of polyominoes. Discrete Appl. Math. **125**(1), 115–133 (2003)
7. Debled-Rennesson, I., Reveillès, J.: A linear algorithm for segmentation of digital curves. Int. J. Pattern Recognit Artif Intell. **9**(4), 635–662 (1995)
8. Dorksen-Reiter, H., Debled-Rennesson, I.: Convex and concave parts of digital curves. J. Geom. Prop. Incomplete Data **31**, 145–159 (2004)
9. Duval, J.: Factorizing words over an ordered alphabet. J. Algorithms **4**(4), 363–381 (1983)

10. Eckhardt, U.: Digital lines and digital convexity. In: Bertrand, G., Imiya, A., Klette, R. (eds.) Digital and Image Geometry. LNCS, vol. 2243, pp. 209–228. Springer, Heidelberg (2001). doi:10.1007/3-540-45576-0_13

11. Feschet, F., Tougne, L.: Optimal time computation of the tangent of a discrete curve: application to the curvature. In: Bertrand, G., Couprie, M., Perroton, L. (eds.) DGCI 1999. LNCS, vol. 1568, pp. 31–40. Springer, Heidelberg (1999). doi:10.1007/3-540-49126-0_3

12. Kim, C.E.: On the cellular convexity of complexes. IEEE Trans. Pattern Anal. Mach. Intell. **3**(6), 617–625 (1981)

13. Kim, C.E., Rosenfeld, A.: Digital straight lines and convexity of digital regions. IEEE Trans. Pattern Anal. Mach. Intell. **4**(2), 149–153 (1982)

14. Klette, G.: Recursive computation of minimum-length polygons. Comput. Vis. Image Underst. **117**(4), 386–392 (2013)

15. Klette, R., Rosenfeld, A.: Digital Geometry: Geometric Methods for Digital Picture Analysis. Elsevier, Amsterdam (2004)

16. Latecki, L.J., Conrad, C., Gross, A.: Preserving topology by a digitization process. J. Math. Imaging Vis. **8**(2), 131–159 (1998)

17. Melkman, A.A.: On-line construction of the convex hull of a simple polyline. Inform. Process. Lett. **25**(1), 11–12 (1987)

18. Minsky, M., Papert, S.: Perceptrons: An Introduction to Computational Geometry. MIT Press, Reading (1969)

19. Ngo, P., Passat, N., Kenmochi, Y., Talbot, H.: Topology-preserving rigid transformation of 2D digital images. IEEE Trans. Image Process. **23**(2), 885–897 (2014)

20. Pavlidis, T.: Algorithms for Graphics and Image Processing. Berlin: Springer, and Rockville: Computer Science Press (1982)

21. Roussillon, T., Sivignon, I.: Faithful polygonal representation of the convex and concave parts of a digital curve. Pattern Recogn. **44**(10–11), 2693–2700 (2011)

22. Sklansky, J.: Recognition of convex blobs. Pattern Recogn. **2**, 3–10 (1970)

23. Sklansky, J.: Measuring concavity on a rectangular mosaic. IEEE Trans. Comput. C-21(12), 1355–1364 (1972)

Weighted Distances on the Trihexagonal Grid

Gergely Kovács[1]([✉]), Benedek Nagy[2], and Béla Vizvári[3]

[1] Edutus College, Tatabánya, Hungary
kovacs.gergely@edutus.hu
[2] Department of Mathematics, Faculty of Arts and Sciences,
Eastern Mediterranean University, Mersin-10, Famagusta, North Cyprus, Turkey
nbenedek.inf@gmail.com
[3] Department of Industrial Engineering, Eastern Mediterranean University,
Mersin-10, Famagusta, North Cyprus, Turkey

Abstract. Recently chamfer distances have been developed not only on the usual integer grids, but also on some non traditional grids including grids which are not lattices. In this paper the trihexagonal grid is considered which is a kind of mix of the hexagonal and triangular grids: its pixels are hexagons and two shaped (oriented) triangles. Three types of 'natural' neighborhood relations are considered on the grid, consequently three weights are used to describe the chamfer distances. Formulae to compute the minimal weights of a connecting path, i.e., the distance of any two pixels, are provided to various cases depending on the relative ratio of the weights. Some properties of these distances, including metricity are also analysed.

1 Introduction

Digital geometry is an important theoretical part of digital image processing. Discrete, digital spaces have different properties than the Euclidean space, e.g., neighborhood of points play important role. Consequently, digital (path based) distance functions have various advantages, see, e.g., [4]. Various digital distance functions are developed since the 1960's, where the two basic digital distances based on the two usual neighborhood on the square grid were investigated [16]. Various grids have various properties and various advantages and disadvantages in applications. E.g., the square grid is easy to use, it has hardware and software support, other grids have more symmetries, may provide better topological properties. Digital geometry and distance functions are developed for various traditional and non traditional grids, both in 2D and in higher dimensions. To move some of the results from a point lattice to other point lattice may not be trivial. In case of at least one of the grids is not a point lattice, we need to find newer and newer approaches, the translation of the results cannot go automatically. Even some results seem to be similar, the details could be very different. A nice symmetric coordinate frame for the hexagonal grid was presented in [3], while the simplest digital distances are investigated in [7] for that grid. The triangular grid was also described by symmetric coordinate system with integer

© Springer International Publishing AG 2017
W.G. Kropatsch et al. (Eds.): DGCI 2017, LNCS 10502, pp. 82–93, 2017.
DOI: 10.1007/978-3-319-66272-5_8

triplets, see, e.g., [9,11]. The hexagonal and triangular grids can be seen as one and two parallel oblique planes of the cubic grid [10], and with three such planes another 'triangular grid' can be obtained [12]. This grid is called trihexagonal grid in this paper since it is mixing the properties of the hexagonal and triangular grids (see Fig. 1). Each node has the same rank and each node is surrounded by the same set of regular polygons in the same order. By its symmetric properties this grid is denoted by T(6, 3, 6, 3) in [15], where 4 is the nodal rank and 6 or 3 is the number of edges of the i-th polygon.

Chamfer (or weighted) distances are providing a relatively good approximation of the Euclidean distance with good algorithmic properties [1]. The concept was also investigated on some non-traditional 2D grids including the triangular grid [13], Khalimsky grid [5,6]; and various 3D grids [17,18]. In this paper chamfer distances on the 2D trihexagonal grid are investigated; our main motivation is to show the basic properties.

2   Description of Trihexagonal Tiling (6, 3, 6, 3)

Figure 1 shows a usual representation of the trihexagonal tiling. The grid is T(6, 3, 6, 3) if dual notation is used, see, *e.g.*, [15]. In this representation hexagons represent the points for which 6-neighborhood is used and triangles the points for which 3-neighborhood is used.

Each pixel of the trihexagonal grid (6, 3, 6, 3) is a hexagon or a triangle, we also call it as a point of the grid. A hexagon has 6 neighbors with common sides (6 triangles). A triangle has 3 neighbors with common sides (3 hexagons).

Similarly to the hexagonal grid or to the triangular grid, each pixel of the trihexagonal grid (6, 3, 6, 3) can also be described by a unique coordinate-triplet.

There is a triangle having coordinate triplet (0, 0, 0), and the directions of the axes can be seen on the Fig. 1. At every time when we step from a triangle to a hexagon (or from a hexagon to a triangle) crossing their common side, the step is done parallel to one of the axes. If the step is in the direction of an axis, then the respective coordinate value is increased by 1, while in case the step is in opposite direction to an axis, the respective coordinate value is decreased by 1.

In this way every point gets a unique coordinate triplet with integer values. However the three values are not independent (we are in a two dimensional space, *i.e.*, plane). Their sum is either 0, 1 or 2. There are two orientations of the used triangles: there are triangles of shape $\triangle$ and there are triangles of shape $\triangledown$. The sum of the coordinate values that address a triangle is 0 or 2 depending the orientation (shape) of the triangle. The sum of the coordinate values of a hexagon is 1.

There are two types of (commonly used) neighborhood on this grid. Two pixels are neighbors if they share a side. Two pixels are semi-neighbors if they share at least a point on their boundaries (e.g., a corner point). Using the coordinate triplets one can give the neighborhood relations in the following formal form. The hexagon $p(p(1), p(2), p(3))$ and the triangle $q(q(1), q(2), q(3))$ are neighbors if $|p(1) - q(1)| + |p(2) - q(2)| + |p(3) - q(3)| = 1$. (See also Fig. 1.)

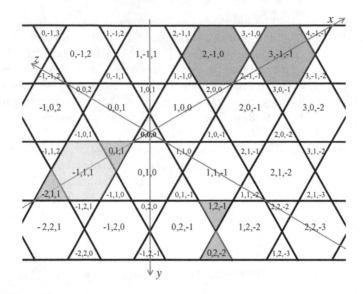

Fig. 1. Coordinate axes and coordinate values assigned to cells of a segment of the trihexagonal grid. The yellow hexagon and the orange triangles are neighbors (α), the pink hexagons are semi-neighbors (β), the blue triangles are semi-neighbors (γ). (Color figure online)

There are no triangles with common side, but every triangle p has 3 semi-neighbor triangle, for example q, for which $|p(1) - q(1)| + |p(2) - q(2)| + |p(3) - q(3)| = 2$.

There are no hexagons with common side, but every hexagon p has 6 semi-neighbor hexagon, for example q, for which $|p(1) - q(1)| + |p(2) - q(2)| + |p(3) - q(3)| = 2$.

3   Definition of Weighted Distances

Let $\alpha, \beta, \gamma \in \mathbb{R}^+$ be positive real weights. The simplest weighted distances allow to step to a neighbor (from a triangle to a hexagon and vice versa) by changing only one coordinate value by ± 1 with weight α. Let the weight of a step from a hexagon to a semi-neighbor hexagon be β. This step changes two coordinates, one by $+1$, and another one by -1. Let the weight of a step from a triangle to a semi-neighbor triangle be γ. This step changes two coordinates, both by $+1$, or both by -1. (See also Fig. 1.)

We can define the weighted distance of any two points (hexagons or triangles) of the grid. Let p and q be two points of the grid. A finite point sequence of points of the form $p = p_0, p_1, \ldots, p_m = q$, where p_{i-1}, p_i are neighbors or semi-neighbors for $1 \le i \le m$, is called a path from p to q. A path can be seen as consecutive steps to neighbors or semi-neighbors. Then the cost of the path is the sum of the weights of its steps.

Finally, let the weighted distance $d(p, q; \alpha, \beta, \gamma)$ of p and q by the weights α, β, γ be the cost of the minimal (basic) weighted paths between p and q.

Usually, there can be several shortest paths from a point to another. The order of steps can be varied, e.g., if a shortest path from $(0, 0, 1)$ to $(2, -1, 0)$ contains two β steps containing the point $(1, 0, 0)$, then the path having two β steps through $(1, -1, 1)$ is also a shortest path. However, since we are not on a lattice, we are not free to use any order of the steps. One need to take care about the following conditions:

- α steps are allowed on any point, and a point of the opposite type is reached by the step (where hexagon and triangle are the types).
- β step can only be used from a hexagon (the sum of coordinates is equal to 1), and it goes to a hexagon.
- γ steps are valid only from a triangle (the sum of the coordinates is equal to 0, or 2), and another triangle is reached (but the sum of the coordinates of the new triangle is different from the original one: the triangles have opposite orientation).

A technical definition is used through the paper. The difference $w_{p,q} = (w(1), w(2), w(3))$ of two points p and q is defined by $w(i) = q(i) - p(i)$, where $i \in \{1, 2, 3\}$.

4   Minimal Weighted Paths

There are various paths with various sums of weights that can be found between any two points. When the weights α, β and γ are known the optimal search (the Dijkstra algorithm) can be used. However depending on the actual ratios and values of the weights α, β, γ, one can compute a minimum weighted paths in a more direct way. Using a combinatorial approach, we give methods for these computations for each possible case.

We use the natural condition $0 \leq \alpha \leq \beta, \gamma$ for the used weight values. We know that with a neighbor step (by weight α) only 1 of the coordinates changes by ± 1; with a semi-neighbor step (by weight β or γ) exactly 2 of the coordinates change by 1 and/or -1, respectively. Therefore it is important to measure the relative weight of a step, that is the cost of the change of a coordinate value by ± 1. These relative weights give the separation of the possible cases.

4.1   Case $2\alpha \leq \beta$ and $2\alpha \leq \gamma$

Lemma 1. *If $2\alpha \leq \beta$ and $2\alpha \leq \gamma$, then the length of the minimal path between the points p and q is*

$$d(p, q; \alpha, \beta, \gamma) = \alpha(|w(1)| + |w(2)| + |w(3)|). \tag{1}$$

Proof. In this case the use of β steps and γ steps is not efficient (their usage do not lead to shorter paths than we have without them). A minimal path can be constructed only by α steps. In every step the absolute value of a coordinate difference is decreasing by 1 implying the formula.   □

For example the distance of the triangle $(0, 0, 0)$ and the hexagon $(1, -1, 1)$ is 3α.

4.2 Case $\gamma \leq 2\alpha \leq \beta$

Lemma 2. *If $\alpha \leq \gamma \leq 2\alpha \leq \beta$, then the length of the minimal path between the points p and q is*

$$d(p, q; \alpha, \beta, \gamma) = \gamma \cdot \min\{|w(i)|\} + \alpha \cdot \left(\sum_{i=1}^{3} |w(i)| - 2 \cdot \min\{|w(i)|\} \right), \quad (2)$$

but if $w_{p,q}$ contains one 0 and two 1 values, or one 0 and two -1 values, then

$$d(p, q; \alpha, \beta, \gamma) = \gamma. \quad (3)$$

Proof. The value of the sum of coordinate differences $\sum_{i=1}^{3} w(i)$ is equal to $0, \pm 1$ or ± 2. (Moreover it is ± 2 only if the two points are different shaped triangles.)

Clearly, in this case β steps cannot appear (when $2\alpha \leq \beta$) in any shortest path or substituting them by two α steps a path with same weight is obtained, we deal with paths containing only α and γ steps. It follows from Lemma 1 there is always a path between p and q consisting of α steps only and their number is $|w(1)| + |w(2)| + |w(3)|$. Obviously, a path with this property has the minimal length among all paths using only α steps as each such step changes the absolute values of differences of the coordinates by 1. If there are two coordinates where the difference has the same sign, then it is possible to make a γ step at the beginning.

If $\min\{|w(i)|\} > 0$, then the signs of two differences (from three) are the same, and one of them has the smallest absolute value, because the sum of differences is between -2 and 2. It means that we are able to use γ steps $\min\{|w(i)|\}$ times instead of twice many α steps, but no more. The path can be constructed that between two γ steps come two α steps (for example between two $(1, 1, 0)$ steps come two $(0, 0, -1)$ steps).

If $\min\{|w(i)|\} = 0$ and the sign of the other two differences are different, then the use of a γ step does not decrease the sum of the absolute values of the coordinate differences, i.e. the use of γ step is not efficient (there is a shortest path without it).

If $\min\{|w(i)|\} = 0$, then the signs of the other two differences are the same only if both values are equal to 1 or both values are equal to -1, for example $(1, 1, 0)$. In this case the distance is γ. □

For example the distance of the triangle $(0, 0, 0)$ and the hexagon $(1, -1, 1)$ is $\gamma + \alpha$, the distance of the triangles $(0, 0, 0)$ and $(2, 0, -2)$ is 4α, but the distance of the triangles $(0, 0, 0)$ and $(1, 0, 1)$ is γ with the given conditions on the weights.

4.3   Case $\beta \leq 2\alpha \leq \gamma$

The use of γ steps instead of α steps is not efficient, but one β step can be better than two α steps.

Lemma 3. *If* $\beta \leq 2\alpha \leq \gamma$, *then the length of the minimal path between the hexagons* p *and* q *is*

$$d(p,q;\alpha,\beta,\gamma) = \beta \cdot \frac{\sum\limits_{i=1}^{3} |w(i)|}{2}. \tag{4}$$

Proof. If p and q are hexagons, then the sum of the coordinate differences is 0, and every β step is a 0-sum step. In this case a minimal path can be constructed only by β steps. In every step the sum of the absolute values of the coordinate differences is decreasing by 2, and this is better than the use of two α steps.   □

For example the distance of the hexagons (0, 0, 1) and (2, −1, 0) is 2β.

If every point of a path is a hexagon, then it is called a hexagonal path in the following. The next lemmas can be proven in similar manner as Lemma 3.

Lemma 4. *If* $\beta \leq 2\alpha \leq \gamma$, *and* p *and* q *are a triangle and a hexagon, then*

$$d(p,q;\alpha,\beta,\gamma) = \beta \cdot \frac{\sum\limits_{i=1}^{3} |w(i)| - 1}{2} + \alpha. \tag{5}$$

Lemma 5. *If* $\beta \leq 2\alpha \leq \gamma$, *and* p *and* q *are triangles, then*

$$d(p,q;\alpha,\beta,\gamma) = \beta \cdot \frac{\sum\limits_{i=1}^{3} |w(i)| - 2}{2} + 2\alpha. \tag{6}$$

For example the distance of the triangle $(0,0,0)$ and the hexagon $(1,−1,1)$ is $\beta + \alpha$, and the distance of the triangles $(0,0,0)$ and $(2,0,−2)$ is $\beta + 2\alpha$.

4.4   Case $\beta \leq \gamma \leq 2\alpha$

Lemma 6. *If* $\beta \leq \gamma \leq 2\alpha$, *then the length of the minimal path between* p *and* q *is given in* (4), (5) *and* (6), *but if* p *and* q *are adjacent triangles, then* $d(p,q;\alpha,\beta,\gamma) = \gamma$.

Proof. In every β step and γ step the sum of the absolute values of the coordinate differences is decreasing by 2. But $\beta \leq \gamma$ means that the use of γ steps is not efficient if one of the points is a hexagon. Then this case is the same as the previous case.

But if p and q are triangles and in the minimal path of the above case the α steps are consecutive steps, then it is feasible and efficient to change them with a γ step. The α steps are consecutive only if there are no β steps, and the triangles are semi-neighbors.   □

For example the distance of the triangles $(0,0,0)$ and $(2,0,−2)$ is $\beta + 2\alpha$, but the distance of the triangles $(0,0,0)$ and $(1,0,1)$ is γ.

4.5   Case $\gamma \leq \beta \leq 2\alpha$

Subcase p and q are Hexagons

Lemma 7. *If $\alpha \leq \gamma \leq \beta \leq 2\alpha$ and p, q are hexagons, then the minimal path contains at least one γ step if and only if $w_{p,q}$ has two coordinates with the same sign and $\gamma + 2\alpha \leq 2\beta$.*

Proof. Of course we cannot use a γ step instead of a β step, or two γ steps instead of two β steps. But we can use a γ step and two α steps instead of two appropriate β steps.

If p and q are hexagons, then the sum of coordinate differences is equal to 0. In this case $\min\{|w(i)\} = 0$ if and only if the sign of the other two differences are different. Then the use of a γ step does not decrease the sum of the absolute values of the coordinate differences, thus the use of γ step is not efficient. The minimal path is the hexagonal path between the two hexagons, which uses only β steps. $\min\{|w(i)\} > 0$ if and only if w_{pq} has two coordinates with the same sign. The hexagonal path is minimal only if $2\beta < \gamma + 2\alpha$, when the use of γ step is not efficient. But if $2\beta \geq \gamma + 2\alpha$, then the minimal path uses γ step: we can begin the minimal path with α, γ, α steps, and the sum of the absolute value of the coordinate differences is decreased by 4. $\qquad\square$

Lemma 8. *If $\alpha \leq \gamma \leq \beta \leq 2\alpha$ and p, q are hexagons, then*

$$d(p,q;\alpha,\beta,\gamma) = \beta \cdot \frac{\sum_{i=1}^{3} |w(i)|}{2} \tag{7}$$

if $2\beta \leq \gamma + 2\alpha$, and

$$d(p,q;\alpha,\beta,\gamma) = \gamma \cdot \min\{|w(i)|\} + 2\alpha \cdot \min\{|w(i)|\} + \beta \cdot \frac{\sum_{i=1}^{3} |w(i)| - 4\min\{|w(i)|\}}{2} \tag{8}$$

if $\gamma + 2\alpha \leq 2\beta$.

Proof. If p and q are hexagons, then the sum of the coordinate differences is 0.

If $\min\{|w(i)|\} = 0$, then the signs of the other two differences are different. It means that the use of γ steps is not efficient.

But if $\min\{|w(i)|\} > 0$, then $w_{p,q}$ has two coordinates with the same sign, for example positive coordinates: the first and the second one. There exists a β-path between the two hexagons, which first $2\min\{|w(i)|\}$ steps are the alternate β steps $(1,0,-1)$ and $(0,1,-1)$. Changing two appropriate β steps to one γ step and two α-steps may be efficient $\min\{|w(i)|\}$ times if $\gamma + 2\alpha \leq 2\beta$.

Of course $4\min\{|w(i)|\} \leq \sum_{i=1}^{3} |w(i)|$ holds, because of $\sum_{i=1}^{3} w(i) = 0$. $\qquad\square$

For example the distance of the hexagons $(0,0,1)$ and $(2,-1,0)$ is 2β or $2\alpha + \gamma$.

Subcase p and q are a Hexagon and a Triangle. If p and q are a hexagon and a triangle, then there exists the above mentioned minimal path between the hexagon and another hexagon, which is the closest hexagon neighbor of the triangle, and there is a final α step between the closest hexagon and the triangle. If the final step of the path between the two hexagons is a β step, then it can be efficient to change the final β and α steps to an α step and a γ step. This way we can increase the number of γ steps by 1. For example we can change the steps $(0, 1, -1)$ and $(1, 0, 0)$ to $(0, 0, -1)$ and $(1, 1, 0)$. When are we able to do this substitution?

In this case the sum of the coordinate differences is ± 1.

Lemma 9. *If $\alpha \leq \gamma \leq \beta \leq 2\alpha$, p and q are a hexagon and a triangle, and $w_{p,q}$ contains only one coordinate with the sign of $\sum\limits_{i=1}^{3} w(i)$, then*

$$d(p, q; \alpha, \beta, \gamma) = \beta \cdot \frac{\sum\limits_{i=1}^{3} |w(i)| - 1}{2} + \alpha \tag{9}$$

if $2\beta \leq \gamma + 2\alpha$, and

$$d(p, q; \alpha, \beta, \gamma) = \gamma \cdot \min\{|w(i)|\}$$

$$+ \alpha \cdot (2\min\{|w(i)|\} + 1) + \beta \cdot \frac{\sum\limits_{i=1}^{3} |w(i)| - 4\min\{|w(i)|\} - 1}{2} \tag{10}$$

if $\gamma + 2\alpha \leq 2\beta$.

Proof. Let us assume, that $\sum\limits_{i=1}^{3} w(i) = 1$ and p is a hexagon and q is a triangle. (The proof of the case -1, or the case of a triangle and a hexagon are similar.) In this case $w_{p,q}$ contains only one positive coordinate, for example the first one, then the final α step of the above mentioned path (based on the closest neighbor hexagon of the triangle) is $(1, 0, 0)$. (For example if the starting hexagon is $(1, 0, 0)$ and the triangle is $(7, -3, -2)$, then the closest neighbor hexagon of the triangle is $(6, -3, -2)$ and not $(7, -4, -2)$ or $(7, -3, -3)$, because the sum of the absolute value of the coordinate differences are here 10, 12 and 12.)

The use of a γ step, which first coordinate is $+1$ is not efficient, because this γ step does not decrease the sum of the absolute values of the coordinate differences. Thus the use of the above mentioned substitution is not efficient, *i.e.* the number of γ steps is equal to the number of γ steps of the path based on the closest neighbor hexagon (in the previous lemma).

If this path uses an α step, then it's coordinates are $(1, 0, 0)$. These are the same as the coordinates of the final α step. Thus we are not able to increase the number of β steps. □

For example the distance of the triangle $(0, 0, 0)$ and the hexagon $(3, -1, -1)$ is $2\beta + \alpha$ or $3\alpha + \gamma$.

In a similar manner one can also prove the following lemma.

Lemma 10. *If $\alpha \leq \gamma \leq \beta \leq 2\alpha$, p and q are a hexagon and a triangle, and $w_{p,q}$ contains two coordinates with the sign of $\sum\limits_{i=1}^{3} w(i)$, then*

$$d(p, q; \alpha, \beta, \gamma) = \beta \cdot \frac{\sum\limits_{i=1}^{3} |w(i)| - 3}{2} + \alpha + \gamma \tag{11}$$

if $2\beta \leq \gamma + 2\alpha$, and

$$d(p, q; \alpha, \beta, \gamma) = \gamma \cdot \min\{|w(i)|\}$$

$$+ \alpha \cdot (2 \min\{|w(i)|\} - 1) + \beta \cdot \frac{\sum\limits_{i=1}^{3} |w(i)| - 4 \min\{|w(i)|\} + 1}{2} \tag{12}$$

if $\gamma + 2\alpha \leq 2\beta$.

For example the distance of the triangle $(0, 0, 0)$ and the hexagon $(2, 1, -2)$ is $\beta + \alpha + \gamma$ in both cases.

Subcase p and q are Triangles. If p and q are triangles, then there exists the above mentioned minimal path between the closest hexagon neighbors of the triangles, and there are starting and final α steps between the closest hexagons and the triangles. This case is similar to the previous one, but sometimes we are able to do the above mentioned changes (at the end of the path) two times (at the beginning, too).

Lemma 11. *If $\alpha \leq \gamma \leq \beta \leq 2\alpha$, p and q are triangles, $\sum\limits_{i=1}^{3} w(i) \neq 0$, and $w_{p,q}$ contains only one coordinate with the sign of $\sum\limits_{i=1}^{3} w(i)$, then*

$$d(p, q; \alpha, \beta, \gamma) = \beta \cdot \frac{\sum\limits_{i=1}^{3} |w(i)| - 2}{2} + 2\alpha \tag{13}$$

if $2\beta \leq \gamma + 2\alpha$, and

$$d(p, q; \alpha, \beta, \gamma) = \gamma \cdot \min\{|w(i)|\}$$

$$+ \alpha \cdot (2 \min\{|w(i)|\} + 2) + \beta \cdot \frac{\sum\limits_{i=1}^{3} |w(i)| - 4 \min\{|w(i)|\} - 2}{2} \tag{14}$$

if $\gamma + 2\alpha \leq 2\beta$.

The proof of this lemma is similar to the proof of Lemma 9.

For example the distance of the triangles $(0,0,0)$ and $(3,-1,0)$ is $2\alpha + \beta$ in both cases.

Lemma 12. *If* $\alpha \leq \gamma \leq \beta \leq 2\alpha$, p *and* q *are triangles,* $\sum_{i=1}^{3} w(i) \neq 0$ *and* $w_{p,q}$ *contains two coordinates with the sign of* $\sum_{i=1}^{3} w(i)$, *then*

$$d(p,q;\alpha,\beta,\gamma) = \beta \cdot \frac{\sum_{i=1}^{3} |w(i)| - 6}{2} + 2\alpha + 2\gamma \tag{15}$$

if $2\beta \leq \gamma + 2\alpha$, *and*

$$d(p,q;\alpha,\beta,\gamma) = \gamma \cdot \min\{|w(i)|\}$$

$$+ \alpha \cdot (2\min\{|w(i)|\} - 2) + \beta \cdot \frac{\sum_{i=1}^{3} |w(i)| - 4\min\{|w(i)|\} + 2}{2} \tag{16}$$

if $\gamma + 2\alpha \leq 2\beta$. *But if moreover* $\sum_{i=1}^{3} |w(i)| = 2$, *then* $d(p,q;\alpha,\beta,\gamma) = \gamma$, *and if* $\sum_{i=1}^{3} |w(i)| = 4$, *then* $d(p,q;\alpha,\beta,\gamma) = \gamma + 2\alpha$.

The proof of this lemma is similar to the proof of Lemma 10.

For example the distance of the triangles $(0,0,0)$ and $(2,2,-2)$ is $2\alpha + 2\gamma$ in both cases.

Lemma 13. *If* $\alpha \leq \gamma \leq \beta \leq 2\alpha$, p *and* q *are triangles, and* $\sum_{i=1}^{3} w(i) = 0$, *then*

$$d(p,q;\alpha,\beta,\gamma) = \beta \cdot \frac{\sum_{i=1}^{3} |w(i)| - 2}{2} + 2\alpha \tag{17}$$

if $2\beta \leq \gamma + 2\alpha$ *or* $\min\{|w(i)|\} = 0$ *and*

$$d(p,q;\alpha,\beta,\gamma) = \gamma \cdot \min\{|w(i)|\}$$

$$+ \alpha \cdot \min\{|w(i)|\} + \beta \cdot \frac{\sum_{i=1}^{3} |w(i)| - 4\min\{|w(i)|\}}{2} \tag{18}$$

if $\gamma + 2\alpha \leq 2\beta$ *and* $\min\{|w(i)|\} > 0$.

The proof of this lemma is similar to the proof of Lemma 10.

For example the distance of the triangles $(0,0,0)$ and $(0,2,-2)$ is $2\alpha + \beta$.

5   Properties of Distances

Minimal weighted paths can be obviously computed by Dijkstra algorithm [2]. However, based on the regular structure of the grid and by the help of an appropriate coordinate system direct formulae are provided to compute them. There are various cases depending on the relation of the used weights; a summary of the results is shown in Table 1.

Table 1. Value of $d(p, q; \alpha, \beta, \gamma)$ depending on the cases of respective relations of the weights

Cases	Between		
	Two hexagons	A hexagon and a triangle	Two triangles
4.1: $2\alpha \leq \beta, \gamma$	$\alpha \sum \lvert w(i) \rvert$		
4.2: $\gamma \leq 2\alpha \leq \beta$	$\gamma \min\{\lvert w(i) \rvert\} + \alpha \left(\sum \lvert w(i) \rvert - 2\min\{\lvert w(i) \rvert\} \right)$		Subcases
4.3, 4.4: $\beta \leq \gamma, 2\alpha$	$\beta \frac{\sum \lvert w(i) \rvert}{2}$	$\beta \frac{\sum \lvert w(i) \rvert - 1}{2} + \alpha$	Subcases
4.5: $\gamma \leq \beta \leq 2\alpha$	Subcases	Subcases	Subcases

For the sake of completeness, we recall the definition of metricity for digital distances. A distance function $d(\cdot, \cdot)$ is a metric if the three properties, the positive definiteness, the symmetry and the triangular inequality, are fulfilled for any points p, q, r of a space, that is, in this paper the trihexagonal grid (6, 3, 6, 3). There are some non-metrical digital distances, e.g., distances based on neighborhood sequences [9], weight sequences [14]. However, in some applications it is important to use metric distances, therefore, it is important to note that, as usual at weighted distances, for any values of $\alpha, \beta, \gamma \in \mathbb{R}^+$, the distance function $d(\cdot, \cdot; \alpha, \beta, \gamma)$ is a metric.

One can also prove the following about the translation invariance of our distance functions.

Theorem 1. *Let $t = (x, y, z)$ be a grid vector, i.e., the difference of the coordinates of two points of the grid. Then, the distance $d(p, q; \alpha, \beta, \gamma) = d(p + t, q + t; \alpha, \beta, \gamma)$ for all pairs of points p, q of the grid and for any $\alpha, \beta, \gamma \in \mathbb{R}^+$ if and only if t is an integer triplet with 0-sum.*

We can conclude that we have done the first steps to include the trihexagonal grid in digital geometry, namely we have studied chamfer distances. Digital or path-based distance functions are well known and widely used. However, the properties of these functions depend highly on the underlying grid: while chamfer polygons on the square grid are well known any relatively easy to describe, it is not the case on the triangular grid (based on the three classical neighborhood, e.g., 63-gons can also be obtained [8]). Future works and possible applications include cases when, e.g., $\gamma < \alpha$, distance transforms, and also studies about the digital disks (chamfer polygons) and their interesting phenomena: conditions for concavities, holes and islands (somewhat similarly to [6]).

References

1. Borgefors, G.: Distance transformations in digital images. Comput. Vis. Graph. Image Process. **34**, 344–371 (1986)
2. Dijkstra, E.W.: A note on two problems in connexion with graphs. Numerische Mathematik **1**, 269–271 (1959). doi:10.1007/BF01386390
3. Her, I.: A symmetrical coordinate frame on the hexagonal grid for computer graphics and vision. ASME J. Mech. Design **115**, 447–449 (1993)
4. Klette, R., Rosenfeld, A.: Digital Geometry: Geometric Methods for Digital Picture Analysis. Elsevier, Amsterdam (2004)
5. Kovács, G., Nagy, B., Vizvári, B.: On weighted distances on the Khalimsky grid. In: Normand, N., Guédon, J., Autrusseau, F. (eds.) DGCI 2016. LNCS, vol. 9647, pp. 372–384. Springer, Cham (2016). doi:10.1007/978-3-319-32360-2_29
6. Kovács, G., Nagy, B., Vizvári, B.: Weighted distances and digital disks on the Khalimsky grid: disks with holes and islands. J. Math. Imaging Vis. **59**, 2–22 (2017). doi:10.1007/s10851-016-0701-5
7. Luczak, E., Rosenfeld, A.: Distance on a hexagonal grid. Trans. Comput **C−25**(5), 532–533 (1976)
8. Mir-Mohammad-Sadeghi, H., Nagy, B.: On the chamfer polygons on the triangular grid. In: Brimkov, V.E., Barneva, R.P. (eds.) IWCIA 2017. LNCS, vol. 10256, pp. 53–65. Springer, Cham (2017). doi:10.1007/978-3-319-59108-7_5
9. Nagy, B.: Metrics based on neighbourhood sequences in triangular grids. Pure Math. Appl. - PU.M.A **13**, 259–274 (2002)
10. Nagy, B.: A family of triangular grids in digital geometry. In: 3rd International Symposium on Image and Signal Processing and Analysis (ISPA 2003), Rome, Italy, pp. 101–106 (2003)
11. Nagy, B.: Characterization of digital circles in triangular grid. Pattern Recognit. Lett. **25**, 1231–1242 (2004)
12. Nagy, B.: Generalized triangular grids in digital geometry. Acta Mathematica Academiae Paedagogicae Nyíregyháziensis **20**, 63–78 (2004)
13. Nagy, B.: Weighted distances on a triangular grid. In: Barneva, R.P., Brimkov, V.E., Šlapal, J. (eds.) IWCIA 2014. LNCS, vol. 8466, pp. 37–50. Springer, Cham (2014). doi:10.1007/978-3-319-07148-0_5
14. Nagy, B., Strand, R., Normand, N.: A weight sequence distance function. In: Hendriks, C.L.L., Borgefors, G., Strand, R. (eds.) ISMM 2013. LNCS, vol. 7883, pp. 292–301. Springer, Heidelberg (2013). doi:10.1007/978-3-642-38294-9_25
15. Radványi, A.G.: On the rectangular grid representation of general CNN networks. Int. J. Circuit Theory Appl. **30**, 181–193 (2002)
16. Rosenfeld, A., Pfaltz, J.L.: Distance functions on digital pictures. Pattern Recognit. **1**, 33–61 (1968)
17. Strand, R., Nagy, B.: Weighted neighbourhood sequences in non-standard three-dimensional grids – metricity and algorithms. In: Coeurjolly, D., Sivignon, I., Tougne, L., Dupont, F. (eds.) DGCI 2008. LNCS, vol. 4992, pp. 201–212. Springer, Heidelberg (2008). doi:10.1007/978-3-540-79126-3_19
18. Strand, R., Nagy, B., Borgefors, G.: Digital distance functions on three-dimensional grids. Theor. Comput. Sci. **412**, 1350–1363 (2011)

An Integer Programming Approach to Characterize Digital Disks on the Triangular Grid

Gergely Kovács[1]([✉]), Benedek Nagy[2], and Béla Vizvári[3]

[1] Edutus College, Tatabánya, Hungary
kovacs.gergely@edutus.hu
[2] Department of Mathematics, Faculty of Arts and Sciences,
Eastern Mediterranean University, Mersin-10, Famagusta, North Cyprus, Turkey
nbenedek.inf@gmail.com
[3] Department of Industrial Engineering, Eastern Mediterranean University,
Mersin-10, Famagusta, North Cyprus, Turkey

Abstract. Generally, the integer hull of a polyhedral set is the convex hull of the integer points of the set. In most of the cases, for example when the set is bounded, the integer hull is a polyhedral set, as well. The integer hull can be determined in an iterative way by Chvátal cuts. Weighted (or chamfer) distances are popular digital distances used in various grids. They are based on the weights assigned to steps to various neighborhood. In the triangular grid there are three usually used neighborhood, consequently, chamfer distances based on three weights are defined. A digital disk (or a chamfer ball) of a grid is the set of the elements which are not on a longer distance from the origin than a given finite bound, radius. These disks are well known and well characterized on the square grid (with even larger neighborhood than the usual 3 × 3), and recently they become a topic of a current research on the triangular grid. The shapes of the disks in the latter case have a great variability. In this paper, the inequalities satisfied by the elements of a disk are analyzed if their Chvátal rank is 1. The most popular coordinate system of the triangular grid uses three coordinates. Individual bounds are described completely. It also gives the complete description of some disks. Further inequalities having Chvátal rank 1 are also discussed.

Keywords: Weighted distances · Chamfer balls · Non-traditional grids · Integer programming · Optimization

1 Introduction

A grid consists of several tiles/pixels. A step is moving from one pixel to another, neighbor one. Each step has a positive length depending on the two pixels and their relative positions. The *distance of two pixels* of the grid is measured by the length of the minimal path between them. A *disk* of a grid is a set of pixels such

© Springer International Publishing AG 2017
W.G. Kropatsch et al. (Eds.): DGCI 2017, LNCS 10502, pp. 94–106, 2017.
DOI: 10.1007/978-3-319-66272-5_9

that their distance from a fixed pixel, say P is not greater than a given value. In most of the cases P is considered to be the origin of grid.

The *shape of a disk* can be defined as follows. Assume that the pixels of the grid are symmetric and have a center point. Then the shape of the disk is the shape of the convex hull of the center points of the pixels which are the elements of the grid. It is well known that the shape of the disk is always an octagon on the square grid based on the usual two neighborhood [1,2]. There is a wide variety of shapes of the disks on the triangular grid [8,12]. On Fig. 1 we show an example.

The pixels of a grid are identified by integer coordinates depending on the type of the grid, e.g., each pixel has its 2 or 3 coordinates which uniquely identify the pixels, in the square and the triangular grids, respectively. These coordinates can be considered as the coordinates of the center point of the pixel. The disk is determined by some inequalities and equations. One inequality restricts the total distance. There are grids where the coordinates must also satisfy some conditions. For example, the sum of the coordinates can be either 0 or 1 in the case of the triangular grid. Finally, the path from the origin to the pixel must be described as well. These constraints are satisfied by many values including even non-integer ones. The problem which is the main topic of this paper is, how can be these constraint used to determine the convex hull of the integer points satisfying the constraints. It is a general problem in integer programming (optimization).

2   The Triangular Grid

The triangular grid is a complete, non-overlapping coverage of the plane by regular triangles [9,10]. Each pixel of the grid can be addressed by an integer coordinate triplet having zero or one sum. These two types of vectors differentiate the two types of the orientations of the pixels of the grid. Zero sum vectors address the even pixels (they have shape $\triangle$ in this paper), while one sum triplets address the odd pixels (their shape is $\triangledown$). As Fig. 1 shows, the coordinate axes have angle 120° pairwise. The set of pixels where one of the coordinate values is fixed, is called a lane and it is orthogonal to the axis with fixed value. For example, the pixels of the top of the Fig. 1 belong to the lane $y = -4$. This symmetric coordinate system captures well the well-known neighborhood relations [3] of the grid (Fig. 1, right). Formally, the points (*i.e.*, pixels) $p = (p(1), p(2), p(3))$ and $q = (q(1), q(2), q(3))$ of the triangular grid are

- *m-neighbors* $(m = 1, 2, 3)$, if
    (i) $|p(i) - q(i)| \leq 1$, for $i = 1, 2, 3$, and
    (ii) $|p(1) - q(1)| + |p(2) - q(2)| + |p(3) - q(3)| \leq m$.
- *strict m-neighbors*, if there is an equality in (ii) for the value of m.

Let $p = (p(1), p(2), p(3))$ and $q = (q(1), q(2), q(3))$ be two points of the triangular grid. A finite sequence of points of the form $p = p_0, p_1, \ldots, p_m = q$, where p_{i-1}, p_i are 3-neighbor points for $1 \leq i \leq m$, is called a *path* from p to q. The term

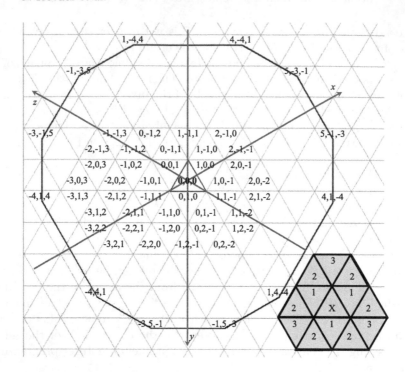

Fig. 1. A part of the triangular grid with the symmetric coordinate system, an example to (the convex hull of) a disk and a pixel (X) with its various neighbors (right left corner).

k-step will be used to abbreviate the sentence 'step to a strict *k*-neighbor point' ($k \in \{1, 2, 3\}$).

There are three types of steps in the problem which are called according to their length 1-step, 2-step, and 3-step. We use the notation given in Table 1 for various possible *k*-steps.

Table 1. 1-steps, 2-steps, and 3-steps on the triangular grid.

1-steps						2-steps						3-steps					
u_1	u_2	u_3	u_4	u_5	u_6	v_1	v_2	v_3	v_4	v_5	v_6	w_1	w_2	w_3	w_4	w_5	w_6
1	0	0	−1	0	0	1	1	0	−1	−1	0	1	1	−1	−1	−1	1
0	1	0	0	−1	0	−1	0	1	1	0	−1	1	−1	1	−1	1	−1
0	0	1	0	0	−1	0	−1	−1	0	1	1	−1	1	1	1	−1	−1

In this paper without loss of generality, we can deal with shortest paths from the origin $(0, 0, 0)$ to point (x, y, z).

Chamfer distances on the triangular grid were investigated in [11] based on various weights for the 3 types of steps on the grid.

3 Chamfer Distances

Using the 3 types of neighbors there could be various paths from a point p to a point q in the triangular grid. Let the weights for the various steps be fixed as a, b, c for 1-, 2-, and 3-steps, respectively, with the condition $c \geq b \geq a > 0$. Then, the sum of the weights of the steps of a path is considered as a weighted path. The chamfer distance (or, in another term, the weighted distance) of two points is the weight of the weighted shortest path between them, $i.e.$, the minimal value among weights of the paths between the points: $d((x_1, y_1, z_1), (x_2, y_2, z_2); a, b, c)$. In [11] it is proven, that this distance function satisfies the metric conditions: it is positive definite, symmetric and the triangle inequality holds.

Let the chamfer ball, $disk(r)$ be the set of pixels (x, y, z) for which the weighted distance between the origin and (x, y, z) is at most r, formally,

$$disk(r) = \{(x, y, z) \mid d((0, 0, 0), (x, y, z); a, b, c) \leq r\}.$$

Chamfer distances are discussed for some other grid in [13]. They are connected to the Frobenius problem of three variables [4, 13].

4 Integer Hull and Chvátal Cuts

Let m and n be two positive integers, A an $m \times n$ matrix and b an m-dimensional vector. The set $P = \{x \mid Ax \leq b\}$ is a polyhedron in the n-dimensional space. The *integer hull* of P is the convex hull of its integer points, $i.e.$ the set $int(P) = conv(P \cap Z^n)$, where Z^n is the lattice of n-dimensional integer vectors.

The set $int(P)$ is not necessarily a polyhedral set [5]. However, it is a polyhedral set in many cases including the case when P is bounded or A and b have rational elements [7]. It means that the disks are always polygons.

The set $int(P)$ can be determined by an iterative procedure. The key tool of the algorithm is the *Chvátal cut*. Assume that $\lambda \in (\mathbb{R}^{\geq 0})^m$ is an m-dimensional vector such that the product $\lambda^T A$ is an integer vector. Then all integer vector x of P must satisfy the inequality

$$\lambda^T Ax \leq \lfloor \lambda^T b \rfloor, \tag{1}$$

where the usual floor function $\lfloor \cdot \rfloor$ is used. If $\lambda^T b \neq \lfloor \lambda^T b \rfloor$ then (1) is not an algebraic consequence of the original inequalities defining P. This inequality is the Chvátal cut which can be added to other inequalities without cutting any integer point from P. Further details of the method can be obtained from [14]. What is important here is that for there are finitely many significantly different Chvátal cuts of every inequality system. One iteration is to generate all of them. When the inequality system is enlarged by the generated inequalities, the procedure can be repeated. The integer hull is obtained after finitely many iterations.

The Chvátal rank of an inequality is the number of the iteration in which it was generated. The rank of the inequalities defining P is 0. The rank of the inequalities generated by rank 0 inequalities is 1, etc. In this paper, only inequalities having rank 1 are investigated. In one iteration the number of Chvátal cuts could be even exponential on the number of original equations (Chap. 23 in [14]), therefore their analysis gives already becomes very time consuming and also gives some nice results. The Chvátal cuts are analyzed systematically only in few papers. [6] discusses the Chvátal cuts of the knapsack polytope of rank 1.

5   Linear Programming Model

There are several methods to describe a chamfer disk. A new approach is discussed in this paper. The polygon of a disk for a given radius r consists of the feasible solutions of an integer programming problem. It means that the coordinate vectors must satisfy certain linear inequalities and must have integer components. The size of the model, *i.e.* number of rows and columns, is fixed. What is changing is only the right-hand side. The theory of linear programming gives a complete description of the potential optimal solutions. The form of the disks can be one of the elements of a finite set according to our empirical observation. Thus it seems possible to give a complete structural description of the disks by uncovering the integer hull of the feasible sets of the related LP problems. The structural description eliminate the necessity of any algorithm.

A 1-step changes one coordinate by 1 unit. A 2-step changes two coordinates into the opposite direction by 1 unit. Finally, a 3-step changes all coordinates by 1 unit, however the direction of the three changes are not the same. Thus, the matrix of the steps is specified in Table 1. Let u_i, v_i and w_i (for $i = 1, \ldots, 6$) are the numbers of various steps. The disk constraint is given in the following form:

$$au_1 + \cdots + au_6 + bv_1 + \cdots + bv_6 + cw_1 + \cdots + cw_6 \leq r,$$

where $c \geq b \geq a > 0$. It should be note here that the triangular grid is not a lattice, therefore the steps of a path are usually not free to permute. However, when a multiset of steps is specified corresponding to a path connecting the origin to another pixel, then because of the constraint on the sum of coordinate values, there exists always a path on the grid which built up by those steps if the sum of the coordinate changes is equal to 0 or 1.

Now, we are ready to continue to show other constraints. All step variables are non-negative:

$$u_1, \ldots, u_6, v_1, \ldots, v_6, w_1, \ldots, w_6 \geq 0.$$

If a point $(x, y, z)^T$ is reached from the origin by the steps, then the step numbers satisfy the equation system

$$u_1 - u_4 + v_1 + v_2 - v_4 - v_5 + w_1 + w_2 - w_3 - w_4 - w_5 + w_6 = x,$$
$$u_2 - u_5 - v_1 + v_3 + v_4 - v_6 + w_1 - w_2 + w_3 - w_4 + w_5 - w_6 = y,$$
$$u_3 - u_6 - v_2 - v_3 + v_5 + v_6 - w_1 + w_2 + w_3 + w_4 - w_5 - w_6 = z.$$

The sum of the coordinates x, y, and z are between 0 and 1:

$$x + y + z \leq 1,$$
$$x + y + z \geq 0.$$

The polyhedral set is defined by the system of constraints shown in Table 2, where the index of the (in)equality is used in further analysis as is indicated there. Using the same index set, the multipliers of the inequalities of the original constraint set are denoted by $\lambda_0, \ldots, \lambda_{23}$, i.e., they are the elements of vector λ.

The facet defining inequalities concern to the coordinates of the points only, *i.e.* to the variables x, y, and z. It is assumed that the facet-defining inequality of the disk is given in the form

$$ex + fy + gz \leq h \tag{2}$$

with $e, f, g, h \in \mathbb{Z}$. The rank of the defining inequalities is 0 by definition. If (2) is not among the defining inequalities of the disk polytope, then its Chvátal rank is 1 only if multipliers $\lambda_0, \ldots, \lambda_{23}$ can be chosen such that $\lambda_0, \ldots, \lambda_{18}, \lambda_{22}, \lambda_{23}$ are nonnegative, and the coefficients on the left-hand side of the generated inequality are $e, f, g, 0, \ldots, 0$, while the right-hand side is less than $h + 1$. If there are several options that can be employed to generate the left-hand side, then the preferred choice is that which gives the minimal right-hand side result. This observation leads to the following linear programming model:

Table 2. The inequalities defining the disk polytope, RHS stands for Right-Hand Side.

index	Left-Hand Side (LHS)						RHS
0		$au_1 + \cdots + au_6$	$+bv_1 + \cdots + bv_6$	$+cw_1 + \cdots + cw_6$			$\leq r$
1		$-u_1$					≤ 0
$\vdots$		$\ddots$					$\vdots \ \vdots$
6		$-u_6$					≤ 0
7			$-v_1$				≤ 0
$\vdots$			$\ddots$				$\vdots \ \vdots$
12				$-u_6$			≤ 0
13				$-w_1$			≤ 0
$\vdots$					$\ddots$		$\vdots \ \vdots$
18						$-w_6$	≤ 0
19	$-x$	$+u_1 \cdots$	$+v_1 \cdots$	$+w_1 \cdots$	$+w_6$		$= 0$
20	$-y$	$\cdots$	$-v_1 \cdots$	$-v_6 +w_1 \cdots$	$-w_6$		$= 0$
21	$-z$	$\cdots -u_6$	$\cdots$	$+v_6 -w_1 \cdots$	$-w_6$		$= 0$
22	$x +y +z$						≤ 1
23	$-x -y -z$						≤ 0

$$\min r\lambda_0 + \lambda_{22}$$
$$-\lambda_{19} + \lambda_{22} - \lambda_{23} = e \tag{3}$$
$$-\lambda_{20} + \lambda_{22} - \lambda_{23} = f \tag{4}$$
$$-\lambda_{21} + \lambda_{22} - \lambda_{23} = g \tag{5}$$
$$a\lambda_0 - \lambda_1 + \lambda_{19} = 0 \tag{6}$$
$$\vdots$$
$$a\lambda_0 - \lambda_6 - \lambda_{21} = 0 \tag{7}$$
$$b\lambda_0 - \lambda_7 + \lambda_{19} - \lambda_{20} = 0 \tag{8}$$
$$\vdots$$
$$b\lambda_0 - \lambda_{12} - \lambda_{20} + \lambda_{21} = 0 \tag{9}$$
$$c\lambda_0 - \lambda_{13} + \lambda_{19} + \lambda_{20} - \lambda_{21} = 0 \tag{10}$$
$$\vdots$$
$$c\lambda_0 - \lambda_{18} + \lambda_{19} - \lambda_{20} - \lambda_{21} = 0 \tag{11}$$
$$\lambda_0, \ldots \lambda_{18}, \lambda_{22}, \lambda_{23} \geq 0. \tag{12}$$

6   Construction of Chvátal Cuts

In general, a cut (1) is not necessarily facet defining cut as even stronger cuts might be generated in farther iterations. In this section, some cuts of rank 1 are generated.

The next lemma goes independently of the fact that we are using the triangular grid, that is, we do not take into account about the constraints $i = 22, 23$, we set their weights $\lambda_{22} = \lambda_{23} = 0$.

Lemma 1. *Chvátal rank of (2) is 1 if*

$$r \cdot \max\left\{ \max_{i \in \{e,f,g\}} \left\{ \frac{|i|}{a} \right\} ; \max_{i,j \in \{e,f,g\}} \left\{ \frac{|i-j|}{b} \right\} ; M_3 \right\} < h + 1,$$

where $M_3 = \max\limits_{i,j,k \in \{e,f,g\}} \left\{ \frac{|i-j-k|}{c} ; \frac{|i+j-k|}{c} \right\}.$

Proof. If $\lambda_{22} = \lambda_{23} = 0$, then we must minimize λ_0 for the best RHS.

In this case from (3), (4), and (5) we get, that $\lambda_{19} = -e$, $\lambda_{20} = -f$, and $\lambda_{21} = -g$.

The minimal value of λ_0 for the Eqs. (6) and (7) is

$$\frac{\max\{|e|, |f|, |g|\}}{a} = \max_{i \in \{e,f,g\}} \left\{ \frac{|i|}{a} \right\} \leq \lambda_0,$$

because λ_i's are nonnegative for $i = 1, \ldots, 6$.

Similarly, the minimal value of λ_0 for the Eqs. (8) and (9) is

$$\frac{\max\{|e-f|,|e-g|,|f-g|\}}{b} = \max_{i,j\in\{e,f,g\}}\left\{\frac{|i-j|}{b}\right\} \leq \lambda_0,$$

because λ_i's are nonnegative for $i = 7, \ldots, 12$.

Finally, the minimal value of λ_0 for the Eqs. (10) and (11) is

$$\frac{\max\{|e+f-g|,|e-f+g|,|-e+f+g|\}}{c} = M_3 \leq \lambda_0,$$

because λ_i's are nonnegative for $i = 13, \ldots, 18$.   □

Corollary 1. *If* $2a \leq b$, $3a \leq c$ *and*

$$r \cdot \max_{i\in\{e,f,g\}}\left\{\frac{|i|}{a}\right\} \leq h+1,$$

then the Chvátal rank of (2) *is* 1.

Theorem 1. *Chvátal rank of the inequality* $x \leq k$ *is* 1 *if and only if*

$$\frac{r+a}{2a} < k+1 \quad \text{if } 2a \leq b \text{ and } 3a \leq c,$$

$$\frac{r+b-a}{b} < k+1 \quad \text{if } b \leq 2a \text{ and } 2b \leq a+c,$$

$$\frac{2r+c-a}{c+a} < k+1 \quad \text{if } c \leq 3a \text{ and } a+c \leq 2b.$$

Proof. In this case, let $e = 1$, $f = g = 0$ in (2). We need to minimize $r\lambda_0 + \lambda_{22}$, where (3)–(12) hold. Only (3)–(5) contain λ_{23} and always in the form $\lambda_{22} - \lambda_{23}$ not counting (12). It means that because of minimizing λ_{22} we need to choose $\lambda_{23} = 0$, because λ_{22} is nonnegative. Let us denote $\lambda_{22} = s$. Thus from (3) we get, that $\lambda_{19} = s - 1$ (it may be negative). From (4) and (5) we get, that $\lambda_{20} = \lambda_{21} = s$.

Then the minimal value for λ_0 satisfies (6) and (7) is the following:

$$\frac{\max\{s,-s,s-1,1-s\}}{a} = \max\left\{\frac{s}{a},\frac{1-s}{a}\right\} \leq \lambda_0. \tag{13}$$

The minimal value for λ_0 satisfies (8)–(9) is

$$\frac{\max\{s-s,s-(s-1),(s-1)-s\}}{b} = \frac{1}{b} \leq \lambda_0. \tag{14}$$

Finally, the minimal value for λ_0 satisfies (10)–(11) is

$$\frac{\max\{s-s-(s-1),s-s+(s-1),-s-s+(s-1),s+s-(s-1)\}}{c} =$$

$$= \frac{s+1}{c} \leq \lambda_0. \tag{15}$$

Then from (13)–(15) we get

$$\max\left\{\frac{s}{a}, \frac{1-s}{a}, \frac{1}{b}, \frac{s+1}{c}\right\} \le \lambda_0.$$

It means that to minimize $r\lambda_0 + s$ we need to minimize the following in s:

$$\max\left\{\frac{rs}{a}+s, \frac{r(1-s)}{a}+s, \frac{r}{b}+s, \frac{r(s+1)}{c}+s\right\}.$$

There are four functions of s in the above maximum, let us denote them by $f_1(s), \ldots, f_4(s)$.

If $r < a$, then our disk contains only the origin. We can assume that $a \le r$. In this case $f_2(s)$ is decreasing, but other three functions are increasing. If $s = 0$, then the value of the second function is greater than the others: $f_2(0) = \frac{r}{a} \ge f_i(0)$, where $i = 1, 3, 4$, because we assumed that $a \le b \le c$.

The minimal value of the maximum of the above functions is obtained when the decreasing one $(f_2(s))$ has the same value as one of the increasing functions (we say that the decreasing function and one of the increasing functions are intersecting each other).

Case 1. The intersection of $f_1(s)$ and $f_2(s)$ is $s = \frac{1}{2}$. In this case the value of the functions is

$$f_1\left(\frac{1}{2}\right) = f_2\left(\frac{1}{2}\right) = \frac{r+a}{2a}.$$

It is maximal for all f_i in $s = \frac{1}{2}$ if $f_1(s) = f_2(s) \ge f_3(s)$, thus

$$\frac{r}{2a} + \frac{1}{2} \ge \frac{r}{b} + \frac{1}{2},$$

and this holds if $2a \le b$. $f_1(s) = f_2(s) \ge f_4(s)$ means that

$$\frac{r}{2a} + \frac{1}{2} \ge \frac{\frac{3}{2}r}{c} + \frac{1}{2},$$

and this holds if $3a \le c$.

Case 2. The intersection of $f_2(s)$ and $f_3(s)$ is $s = \frac{b-a}{b}$ from $\frac{1-s}{a} = \frac{1}{b}$. In this case the value of the functions is

$$f_2\left(\frac{b-a}{b}\right) = f_3\left(\frac{b-a}{b}\right) = \frac{r+b-a}{b}.$$

It is maximal for all f_i in $s = \frac{b-a}{b}$ if $f_2(s) = f_3(s) \ge f_1(s)$, thus

$$\frac{r}{b} + \frac{b-a}{b} \ge \frac{r(b-a)}{ba} + \frac{b-a}{b},$$

and this holds if $b \le 2a$. $f_2(s) = f_3(s) \ge f_4(s)$ means that

$$\frac{r}{b} + \frac{b-a}{b} \ge \frac{r(2b-a)}{bc} + \frac{b-a}{b},$$

and this holds if $2b \le a + c$.

Case 3. The intersection of $f_2(s)$ and $f_4(s)$ is $s = \frac{c-a}{c+a}$ from $\frac{1-s}{a} = \frac{s+1}{c}$. In this case the value of the functions is

$$f_2\left(\frac{c-a}{c+a}\right) = f_4\left(\frac{c-a}{c+a}\right) = \frac{2r+c-a}{c+a}.$$

It is maximal for all f_i in $s = \frac{c-a}{c+a}$ if $f_2(s) = f_4(s) \geq f_1(s)$, thus

$$\frac{2r}{c+a} + \frac{c-a}{c+a} \geq \frac{r(c-a)}{a(c+a)} + \frac{c-a}{c+a},$$

and this holds if $c \leq 3a$. $f_2(s) = f_4(s) \geq f_3(s)$ means that

$$\frac{2r}{c+a} + \frac{c-a}{c+a} \geq \frac{r}{b} + \frac{c-a}{c+a},$$

and this holds if $a + c \leq 2b$.   □

The conditions of the above three cases contain all possibilities. If $2a \leq b$ and $3a \leq c$, then this is Case 1. If $b \leq 2a$ and $3a \leq c$, then $2b \leq 4a \leq a + c$, thus this subcase is part of Case 2. If $2a \leq b$ and $c \leq 3a$, then $c + a \leq 4a \leq 2b$, thus this subcase is part of Case 3. If $b \leq 2a$ and $c \leq 3a$, then $a + c \leq 2b$ or $2b \leq a + c$ is possible, thus one part of this subcase belongs to Case 2, another part of this subcase belongs to Case 3. Further we will refer to these cases as we have described them here.

Theorem 2. *Chvátal rank of the inequality $-x \leq l$ is 1 if and only if*

$$\frac{r}{2a} < l + 1 \quad \text{if } 2a \leq b \text{ and } 3a \leq c,$$

$$\frac{r}{b} < l + 1 \quad \text{if } b \leq 2a \text{ and } 2b \leq a + c,$$

$$\frac{2r}{c+a} < l + 1 \quad \text{if } c \leq 3a \text{ and } a + c \leq 2b.$$

The proof follows the same idea as the previous one with $e = -1$, $f = g = 0$ in (2).

7   Facet-Defining Inequalities

Theorem 3. *The above mentioned inequalities $x \leq k$ and $-x \leq l$ are facet-defining in Case 1 and in Case 2.*

Proof. **Case $2a \leq b$ and $3a \leq c$.** This case belongs to Case 1 of the above two Theorems. It is in [11] that the distance function between $(0,0,0)$ and (x, y, z) is $d(a, b, c) = a(|x| + |y| + |z|)$. If we want to create the point of the disk, which has minimal or maximal value in x, we need to solve the following problem:

$$\min(\text{or max})x$$
$$d(a, b, c) \leq r \tag{16}$$

$$x + y + z \le 1$$
$$-x - y - z \le 0$$
$$x, y, z \in \mathbb{Z}$$

If $x + y + z = 0$, then the maximal value of $|x|$ can be equal to $\frac{|x|+|y|+|z|}{2}$. In this case $2|x| = |x| + |y| + |z| \le \frac{r}{a}$ and it means that $-x \le \lfloor \frac{r}{2a} \rfloor$ is a facet-defining equation.

If $x + y + z = 1$, then the maximal value of x can be equal to $\frac{|x|+|y|+|z|+1}{2}$. In this case $2x - 1 = |x| + |y| + |z| \le \frac{r}{a}$ and $x \le \lfloor \frac{r+a}{2a} \rfloor$ is a facet-defining equation.

Case $b \le 2a$ and $3a \le c$. In this case $2b \le a + c$, and this case belongs to Case 2 of the above two Theorems. The distance function between $(0,0,0)$ and (x, y, z) is in [11]:

$$d(a,b,c) = \begin{cases} b\frac{|x|+|y|+|z|}{2}, & \text{if } x + y + z = 0; \\ a + b\frac{|x|+|y|+|z|-1}{2}, & \text{if } x + y + z = 1. \end{cases}$$

If $x + y + z = 0$, then from (16) and from the maximal value of $|x|$ we get that $2|x| = |x| + |y| + |z| \le \frac{2r}{b}$ and it means that $-x \le \lfloor \frac{r}{b} \rfloor$ is a facet-defining equation.

If $x + y + z = 1$, then from (16) and from the maximal value of x we get that $2x - 1 = |x| + |y| + |z| \le \frac{2r-2a+b}{b}$ and $x \le \lfloor \frac{r+b-a}{b} \rfloor$ is a facet-defining equation.

Case $b \le 2a$, $c \le 3a$ and $a + b \le c$. In this case $2b \le a + c$, and this case belongs to Case 2. In this case function d is the same as in the previous case, i.e., our statements hold.

Case $b \le 2a$, $c \le 3a$, $c \le a + b$ and $2b \le a + c$. This case belongs to Case 2, too. If $x + y + z = 0$, then function d is the same as in the previous case. If $x + y + z = 1$ and we want to maximize the value of $|x|$, then x is positive and y and z are negative. In this case [11] uses the same distance function as in the previous case. There is a third distance function in [11] in the case of $x + y + z = 1$ for the subcase a negative and two positive coordinates, but in this subcase the value of $|x|$ is not maximal.

Case $b \le 2a$, $c \le 3a$, $c < a+b$ and $a+c \le 2b$ and Case $2a \le b$ and $c \le 3a$. These cases belong to Case 3.   □

The result of the theorem can give at most 6 of the sides of the chamfer disk, that is, actually, the embedded hexagon of the disk.

Example. The inequalities provided by the theorem can be both facet defining and non-facet defining in Case 3. If $a = 4$, $b = 7$, $c = 8$ and $r = 30$ (see Fig. 1), then this case belongs to Case 3.

(a) In this case the Chvátal rank of the inequality of $-x \le l$ is 1 if and only if $\lfloor \frac{60}{12} \rfloor = 5 \le l$. Let $v(i)$ be the sorted coordinate values of (x, y, z) in a non-increasing way by their absolute values, i.e., $|v(3)| \le |v(2)| \le |v(1)|$. The distance function between $(0, 0, 0)$ and (x, y, z) is in [11]:

$$d(a,b,c) = |v(1)|\frac{b}{2} + |v(2)|\frac{b}{2} + |v(3)|(a+c - \frac{3}{2}b), \tag{17}$$

if $x+y+z=0$. In this case $a+c-\frac{3}{2}b \leq \frac{b}{2}$, thus the value of $d(a,b,c)$ is minimal for a given x if $|v(3)|$ is close to $|v(2)|$, i.e., if x is even, then $|v(2)| = |v(3)| = \frac{|x|}{2}$; and if x is odd, then $|v(2)| = |v(3)| + 1 = \frac{|x|+1}{2}$. For example if $(x,y,z) = (-5,3,2)$, then $v(1) = -5$, $v(2) = 3$, $v(3) = 2$, and in this case $d(4,7,8) = 31$, thus there is no point with $x = -5$ for this distance function to satisfy $d(.) \leq r$.

If $x+y+z=1$, then the distance function between $(0,0,0)$ and (x,y,z) is different from (17). If $x+y+z=1$ and $x=-5$, then by decreasing one of the positive coordinates by 1, the new point has the same x value and the distance of the new point from $(0,0,0)$ is less than the distance of the original one, thus there is no point with $x=-5$ for the distance functions of case $x+y+z=1$ to satisfy $d(.) \leq r$. It means that the disk of the value $r=30$ has no point with $x=-5$, i.e., $-x \leq 5$ is not facet-defining.

(b) The Chvátal rank of the inequality of $x \leq 5$ is 1. If $(x,y,z) = (5,-1,-3)$, then $d(4,7,8) = 30$ based on [11], thus $x \leq 5$ is a facet-defining equation.

(c) $(5,-1,-3)$ and $(4,1,-4)$ are points of the disk, the inequality $x-z \leq 8$ holds for these pixels of the disk. The Chvátal rank of this inequality is 1 if $h=8$ in Lemma 1, thus $x-z \leq 8$ is a facet-defining equation with Chvátal rank of 1.

The disk of Fig. 1 has 12 facet-defining equation: there are three 1-ranked inequalities similar to $x \leq 5$; six 1-ranked inequalities similar to $x-z \leq 8$; and three not 1-ranked inequalities similar to $-x \leq 4$.

A further analysis of 1-ranked conditions is planned. We believe that we will find connection between the chamfer radius of the disk and the Eucledian radius by using the theory of 1-ranked conditions.

References

1. Borgefors, G.: Distance transformations in digital images. Comput. Vis. Graph. Image Process. **34**(3), 344–371 (1986)
2. Butt, M.A., Maragos, P.: Optimum design of chamfer distance transforms. IEEE Trans. Image Process. **7**(10), 1477–1484 (1998)
3. Deutsch, E.S.: Thinning algorithms on rectangular, hexagonal and triangular arrays. Comm. ACM **15**, 827–837 (1972)
4. Hujter, M., Vizvári, B.: The exact solutions to the Frobenius problem with three variables. J. Ramanujan Math. Soc. **2**, 117–143 (1987)
5. Jeroslow, R.G.: Comments on integer hull of two linear constraints. Oper. Res. **19**, 1061–1069 (1971)
6. Kovács, G., Vizvári, B.: On Chvátal complexity of knapsack problems, RUTCOR, Rutgers University, Research report, 15–2008 (2008)
7. Meyer, R.R.: On the existence of optimal solutions to IP and MIP problems. Math. Program. **7**, 223–235 (1974)
8. Mir-Mohammad-Sadeghi, H., Nagy, B.: On the chamfer polygons on the triangular grid. In: Brimkov, V.E., Barneva, R.P. (eds.) IWCIA 2017. LNCS, vol. 10256, pp. 53–65. Springer, Cham (2017). doi:10.1007/978-3-319-59108-7_5
9. Nagy, B.: Shortest path in triangular grids with neighbourhood sequences. J. Comput. Inf. Technol. **11**, 111–122 (2003)
10. Nagy, B.: Characterization of digital circles in triangular grid. Pattern Recogn. Lett. **25**, 1231–1242 (2004)

11. Nagy, B.: Weighted distances on a triangular grid. In: Barneva, R.P., Brimkov, V.E., Šlapal, J. (eds.) IWCIA 2014. LNCS, vol. 8466, pp. 37–50. Springer, Cham (2014). doi:10.1007/978-3-319-07148-0_5

12. Nagy, B., Mir-Mohammad-Sadeghi, H.: Digital disks by weighted distances in the triangular grid. In: Normand, N., Guédon, J., Autrusseau, F. (eds.) DGCI 2016. LNCS, vol. 9647, pp. 385–397. Springer, Cham (2016). doi:10.1007/978-3-319-32360-2_30

13. Remy, E., Thiel, E.: Computing 3D medial axis for chamfer distances. In: Borgefors, G., Nyström, I., Baja, G.S. (eds.) DGCI 2000. LNCS, vol. 1953, pp. 418–430. Springer, Heidelberg (2000). doi:10.1007/3-540-44438-6_34

14. Schrijver, A.: Theory of Linear and Integer Programming. Wiley, Chichester (1986)

Discrete Tomography

High-Level Algorithm Prototyping: An Example Extending the TVR-DART Algorithm

Axel Ringh[1]([✉]), Xiaodong Zhuge[2], Willem Jan Palenstijn[2],
Kees Joost Batenburg[2,3], and Ozan Öktem[1]

[1] Department of Mathematics, KTH Royal Institute of Technology,
Stockholm, Sweden
{aringh,ozan}@kth.se

[2] Computational Imaging, Centrum Wiskunde & Informatica (CWI),
Amsterdam, The Netherlands
{x.zhuge,willem.jan.palenstijn,joost.batenburg}@cwi.nl

[3] Mathematical Institute, Leiden University, Leiden, The Netherlands

Abstract. Operator Discretization Library (ODL) is an open-source Python library for prototyping reconstruction methods for inverse problems, and ASTRA is a high-performance Matlab/Python toolbox for large-scale tomographic reconstruction. The paper demonstrates the feasibility of combining ODL with ASTRA to prototype complex reconstruction methods for discrete tomography. As a case in point, we consider the total-variation regularized discrete algebraic reconstruction technique (TVR-DART). TVR-DART assumes that the object to be imaged consists of a limited number of distinct materials. The ODL/ASTRA implementation of this algorithm makes use of standardized building blocks, that can be combined in a plug-and-play manner. Thus, this implementation of TVR-DART can easily be adapted to account for application specific aspects, such as various noise statistics that come with different imaging modalities.

1 Introduction

Inverse problems refer to the task of reconstructing parameters characterizing the system under investigation from indirect observations. Such problems arise in several areas of science and engineering, and in particular for *tomographic imaging*. The idea here is to expose the object to penetrating waves or particles

---

A. Ringh and O. Öktem—The authors are supported by the Swedish Research Council (VR) grant 2014-5870, and the Swedish Foundation of Strategic Research (SSF) grant AM13-0049.

X. Zhuge—The author is supported by the Stichting voor de Technische Wetenschappen (STW) through a personal grant (Veni, 13610).

W.J. Palenstijn—The author is supported by the Stichting voor de Technische Wetenschappen (STW), project 13314.

K.J. Batenburg—The author is supported by the Netherlands Organization for Scientific Research (NWO), project 639.073.506.

© Springer International Publishing AG 2017
W.G. Kropatsch et al. (Eds.): DGCI 2017, LNCS 10502, pp. 109–121, 2017.
DOI: 10.1007/978-3-319-66272-5_10

from different directions. The measured transmission (or emission) data is then used as input to a reconstruction scheme that computes an estimate of the interior structure of the object.

Computed tomography (CT) has a wide range of applications, e.g., X-ray CT [9] in medical imaging and electron tomography (ET) [8,10] in biology and material science. A key element is to model the interaction between the object and the wave/particle probe with sufficient accuracy. The resulting inverse problems are often ill-posed, for example in the sense that small errors in data get amplified. Hence, one must stabilize the reconstruction (regularization) by exploiting a priori knowledge of the unknown interior structure. Discrete tomography considers a specific type of prior knowledge, where it is assumed that the unknown object consists of a small number of different materials, each corresponding to a characteristic, approximately constant grey level in the reconstruction. A variety of reconstruction algorithms have been proposed for discrete tomography problems including primal-dual subgradient algorithms [12], network flow algorithms [3], statistical methods [2,7] and algebraic methods [4,17].

At first sight it may seem that most aspects of a reconstruction method are problem-specific. This is fortunately not the case. In fact, the general theory developed during the last three decades provides a number of generic frameworks that are adaptable to specific ill-posed inverse problems. When properly adapted, the methods derived from the general framework compare favorably with application-specific approaches. Furthermore, using general mathematical tools to address specific problems also provides new insights that may go unnoticed if one uses an entirely problem-specific approach. An example is sparsity promoting regularization, which is a general framework that outperforms, or matches, many application-specific state-of-the-art approaches for inverse problems where data are highly noisy or under-sampled.

Despite the above, most concrete implementations of reconstruction methods are tied to a specific application in the sense that minor mathematical modifications lead to a large amount of low level implementations and modifications, which require substantial dedicated algorithmic and programming efforts. Operator Discretization Library (ODL) [1] and ASTRA toolbox [14] are two open-source software libraries developed to assist fast prototyping of reconstruction algorithms. When used together, a user may implement a generic reconstruction method and use it on different tomographic real-world problems without having to re-implement all necessary parts from the bottom up. This becomes especially useful for complex reconstruction methods. *This paper demonstrates the capabilities of ODL and ASTRA on a recently proposed discrete tomography algorithm, total variation regularized discrete algebraic reconstruction technique (TVR-DART).*

2   Inverse Problems and Tomography

Mathematically, an inverse problem in imaging can be stated as the problem of reconstructing an image $f \in X$ representing the object under study from data

$g \in Y$ where

$$g = \mathcal{A}(f) + \text{"noise"}. \tag{1}$$

Here, $\mathcal{A} : X \to Y$ (forward operator) models how an image gives rise to data in the absence of noise. Moreover, X is a suitable Hilbert space of real valued functions supported on a fixed domain $\Omega \subset \mathbb{R}^n$ whose elements represent attenuation of emission values. Likewise, Y is a Hilbert space of real-valued functions that represent data and that are defined on some manifold $\mathbb{M}$ (data manifold).

In *tomographic imaging*, data can often be modeled as line integrals of the function f that describes the object along a line, i.e., data is a real-valued function on some set of lines $\mathbb{M}$ in $\mathbb{R}^n$. We can now introduce coordinates on this data manifold. A line in $\mathbb{R}^n$ can be described by a directional vector in the unit sphere S^{n-1} and a point that it passes through. This provides coordinates on $\mathbb{M}$ where a line is given by $(\omega, x) \in S^{n-1} \times \mathbb{R}^n$ with $x \in \omega^\perp$. Here, $\omega^\perp \subset \mathbb{R}^n$ is the unique plane through the origin with $\omega \in S^{n-1}$ as its normal vector. The corresponding forward operator $\mathcal{A}$ is the *ray transform*, which is expressible in the aforementioned coordinates as

$$\mathcal{A}(f)(\omega, x) := \int_{-\infty}^{\infty} f(x + t\omega)dt \quad \text{for } f \in X. \tag{2}$$

Tomographic data can then be seen as values of $\mathcal{A}(f)(\omega, x)$ for a sampling of $\omega \in S^{n-1}$ (angular sampling) and $x \in \omega^\perp$ (detector sampling). With slight abuse of terminology, one refers to these data as the "projection" of f along the line given by (ω, x).

3   Overview of ODL and ASTRA

Many reconstruction schemes can be formulated in a generic, yet adaptable, manner by stating them in an abstract coordinate-free setting using the language of functional analysis. The adaptability stems from "parametrizing" the scheme in terms of the forward operator, the data noise model, and the type of a priori information that one seeks to exploit. These can be further broken down into more *generic* components, each representing a well-defined generally applicable mathematical structure or operation. Another advantage that comes with a generic formulation is that it makes the reconstruction scheme more transparent.

These considerations form a natural blueprint for a modular software library for inverse problems where the forward operator, data noise model, and prior model are treated as independent exchangeable components. ODL is such a software library whose *design principles* are *modularity, abstraction*, and *compartmentalization* [1] that is freely available at http://github.com/odlgroup/odl. To realize these design principles, ODL separates *which* mathematical object or operation one seeks to represent from *how* it is implemented using concrete computational routines. Mathematical objects and operations are represented by abstract classes with abstract methods and specific implementations are represented by concrete subclasses. Hence, these abstract classes form a domain specific language for functional analysis and corresponding subclasses couple to relevant numerical libraries. In this way one can express abstraction in a way that allows combining generic and application-specific code. The generic part, such as an optimization method, can be formulated in a coordinate-free manner using abstract classes and methods, whereas application-specific parts, such as evaluating the forward operator, are contained in specific subclasses. Hence, one can express reconstruction schemes using a clean near-mathematical syntax and involved implementation specific details are hidden in concrete subclasses.

A key part of ODL is the notion of an operator between two vector spaces. ODL offers operator calculus for constructing operators from existing ones, typically using composition. Furthermore, an operator may also have a number of additional associated operators, like its (Fréchet) derivative, inverse, and adjoint. Whenever possible, such associated operators are automatically generated when an operator is defined using the operator calculus in ODL, e.g., derivative operators are formed using the chain rule. This is a very powerful part of ODL that reduces the risk for errors and simplifies testing.

Another important part of ODL is its usage of external software libraries for performing specific tasks. When working with tomographic inverse problems, one such task is computing the 2D/3D ray transform and its adjoint. For this ODL employs the ASTRA toolbox [11,13,14], which is a high-performance, GPU accelerated toolbox for tomographic reconstruction freely available from http://www.astra-toolbox.com. The toolbox supports many different data manifolds arising in tomography, including for example circular cone beam, laminography, tomosynthesis, and electron tomography, see [13] for details. It also provides both Matlab and Python interfaces, that expose the core tomographic operations. The latter is used for seamless integration between ODL and ASTRA in the sense that forward and backprojection routines in ASTRA are available as operators in ODL. Likewise, the tomographic data acquisition model in ASTRA is fully reflected by the corresponding data model in ODL. This open up for using ASTRA routines from ODL without unnecessary data copying between GPU and CPU.

The reconstruction methods available in ASTRA are mostly iterative methods. In ODL, on the other hand, one can easily formulate a variational reconstruction algorithm. In this work we will consider ODL/ASTRA to formulate a variational reconstruction algorithm and solve the corresponding optimization problem.

4   Discrete Algebraic Reconstruction

A large class of reconstruction methods for ill-posed inverse problems can be formulated as solving an optimization problem:

$$\min_{f \in X} \Big[\mathcal{L}\big(\mathcal{A}(f), g\big) + \lambda \mathcal{R}(f) \Big]. \tag{3}$$

Here, $\mathcal{R} \colon X \to \mathbb{R}_+$ is the regularization term that accounts for the a priori knowledge by penalizing unfeasible solution candidates, $\mathcal{L} \colon Y \times Y \to \mathbb{R}_+$ is the data-fit term that quantifies how well two points in the data space agree with each other, and $\lambda > 0$ is the regularization parameter that weights the a priori knowledge against the need to minimize the data-fit term.

Discrete tomography (DT) is a class of tomographic reconstruction methods that are based on the assumption that the unknown object f consists of a few distinct materials, each producing a (almost) constant gray value in the reconstruction. The total variation regularized discrete algebraic reconstruction technique (TVR-DART) is a recent approach that is adapted towards discrete tomography [16,17], which has proven to be more robust than the original DART algorithm [4]. In particular, using the TVR-DART method allows one to significantly improve reconstruction quality and to drastically reduce the number of required projection images and/or exposure to the sample. The idea in TVR-DART is that the image to be recovered is step-function like with a transition between regions that is not necessarily sharp. This is typically the case when one images specimens consisting of only a few different material compositions where each material has a distinct gray value in the corresponding image.

TVR-DART aims to capture such a priori information by combining principles from discrete tomography and compressive sensing. The original formulation in [16,17] uses L^2-norm as the data-fit term, which comes from the assumption that noise in data is additive Gaussian. This is however not always the case, e.g., in HAADF-STEM tomography [8] the noise in data is predominantly Poisson distributed, especially under very low exposure (electron dose) [10].

In the following, we will first formulate TVR-DART in an abstract manner using the language of functional analysis. Next, this abstract version is implemented in ODL/ASTRA using the operator calculus in ODL. Thereby, we can encapsulate all application-specific parts and use high-level ODL solvers that usually expect an operator as argument. In this way, the same problem can be solved with different methods by simply calling different solvers. When a new solver or application-specific code is needed, it needs to be written only once at one place, and can be tested separately. At the lower level of ODL, efficient forward and backward projections are performed using GPU accelerated code provided by the ASTRA toolbox.

In summary, we offer a generic, yet adaptable, version of TVR-DART with a plug-and-play structure that allows one to change the forward operator and the noise model. We demonstrate this flexibility by switching the data-fit term to one that better matches data with Poisson noise, which is more appropriate for tomographic data under low dose conditions.

5   ODL Implementation of TVR-DART

Bearing in mind the functional analytic viewpoint in ODL, our starting point is to formulate the TVR-DART scheme in an abstract setting (Sect. 5.1, Eq. (7)). In the following four sections (Sects. 5.2, 5.3, 5.4 and 5.5) we show how the abstract TVR-DART scheme is implemented using the ODL operator calculus with ASTRA as computational backend for computing projections and corresponding backprojections. In order to emphasis the important points, and due to space limitations, we have left out some of the code indicated with "[...]". The full source code is available at http://github.com/aringh/TVR-DART. We conclude by showing some reconstructions. Note however that the goal of the paper is not to evaluate TVR-DART, for that see [17]. It is to show the flexibility in using ODL/ASTRA as a prototyping tool. Here TVR-DART merely severs as an example of a complex reconstruction method.

5.1   Abstract Formulation

The key assumption in TVR-DART is that the image we seek consists of n gray-scale levels that are separated by a narrow, but smooth, transition layer. Thus, we introduce a (parametrized) segmentation operator $\mathcal{T} : X \times \Theta \to X$ that acts as a kind of segmentation map. It is here given as

$$\mathcal{T}(f, \theta)(x) = \sum_{i=1}^{n-1} (\rho_i - \rho_{i-1}) u_{k_i} (f(x) - \tau_i) \quad \text{for } x \in \Omega \text{ and } \theta \in \Theta. \quad (4)$$

The parameter space $\Theta := (\mathbb{R} \times \mathbb{R} \times \mathbb{R})^n$ defines the transition characteristics of the n layers (the background ρ_0 is often set to 0). Concretely, $\theta = (\theta_1, \ldots, \theta_n) \in \Theta$ with $\theta_i := (\rho_i, \tau_i, K_i)$ where ρ_i is the gray-scale level of the i:th level, τ_i is the mid-point gray-scale value, $k_i := K_i/(\rho_i - \rho_{i-1})$ is the sharpness of the smooth gray-scale transition, and $u \colon \mathbb{R} \to [0, 1]$ is the logistic function that models the transition itself

$$u_k(s) := \frac{1}{1 + e^{-2ks}} \quad \text{for } s \in \mathbb{R}. \quad (5)$$

The TVR-DART algorithm for solving (1) is now defined as a method that yields a minimizer to

$$\min_{f \in X, \theta \in \Theta} \left[\mathcal{L} \left([\mathcal{A} \circ \mathcal{T}](f, \theta), g \right) + \lambda [\mathcal{S} \circ \mathcal{T}](f, \theta) \right]. \quad (6)$$

In the above, $\mathcal{L} \colon Y \times Y \to \mathbb{R}_+$ is an appropriate data-fit term and $\mathcal{S} \colon X \to \mathbb{R}_+$ is the regularization. The variant considered in [17] uses a data-fit term $\mathcal{L}(\cdot, g) = \| \cdot - g \|_2^2$ and a regularizing functional $\mathrm{TV}_\varepsilon = [\mathcal{H}_\varepsilon \circ \nabla]$, where $\mathcal{H}_\varepsilon$ is the Huber norm and ∇ is the spatial gradient operator. The Huber norm is a smooth surrogate functional for the L^1-norm, and $\mathcal{H}_\varepsilon \circ \nabla$ is thus a smoothed version of TV. Hence, (6) becomes

$$\min_{f \in X, \theta \in \Theta} \left[\left\| [\mathcal{A} \circ \mathcal{T}](f, \theta) - g \right\|_2^2 + \lambda [\mathcal{H}_\varepsilon \circ \nabla \circ \mathcal{T}](f, \theta) \right]. \quad (7)$$

Gradient based methods can be used to solve (7) since its objective functional is smooth. In [17] one such solution method was presented, based on an alternating optimization over f and θ. We take a similar approach here, and to this end define the two operator $\mathcal{T}_\theta : X \to X$, defined by $\mathcal{T}_\theta(f) = \mathcal{T}(f, \theta)$, and $\mathcal{T}_f : \Theta \to X$, defined $\mathcal{T}_f(\theta) = \mathcal{T}(f, \theta)$, where θ and f are seen as fix parameters, respectively. In the current implementation we view the sharpness parameter as fixed, but optimize over gray-scale value and mid-point.

5.2   Defining the Inverse Problem

We begin by defining the reconstruction space $X = L_2(\Omega)$ assuming digitization by uniform sampling in Ω with 320×320 pixels:

```
X = odl.uniform_discr(min_pt=[−200, −200],
      max_pt=[200, 200], shape=[320, 320])
```

Next is to define the forward operator as the ray transform $\mathcal{A} : X \to Y$ in (2).

```
M_angle_part = odl.uniform_partition(0, np.pi, 18, nodes_on_bdry=
      True)
M_detector_part = odl.uniform_partition(−200, 200, 500)
M = odl.tomo.Parallel2dGeometry(M_angle_part, M_detector_part)
A = odl.tomo.RayTransform(X, M, impl='astra_cuda')
```

Note that there is no need to explicitly specify the range Y of the ray transform $\mathcal{A}$, which are functions defined on $\mathbb{M}$ (data manifold). Y is given indirectly by the **geometry**-object, which defines $\mathbb{M}$ through M_angle_part for the angular sampling of the lines and M_detector_part for the detector sampling. This information is typically provided by the experimental setup.

5.3   Defining the Objective Functional

To define the objective functional in (7), we begin by setting up the soft segmentation operator $\mathcal{T}_\theta : X \to X$ (see Sect. 5.5 for its implementation):

```
T_theta = SoftSegmentationOperator(X, base_value, thresholds,
      values, sharpness)
```

The regularization term $\mathcal{R} : X \to X$ in (7), when optimizing over f, is given by $\mathcal{R}_\theta := \mathcal{H}_\varepsilon \circ \nabla \circ \mathcal{T}_\theta$, which can be implemented using the operator calculus in ODL:

```
gradient = odl.Gradient(X)
gradient = odl.PointwiseNorm(gradient.range) * gradient
H = HuberNorm(X, 0.0001)
R_theta = H * gradient * T_theta
```

In the above, `PointwiseNorm` is used to define the isotropic TV-like term and a description of how to implement the Huber norm is given in Sect. 5.5. Next the data-fit term in (7) as a function $f \mapsto \left\| [\mathcal{A} \circ \mathcal{T}_\theta](f) - g \right\|_2^2$ is implemented as

```
l2_norm = odl.solvers.L2NormSquared(A.range)
l2_norm = l2_norm.translated(data)
data_fit_theta = l2_norm * A * T_theta
```

The `l2_norm.translated(data)` command shifts the origin of the L_2-norm functional, i.e., it changes the functional $\| \cdot \|_2^2$ into $\| \cdot -g \|_2^2$. Hence, the complete objective functional in (7), when optimizing over f, can be assembled as

```
obj_theat = data_fit_theta + reg_param * R_theta
```

The implementation for optimizing over θ is analogous.

5.4 Solving the Optimization Problem

Since the objective functional in (7), both when seen as a functional over f and over θ, is smooth we can use a smooth solver such as limited-memory BFGS [6, Sect. 13.5] with backtracking line-search [6, Sect. 11.5] in the alternating optimization. A BFGS solver and backtracking line-search is built into ODL, and the alternating optimization can thus be implemented as follows.

```
reco = fbp
theta = theta_init
for i in range(10):
  [...]
  linesearch = odl.solvers.BacktrackingLineSearch(obj_theta)
  odl.solvers.bfgs_method(f=obj_theta, x=reco, line_search=
      linesearch, maxiter=10, tol=1e-8, num_store=10)
  [...]
  linesearch = odl.solvers.BacktrackingLineSearch(obj_f)
  odl.solvers.bfgs_method(f=obj_f, x=theta, line_search=
      linesearch, maxiter=2, tol=1e-8, num_store=2)
```

The command x=reco specifies the initial iterate for the BFGS solver, and in the first outer iteration is taken as a reconstruction fbp obtained from standard filtered backprojection (FBP) using a Hann filter. Example reconstructions using this algorithm are shown later in Fig. 1.

5.5 Implementing the Huber Norm and Soft Segmentation Operator

The Huber norm and soft segmentation operator are not part of ODL and need to be added. They are implemented as an ODL Functional and Operator object, respectively.

Starting with the Huber norm, its mathematical definition is

$$\mathcal{H}_\varepsilon(f) = \int_\Omega f_\varepsilon(x)dx \quad \text{where} \quad f_\varepsilon(x) = \begin{cases} |f(x)| - \dfrac{\varepsilon}{2} & \text{if } |f(x)| \geq \varepsilon \\ \dfrac{f(x)^2}{2\varepsilon} & \text{if } |f(x)| < \varepsilon. \end{cases}$$

In the ODL implementation below we uses q_part to denote the quadratic region, i.e., where $|f(x)| < \varepsilon$.

```
class HuberNorm(Functional):
  [...]
  def _call(self, f):
    """Evaluating the functional."""
    q_part = f.ufuncs.absolute().asarray() < self.epsilon
    q_part = np.float32(q_part)
    f_eps = ((f * q_part)**2 / (2.0 * self.epsilon) +
             (f.ufuncs.absolute() - self.epsilon / 2.0) *
             (1-q_part))
    # This line takes the inner product with the one-function.
    return f_eps.inner(self.domain.one())
```

Since we use a smooth solver, we also need to provide the gradient associated with the Huber norm, which is an element $\nabla \mathcal{H}_\varepsilon(f) \in X$ that satisfies

$$\mathcal{H}'_\varepsilon(f)(h) = \langle \nabla \mathcal{H}_\varepsilon(f), h \rangle_X.$$

In the above, the bounded linear operator $\mathcal{H}'_\varepsilon(f) : X \to \mathbb{R}$ is the Fréchet derivative of $\mathcal{H}_\varepsilon$ at f. For the Huber norm, the gradient at $f \in X$ is

$$\nabla \mathcal{H}_\varepsilon(f)(x) = \frac{\partial}{\partial f} f_\varepsilon(x) = \begin{cases} 1 & \text{if } f(x) \geq \varepsilon \\ -1 & \text{if } f(x) \leq -\varepsilon \\ \dfrac{f(x)}{\varepsilon} & \text{if } |f(x)| < \varepsilon. \end{cases}$$

The above is implemented in ODL as a property of the Huber norm functional:

```
@property
def gradient(self):
  """Gradient operator of the functional."""
  func = self
  class HuberNormGradient(Operator):
  [...]
    def _call(self, f):
      q_part = f.ufuncs.absolute().asarray() < func.epsilon
      q_part = np.float32(q_part)
      f_eps_diff = ((f * q_part) / (func.epsilon) +
                    f.ufuncs.sign() * (1-q_part))
      return f_eps_diff
  return HuberNormGradient()
```

The soft segmentation operator $\mathcal{T}_\theta : X \to X$, implicitly defined in (4), can be implemented in a similar way. Below we compute the Fréchet derivative with respect to f (since this operator is not a functional, it does not have a gradient) that can be implemented in ODL as a property. In ODL the derivative of $\mathcal{T}_\theta$ with respect to the function, at $f \in X$, is itself a linear operator $\mathcal{T}'_\theta(f) : X \to X$ given by

$$\mathcal{T}'_\theta(f)(h)(x) = \sum_{i=1}^{n-1} h(x)(\rho_i - \rho_{i-1})u'_{k_i}\big(f(x) - \tau_i\big) \quad \text{for } h \in X,$$

where u'_k is the derivative of the logistic function (5)

$$u'_k(t) = \frac{2ke^{-2kt}}{(1 + e^{-2kt})^2}.$$

The operator $\mathcal{T}_f : \Theta \to X$ is implemented analogously, where the Fréchet derivative of $\mathcal{T}_f$ with respect to θ can be derived similarly.

5.6   Extension to Handle Data with Poisson Noise

It is well-known that minimizing the Kullback-Leibler (KL) divergence of count data is equivalent to *maximum likelihood* estimation when the noise in data g is Poisson distributed [5]. The original notion of KL divergence comes from information theory, and is thus defined for probability measures. The one used in inverse problems is the generalization below to nonnegative functions:

$$\mathbb{D}_{\mathrm{KL}}(g \mid h) = \begin{cases} \int_\Omega \left(g(y) \log\left(\frac{g(y)}{h(y)}\right) + h(y) - g(y) \right) dy & g(y) \geq 0, \; h(y) > 0 \\ +\infty & \text{else.} \end{cases} \quad (8)$$

Noise in low count data is often better modeled using a Poisson distribution rather than an additive Gaussian distribution, and many electron tomography applications are low count data [10]. Hence, using (8) as data-fit term in (6), i.e.,

$$\mathcal{L}([\mathcal{A} \circ \mathcal{T}_\theta](f), g) = \mathbb{D}_{\mathrm{KL}}\big(g \mid [\mathcal{A} \circ \mathcal{T}_\theta](f)\big),$$

is of interest to applications where the TVR-DART algorithm will be used. Since the KL divergence is already available as a functional in ODL, in order to use KL instead of L_2 we only need to change the data-fit functional:

```
kl = odl.solvers.KullbackLeibler(A.range, prior=data)
data_fit_theta = kl * A * T_theta
```

5.7   Reconstructions

The resulting reconstructions from running the TVR-DART is summarized in Fig. 1, which compares TVR-DART and TV reconstructions, both approaches using L^2-norm and the KL as data-fit term, the former more suitable for data with additive Gaussian noise and the latter more suitable for data with Poisson noise. Tomographic data used for the tests is simulated using both Gaussian and Poisson noise, and for both TV and TVR-DART we have used an FBP reconstruction as an initial starting iterate. In Fig. 1 we also give some figure of merits for the reconstructions, namely Relative Mean Error (RME) (see, e.g., [17, p. 460]) and Structural Similarity index (SSIM) [15].

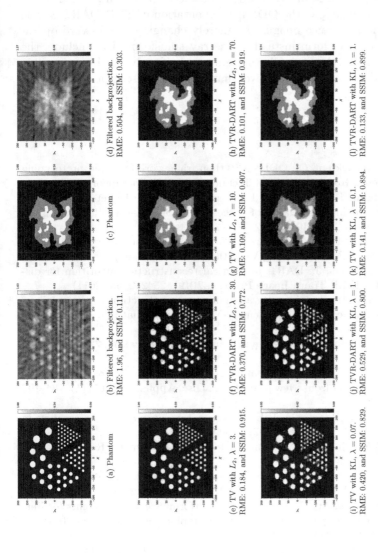

Fig. 1. Phantoms are shown in 1a and 1c. Data is generated from the phantoms and then perturbed by noise. In 1b and 1d, FBP reconstructions from the data with Poisson noise are shown. The reconstructions in the second row, 1e through 1h, are from data with white Gaussian noise, and the reconstructions in the third row, 1i through 1l, are from data with Poisson noise. The images are of size 320×320, and data is acquired from a parallel beam geometry, with 18 equidistant angles between 0 and π and with 500 discretization points on the detector. For data with white Gaussian noise, the noise has a norm that is 5% of the norm of data, and for data with Poisson noise, the data is the outcome of a Poisson distributed variable with parameter given by the noise-free data.

6 Conclusions

We have shown how TVR-DART can be implemented in ODL/ASTRA, and utilizing the modularity and flexibility of ODL we extended the algorithm by changing the data-fit functional $\mathcal{L}$ in (6). In the same way, it is straightforward to change to forward operator $\mathcal{A}$ in (6) in order to use the algorithm in other imaging modalities. As an example, the ODL implementation of TVR-DART can be applied to magnetic resonance imaging by merely changing the forward operator $\mathcal{A}$ to the Fourier transform instead of the ray transform. To conclude, the combination of ODL and ASTRA allows users to specify advanced tomographic reconstruction methods using a high-level mathematical description that facilitates rapid prototyping.

References

1. Adler, J., Kohr, H., Öktem, O.: ODL - a Python framework for rapid prototyping in inverse problems. Royal Institute of Technology (2017) (in Preparation)
2. Alpers, A., Poulsen, H.F., Knudsen, E., Herman, G.T.: A discrete tomography algorithm for improving the quality of three-dimensional X-ray diffraction grain maps. J. Appl. Crystallogr. **39**(4), 582–588 (2006)
3. Batenburg, K.J.: A network flow algorithm for reconstructing binary images from continuous X-rays. J. Math. Imaging Vis. **30**(3), 231–248 (2008)
4. Batenburg, K.J., Sijbers, J.: DART: a practical reconstruction algorithm for discrete tomography. IEEE Trans. Image Process. **20**(9), 2542–2553 (2011)
5. Bertero, M., Lantéri, H., Zanni, L.: Iterative image reconstruction: a point of view. Math. Methods Biomed. Imaging Intensity-Modulated Radiat. Ther. (IMRT) **7**, 37–63 (2008)
6. Griva, I., Nash, S.G., Sofer, A.: Linear and Nonlinear Optimization, 2nd edn. SIAM (2009)
7. Liao, H.Y., Herman, G.T.: A coordinate ascent approach to tomographic reconstruction of label images from a few projections. Disc. Appl. Math. **151**(1), 184–197 (2005)
8. Midgley, P.A., Dunin-Borkowski, R.E.: Electron tomography and holography in materials science. Nat. Mater. **8**(4), 271 (2009)
9. Natterer, F., Wübbeling, F.: Mathematical Methods in Image Reconstruction. SIAM (2001)
10. Öktem, O.: Mathematics of electron tomography. In: Scherzer, O. (ed.) Handbook of Mathematical Methods in Imaging, pp. 937–1031. Springer, New York (2015)
11. Palenstijn, W.J., Batenburg, K.J., Sijbers, J.: Performance improvements for iterative electron tomography reconstruction using graphics processing units (GPUs). J. Struct. Biol. **176**(2), 250–253 (2011)
12. Schüle, T., Schnörr, C., Weber, S., Hornegger, J.: Discrete tomography by convex-concave regularization and DC programming. Disc. Appl. Math. **151**(1), 229–243 (2005)
13. van Aarle, W., Palenstijn, W.J., Cant, J., Janssens, E., Bleichrodt, F., Dabravolski, A., De Beenhouwer, J., Batenburg, K.J., Sijbers, J.: Fast and flexible X-ray tomography using the astra toolbox. Opt. Express **24**(22), 25129–25147 (2016)

14. van Aarle, W., Palenstijn, W.J., De Beenhouwer, J., Altantzis, T., Bals, S., Batenburg, K.J., Sijbers, J.: The ASTRA Toolbox: A platform for advanced algorithm development in electron tomography. Ultramicroscopy **157**, 35–47 (2015)
15. Wang, Z., Bovik, A.C., Sheikh, H.R., Simoncelli, E.P.: Image quality assessment: from error visibility to structural similarity. IEEE Trans. Image Process. **13**(4), 600–612 (2004)
16. Zhuge, X., Jinnai, H., Dunin-Borkowski, R.E., Migunov, V., Bals, S., Cool, P., Bons, A.J., Batenburg, K.J.: Automated discrete electron tomography - towards routine high-fidelity reconstruction of nanomaterials. Ultramicroscopy **175**, 87–96 (2017)
17. Zhuge, X., Palenstijn, W.J., Batenburg, K.J.: TVR-DART: a more robust algorithm for discrete tomography from limited projection data with automated gray value estimation. IEEE Trans. Image Process. **25**(1), 455–468 (2016)

A Parametric Level-Set Method for Partially Discrete Tomography

Ajinkya Kadu[1]([✉]), Tristan van Leeuwen[1], and K. Joost Batenburg[2]

[1] Mathematical Institute, Utrecht University, Utrecht, The Netherlands
a.a.kadu@uu.nl
[2] Centrum Wiskunde & Informatica, Amsterdam, The Netherlands

Abstract. This paper introduces a parametric level-set method for tomographic reconstruction of partially discrete images. Such images consist of a continuously varying background and an anomaly with a constant (known) grey-value. We express the geometry of the anomaly using a level-set function, which we represent using radial basis functions. We pose the reconstruction problem as a bi-level optimization problem in terms of the background and coefficients for the level-set function. To constrain the background reconstruction, we impose smoothness through Tikhonov regularization. The bi-level optimization problem is solved in an alternating fashion; in each iteration we first reconstruct the background and consequently update the level-set function. We test our method on numerical phantoms and show that we can successfully reconstruct the geometry of the anomaly, even from limited data. On these phantoms, our method outperforms Total Variation reconstruction, DART and P-DART.

Keywords: Discrete tomography · Level-set method · Model splitting · Geometric inversion

1 Introduction

The need to reconstruct (quantitative) images of an object from tomographic measurements appears in many applications. At the heart of many of these applications is a projection model based on the Radon transform. Characterizing the object under investigation by a function $u(\mathbf{x})$ with $\mathbf{x} \in \mathcal{D} = [0,1]^2$, tomographic measurements are modeled as

$$p_i = \int_{\mathcal{D}} u(\mathbf{x})\delta(s_i - \mathbf{n}(\theta_i) \cdot \mathbf{x})\,\mathrm{d}\mathbf{x}, \tag{1}$$

where $s_i \in [0,1]$ denotes the shift, $\theta_i \in [0, 2\pi)$ denotes the angle and $\mathbf{n}(\theta) = (\cos\theta, \sin\theta)$. The goal is to retrieve u from a number, m, of such measurements for various shifts and directions.

If the shifts and angles are regularly and densely sampled, the transform can be inverted directly by Filtered back-projection or Fourier reconstruction [9].

© Springer International Publishing AG 2017
W.G. Kropatsch et al. (Eds.): DGCI 2017, LNCS 10502, pp. 122–134, 2017.
DOI: 10.1007/978-3-319-66272-5_11

A common approach for dealing with non-regularly sampled or missing data, is *algebraic reconstruction*. Here, we express u in terms of a basis

$$u(\mathbf{x}) = \sum_{j=1}^{n} u_j b(\mathbf{x} - \mathbf{x}_j),$$

where b are piece-wise polynomial basis functions and $\{\mathbf{x}_j\}_{j=1}^{n}$ is a regular (pixel) grid. This leads to a set of m linear equations in n unknowns

$$\mathbf{p} = W\mathbf{u},$$

with $w_{ij} = \int_{\mathcal{D}} b(\mathbf{x} - \mathbf{x}_j)\delta(s_i - \mathbf{n}(\theta_i) \cdot \mathbf{x})\,d\mathbf{x}$. Due to noise in the data or errors in the projection model the system of equations is typically inconsistent, so a solution may not exist. Furthermore, there may be many solutions that fit the observations equally well when the system is underdetermined. A standard approach to mitigate these issues is to formulate a regularized least-squares problem

$$\min_{\mathbf{u}} \tfrac{1}{2}\|W\mathbf{u} - \mathbf{p}\|_2^2 + \tfrac{\lambda}{2}\|R\mathbf{u}\|_2^2,$$

where R is the regularization operator with parameter λ balancing the data-misfit and regularity of the solution. Such a formulation is popular mainly because very efficient algorithms exist for solving it. Depending on the choice of R, however, this formulation forces the solution to have certain properties which may not reflect the truth. For example, setting R to be the discrete Laplace operator will produce a smooth reconstruction, whereas setting R to be the identity matrix forces the individual coefficients u_i to be small. In many applications such quadratic regularization terms do not reflect the characteristics of the object we are reconstructing. For example, if we expect u to be piecewise constant, we could use a Total Variation regularization term $\|R\mathbf{u}\|_1$ where R is a discrete gradient operator [13]. Recently, a lot of progress has been made in developing efficient algorithms for solving such non-smooth optimization problems [6]. If the object under investigation is known to consist of only two distinct materials, the regularization can be formulated in terms of a non-convex constraint $\mathbf{u} \in \{u_0, u_1\}^n$. The latter leads to a combinatorial optimization problem, solutions to which can be approximated using heuristic algorithms [3].

In this paper, we consider tomographic reconstruction of *partially discrete* objects that consist of a region of constant density embedded in a continuously varying background. In this case, neither the quadratic, Total Variation nor non-convex constraints by themselves are suitable. We therefore propose the following parametrization

$$u(\mathbf{x}) = \begin{cases} u_1 & \text{if } \mathbf{x} \in \Omega, \\ u_0(\mathbf{x}) & \text{otherwise.} \end{cases}$$

The inverse problem now consists of finding $u_0(\mathbf{x})$, u_1 and the set Ω. We can subsequently apply suitable regularization to u_0 separately. To formulate a tractable optimization algorithm, we represent the set Ω using a level-set function $\phi(\mathbf{x})$ as follows:

$$\Omega = \{\mathbf{x} \mid \phi(\mathbf{x}) > 0\}.$$

In the following sections, we assume knowledge of u_1 and discuss how to formulate a variational problem to reconstruct Ω and u_0 based on a parametric level-set representation of Ω.

The outline of the paper is as follows. In Sect. 2 we discuss the parametric level-set method and propose some practical heuristics for choosing various parameters that occur in the formulation. A joint background-anomaly reconstruction algorithm for partially discrete tomography is discussed in Sect. 3. The results on a few moderately complicated numerical phantoms are presented in Sect. 4. We provide some concluding remarks in Sect. 5.

2   Level-Set Methods

In terms of the level-set function, we can express u as

$$u(\mathbf{x}) = (1 - h(\phi(\mathbf{x})))u_0(\mathbf{x}) + h(\phi(\mathbf{x}))u_1,$$

where h is the Heaviside function.

Level-set methods have received much attention in geometric inverse problems, interface tracking, segmentation and shape optimization. In the classical level-set method, introduced by Sethian and Osher [11], the level-set is evolved according to the Hamilton-Jacobi equation

$$\frac{\partial \phi}{\partial t} + v|\nabla \phi| = 0,$$

where $\phi(\mathbf{x}, t)$ now denotes the level-set function as a time-dependent quantity for representing the shape and v denotes the normal velocity at the boundary of the shape. In the inverse-problems setting, the velocity v is often derived from the gradient of the cost function with respect to the model parameter [5,7]. There are various numerical issues associated with the numerical solution of level-set equation, e.g. reinitialization of the level-set. We refer the interested reader to a seminal paper in level-set methods [11] and its application to computational tomography [10].

Instead of taking this classical level-set approach, we employ a parametric level-set approach, first introduced by Aghasi et al. [1]. In this method, the level-set function is parametrized using radial basis functions (RBF)

$$\phi(\mathbf{x}) = \sum_{j=1}^{n'} \alpha_j \Psi(\beta_j \|\mathbf{x} - \boldsymbol{\chi}_j\|_2),$$

where $\Psi(\cdot)$ is a radial basis function, $\{\alpha_j\}_{j=1}^{n'}$ and $\{\boldsymbol{\chi}_j\}_{j=1}^{n'}$ are the amplitudes and nodes respectively, and the parameters $\{\beta_j\}_{j=1}^{n'}$ control the widths. Introducing the kernel matrix $A(\boldsymbol{\chi}, \boldsymbol{\beta})$ with elements

$$a_{ij} = \Psi(\beta_j \|\mathbf{x}_i - \boldsymbol{\chi}_j\|_2),$$

we can now express $\mathbf{u}$ on the computational grid $\{\mathbf{x}_i\}_{i=1}^n$ as

$$\mathbf{u} = (1 - h(A(\boldsymbol{\chi}, \boldsymbol{\beta})\boldsymbol{\alpha})) \odot \mathbf{u}_0 + h(A(\boldsymbol{\chi}, \boldsymbol{\beta})\boldsymbol{\alpha})u_1, \qquad (2)$$

where h is applied element-wise to the vector $A(\boldsymbol{\chi}, \boldsymbol{\beta})\boldsymbol{\alpha}$ and $\odot$ denotes the element-wise (Hadamard) product. By choosing the parameters $(\boldsymbol{\chi}, \boldsymbol{\beta}, \boldsymbol{\alpha})$ appropriately we can represent any (smooth) shape. To simplify matters and make the resulting optimization problem more tractable, we consider a fixed regular grid $\{\boldsymbol{\chi}_j\}_{j=1}^{n'}$ and a fixed width $\beta_j \equiv \beta$. In the following we choose β in accordance with the grid spacing $\Delta\chi$ as $\beta = 1/(\eta\Delta\chi)$, where η corresponds to the width of the RBF in grid points.

Example. To illustrate the representation of a shape using finitely many radial basis functions, we consider the green shape shown in Fig. 1(a). Starting from an initial guess (Fig. 1(a), red) we obtain the coefficients $\boldsymbol{\alpha}$ by solving a non-linear least-squares problem $\min_{\boldsymbol{\alpha}} \|h(A\boldsymbol{\alpha}) - \mathbf{y}\|_2^2$, where $\mathbf{y} \in \{0,1\}^n$ indicates the true shape. This leads to the representation shown in Fig. 1(b). With $n' = 196$ RBFs, it is possible to reconstruct a smooth shape discretized on a grid with $n = 256 \times 256$ pixels.

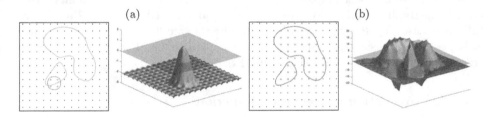

$$(a) \qquad\qquad\qquad\qquad\qquad (b)$$

Fig. 1. Any (sufficiently) smooth level-set can be reconstructed from radial basis functions. (a) The shape to be reconstructed is denoted in *green*. The initial shape (*dash-dotted* line) is generated by some positive RBF coefficients (denoted by *red pluses*) near the center and negative coefficients elsewhere (denoted by *blue dots*). Also shown is the corresponding initial level-set function. The reconstructed shape denoted with a *dash-dotted* line, the sign of the RBF-coefficients as well as the corresponding level-set function are shown in (b). (Color figure online)

Finally, the discretized reconstruction problem for determining the shape is now formulated as

$$\min_{\boldsymbol{\alpha}} \left\{ f(\boldsymbol{\alpha}) = \|W[(u_1 \mathbf{1} - \mathbf{u}_0) \odot h_\epsilon(A\boldsymbol{\alpha})] - (\mathbf{p} - W\mathbf{u}_0)\|_2^2 \right\}, \qquad (3)$$

where h_ϵ is a smooth approximation of the Heaviside function. The gradient and Gauss-Newton Hessian of $f(\boldsymbol{\alpha})$ are given by

$$\begin{aligned} \nabla f(\boldsymbol{\alpha}) &= A^T D_{\boldsymbol{\alpha}}^T W^T \mathbf{r}(\boldsymbol{\alpha}), \\ H_{GN}(f(\boldsymbol{\alpha})) &= A^T D_{\boldsymbol{\alpha}}^T W^T W D_{\boldsymbol{\alpha}} A, \end{aligned} \qquad (4)$$

where the diagonal matrix and residual vectors are given by

$$D_\alpha = \text{diag}((u_1 \mathbf{1} - \mathbf{u}_0) \odot h'_\epsilon(A\alpha)),$$
$$\mathbf{r}(\alpha) = W[(u_1 \mathbf{1} - \mathbf{u}_0) \odot h_\epsilon(A\alpha)] - (\mathbf{p} - W\mathbf{u}_0).$$

Using a Gauss-Newton method, the level-set parameters are updated as

$$\alpha^{(k+1)} = \alpha^{(k)} - \mu_k \left(H_{GN}(f(\alpha^{(k)})) \right)^{-1} \nabla f(\alpha^{(k)}), \qquad (5)$$

where μ_k is a stepsize chosen to satisfy the weak Wolfe conditions [15] and $\alpha^{(0)}$ is a given initial estimate of the shape. The weak Wolfe conditions consist of sufficient decrease and curvature conditions and ensure global convergence to a local minimum.

From Eq. (4), it can be observed that the ability to update the level-set parameters depends on two main factors: (1) The difference between $\mathbf{u}_0$ and u_1, and (2) the derivative of the Heaviside function. Hence, the support and smoothness of h'_ϵ plays a crucial role in the sensitivity. More details on the choice of h_ϵ are discussed in Sect. 2.1.

Example. We demonstrate the parametric level-set method on a (binary) discrete tomography problem. We consider the model described in Fig. 2(a). For a full-angle case ($0 \leq \theta \leq \pi$) with a large number of samples, Fig. 2(c) shows that it is possible to accurately reconstruct a complex shape. The model is reconstructed by iteratively updating α using Eq. (5).

2.1   Approximation to Heaviside Function

The update of the level-set function depends crucially on the choice of the Heaviside function. In Eq. (4) we see that h'_ϵ acts as a windowing function that controls which part of the level-set function is updated. The windowing function should achieve the following: (i) limit the update to a small region around the boundary of the shape; (ii) have a uniform amplitude in the boundary region; and (iii) guarantee a minimum width of the boundary region. Failure to meet these requirements may result in poor updates for the level-set parameter α and premature break-down of the algorithm.

Requirement (i) is easily fulfilled as any smooth approximation of the Heaviside will have a rapidly decaying derivative. To satisfy the second requirement we construct the Heaviside function by smoothing the piece-wise linear function $\frac{1}{2} + \frac{x}{2\epsilon}$ for $|x| \leq \epsilon$. This approximation is shown in Fig. 3 alongside two common smooth approximations of the Heaviside. We now discuss how we can satisfy the third requirement, starting with a formal definition of the width of the level-set boundary layer as shown in Fig. 3(c).

Definition 1. *In accordance with the compact approximation of the Heaviside function with width ϵ, we define the minimum width of the level-set boundary layer as $\Delta = \min_{\mathbf{x}_0, \mathbf{x}_1} \|\mathbf{x}_0 - \mathbf{x}_1\|_2$ such that $\phi(\mathbf{x}_0) = 0$ and $|\phi(\mathbf{x}_1)| = \epsilon$.*

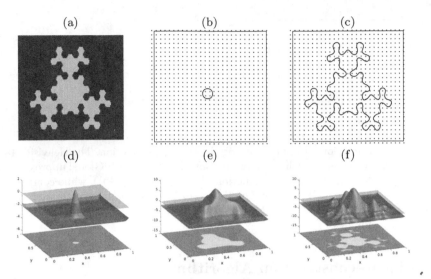

Fig. 2. Parametric level-set method for Discrete tomography problem. (a) True model ($n = 256 \times 256$) (b) RBF grid ($n' = 27 \times 27$) with initial level-set denoted by *green line*, positive and negative RBFs are denoted by *red pluses* and *blue dots* respectively (c) Final level-set denoted by the *green line*, and the corresponding positive and negative RBFs (d) Initial level-set function (e) level-set function after 10 iterations (f) final level-set function after 25 iterations. (Color figure online)

Lemma 1. *For any smooth and compact approximation of the Heaviside function with finite width ϵ, the width of the level-set boundary layer, Δ, satisfies*

$$\Delta \geq \epsilon / \|\nabla \phi\|_{\infty}.$$

Proof. A Taylor series expansion of $\phi(\mathbf{x})$ around $\mathbf{x}_0$ for which $\phi(\mathbf{x}_0) = 0$, we get

$$\phi(\mathbf{x}) = (\mathbf{x} - \mathbf{x}_0)^T \nabla \phi(\boldsymbol{\xi}),$$

with $\boldsymbol{\xi} = t\mathbf{x}_0 + (1 - t)\mathbf{x}$ for some $t \in [0, 1]$. This leads to

$$|\phi(\mathbf{x})| \leq \|\mathbf{x} - \mathbf{x}_0\|_2 \cdot \|\nabla \phi(\boldsymbol{\xi})\|_2 \leq \|\mathbf{x} - \mathbf{x}_0\|_2 \cdot \|\nabla \phi\|_{\infty}.$$

Choosing $\mathbf{x} = \mathbf{x}_1$ with $|\phi(\mathbf{x}_1)| = \epsilon$, we have $\|\mathbf{x}_1 - \mathbf{x}_0\|_2 \geq \epsilon / \|\nabla \phi\|_{\infty}$. Since this holds for all $\mathbf{x}_0, \mathbf{x}_1$ we obtain the desired result.   □

To ensure a minimum width of the boundary layer, Lemma 1 suggest to choose ϵ proportional to $\|\nabla \phi\|_{\infty}$. For computational simplicity, we approximate this using upper and lower bounds [8] and set:

$$\epsilon = \kappa \left(\frac{\max(\phi(\mathbf{x})) - \min(\phi(\mathbf{x}))}{\Delta x} \right) = \kappa \left(\frac{\max(A\boldsymbol{\alpha}) - \min(A\boldsymbol{\alpha})}{\Delta x} \right), \qquad (6)$$

where κ controls the width of level-set boundary in terms of the underlying computational grid. A small value of κ leads to the narrow boundary while big value leads a wide boundary.

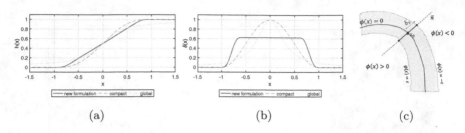

Fig. 3. New formulation for approximating the Heaviside function. The Heaviside functions (a) and corresponding Dirac-Delta functions (b) with $\epsilon = 1$. Global approximation is constructed from inverse tangent function ($\frac{1}{2}(1 + \frac{2}{\pi}\arctan(\pi\frac{x}{\epsilon}))$), while compact one is composed of linear and sinusoid functions. (c) level-set boundary (*orange* region) around zero level-set denoted by *blue* line, n represents the normal direction at $\mathbf{x}_0$. (Color figure online)

3　Joint Reconstruction Algorithm

Reconstructing both the shape and the background parameter can be cast as a bi-level optimization problem

$$\min_{\mathbf{u}_0, \boldsymbol{\alpha}} \left\{ f(\boldsymbol{\alpha}, \mathbf{u}_0) := \tfrac{1}{2}\|W[(1 - h(A\boldsymbol{\alpha}))\mathbf{u}_0 + h(A\boldsymbol{\alpha})u_1] - \mathbf{p}\|_2^2 + \tfrac{\lambda}{2}\|L\mathbf{u}_0\|_2^2 \right\}, \quad (7)$$

where L is of form $[L_x^T \; L_y^T]^T$ where L_x and L_y is the second-order finite-difference operators in the x and y direction, respectively. This optimization problem is *separable*; it is quadratic in $\mathbf{u}_0$ and non-linear in $\boldsymbol{\alpha}$. In order to exploit the fact that the problem has a closed-form solution in $\mathbf{u}_0$ for each $\boldsymbol{\alpha}$, we introduce a reduced objective

$$\overline{f}(\boldsymbol{\alpha}) = \min_{\mathbf{u}_0} f(\boldsymbol{\alpha}, \mathbf{u}_0).$$

The gradient and Hessian of this reduced objective are given by

$$\nabla \overline{f}(\boldsymbol{\alpha}) = \nabla_{\boldsymbol{\alpha}} f(\boldsymbol{\alpha}, \overline{\mathbf{u}}_0), \quad (8)$$

$$\nabla^2 \overline{f}(\boldsymbol{\alpha}) = \nabla_{\boldsymbol{\alpha}}^2 f - \nabla_{\boldsymbol{\alpha}, \mathbf{u}_0}^2 f \left(\nabla_{\mathbf{u}_0}^2 f\right)^{-1} \nabla_{\boldsymbol{\alpha}, \mathbf{u}_0}^2 f, \quad (9)$$

where $\overline{\mathbf{u}}_0 = \mathrm{argmin}_{\mathbf{u}_0} f(\boldsymbol{\alpha}, \mathbf{u}_0)$ [2].

Using a modified Gauss-Newton algorithm to find a minimizer of $\overline{f}$, it leads to the following alternating algorithm

$$\mathbf{u}_0^{(k+1)} = \arg \min_{\mathbf{u}_0} f(\boldsymbol{\alpha}^{(k)}, \mathbf{u}_0) \quad (10)$$

$$\boldsymbol{\alpha}^{(k+1)} = \boldsymbol{\alpha}^{(k)} - \mu_k \left(H_{GN}(f(\boldsymbol{\alpha}^{(k)}))\right)^{-1} \nabla_{\boldsymbol{\alpha}} f(\boldsymbol{\alpha}^{(k)}, \mathbf{u}_0^{(k+1)}), \quad (11)$$

where the expressions for the gradient and Gauss-Newton Hessian are given by (4). Convergence of this alternating approach to a local minimum of (7) is guaranteed as long as the step-length satisfies the strong Wolfe conditions [15].

The reconstruction algorithm based on this iterative scheme is presented in Algorithm 1. We use the LSQR method in step 3, with a pre-defined maximum number of iterations (typically 200) and a tolerance value. A trust-region method is applied to compute $\alpha^{(k+1)}$ in step 4 restricting the conjugate gradient to *only* 10 iterations. We perform a total of $K = 50$ iterations to reconstruct the model.

Algorithm 1. Joint reconstruction algorithm

Require: $\mathbf{p}$ - data, W - forward modeling operator, u_1 - anomaly property, A - RBF Kernel matrix, α_0 - initial RBF weights, κ - Heaviside parameter
Ensure: α_K - final weights, $\mathbf{u}$ - corresponding model
1: **for** $k = 0$ to $K - 1$ **do**
2:     compute Heaviside ϵ from Equation (6)
3:     compute background parameter $\mathbf{u}_0^{(k+1)}$ by solving Eq. (10)
4:     compute level-set parameter $\alpha^{(k+1)}$ from Eq. (11)
5: **end for**
6: compute $\mathbf{u}$ from Eq. (2).

---

4   Numerical Experiments

The numerical experiments are performed on 4 phantoms shown in Fig. 4. We scale the phantoms such that $u_1 = 1$. For the first two phantoms, the background varies from 0 to 0.5, while for the next two, it varies from 0 to 0.8. In order to avoid the *inverse crime*, we use two different discretization schemes for Eq. (1) (namely, line kernel [4] for data generation, and Joseph kernel [4] for forward modeling). We use ASTRA toolbox to compute the forward and backward projections [4]. First, we show the results on the full-view data and later we compare various methods on a limited-angle case.

For the parametric level-set method, we use Wendland compactly supported radial basis functions [8]. The RBF nodes are placed on a 5 times coarser grid than the model grid, with an extension of two points outside the model grid to avoid boundary effects. To constrain the initial level-set boundary to 4 gridpoints, the Heaviside width parameter κ is set to be 0.01.

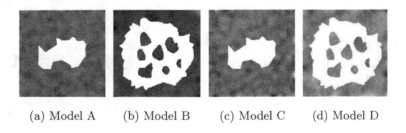

(a) Model A   (b) Model B   (c) Model C   (d) Model D

Fig. 4. Phantoms for Simulations. All the models have resolution of 256×256 pixels.

The level-set parameter α is optimized using the *fminunc* package (trust-region algorithm) in MATLAB. A total of 50 iterations are performed for predicting the α, while 200 iterations are performed for predicting $\mathbf{u}_0(x)$ using LSQR at each step.

4.1 Regularization Parameter Selection

The reconstruction with the proposed algorithm is influenced by the regularization parameter λ (cf. (7)). In general, there are various strategies to choose this parameter, e.g., [14]. As our problem formulation is non-linear, many of these strategies do not apply. Instead we analyze the influence of the regularization parameter numerically as follows.

We define two measures (in the least-squares sense) to quantify the residuals: the data residual (DR), which determines the fit between the true data and reconstructed data, and the model residual (MR), which determines the fit between reconstructed model and true model. Finally, we use the Jaccard index (JI), defined as a similarity coefficient between two sets, to capture the error in the reconstructed shape. In practice, one can only use the data residual to select the regularization parameter λ. It is evident from Fig. 5 that there exists a sufficiently large region of λ for which the reconstructions are equally good. Moreover, this region is easily identifiable from the data residual plot for various λ values.

4.2 Full-View Test

For the full-view case, the projection data is generated on a 256×256 grid with 256 detectors and 180 equidistant projections ($0 \leq \theta \leq \pi$). The Gaussian noise of 10 dB Signal-to-Noise ratio (SNR) is added to the data. The results on phantoms A, B, C and D with the full-view data are shown in Fig. 6. The geometry of the

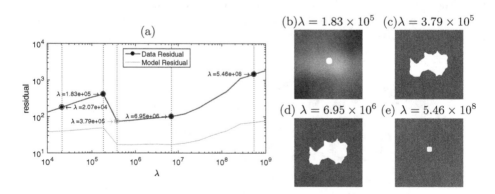

Fig. 5. Variation of residuals with regularization parameter for Tikhonov. Appropriate region for choosing λ exists between 3.79×10^5 and 6.95×10^6. (a) behavior of DR and MR over λ for model A with noisy limited-angle data. (b), (c), (d), (e) show reconstructions for various λ values.

anomalies in all of these reconstructed models are very close to the ground truth, as is indicated by the Jaccard index shown above the figures. The background, though, has been smoothened out with the Tikhonov regularization.

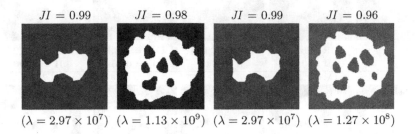

$$JI = 0.99 \qquad JI = 0.98 \qquad JI = 0.99 \qquad JI = 0.96$$

$$(\lambda = 2.97 \times 10^7) \quad (\lambda = 1.13 \times 10^9) \quad (\lambda = 2.97 \times 10^7) \quad (\lambda = 1.27 \times 10^8)$$

Fig. 6. Full-view test: reconstructions with full-view data for the regularization parameter λ shown below it.

4.3 Limited-Angle Test

In this case, we generate synthetic data using *only* 5 projections, namely, $\theta = \{0, \pi/6, \pi/3, \pi/2, 2\pi/3\}$. We add Gaussian noise of 10 dB SNR to the synthetic data. To check the performance of the proposed method, we compare it to Total Variation method [4], DART [3] and its modified version for partially discrete tomography, P-DART [12]. For Total Variation, we determine the shape from the final reconstruction via a simple segmentation step (thresholding). A total of 200 iterations were performed with the Total Variation method and the regularization parameter determined such that it optimally reconstructs the shape. In DART, the background is modeled using 20 discrete grey-values for model A and B, while 30 discrete grey-values for model C and D. True model has been segmented per mentioned grey-values to generate data for DART. 40 DART iterations were performed in each case. For P-DART, a total of 150 iterations were performed.

The results on limited-angle data are presented in Fig. 7. The proposed method is able to capture most of the fine details (evident from the Jaccard Index) in the phantoms even with the very limited data with moderate noise. The P-DART method achieves the least amount of data residual in all the cases, but fails to capture the complete geometry of the anomaly. The Total variation method gives surprisingly good reconstructions of the shape. However, we obtained these results by selecting the best over a large range of regularization parameters. The level-set method consistently gives the best reconstruction of the shape.

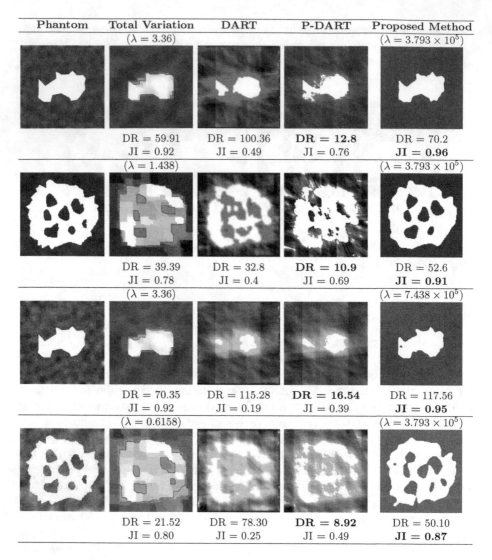

Phantom	Total Variation	DART	P-DART	Proposed Method
	($\lambda = 3.36$)			($\lambda = 3.793 \times 10^5$)
	DR = 59.91 JI = 0.92	DR = 100.36 JI = 0.49	**DR = 12.8** JI = 0.76	DR = 70.2 **JI = 0.96**
	($\lambda = 1.438$)			($\lambda = 3.793 \times 10^5$)
	DR = 39.39 JI = 0.78	DR = 32.8 JI = 0.4	**DR = 10.9** JI = 0.69	DR = 52.6 **JI = 0.91**
	($\lambda = 3.36$)			($\lambda = 7.438 \times 10^5$)
	DR = 70.35 JI = 0.92	DR = 115.28 JI = 0.19	**DR = 16.54** JI = 0.39	DR = 117.56 **JI = 0.95**
	($\lambda = 0.6158$)			($\lambda = 3.793 \times 10^5$)
	DR = 21.52 JI = 0.80	DR = 78.30 JI = 0.25	**DR = 8.92** JI = 0.49	DR = 50.10 **JI = 0.87**

Fig. 7. Reconstructions with noisy limited data. The first column shows the true models, while the last 4 columns show the reconstructions with various methods. Red dotted line shows the contour of the segmented model in Total Variation method. Different measures are also shown below each reconstructed model. (Color figure online)

5   Conclusions and Discussion

We discussed a parametric level-set method for partially discrete tomography. We model such objects as a constant-valued shape embedded in a continuously varying background. The shape is represented using a level-set function, which in turn is represented using radial basis functions. The reconstruction problem is posed as a bi-level optimization problem for the background and level-set parameters. This reconstruction problem can be efficiently solved using a variable projection approach, where the shape is iteratively updated. Each iteration requires a full reconstruction of the background. The algorithm includes some practical heuristics for choosing various parameters that are introduced as part of the parametric level-set method. Numerical experiments on a few numerical phantoms show that the proposed approach can outperform other popular methods for (partially) discrete tomography in terms of the reconstruction error. As the proposed algorithm requires repeated full reconstructions, it is currently an order of magnitude slower than the other methods. Future research is directed at making the method more efficient.

Acknowledgments. This work is part of the Industrial Partnership Programme (IPP) 'Computational sciences for energy research' of the Foundation for Fundamental Research on Matter (FOM), which is part of the Netherlands Organisation for Scientific Research (NWO). This research programme is co-financed by Shell Global Solutions International B.V. The second and third authors are financially supported by the NWO as part of research programmes 613.009.032 and 639.073.506 respectively.

References

1. Aghasi, A., Kilmer, M., Miller, E.L.: Parametric level set methods for inverse problems. SIAM J. Imaging Sci. **4**(2), 618–650 (2011)
2. Aravkin, A.Y., Van Leeuwen, T.: Estimating nuisance parameters in inverse problems. Inverse Prob. **28**(11), 115016 (2012)
3. Batenburg, K.J., Sijbers, J.: Dart: a practical reconstruction algorithm for discrete tomography. IEEE Trans. Image Process. **20**(9), 2542–2553 (2011)
4. Bleichrodt, F., van Leeuwen, T., Palenstijn, W.J., van Aarle, W., Sijbers, J., Batenburg, K.J.: Easy implementation of advanced tomography algorithms using the astra toolbox with spot operators. Numer. Algorithms **71**(3), 673–697 (2016)
5. Burger, M.: A level set method for inverse problems. Inverse Prob. **17**(5), 1327 (2001)
6. Chambolle, A., Pock, T.: A first-order primal-dual algorithm for convex problems with applications to imaging. J. Math. Imaging Vis. **40**(1), 120–145 (2011)
7. Dorn, O., Lesselier, D.: Level set methods for inverse scattering. Inverse Prob. **22**(4), R67 (2006)
8. Kadu, A., Van Leeuwen, T., Mulder, W.A.: Salt reconstruction in full waveform inversion with a parametric level-set method. IEEE Trans. Comput. Imaging **3**(2), 305–315 (2016)
9. Kak, A.C., Slaney, M.: Principles of Computerized Tomographic Imaging. SIAM, Philadelphia (2001)
10. Klann, E., Ramlau, R., Ring, W.: A mumford-shah level-set approach for the inversion and segmentation of SPECT/CT data. Inverse Probl. Imaging **5**(1), 137–166 (2011)
11. Osher, S., Fedkiw, R.: Level Set Methods and Dynamic Implicit Surfaces, vol. 153. Springer, New York (2006). doi:10.1007/b98879
12. Roelandts, T., Batenburg, K., Biermans, E., Kübel, C., Bals, S., Sijbers, J.: Accurate segmentation of dense nanoparticles by partially discrete electron tomography. Ultramicroscopy **114**, 96–105 (2012)

13. Sidky, E.Y., Pan, X.: Image reconstruction in circular cone-beam computed tomography by constrained, total-variation minimization. Phys. Med. Biol. **53**(17), 4777 (2008)
14. Thompson, A.M., Brown, J.C., Kay, J.W., Titterington, D.M.: A study of methods of choosing the smoothing parameter in image restoration by regularization. IEEE Trans. Pattern Anal. Mach. Intell. **13**(4), 326–339 (1991)
15. Wright, S., Nocedal, J.: Numerical Optimization. Springer Series in Operations Research and Financial Engineering. Springer, New York (1999). doi:10.1007/b98874

Maximal N-Ghosts and Minimal Information Recovery from N Projected Views of an Array

Imants Svalbe and Matthew Ceko[✉]

School of Physics and Astronomy, Monash University, Melbourne, Australia
{imants.svalbe,matthew.ceko}@monash.edu

Abstract. Digital data is now frequently stored privately and securely in the "cloud". One repository stores several different sets of projections of the original data. Each set is kept on a separate, remote server. The information residing on any local server, purposefully, is insufficient to exactly reconstruct the full data. Here we ask: how much useful information can be gleaned from one local projection set? We answer that question by examining projection ghosts. A ghost is an assembly of signed pixels positioned to have zero sums along chosen discrete directions. The shape of each ghost is defined uniquely by its distinct set of N directions. An N-ghost with a shape that fits snuggly inside the boundary of an array defines precisely all of the array locations that cannot be exactly reconstructed from those N projected views. Minimal N-ghosts contain $2N$ elements: one $(-1/+1)$ pair is needed for zero-sums along each of the N directions. Maximal N-ghosts contain 2^N elements: the number of $(-1/+1)$ elements can double N times, once for each of the N directions. Here we construct maximal N-ghosts that cover a large area of their bounding array. By maximising the number of unrecoverable ghosted pixels, we minimise the information that can be reconstructed from N projected views. We show that at least 60% of the data in an $m \times m$ array, for $m \approx N^2/4$, can be masked or made "unreadable", for a maximal set of N noise-free projections of the original $m \times m$ data.

Keywords: Discrete projection · Cloud storage · Data security · Mojette transform · Tomographic reconstruction

1 Introduction

In tomography, absorption of beams with finite width are measured passing through a continuum of absorbing atoms. Absorption profiles are recorded in separate bins to be reconstructed on discrete lattices, and therefore the reconstructions must be approximated. For projections on a discrete array, each projected ray can simply sum the intensity at each array site. On a discrete array, projection angles are not continuous. The rays follow paths that link regularly staggered lattice sites. For a square lattice we take p steps across and q steps down to define angle $p{:}q$, where p and q are relatively prime, giving angle $\theta_{p:q} = \tan^{-1}(q/p)$. Sums of pixels that lie on $p{:}q$ lines are known as Dirac-Mojette projections

© Springer International Publishing AG 2017
W.G. Kropatsch et al. (Eds.): DGCI 2017, LNCS 10502, pp. 135–146, 2017.
DOI: 10.1007/978-3-319-66272-5_12

(also called X-rays). If p takes signed integer values, $0° \leq \theta_{p:q} < 180°$, the set of discrete angles native to a $P \times Q$ pixel array can be found using the Farey-Harros set of fractions, as F_{PQ} [10].

An N-ghost is a set of signed elements placed in a particular pattern on a digital array. For each of N chosen discrete angles, any parallel rays that pass across these patterns will always intersect pairs of ghost elements that sum to exactly zero. By design, ghosts remain invisible when viewed from any of these N angles. Ghosts are also known as switching elements or the null set. A ghost whose $(+/-)$ values are scaled by a constant, or a ghost with its signs reversed, remains a ghost. Exact reconstruction of any array of elements from N (noise-free) projected views is possible if and only if their N-ghost extends beyond the array boundary when the ghost is superimposed over the array. This is the Katz criterion [12]. An N-ghost with any of its elements positioned outside the image space cannot be superimposed on the image and also remain invisible in all N directions. For a set of projections $\{p_i{:}q_i\}$, $1 \leq i \leq N$, on a $P \times Q$ image, the Katz criterion permits exact reconstruction of that array if and only if

$$\sum |p_i| \geq P \text{ or } \sum |q_i| \geq Q. \tag{1}$$

This paper considers methods to make tomographic reconstruction as difficult as possible. This inverts the usual problem of choosing sets of projection angles to facilitate image recovery. A ghost projection contains no new information about internal parts of an object, because the N projections of those parts, by design, have zero sums. An N-ghost adds nothing to the content for any object viewed in those N directions.

The aim here is to minimise the information that can be recovered from a partial set of array projections. The motivation of this work is to improve the security and privacy of information that is stored as partial sets of projected data in a distributed manner across multiple servers. For example, the RozoFS system [18], or for digital data communications via packet networks [17].

This paper is structured as follows: Sect. 2 provides a short review of the theory and construction of ghost projections, including methods to make minimal ghosts and methods for the exact reconstruction of images from discrete noise-free projections. Section 3 presents techniques to construct maximal N-ghosts and discusses their properties. It shows how well these ghosts mask given locations to inhibit exact array recovery. Section 4 discusses how this work might be extended to cover a wider range of practical applications, whilst Sect. 5 provides a brief summary and conclusion.

2   Ghost Review

2.1   Ghosts and Tomography

The notion of ghosts arose with the development of practical schemes to implement the inverse Radon transform. The central slice theorem makes it possible to recover internal object details from the intensity profiles of beams transmitted through the object at several angles. Tomography has found rapid application in diverse areas such as astrophysics, archaeology, biology, medical imaging, geology, part inspection for manufacturing and security scanning.

Early work on ghosts for image reconstruction in the digital domain (using discrete data from finite detector arrays) was presented by [12,13,15]. The non-unique mapping between arrays and their projections, in the mathematical sense, was flagged by G.G. Lorenz as early as 1949 [14], as did related work on the recovery of matrices from their row and column sums. It was soon realised that the uniqueness of a reconstructed image depends on the absence of ghosts (or switching elements). The presence of a ghost means the constraints of any finite projection set may be equally well satisfied by arbitrary multiple (i.e. differently ghosted) objects.

Recent work on ghosts has aimed to minimise the redundant information in sets of projections. The objective was to achieve shorter acquisition times and require fewer probing rays [5,11]. Along the way, many valuable theoretical links have been established, such as the use of polynomials, cyclotomic equations and the Vandermonde matrix to represent discrete projection data. A wide variety of methods to reconstruct images from projections have been developed [1,4,16]. Work has been done to recover images from sets with several missing projections that has resulted in several image de-ghosting algorithms [6], or to recover data from lossy transmission of packets across communication networks [16].

2.2   Constructing N-Ghosts

An elementary ghost for a discrete angle p:q, denoted $g_{p:q}$, is a pair of signed points separated by p columns and q rows. This is defined as

$$g_{p:q}(i,j) = \begin{cases} +1 & \text{if } (i,j) = (0,0) \\ -1 & \text{if } (i,j) = (q,p) \\ 0 & \text{elsewhere.} \end{cases} \tag{2}$$

A ghost G, over N directions can then be constructed through $N-1$ discrete convolutions of N elementary ghosts.

$$G = g_{p_1:q_1} * \cdots * g_{p_N:q_N} \tag{3}$$

Geometrically, this corresponds to starting with a pair of oppositely signed points that define a ghost in the first chosen direction and then dilating, (add a translated copy with reversed signs) that ghost in the second selected direction. The result for the composite 2-ghost is then dilated in the third direction, and so on, for N directions. Figure 1 shows this process to make the 4-ghost with directions $\{1{:}0, 0{:}1, 1{:}1, -1{:}1\}$.

The bounding polygon of the N-ghost, given by the convex hull of ghost points, is then comprised of $2N$ vectors from the angles p:q (including reflections). Therefore the convex hull of all N-ghosts built through convolutions of elementary ghosts are $180°$ symmetric. Ghosts defined through Eq. (3) are the

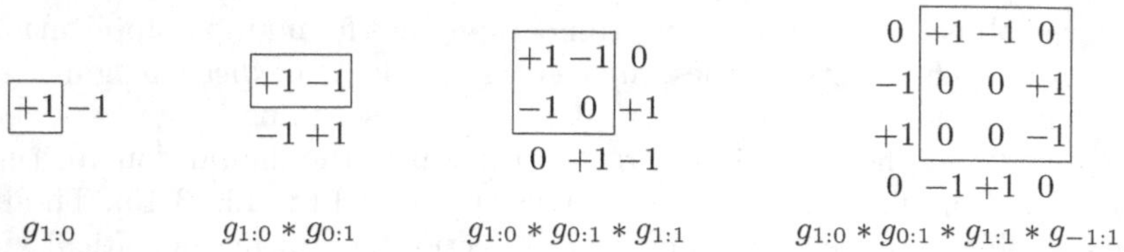

$$
\begin{array}{cc} \boxed{+1}\,-1 \end{array}
\qquad
\begin{array}{cc} \boxed{+1\ -1} \\ -1\ +1 \end{array}
\qquad
\begin{array}{ccc} \boxed{+1\ -1}\ \ 0 \\ -1\ \ 0\ \boxed{+1} \\ 0\ +1\ -1 \end{array}
\qquad
\begin{array}{cccc} 0\ \boxed{+1\ -1\ \ 0} \\ -1\ \boxed{0\ \ 0\ +1} \\ +1\ \boxed{0\ \ 0\ -1} \\ 0\ -1\ +1\ \ 0 \end{array}
$$

$$g_{1:0} \qquad\qquad g_{1:0} * g_{0:1} \qquad g_{1:0} * g_{0:1} * g_{1:1} \qquad g_{1:0} * g_{0:1} * g_{1:1} * g_{-1:1}$$

Fig. 1. Construction of an N-ghost by sequential dilation. Left to right: the 1D ghost with direction 1:0, dilated in direction 0:1, then 1:1 followed by -1:1 to form the $N = 4$ (minimal) ghost with 8 non-zero elements. Boxes show the location of the previous ghost before sign reversal and shift by $p{:}q$.

minimal ghost configurations for given directions $\{p_i{:}q_i\}$ [11]. It is also possible to construct ghosts through other methods, such as using U-polygons [8]. However, in this work we consider only ghosts defined by (3) as these are the ghosts that arise as errors in Mojette reconstruction.

2.3   Reconstruction from Projections

The Dirac-Mojette projections of known, fixed elements of a discrete array create noise-free projections. Such arrays can be reconstructed exactly from a sufficient set of their Mojette projections using the Corner Based Inversion (CBI) method of [16]. This method begins by matching pixels to pairs of projection bins from the corners of the image. These rays contain one or two summed elements. We can iteratively subtract the un-summed pixel values from the summed bins to progressively unpack the content of projected values, working from the array edges into the array centre. CBI is not robust even for small, isolated levels of noise. Any error, even at one bin location, will propagate into all subsequent unpacked values, as studied in [2].

For a set of N sub-Katz projections, where $\sum|p_i| < P$ and $\sum|q_i| < Q$, the CBI method correctly unpacks as many image pixels as possible, leaving unknown values for the remaining pixels. The unrecovered data corresponds to the $(P - \sum|p_i|)(Q - \sum|q_i|)$ possible translations of N-ghost pixels inside the reconstructed image [3,7]. Figure 2 demonstrates the simplest case where the N-ghost is the same size as the image, and therefore the errors are simply an embedded N-ghost with scaled values.

There are also additional "shadow" pixels located along the N directions that are blocked by a ghost element and cannot be further unpacked (see Fig. 4). All values that lie within the convex hull of translated ghost pixels are initially unsolved. Using ghost solving [6] or by assigning an arbitrary (hence usually incorrect) value to one of these ghost pixels, it is possible to re-apply CBI to remove many of the shadow pixels (Fig. 2d). The remaining unreconstructed pixels are, at least, all of the non-zero positions of the N-ghost. We wish to choose a set of N directions that maximises the number of these N-ghost pixels and thus maximise the number of unrecoverable array pixels.

$$
\begin{array}{r|l}
\mathbf{0\!:\!1} & 84 \ \ 110 \ \ 83 \ \ 83 \ \ 105 \\
\mathbf{1\!:\!0} & 72 \ \ 58 \ \ \ \ 53 \ \ 81 \ \ 106 \ \ 95 \\
\mathbf{1\!:\!1} & 4 \ \ \ \ 27 \ \ \ \ 46 \ \ 54 \ \ 89 \ \ \ \ \ 48 \ \ 87 \ \ 48 \ \ 42 \ \ 20 \\
\mathbf{-1\!:\!1} & 25 \ \ 57 \ \ \ \ 30 \ \ 76 \ \ 42 \ \ \ \ \ 79 \ \ 76 \ \ 48 \ \ 24 \ \ 8 \\
\mathbf{1\!:\!2} & 4 \ \ \ \ 3 \ \ \ \ \ \ 31 \ \ 37 \ \ 49 \ \ \ \ \ 72 \ \ 37 \ \ 38 \ \ 45 \ \ 27 \ \ 37 \ \ 37 \ \ 28 \ \ 20
\end{array}
$$

a)

8	19	17	24	4
5	10	18	22	3
21	11	12	2	7
23	13	1	29	15
28	16	26	6	30
20	14	9	27	25

b)

8	0	0	24	4
0	0	0	0	3
0	0	0	0	7
23	0	0	0	0
28	0	0	0	0
20	14	0	0	25

c)

8	10	26	24	4
14	10	18	13	3
12	11	21	2	7
23	13	−8	29	24
28	25	26	6	21
20	14	0	36	25

d)

Fig. 2. CBI for a 5×6 array. (a) $N = 5$ projections $\{0\!:\!1, 1\!:\!0, 1\!:\!1, -1\!:\!1, 1\!:\!2\}$. (b) The original array values. (c) The raw CBI result using the array projections. Initially only the corner values of the array are solved. The ghost pixels for this $N = 5$ set are underlined. (d) Setting (arbitrarily) the middle ghost pixel on the bottom row to $\underline{0}$ and re-applying CBI yields the remaining array values. Note the underlined values at the locations of the ghost elements are all wrong (here by either $+9$ or -9). Here $18/30 = 60\%$ of the array can be recovered, 40% cannot.

2.4 Minimal Ghosts

A projection ghost increases the pixel sum (by $+v$) of an object at one point along a ray, whilst equally decreasing the object sum (by $-v$) at some other point along the same ray, leaving the total number of pixels unchanged. Clearly the object has been changed, but the net projected view remains the same. To do this in N-directions requires at least N increments and N decrements, one pair for each of the N directions.

A 2D ghost has a finite area and enclosing perimeter. Each ghost element located on the perimeter of the ghost area must be linked to at least one other perimeter ghost element of the opposite sign, by a line $p_i\!:\!q_i$ in one of the N directions. The same must happen for the parallel ray passing through the opposite edge of the perimeter. The polygonal boundary formed by the vertices of the convex hull of any ghost is then symmetric and always contains at least $2N$ points. A minimal N-ghost thus traces the perimeter of a symmetric polygon, with $2N$ sides, made from just N $(+1)$ and N (-1) entries. The area inside the perimeter of a minimal ghost polygon is free of ghost elements (Fig. 3a).

Minimal ghosts in 2D only exist for $N = 1, 2, 3, 4, 6$ [9]. All other N-ghosts contain additional points on their polygon interiors. Near-minimal ghosts for $N = 5$ and all $N > 6$ contain as close as possible to $2N$ points. Near-minimal ghosts are comprised of a symmetric bounding perimeter points, but they always contain interior points, usually most of these have value ± 1. Fig. 3b shows the near-minimal case for $N = 5$, this is the same ghost seen in Fig. 2.

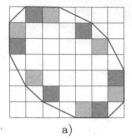 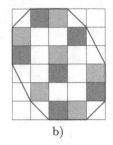 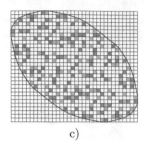

a)   b)   c)

Fig. 3. (a) A minimal ghost for $N = 6$ has 6 (+1) and 6 (−1) points on its boundary and an empty interior. (b) A near-minimal ghost for $N = 5$ contains 12 non-zero points, made up of 10 perimeter points and 2 points interior to the bounding polygon. (c) A 214 $N = 16$ point near-minimal ghost for in a 34×30 array, 108 (−1), 104 (+1) and 2 (+2) values, ghost volume 655.

Ghost elements that have the same sign when overlapped at any location by the dilation process can be summed to produce "grey" ghosts, with internal integer entries of $\pm 2, ..., \pm n$. The value of n for near-minimal ghosts grows rapidly with N [19]. Grey values where summed ghost points overlap are known as multiple points. For grey ghosts, we can find sets of angles that minimise the N-ghost "volume". Here, the volume is computed as the area of the bounding $2N$-sided polygon multiplied by its height. The height can be taken as the mean absolute grey scale of all the ± 1 to $\pm n$ entries (Fig. 4a). There is experimental evidence that ghost volumes have an asymptotic limiting density [19], i.e. that all 2D ghosts have a minimum 3D volume. The height of a grayscale ghost can be traded against the area of its footprint. Ghost elements that have opposite signs at any location when they are overlapped by the dilation process will cancel and thus can reduce the count of internal points (as shown in Fig. 3b and c).

The theoretical design and practical construction of projection sets that produce minimal ghosts is important. It helps to minimise the size and number of projections required to meet the Katz criterion and thus permit exact image reconstruction. Any set of N projections will fail to exactly reconstruct any image that is larger than the size of the corresponding N-ghost. As a minimal (or near-minimal) N-ghost contains at least $2N$ elements, at least $2N$ pixels of any image larger than this ghost cannot be (exactly) reconstructed. As the aim here is to maximise the number of un-reconstructed pixels, we need to build N-ghosts that contain not as few but as many ghost elements as possible.

3   Maximal Ghosts

Minimal (and near-minimal) ghosts are important for designing tomographic systems in which we would like to make reconstruction as easy as possible. However, for applications such as data fragmentation for distributed cloud storage, we wish to do the opposite. If any server is compromised, and insufficient projection data is obtained for an object, the amount of data that can be reconstructed should be minimised. Maximal ghosts contain exactly 2^N ghost points, including multiple points. Therefore, they create a maximum amount of error in reconstructed data. These 2^N points should also cover the maximum area of their bounding array, to minimise the regions of uniqueness that can be exactly reconstructed. Although the same amount of information is missing for any set of projections where $\sum |p_i|$ and $\sum |q_i|$ are equal, the data that can be exactly reconstructed without said information depends on the ghost constructed by the set of directions.

3.1   Maximal Ghost Examples

The maximal and minimal ghosts are identical for $N = 1$ (for example $\{0{:}1\}$) and $N = 2$ (for example $\{0{:}1, 1{:}0\}$). Any affine rotation of these ghosts is also a ghost. For $N = 3$, a minimal ghost set is $\{0{:}1, 1{:}0, 1{:}1\}$ or $\{0{:}1, 1{:}0, -1{:}1\}$, with each 3-ghost having $2 \cdot 3 = 6$ boundary points and no interior points. In contrast, the set $\{0{:}1, 1{:}1, -1{:}1\}$ forms a compact 3-maximal ghost comprised of $2^3 = 8$ points, with no holes in its interior.

We assume here that the original data is an array that fits inside an approximately square $m \times m$ or $m \times (m + 1)$ region, where the more valuable content (when the data is image-like) can be reasonably expected to lie mostly across the centre region of the array. The maximal N-ghosts are made here to have matching size $m \times m$, to overlap the given data as much as possible.

When the data is of text or other formats, long, thin, rectangular boxes, of size $n \times m$, with $n \ll m$ may be more appropriate. The shape of the maximal ghosts then needs to be adapted to accommodate the appropriate image shape. Any discrete angle has four-fold symmetric angles, $p{:}q$, $-p{:}q$, $q{:}p$ and $-q{:}p$, except for the pairs $\{0{:}1, 1{:}0, 1{:}1, -1{:}1\}$. Changing the symmetry of the $p{:}q$ angles that are selected within the set N can be used to deliberately increase or shorten the growth of a ghost preferentially along the x- or y-axis, thus changing the aspect ratio of the part of the image they reconstruct.

The number of ghost elements double after each dilation into a new $p{:}q$ direction. We can easily achieve the maximal count of ghost elements (2^N) by choosing p and q to be large enough that the ghosts never overlap under successive dilations. However, this approach also introduces large gaps between the dilated patterns. These gaps will reduce the masking efficiency as many unghosted internal pixels will be recoverable using simple CBI reconstruction. Our refined aim is then to have the maximal number of ghost elements ghost elements closely packed, with few or no holes inside the ghost perimeter.

3.2   Constructing Maximal N-Ghosts

The projection sets $\{0{:}1, 1{:}0\}$ and $\{1{:}0, 1{:}1, -1{:}1\}$ construct maximal N-ghosts for $N = 2$ and $N = 3$ respectively, as shown in Fig. 4. Therefore, we will use these ghost as starting points from which we can build larger N-ghosts such that overlapping points never cancel over successive dilation directions.

a)
$$\begin{array}{cc} +1 & -1 \\ -1 & +1 \end{array}$$

b)
$$\begin{array}{cccc} 0 & +1 & -1 & 0 \\ -1 & +1 & -1 & +1 \\ 0 & +1 & -1 & 0 \end{array}$$

Fig. 4. (a) Maximal ghost for $N = 2$ given by $\{0{:}1, 1{:}0\}$. (b) Maximal ghost for $N = 3$ given by $\{1{:}0, 1{:}1, -1{:}1\}$.

Notice that the pattern of positive and negative values for the $N = 2$ ghost form a checkerboard. We now look for another angle, such that convolving its elementary ghost does not result in the cancelling of any ghost points. This corresponds to finding the translation for which placing the $N = 2$ ghost with reversed signs never aligns positive with negative points. If we choose p to be even, this occurs when q is odd. Likewise, if q is even, then p must be odd. Due to the alignment of positive and negative ghost points over the new dilation, the resulting $N = 3$ ghost is also a checkerboard pattern (excluding zero points). Hence this can be repeated to obtain any maximal N-ghost with 2^N points.

Similar reasoning can be used to construct N-ghosts starting from the set $\{1{:}0, 1{:}1, -1{:}1\}$. For this set, the columns alternate between negative and positive. Therefore, to construct any maximal N-ghost starting from this set, we must ensure the reversed, translated columns align with similar signs. As long as p is chosen to be odd, then ghost points will never cancel. Equivalently, this can also be performed using $\{0{:}1, 1{:}1, -1{:}1\}$, by choosing odd q.

For maximally inefficient reconstruction, we wish to cover as many possible pixel sites with non-zero ghost values. We have seen that the ghost from the projection set $\{0{:}1, 1{:}0\}$ covers its entire bounding array with non-zero values due to the directions aligning with the lattice. Since for $N > 2$ we can no longer choose directions that align with the grid, we must choose angles that approximate it. Therefore we choose $\pm p{:}1 \pm 1{:}q$ for the largest valid p and q possible without disconnecting ghost points. Then maximal ghosts for even N can be constructed using symmetric discrete projection sets

$$S_E = \{0{:}1, 1{:}0, 1{:}2n, -1{:}2n, 2n{:}1, -2n{:}1\} \tag{4}$$

or subsets thereof for $n \in \mathbb{N}$. Similarly, odd N maximal ghosts can be built using the projection set

$$S_O = \{0{:}1, 1{:}1, -1{:}1, 1{:}(2n+1), -1{:}(2n+1), (2n+1){:}1, -(2n+1){:}1\} \tag{5}$$

or any subsets with $n \in \mathbb{N}$. S_E and S_O in their entirety produce ghosts that are extremely compact and dense. They also result in some ambiguity in choice of angles when $N \neq 4n + 2$ for even N, and $N \neq 4n + 3$ for odd N. Therefore we offer some subsets G_N as examples, to produce approximately square ghosts. For even N, we can choose

$$G_N = \{0{:}1, 1{:}0, 1{:}2, 2{:}1, 1{:}4, 4{:}1, ..., 1{:}(N-2), (N-2){:}1\} \tag{6}$$

$$\text{or } G_N = \{0{:}1, 1{:}0, 1{:}2, -2{:}1, 1{:}4, -4{:}1, ..., 1{:}(N-2), -(N-2){:}1\}. \tag{7}$$

Note (6) and (7) result in square $m \times m$ ghosts. For odd N maximal ghosts, we choose a similar symmetric pattern that give $m \times (m+1)$ ghosts.

$$G_N = \{0{:}1, 1{:}1, -1{:}1, 1{:}3, 3{:}1, 1{:}5, 5{:}1, ..., 1{:}(N-2), (N-2){:}1\} \tag{8}$$

$$\text{or } G_N = \{0{:}1, 1{:}1, -1{:}1, 1{:}3, -3{:}1, 1{:}5, -5{:}1, ..., 1{:}(N-2), -(N-2){:}1\} \tag{9}$$

The choice (6) constructs ghosts that have no holes inside the $2N$ point symmetric polygon that forms the boundary of any N-ghost. All interior points are filled by non-zero ghost elements. Using choice (7) inserts just four interior holes, these holes always lie near the corners of the polygon. Examples of these maximal ghosts, for $N = 8$ and $N = 9$, are shown in Fig. 5. For odd N, the situation is similar, although the boundaries are less sharply defined, with the arrays from option (9) having 8 interior holes. As N grows, the largest angle in the $2N$-polygon perimeter closer approximates 1:0 (or 0:1). Hence the ghost boundary will approach the array boundary, and the fill will approach 100%.

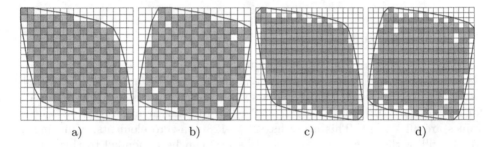

Fig. 5. Maximal 8-ghosts as 17×17 arrays, and maximal 9-ghosts as 21×22 arrays with. (a) G_8 using Eq. (6), has 191 non-zero elements. (b) G_8 using Eq. (7), has 205 non-zero elements, with 4 holes. (c) G_9 using Eq. (8), contains 312 non-zero elements. (d) G_9 using Eq. (9), contains 322 non-zero elements, with 8 interior holes near the corners of the array.

Tables 1 and 2 give the image sizes and percentage of the array that the N-ghost fills for even and odd N, respectively. The size of the array can be computed from the sum of the absolute values of p and q values in the angle set. For N even (6), (7), this sum gives array size $m = N(N-2)/4 + (N-2)/2 + 2 = (N^2 + 4)$. Similarly $m = (N^2 + 5)/4$ for N odd (8), (9). The entry "count" in Tables 1 and 2 is the number of non-zero elements in the ghost, the fill factor is the percentage of non-zero pixels in the full array. The pair of rows in the table at each N show results using choice (6) then (7) for even N, and (8) then (9) for odd N.

To maintain closely spaced elements with few holes, the overlap of ghost elements will also increase the grey level of the ghost. The range of greys in a ghost must be less than the grey values in the image. The sum of the ghost values is always zero. The sum of the absolute values on the ghost array is 2^N as elements at each location accumulate from the overlap of N successive dilations.

Table 1. N-ghost array sizes, redundancy and fill factors for even $N = 4$ to $N = 14$.

N	Array size	Range	Count	Fill (%)
4	5 × 5 (6)	$\{-1, ..., 2\}$	15	60.00
	(7)	$\{-1, ..., 1\}$	16	64.00
6	10 × 10	$\{-1, ..., 1\}$	60	60.00
		$\{-1, ..., 1\}$	64	64.00
8	17 × 17	$\{-3, ..., 4\}$	191	61.09
		$\{-4, ..., 3\}$	205	70.09
10	26 × 26	$\{-5, ..., 5\}$	484	71.60
		$\{-4, ..., 4\}$	512	75.74
12	37 × 37	$\{-10, ..., 10\}$	1039	75.89
		$\{-9, ..., 9\}$	1085	79.25
14	50 × 50	$\{-26, ..., 26\}$	1980	79.20
		$\{-25, ..., 25\}$	2048	81.92

Table 2. N-ghost array sizes, redundancy and fill factors for odd $N = 5$ to $N = 15$.

N	Array size	Range	Count	Fill (%)
5	7 × 8 (8)	$\{-1, ..., 1\}$	32	57.14
	(9)	$\{-1, ..., 1\}$	32	57.14
7	13 × 14	$\{-2, ..., 2\}$	112	61.54
		$\{-2, ..., 2\}$	112	61.54
9	21 × 22	$\{-4, ..., 4\}$	312	67.53
		$\{-4, ..., 4\}$	322	69.70
11	31 × 32	$\{-7, ..., 7\}$	720	72.58
		$\{-8, ..., 8\}$	744	75.00
13	43 × 44	$\{-17, ..., 17\}$	1448	76.53
		$\{-18, ..., 18\}$	1490	78.75
15	57 × 58	$\{-44, ..., 44\}$	2632	79.61
		$\{-43, ..., 43\}$	2696	81.55

3.3   Binary Maximal Ghosts

Maximal N-ghosts with an alphabet of just ±1 elements can also be constructed. Here the 7 angles were chosen, by eye, to place successive dilations so they do not cause any overlap between ghosts points while not creating holes. This dense set for $N = 7$, $p{:}q = \{0{:}1, 1{:}0, 1{:}2, -1{:}2, -3{:}2, 5{:}2, -1{:}6\}$, has array size 16×13, where values are all $-1/+1$. This array has $2^7 = 128$ non-zero elements, achieving a 61.54% fill of the array area. The angle $-11{:}2$ can be appended to this set to construct a maximal binary 8-ghost. It can be seen in Fig. 6 that this translates the binary 7-ghost to fit with itself in a jigsaw-like fashion to produce a fill of 59.26%. It is not yet known if maximal binary ghosts exist for all N.

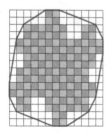 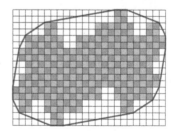

Fig. 6. A maximal 7-ghost, $p{:}q = \{0{:}1, 1{:}0, 2{:}1, -2{:}1, 2{:}3, 6{:}1, -10{:}1\}$ (left), comprised of 128 ± 1 points. The direction $-11{:}2$ is added to give a maximal binary 8-ghost (right).

4   Further Work

We have, so far, only tested how these N-ghosts affect array recovery using the CBI reconstruction method. We next want to apply several statistical/iterative inversion methods to evaluate the potential for "exact" or nearly-exact recoverability of arrays from "ghosted" partial sets of projection data. In many cases, the stored data will have already been encrypted, so any inversion process is unlikely to be aided by using structural or statistical clues from any sections of partially recovered data. However the part played by the range of data values needs further scrutiny, as the constraints on valid reconstructions are tighter (even for random arrays) when the data values are drawn from a restricted alphabet, for example for binary or ternary arrays.

The maximal N-ghost examples given in Fig. 6 was built by hand. It would be useful to devise an algorithm to produce ghosts for any N that are comprised of only $-1/+1$ values and at the same time ensure the lowest possible number of zero elements (holes) occur inside the bounding polygon. It may be possible to also control the location of holes, as they too can be seen as parts of a complementary ghost structure that is enclosed by the same N angles.

When two or more partial sets of projections that are kept on separate servers are pooled, exact and efficient recovery of the exact data must still be possible. Individual sets of N-ghosts must then be distinct, but compatible in the joint reconstruction sense. Maximal ghosts for even N made from Eqs. (6) or (7) would pair well with the odd ghosts for $N \pm 1$ (via Eqs. (8) or (9)), as they have just the 0:1 projection in common and these partial sets are all similarly robust.

5   Summary

This paper considers methods to assemble sets of N discrete projections $\{p_i{:}q_i\}$ that are insufficient for an exact reconstruction, and structured to yield the least possible information about the original array. The unrecoverable pixels of an array are shown to be the elements of the locally-held ghost for those N directions. The geometry of elementary ghost convolutions is used to construct maximal N-ghosts that are comprised of symmetric sets of discrete angles built from 1:n for $n < N$, where n and N are either odd or even. These ghosts maximise the fraction (f) of an array that is not able to be reconstructed from those N projected views. In practice, we can assemble sets where $f \geq 60\%$ for $N \geq 4$.

Acknowledgments. The School of Physics and Astronomy at Monash University, Australia, has supported and provided funds for this work. M.C. has the support of the Australian government's Research Training Program (RTP) and the J.L. William scholarship from the School of Physics and Astronomy at Monash University.

References

1. van Aarle, W., Palenstijn, W.J., De Beenhouwer, J., Altantzis, T., Bals, S., Batenburg, K.J., Sijbers, J.: The ASTRA toolbox: a platform for advanced algorithm development in electron tomography. Ultramicroscopy **157**, 35–47 (2015)
2. Alpers, A., Gritzmann, P.: On stability, error correction, and noise compensation in discrete tomography. SIAM J. Discrete Math. **20**(1), 227–239 (2006)
3. Alpers, A., Larman, D.G.: The smallest sets of points not determined by their X-rays. Bull. Lond. Math. Soc. **47**(1), 171–176 (2015)
4. Batenburg, K.J., Plantagie, L.: Fast approximation of algebraic reconstruction methods for tomography. IEEE Trans. Image Process. **21**(8), 3648–3658 (2012)
5. Brunetti, S., Dulio, P., Hajdu, L., Peri, C.: Ghosts in discrete tomography. J. Math. Imaging Vis. **53**(2), 210–224 (2015)
6. Chandra, S., Svalbe, I., Guédon, J., Kingston, A., Normand, N.: Recovering missing slices of the discrete Fourier transform using ghosts. IEEE Trans. Image Process. **21**(10), 4431–4441 (2012)
7. Dulio, P., Frosini, A., Pagani, S.M.: A geometrical characterization of regions of uniqueness and applications to discrete tomography. Inverse Probl. **31**(12), 125011 (2015)
8. Dulio, P., Peri, C.: On the geometric structure of lattice U-polygons. Discrete Math. **307**(19), 2330–2340 (2007)
9. Gardner, R., Gritzmann, P.: Discrete tomography: determination of finite sets by x-rays. Trans. Am. Math. Soc. **349**(6), 2271–2295 (1997)
10. Guédon, J.: The Mojette Transform: Theory and Applications. ISTE Wiley, New York (2009)
11. Hajdu, L., Tijdeman, R.: Algebraic aspects of discrete tomography. Journal fur die Reine und Angewandte Mathematik **534**, 119–128 (2001)
12. Questions of Uniqueness and Resolution in Reconstruction from Projections. Lecture Notes in Biomathematics, vol. 26. Springer, Heidelberg (1978). doi:10.1007/978-3-642-45507-0
13. Kuba, A.: The reconstruction of two-directionally connected binary patterns from their two orthogonal projections. Comput. Vis. Graph. Image Process. **27**(3), 249–265 (1984)
14. Lorentz, G.: A problem of plane measure. Am. J. Math. **71**(2), 417–426 (1949)
15. Louis, A., Törnig, W.: Ghosts in tomography-the null space of the Radon transform. Math. Methods Appl. Sci. **3**(1), 1–10 (1981)
16. Normand, N., Kingston, A., Évenou, P.: A geometry driven reconstruction algorithm for the Mojette transform. In: Kuba, A., Nyúl, L.G., Palágyi, K. (eds.) DGCI 2006. LNCS, vol. 4245, pp. 122–133. Springer, Heidelberg (2006). doi:10.1007/11907350_11
17. Normand, N., Svalbe, I., Parrein, B., Kingston, A.: Erasure coding with the finite Radon transform. In: Wireless Communications and Networking Conference (WCNC), pp. 1–6. IEEE (2010)
18. Pertin, D.: Mojette erasure code for distributed storage (Ph.D Thesis). University of Nantes (2016)
19. Svalbe, I., Chandra, S.: Growth of discrete projection ghosts created by iteration. In: Debled-Rennesson, I., Domenjoud, E., Kerautret, B., Even, P. (eds.) DGCI 2011. LNCS, vol. 6607, pp. 406–416. Springer, Heidelberg (2011). doi:10.1007/978-3-642-19867-0_34

Ambiguity Results in the Characterization of *hv*-convex Polyominoes from Projections

Elena Barcucci[1], Paolo Dulio[2], Andrea Frosini[1(✉)], and Simone Rinaldi[3]

[1] Dipartimento di Matematica e Informatica "U. Dini", Università di Firenze,
viale Morgagni 65, 50134 Florence, Italy
{elena.barcucci,andrea.frosini}@unifi.it
[2] Dipartimento di Matematica "F. Brioschi", Politecnico di Milano,
Piazza Leonardo da Vinci 32, 20133 Milan, Italy
paolo.dulio@polimi.it
[3] Dipartimento di Ingegneria dell'Informazione e Scienze Matematiche,
Università di Siena, Via Roma, 56, 53100 Siena, Italy
rinaldi@unisi.it

Abstract. In 1997 R. Gardner and P. Gritzmann proved a milestone result for uniqueness in Discrete Tomography: a finite convex discrete set can be uniquely determined by projections taken in any set of seven planar directions. The number of required directions can be reduced to 4, providing their cross-ratio, arranged in order of increasing angle with the positive x-axis, does not belong to the set $\{4/3, 3/2, 2, 3, 4\}$.

Later studies, supported by experimental evidence, allow us to conjecture that a similar result may also hold for the wider class of *hv*-convex polyominoes.

In this paper we shed some light on the differences between these two classes, providing new 4-tuples of discrete directions that do not lead to a unique reconstruction of *hv*-convex polyominoes. We reach our main result by a constructive process. This generates switching components along four directions by a recursive composition of only three of them, and then by shifting the obtained structure along the fourth one.

Furthermore, we stress the role that the horizontal and the vertical directions have in preserving the *hv*-convexity property. This is pointed out by showing that these often appear in the 4-tuples of directions that allow uniqueness.

A final characterization theorem for *hv*-convex polyominoes is still left as open question.

Keywords: Discrete geometry · Discrete tomography · *hv*-convex set · Uniqueness problem · Switching component

1 Introduction

We consider the problem of the characterization of finite discrete set of points of the $2D$ integer lattice from projections, i.e., from the knowledge of the number of

© Springer International Publishing AG 2017
W.G. Kropatsch et al. (Eds.): DGCI 2017, LNCS 10502, pp. 147–158, 2017.
DOI: 10.1007/978-3-319-66272-5_13

points lying on parallel lines along a given set of discrete directions. This research is of great relevance in Discrete Tomography, and it is motivated by the need of a faithful reconstruction of the internal structure of an object that is not directly accessible. The general problem has been studied since the beginning of the 50's, both from theoretical and computational perspectives (see [12] for a survey on the topic). Due to the existence of different sets of points that are consistent with the projections along any set of discrete directions (in [8], Fishburn et al. gave several characterizations of the finite subsets of Z^n that are uniquely determined by their horizontal and vertical projections), constraints have been added to restrict the space of solutions to obtain faithfulness, say uniqueness, in the reconstruction process.

In [10], the authors provided the following milestone result that answers the uniqueness problem in case of convex polyominoes, i.e. the class $\mathcal{C}$ of connected sets of points that match their convex hulls:

Theorem 1. *Let us consider the class $\mathcal{C}$; it holds that*

1. *if U is a set of four discrete directions not having cross ratio $\{4/3, 3/2, 2, 3, 4\}$, then $\mathcal{C}$ is characterized by U (i.e., uniquely determined by the projections along the directions of U);*
2. *$\mathcal{C}$ is characterized by any set of seven mutually nonparallel directions;*
3. *there is a set of six discrete directions not characterizing $\mathcal{C}$;*
4. *$\mathcal{C}$ cannot be characterized by any set of three discrete directions.*

For every set of four directions $U=\{u_1, u_2, u_3, u_4\}$, the cross ratio is defined as

$$\rho(U) = \frac{(h_3 - h_1)(h_4 - h_2)}{(h_3 - h_2)(h_4 - h_1)},$$

where $u_i = (u_{ix}, u_{iy})$ and $h_i = \frac{u_{iy}}{u_{ix}}$, for $i = 1, 2, 3, 4$ and $0 \leq h_1 < h_2 < h_3 < h_4$. If $u_4 = (0, 1)$, say by abuse of notation $h_4 = \infty$, then we only keep the terms not containing h_4, and the cross ratio reduces to $\rho(U) = \frac{(h_3 - h_1)}{(h_3 - h_2)}$.

A major task in proving the above result relies on the analysis of the class of *lattice U-polygons*, consisting of any non-degenerate convex polygon P such that, for any vertex v of P, and for any direction $u \in U$, the line through v and parallel to u meets a different vertex v' of P. The proof of Theorem 1 comes out from a number of lemmas and properties that combine tools from p-adic valuations, projective geometry, algebraic number theory and convexity.

After the paper of Gardner and Gritzmann appeared, the leading idea was that, when a kind of convexity information is available, ambiguous reconstructions are somehow related to the existence of U-polygons. Therefore, understanding the properties of such structures can help in achieving uniqueness results and reconstruction hints (see [1–6] for interesting examples). In [4] it was proved that if $|U| \geq 4$, and the values of the cross ratio of any set of four directions in U, arranged in order of increasing angle with the positive x-axis, are in the set $\{4/3, 3/2, 2, 3, 4\}$, then a lattice U-polygon does exist. All the lattice U-polygons exhibited in [4] are dodecagons, produced by tiling affinely regular hexagons.

Such kind of hexagonal tiling has been investigated and generalized in [6,7], where several results concerning the symmetries of U-polygons, and their geometric structure have been also provided.

In [4], and later on in [9], some classes of polyominoes that are uniquely determined by projections are provided, combining the notion of horizontal and vertical convexity with additional geometrical constraints. The characterization of the class of hv-convex polyominoes, say $\mathcal{HV}$, from projections is still open: again in [4], it was conjectured that Theorem 1 continues to hold for the elements of this class.

Our study sheds some new light on this problem: we restrict the conjecture by showing that ambiguities in the reconstruction do persist for any set of directions whose elements belong to the same quadrant and do not include both the horizontal and the vertical ones. Then, we go further by showing that no characterization is possible for any set of four directions including the two axial ones and such that the remaining two form a sufficiently small angle.

The overall organization of the paper includes, in the next section, a review to the basics of discrete tomography and the related uniqueness problem. Then, Sect. 3 presents our main results, i.e., the construction of specific hv-switchings along a class of 4-tuples of directions. The prominent role of the projections along the horizontal and vertical directions is also underlined. Finally, in Sect. 4 we provide perspectives and open problems related to the definition of hv-convex polyominoes from a generic set of given projections.

2 Definitions and Known Results

A finite discrete subset of points S in the integer lattice is usually represented by a set of cells (unit squares whose centers are the points in the set itself) on a squared surface, and its dimensions are those of the minimal bounding rectangle. We choose to identify the lower leftmost cell of such rectangle with the origin of the integer lattice, so that each set of points will be considered up to translation.

A *polyomino* is a finite union of cells whose interior is connected (see Fig. 1(a)). A *column* (*row*) of a polyomino is the intersection between the polyomino and an infinite strip of cells whose centers lie on a vertical (horizontal) line.

Nevertheless, several subclasses of interest were considered by putting on polyominoes constraints defined by the notion of *convexity* along different directions. In particular, a polyomino is *h-convex* (resp. *v-convex*) if each of its rows (resp. columns) is connected. A polyomino is *hv-convex*, if it is both *h*-convex and *v*-convex (see Fig. 1(b)). Finally, a polyomino is *convex* if it is convex w.r.t. all the discrete directions, i.e., if it equals its discrete convex hull (see Fig. 1(c)).

Let $u = (u_x, u_y)$ be a discrete direction i.e., a couple of coprime integers. To each discrete set S and direction u, we associate an integer vector $X_u(S)$ that stores the number of points of S that lie on each line parallel to u and intersecting the minimal bounding rectangle of S; we indicate such a vector as the (vector of) projections of S along u (see Fig. 1(c)).

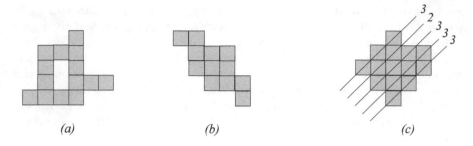

Fig. 1. (*a*) a polyomino; (*b*) an *hv*-convex polyomino; (*c*) a convex polyomino and its projections along the direction $u = (1, 1)$.

Let U be a finite set of discrete directions. We say that S is determined by (the projections along the directions of) U, if $X_u(R) = X_u(S)$ implies $R = S$. We say that S is ambiguous w.r.t. U otherwise. Finally, a class of discrete sets $\mathcal{S}$ is characterized by U if all its elements are non ambiguous w.r.t. U inside the class.

One of the main aims in the field of Discrete Tomography is the achievement of a faithful reconstruction of an unknown object, regarded as a discrete set of points at a certain resolution, from a set of projections. As one can imagine, the existence of different sets of points sharing the same set of projections may dramatically change into meaningless the whole process, so the relevance of the following problem:

Uniqueness $(\mathcal{S}, U)$

Instance: a class of discrete sets $\mathcal{S}$, and a set of discrete directions U.

Question: Does there exist two distinct elements of $\mathcal{S}$ having the same projections along the directions of U?

2.1   The Notion of Switching Components

Many authors studied Uniqueness $(\mathcal{S}, U)$ in terms of the cardinality and the characteristics of the class of discrete sets sharing the same projections along U and, since the very beginning, they relied on the idea of *switching component* (in a very first study, Ryser [14] called it *interchange*) i.e., a rearranging of some elements of a set that preserves the projections along U.

More precisely, a discrete set S contains a *switching component* along U if there is a set of cells S' in the corresponding minimal bounding rectangle such that:

- $S' = S^0 \cup S^1$, $S^0 \cap S^1 = \emptyset$ and $|S^0| = |S^1|$;
- if $a \in S^0$ and $b \in S^1$, then exactly one among a and b belongs to S;
- for each $u \in U$, $X_u(S^0) = X_u(S^1)$ i.e., each line parallel to one direction of U contains the same number of elements in S^0 and S^1.

We call S' a switching component along U.

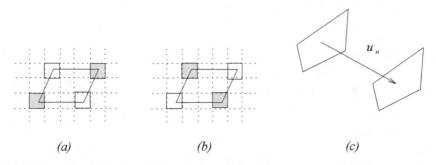

Fig. 2. (a) the switching components S' and its two sets S^0 and S^1 along the directions $(1,0)$ and $(1,2)$. The dark squares are elements of S; (b) the switching component $\overline{S}'$ obtained by S' after changing the values of its elements; (c) the recursive step of the construction of a switching along the n-th direction u_n.

Figure 2(a), shows the two different switching components S^0 and S^1 along two directions, (the light gray cells are not elements of S).

From the definition of switching component, it holds that if a set S has a switching component S' along U, then by changing the values of the elements of S', we obtain another set having the same projections. We denote this discrete set by $\hat{S}$. Furthermore, we call the switching component obtained by changing the values of all the elements of S', its *dual*, and we indicate it by $\overline{S}'$ (see Fig. 2(b)). Obviously the dual operator has period 2. We recall the following result (see [13]):

Theorem 2. *Let $\mathcal{S}$ be a class of discrete sets. An element $S \in \mathcal{S}$ is ambiguous (in $\mathcal{S}$) w.r.t. to a set of directions U, if and only if S has a switching component along U such that $\hat{S} \in \mathcal{S}$.*

The following simple property, rediscovered several times (see [12]), introduces a recursive way to generate switching components along a generic set of discrete directions:

Property 1. Let S' be a switching component along $u_1, u_2, \ldots, u_{n-1}$, and S'' be the dual of the translation of S' along the further direction u_n so that $S' \cap S'' = \emptyset$, i.e., $S'' = \{ku_n + (i,j) : (i,j) \in \overline{S}'\}$, for k sufficiently large integer. The set $S' \cup S''$ is a switching component along the directions $u_1, u_2, \ldots, u_n$.

From this property, two main consequences follow:

(i) we can construct a switching component along a generic set U of directions in the recursive way shown in Fig. 2;

(ii) since we can construct a switching component along every U, then the class of all the possible discrete sets cannot be characterized by any set of discrete directions.

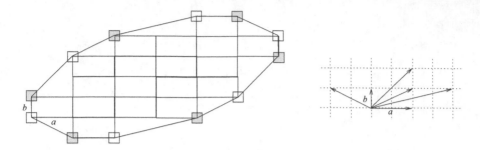

Fig. 3. An example of a (dodecagonal) switching component along the directions $(1,0)$, $(0,1)$, (a,b), $(-a,b)$, $(a,2b)$, and $(2a,b)$.

We stress that the switching components may also be composed in order to achieve complex switching configurations that connect all the elements sharing the same projections to form a connected graph structure.

If we restrict our study to the class $\mathcal{C}$ of convex polyominoes, Theorem 1 states necessary and sufficient conditions in order for a discrete set to be characterized by projections. In particular, any set of seven directions provides uniqueness, while six do not always suffice as the U-dodecagon in Fig. 3 shows, since every set of four directions in U returns a value of the cross ratio belonging to $\{4/3, 3/2, 2, 3, 4\}$.

On the other hand, if we relax too much the constraints on the convexity of the sets, for example preserving convexity only along the horizontal direction, then a negative result holds, as shown in [4], Theorem 3.7:

Theorem 3. *No finite set of directions characterizes the class of horizontally (resp. vertically) convex polyominoes.*

Our intent is to inspect the cut-off line between those results that seems to show up when both the horizontal and vertical convexity are present.
In the same paper, it has been proposed the following

Conjecture 1. The class $\mathcal{HV}$ is characterized by a set of four discrete directions $U = \{(1,0), (0,1), u_3, u_4\}$ such that $\rho(U)$ does not belong to $\{4/3, 3/2, 2, 3, 4\}$.

that has been supported by computational evidence: hv-convex polyominoes were randomly generated and each of them reconstructed using its projections along a set of four directions whose cross ratio does not belong to $\{4/3, 3/2, 2, 3, 4\}$. It was verified that the algorithm uniquely reconstructed the generated hv-convex polyominoes. In the next section, we modify the conjecture by showing a new class of 4-tuples of directions that allow switching components in hv-convex polyominoes.

3   A New Class of Switching Components

The following result from [4,11], provides a step forward to the study of the uniqueness problem on the class $\mathcal{HV}$:

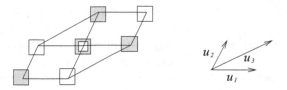

Fig. 4. The construction of the hexagonal switching along three directions u_1, u_2, and u_3.

Property 2. Let u_1, u_2, and u_3 be three discrete directions. There is a hexagonal switching H along $U = \{u_1, u_2, u_3\}$.

Proof. We prove the statement by construction. First of all we construct the switching S along u_1 and u_2, following Property 1. Then, we extend both the sides of S until they equal a k multiple of u_3, and finally we add to S its dual along ku_3, i.e. $H = S \cup ku_3 + \overline{S}$. The two coincident points are deleted (see Fig. 4).   □

Observe that hexagonal switching components have just six distinct points, which is the minimal cardinality for a three directions' switching, while the general construction along three directions defined in Property 1 requires eight points. Furthermore, each hexagonal switching is *convex* in the sense that it is a switching of a convex set.

However, from Property 2 it is easy to realize that hexagonal switching components are not different from those defined in Property 1: as a matter of fact, they can be obtained by translating along the direction u_3 an appropriate magnification of the switching components along u_1 and u_2, so that two opposite of the eight points coincide and so annihilate.

Let $U = \{u_1, u_2, u_3\}$ be a set of three directions in $\mathbb{R}^2$. The group $\mathcal{S}_3^U$ which permutes the indices $\{1, 2, 3\}$ can be represented as a group of symmetries fixing a triangle T with edges parallel to $\{u_1, u_2, u_3\}$. In [7, Theorem 6], the following characterization of lattice U-polygons has been determined

Theorem 4. *Let $U = \{u_1, u_2, u_3\}$ be any set of three lattice directions. Let P be a lattice hexagon. Then P is a U-polygon if and only if $\mathcal{S}_3^U(P) = P$.*

As a consequence, a hexagonal switching along $U = \{u_1, u_2, u_3\}$ always returns a rational magnification of a lattice U-hexagon P, where the triangle T reduces to its barycenter.

Property 3. The class $\mathcal{HV}$ cannot be characterized by a finite set of discrete directions U such that:

(i) at most one among the directions $h = (1, 0)$, and $v = (0, 1)$ belongs to U; and

(ii) all the directions in U belong to the same quadrant.

This property predicts the basic role played by the two convexity directions h and v when they belong to U; for brevity we omit its proof, that is related to the possibility of connecting sufficiently distant parts of a switching component defined as in Property 1 with a path that preserves hv-convexity. It is also easy to verify that the same construction holds when the axes direction is replaced by a generic one:

Property 4. The class $\mathcal{HV}$ cannot be characterized by a finite set of discrete directions U not containing $(1,0)$ and $(0,1)$ and such that all but one lie in the same quadrant.

Relying on these results, we focus on the characterization of $\mathcal{HV}$ by means of a set of four directions $U = \{h, v, u_3, u_4\}$ such that $\rho(U) \notin \{4/3, 3/2, 2, 3, 4\}$.

3.1 Composing Hexagonal Switching Components Along a Diagonal Direction

The definition of hexagonal switching component along three directions suggests a possible construction of hv-convex switching operations along four directions. From [7, Corollary 7] we know that an hexagon H is a lattice U-polygon if and only if for any two diagonals of H, having direction v_1, v_2, there exists a symmetry $\sigma \in \mathcal{S}_3^U$ such that $\sigma(\{u_1, u_2, u_3\}) = \{u_1, u_2, u_3\}$, and $\sigma(v_1) = v_2$. This induces to explore further switchings of a hexagonal one when performed along a diagonal direction. We define such switchings as *diagonal-hexagon switchings*. Also, up to affine transformation, we can always assume that $U = \{(a,0), (0,b), (a,b)\}$.

Property 5. For any three integer numbers a, b, k

(*i*) there exists an hv-convex hexagonal switching H_k along the set of three directions $U = \{(a,0), (0,b), (a,b)\}$;
(*ii*) there exists an hv-convex diagonal-hexagon switching H'_k along $U'_k = U \cup \{(2a, 2b) + k(2a, b)\}$. The same holds if $U'_k = U \cup \{(2a, 2b) + k(a, 2b)\}$.

Proof. Let us consider the set $U = \{(a,0), (0,b), (a,b)\}$ and construct the related hexagonal switching H. Then, we translate it along the direction $(2a, b)$, and we annihilate the incident points having opposite values. The obtained ten points configuration, say $H_1 = H \cup ((2a, b) + H)$, is still an hv-convex switching along U. Now we define $H'_1 = H_1 \cup ((4a, 3b) + \overline{H}_1)$. It is immediate to see that H'_1 is an hv-convex switching along $U' = U \cup \{(4a, 3b)\}$. Now, by iterating the construction for k times, we get

$$H_k = H \bigcup_{i=1}^{k} (k(2a, b) + H) \quad \text{and} \quad H'_k = H_k \cup ((2a, 2b) + k(2a, b) + \overline{H}_k). \quad (1)$$

Then H'_k provides an hv-convex switching component with respect to the set $U'_k = U \cup \{((2a, 2b) + k(2a, b))\}$. In the case $U'_k = U \cup \{((2a, 2b) + k(a, 2b))\}$, the statement follows with a quite analogous argument. □

Theorem 5. *The cross ratio of any diagonal-hexagon switching U'_k is*

$$\rho(U'_k) = 2\frac{k+1}{k}.$$

Proof. Suppose that $U'_k = U \cup \{((2a, 2b) + k(2a, b))\}$. Then, being $a, b, k > 0$, the slope m_k of the k-dependent direction of U'_k satisfies

$$0 < m_k = \frac{(k+2)b}{2(k+1)a} < \frac{b}{a}.$$

Assume the slope as a projective coordinate. Then, by arranging increasingly the four slopes of the directions in U'_k, the computation of their cross ratio $\rho(U'_k)$ provides

$$\rho(U'_k) = (0, \ m_k, \ \frac{b}{a}, \ \infty) = 2\frac{k+1}{k}.$$

In case $U'_k = U \cup \{((2a, 2b) + k(a, 2b))\}$, it results

$$0 < \frac{b}{a} < m_4 = \frac{2(k+1)b}{(k+2)a},$$

so that

$$\rho(U'_k) = (0, \ \frac{b}{a}, \ m_k, \ \infty) = 2\frac{k+1}{k},$$

and the statement follows. $\square$

Remark 1. Note that $\rho(U'_1) = 4$, $\rho(U'_2) = 3$, and

$$\lim_{k \to \infty} \rho(U'_k) = 2.$$

Moreover, up to rearranging the order of the directions, the values $4/3$ and $3/2$ can be turned to 4 and 3, respectively. This points out that the set of cross ratio determined by diagonal-hexagon switching

$$\left\{ 2\frac{k+1}{k}, \ k \in \mathbb{N}, \right\},$$

naturally refines the set $\{4/3, 3/2, 2, 3, 4\}$, that prevents uniqueness in the class $\mathcal{C}$ of convex lattice sets.

Figure 5 shows the above result in the case when $k = 3$ and the related switching along the directions $U' = \{(a, 0), (0, b), (a, b), (8a, 5b)\}$ whose cross ratio is $8/3$.

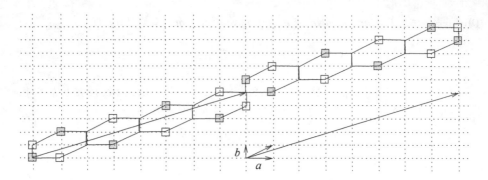

Fig. 5. The composition of four hexagonal switchings, i.e., $k = 3$ w.r.t. the set $U = \{(a, 0), (b, 0), (a, b)\}$ along the direction $(2a, b)$ and the related switching components. The fourth obtained direction is $(8a, 5b)$.

3.2   General Construction of the Switching Components

Finally, we push to the last step our construction of a new class of hv-convex switching components by observing that successive shifts along one of the two directions $u = (2a, b)$ or $v = (a, 2b)$ of the hexagon H can also generate new 4-tuples of elements for the set of directions U'.

In particular, let us consider a sequence of k_1 successive shifts of H along u and k_2 along v: the obtained switching along U, say H_{k_1, k_2} is represented in Fig. 6(a). Following Eq. (1), we define

$$H'_{k_1, k_2} = H_{k_1, k_2} \cup ((2a, 2b) + k_1(2a, b) + k_2(a, 2b) + \overline{H}_{k_1, k_2}).$$

The following holds

Theorem 6. *The set H'_{k_1, k_2} is an hv-convex switching along the directions $U' = U \cup \{u_4\}$, with $u_4 = (2a, 2b) + k_1 u + k_2 v$.*

The proof can be achieved after observing that H_{k_1, k_2} is an hv-convex switching along U, and the union with its dual, once translated along the direction u_4, preserve both the projections along u_4, by definition, and the hv-convexity.

Figure 6(b) shows an example of the switching when $k_1 = 1$ and $k_2 = 3$. We again underline that the cross ratio of the obtained direction $u_4 = (2a + k_1 2a + k_2, 2b + k_1 b + k_2 2b) = (7a, 9b)$ does not belong to the set $\{4/3, 3/2, 2, 3, 4\}$.

After defining the set of discrete directions

$$D = \{(2a + k_1 2a + k_2, 2b + k_1 b + k_2 2b) : k_1, k_2 \in \mathbb{N}\},$$

Theorem 6 allows us to modify Conjecture 1 as follows:

Conjecture 2. The class of hv-convex polyominoes is characterized by a set of four discrete directions $U = \{(1, 0), (0, 1), (a, b), u_4\}$ such that:

– $\rho(U) \notin \{4/3, 3/2, 2, 3, 4\}$;
– $u_4 \notin D$.

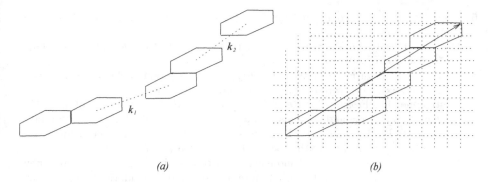

(a)   *(b)*

Fig. 6. *(a)* a generic composition of k_1 hexagonal switchings along the direction $(2a, b)$ and k_2 along the direction $(a, 2b)$; *(b)* an example when $k_1 = 1$ and $k_2 = 3$. The obtained fourth direction is $(7a, 9b)$.

4 Conclusions and Perspectives

In this paper we have addressed the problem of generalizing the results of [10], related to the lattice convex sets, to the wider class $\mathcal{HV}$ of hv-convex polyominoes. We have obtained a few preliminary properties towards a detailed answer to Conjecture 1. In particular we have obtained a complete characterization of diagonal-hexagon switching in $\mathcal{HV}$, and we have determined all the possible values of the cross ratio involved in the corresponding ambiguous reconstructions. These have been explicitly exhibited also from the geometric point of view.

As a further step we wish to investigate different extensions of the presented constructions, in order to get a general description of all switchings preserving hv-convexity. This would allow to get a uniqueness result for hv-convex polyominoes, as well as to get new information on the geometric meaning of the values of the cross ratio related to ambiguous reconstructions

Acknowledgment. This study has been partially supported by INDAM - GNCS Project 2016.

References

1. Alpers, A., Larman, D.G.: The smallest sets of points not determined by their X-rays. Bull. London Math. Soc. **47**, 171–176 (2015)
2. Alpers, A., Tijdeman, R.: The two-dimensional Prouhet-Tarry-Escott problem. J. Number Theor. **123**, 403–412 (2007)
3. Brunetti, S., Dulio, P., Hajdu, L., Peri, C.: Ghosts in discrete tomography. J. Math. Imaging Vis. **53**(2), 210–224 (2015)
4. Barcucci, E., Del Lungo, A., Nivat, M., Pinzani, R.: X-rays characterizing some classes of discrete sets. Linear Algebra Appl. **339**, 3–21 (2001)
5. Cipolla, M., Lo Bosco, G., Millonzi, F., Valenti, C.: An island strategy for memetic discrete tomography reconstruction. Inform. Sci. **257**, 357–368 (2104)

6. Dulio, P.: Convex decomposition of U-polygons. Theor. Comput. Sci. **406**, 80–89 (2008)
7. Dulio, P., Peri, C.: On the geometric structure of lattice U-polygons. Discrete Math. **307**, 2330–2340 (2007)
8. Fishburn, P., Lagarias, J., Reeds, J., Shepp, L.: Sets uniquely determined by projections on axes II. Discrete Appl. Math. **91**, 149–159 (1991)
9. Castiglione, G., Frosini, A., Restivo, A., Rinaldi, S.: Enumeration of L-convex polyominoes by rows and columns. Theor. Comput. Sci. **347**(1-2), 336–352 (2005)
10. Gardner, R.J., Gritzmann, P.: Discrete tomography: determination of finite sets by X-rays. Trans. Amer. Math. Soc. **349**, 2271–2295 (1997)
11. Gardner, R., Gritzmann, P.: Uniqueness and complexity in discrete tomography. In: Herman, G., Kuba, A. (eds.) Discrete Tomography: Foundations, Algorithms, and Applications, pp. 85–113. Birkhäuser, Basel (1999)
12. Herman, G.T., Kuba, A. (eds.): Discrete tomography: Foundations algorithms and applications. Birkhauser, Boston (1999)
13. Kuba, A.: Reconstruction of unique binary matrices with prescribed elements. Acta Cybern. **12**, 57–70 (1995)
14. Ryser, H.: Combinatorial Mathematics, The Carus Mathematical Monographs No. 14, The Mathematical Association of America, Rahway (1963)

Mojette Transform on Densest Lattices in 2D and 3D

Vincent Ricordel$^{(\boxtimes)}$, Nicolas Normand$^{(\boxtimes)}$, and Jeanpierre Guédon$^{(\boxtimes)}$

Université de Nantes, LS2N UMR CNRS 6004 Polytech Nantes,
Rue Christian Pauc, BP 50609, 44306 Nantes Cedex 3, France
{vincent.ricordel,nicolas.normand,jeanpierre.guedon}@univ-nantes.fr

Abstract. The Mojette Transform (MT) is an exact discrete form of the Radon transform. It has been originally defined on the lattice Z^n (where n is the dimension). We propose to study this transform when using the densest lattices for the dimensions 2 and 3, namely the lattice A^2 and the face-centered cubic lattice A^3. In order to compare the legacy MT using Z^n, versus the new MT using A^n, we define a fair comparison methodology between the two MT schemes. In particular we detail how to generate the projection angles by exploiting the lattice symmetries and by reordering the Haros-Farey series. Statistic criteria have been also defined to analyse the information distribution on the projections. The experimental results study shows the specific nature of the information distribution on the MT projections due to the high compacity of the A^n lattices.

Keywords: Mojette Transform · Discrete tomography · Lattices · Densest lattices · Haros-Farey series

1 Objectives of the Study

The Mojette transform is an exact discrete form of the Radon transform [1] defined for specific rational projection angles. Guédon *et al.* originally developed this transform and its corresponding inverse in 1995, in order to represent an image as a set of discrete projections which can be chosen highly redundant (i.e. a frame description). Since 1995, the MT proprieties have been largely explored (spline MT, reconstructability of convex regions, MT in high dimensions, multi-resolution MT, *etc.*), and a lot of applications have been found (data communication and storage, Mojette discrete tomography, Mojette based security, *etc.*). Nevertheless, the MT has been mainly defined, studied and applied using the lattice Z^n (where n is the dimension of the initial lattice to transform). We propose to study this transform when using densest lattices, because we expect that the lattice high compacity will improve the MT performances when representing the data. In the paper we naturally start the study by considering the first dimensions 2 and 3 for which the densest lattices are known.

Thank to Clément Rougale, Jimmy Thomas, Ugo Maury and Maxime Pineau who worked on this project for their MSc at Polytech Nantes.

© Springer International Publishing AG 2017
W.G. Kropatsch et al. (Eds.): DGCI 2017, LNCS 10502, pp. 159–170, 2017.
DOI: 10.1007/978-3-319-66272-5_14

The paper is organised as follows: in the second section, basics on MT and lattices are given. We focus on the MT proprieties (direct/inverse transform, projection matrix, conditions of reconstructability) that are used in the paper, and the densest lattices for the dimensions 2 and 3 (namely the lattice A^2 and the face-centered cubic lattice A^3) are also presented, the lattice density will be also defined at this level. A fair comparison method between the two MT schemes has to be defined, in order to compare the legacy MT using Z^n, versus the new MT using A^n. The comparison methodology is detailed in the third section where we explain in particular how to generate the truncated lattice containing the data to transform, and how to generate the projection angles by exploiting the lattice symmetries and by reordering the Haros-Farey series. The fourth section gives the experimental results, and it explains the statistic criteria used to analyse the specific nature of the information distribution on the MT projections. A conclusion and perspectives are given in the last section.

2   Basics on Mojette Transform and Lattices

2.1   The Mojette Transform

Direct Transform. The Mojette transform is an exact discrete form of the Radon transform defined for specific rational projection angles. Following the work of M. Katz [2], Guédon *et al.* originally developed this transform and its corresponding inverse in 1995, in order to represent an image as a set of discrete projections. The rational projection angles θ_i are defined by a set of discrete vectors (p_i, q_i) as $\theta_i = \tan(q_i, p_i)$, with the condition that q_i and p_i are coprime (i.e. $\gcd(p_i, q_i) = 1$), and q_i is restricted to be positive except for the case $\{p_i, q_i\} = (1, 0)$. The transform domain of an image (or any truncated 2D lattice) is a set of projections where each element (called bin) corresponds to the sum of the pixels centered on the line of projection. This is a linear transform defined for each projection angle by the operator:

$$[\mathcal{M}f](b, p, q) = proj_{p,q}(b) = \sum_{k=-\infty}^{\infty} \sum_{l=-\infty}^{\infty} f(k, l) \Delta(b + kq - lp); \qquad (1)$$

where (k, l) defines the location of an image pixel, b is the index of a bin, and $\Delta(n)$ is the Kronecker delta function, equals to 1 when $n = 0$ and 0 otherwise. The line of projection is represented by $b = kq - lp$, and then $\Delta(b + kq - lp)$ is equal to 1 only for the pixels on this line. The previous Eq. 1 can be rewritten in a matrix form:

$$\mathcal{M}ft(b, p, q) = \sum_k \sum_l f(k, l) \Delta \left(\mathcal{B} - \mathcal{P}_{2 \to 1} \begin{bmatrix} k \\ l \end{bmatrix} \right); \qquad (2)$$

$$= \sum_k \sum_l f(k, l) \Delta \left([b] - [-q\ p] \begin{bmatrix} k \\ l \end{bmatrix} \right);$$

where $\mathcal{P}_{2\rightarrow1}\begin{bmatrix} k \\ l \end{bmatrix}$ is the projection matrix.

Equation (2) can be generalised to higher dimensions. In 3D, a projection plane is defined by a discrete vector (p, q, r) with $gcd(p, q, r) = 1$. In the same way, the projection planes are built from a discrete 3D volume $f(k, l, m)$. Bins are discrete points onto the projected plane, indexed by a vector $\mathcal{B} = \begin{bmatrix} b_1 & b_2 \end{bmatrix}^t$. The 3D Mojette transform can then be defined as [1, Chap. 3], [3]:

$$\mathcal{M}f(b_1, b_2, p, q, r) = \sum_{k,l,m} f(k, l, m)\Delta\left(\mathcal{B} - \mathcal{P}_{3\rightarrow2}\begin{bmatrix} k \\ l \\ m \end{bmatrix}\right);$$

$$= \sum_{k,l,m} f(k, l, m)\Delta\left(\begin{bmatrix} b_1 \\ b_2 \end{bmatrix} - \mathcal{P}_{3\rightarrow2}\begin{bmatrix} k \\ l \\ m \end{bmatrix}\right).$$

(3)

Moreover, in order to obtain a simple and unique index method for the vector of projection, the following conventions are taken [1, Chap. 3]: $r \geq 0$ and $q \geq 0$ if $r = 0$. The projection $\mathcal{P}_{3\rightarrow2}$ matrix can then be defined as following [1, Chap. 3]:

$$\mathcal{P}_{3\rightarrow2} = \begin{cases} \begin{bmatrix} 1 & 0 & -\frac{p}{r} \\ 0 & 1 & -\frac{q}{r} \end{bmatrix} & \text{if } r \neq 0 \text{ and } q \neq 0; \\[2ex] \begin{bmatrix} 1 & -\frac{p}{q} & 0 \\ 0 & 0 & 1 \end{bmatrix} & \text{if } r = 0 \text{ and } q \neq 0; \\[2ex] \begin{bmatrix} 0 & 1 & 0 \\ 0 & 0 & 1 \end{bmatrix} & \text{if } r = 0 \text{ and } q = 0. \end{cases}$$

(4)

This matrix is not optimal as it does not use entire displacement (i.e. ratios are used in this matrix) which creates point with non integer coordinates. Other matrices $\mathcal{P}$ exist and can be generated from the direction projections as presented in DGCI 2005 [4]. The full and detailed explanations in order to generate the matrix $\mathcal{P}$ from the direction projection $(v_1, v_2, \dots, v_n)$ are given in [4,5].

The projection of a 3D regular lattice on a plane with the vector (p, q, r) always produces a 2D regular lattice, according to the vector (p, q, r) [3].

Projection Matrix and Reconstructibility. The reconstructability is the ability to ensure the exact reconstruction of any information using only a set of viewpoints. In other words, a region is reconstructible by a set of projections if a unique correspondence exists between the region and the set of projections [6]. The conditions determining if a set of Mojette projections is sufficient for inverting the transform depends strongly on the discrete shape of the region support under projection. Simple rules exist when the shape is convex [1, Chap. 4]. For rectangular regions, the Katz criterion solves this problem [2]. The reconstruction conditions for any convex region were derived by Normand [1, Chap. 4].

In 2D, each projection direction vector (p_i, q_i) is associated with a two-pixels structuring element B_i (2PSE). Taking G as the region support formed by the successive dilations of the structuring elements B_i, the convex region is reconstructible if and only if it can not contain G [6]. In other words, a convex region (i.e. an image) is not reconstructible if and only if the dilation result by 2PSE is not included in the image support [1, Chap. 4]. This can be also rephrased as: an image of convex support C is reconstructible if and only if the successive erosions of the C formed by the structuring elements B_i gives an empty set [6].

In 3D, the method is extended [3, Chap. 4]: each projection direction vector (p_i, q_i, r_i) is associated with a two-voxels structuring element, and any convex 3D region R is reconstructible by a set of projections S_I, if the dilation of the two-voxels structuring elements produces a form D than is not included in R.

2.2   The Lattices

Lattices. A lattice Λ is a regular arrangement of points in a n-dimensional space. Λ is characterised by its basis [7, (Chap. 1)] or correspondingly by its *generator matrix*:

$$M_\Lambda = \begin{bmatrix} v_{11} & v_{12} & \cdots & v_{1m} \\ v_{21} & v_{22} & \cdots & v_{2m} \\ \vdots & \vdots & \vdots & \vdots \\ v_{n1} & v_{n2} & \cdots & v_{nm} \end{bmatrix}. \tag{5}$$

By combination of the basis vectors, the lattice *fundamental parallelotope* is constructed. This parallelotope, when repeated, can fill the whole space with just one lattice point in each copy.

Different lattices have been studied to solve different problems as sphere packing problem, sphere covering problem, kissing number, fast quantization, *etc.* In the paper we focus on the Z^n lattice, and on the densest lattices for the dimensions 2 and 3, respectively A^2 and A^3.

A sphere packing is an arrangement of non-overlapping identical spheres within a containing space. The lattice is then constituted with the spheres centers, and the densest packing maximises the volume occupied by the spheres. The lattice density can be defined by:

$$\Delta = \frac{\text{vol. of one sphere}}{\text{vol. of the fundamental region}} = \frac{\text{vol. of one sphere}}{det(MM^{tr})^{\frac{1}{2}}}. \tag{6}$$

Z^n **lattices.** The integer lattice is defined as [7, (Chap. 4)]:

$$Z^n = \{(x_1, \ldots, x_n) | x_i \in \mathbb{Z}\}. \tag{7}$$

Its generator matrix is the identity matrix. The densities of the lattices Z^2 and Z^3 are respectively $\Delta_{Z^2} = \frac{\pi}{4} = 0.785...$ and $\Delta_{Z^3} = \frac{\pi}{6} = 0.524...$

We will exploit the lattice symmetries for the MT. The Z^n automorphism group consists of all possible symmetries that are obtained by vector coordinate permutation and/or sign change, the set of permutations has a cardinality of $(2^n n!)$. So, after removing the sign changes (e.g. (a, b) and $(-a, -b)$ are the same vector but with opposite direction), Z^2 and Z^3 counts respectively 4 (see Fig. 1) and 24 symmetries (see Fig. 2).

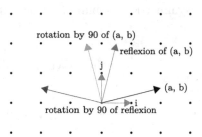

Fig. 1. Symmetries in Z^2 (without sign change).

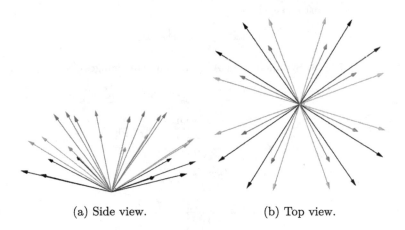

(a) Side view.   (b) Top view.

Fig. 2. Symmetry in Z^3 (without sign change).

A^n lattices. The A^n lattice (for $n \geq 1$) can be defined as [7, Chap. 4]:

$$A^n = \{(x_0, x_1, \ldots, x_n) \in Z^{n+1} | x_0 + x_1 + \cdots + x_n = 0\}. \tag{8}$$

The generator matrix is

$$M_{A^n} = \begin{bmatrix} -1 & 1 & 0 & 0 & \ldots & 0 & 0 \\ 0 & -1 & 1 & 0 & \ldots & 0 & 0 \\ 0 & 0 & -1 & 1 & \ldots & 0 & 0 \\ \vdots & \vdots & \vdots & \vdots & \ddots & \vdots & \vdots \\ 0 & 0 & 0 & 0 & \ldots & -1 & 1 \end{bmatrix}. \tag{9}$$

A^n is the densest lattice for dimensions 2 and 3. A^3 is also known as the face-centered cubic lattice (FCC). The densities of the lattices A^2 and A^3 are respectively $\Delta_{A^2} = \frac{\pi}{\sqrt{12}} = 0.9069...$ and $\Delta_{A^3} = \frac{\pi}{\sqrt{18}} = 0.7405....$

The set of permutations of the automorphism group of the lattice A^2 (respect. A^3) has a cardinality of 16 (respect. 48). After removing the sign changes, 6 symmetries (respect. 12 symmetries) remain [7, Chap. 4] (see also the Figs. 3 and 4).

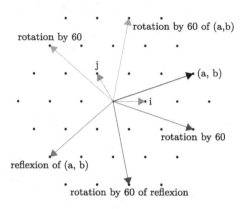

Fig. 3. Symmetries in A^2 (without sign changes).

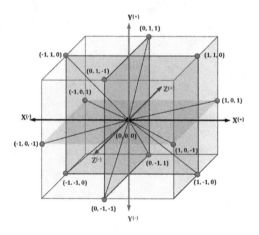

Fig. 4. Neighbors of the point $(0,0,0)$ in A^3 [8]

2.3   Projections and Haros-Farey Sequences

The Haros-Farey sequence gives the set of rational angles in a centered square or cube, this sequence is used to enumerate the MT projections (up to the reconstructability conditions).

In 2D, the Haros-Farey sequence of order N, denoted F_N, is the ordered sequence of irreducible ratios included between 0 and 1, where the denominator is less than or equal to N. In order to get F_{N+1} from F_N, a median ratio $\frac{a_{12}}{b_{12}}$ is inserted between each ratio $\frac{a_1}{b_1}$ and $\frac{a_2}{b_2}$ of F_N, such as $a_{12} = a_1 + a_2$ and $b_{12} = b_1 + b_2$ if $a_{12} < N + 1$ and $b_{12} < N + 1$ [5]. F_1, F_2, F_3 are given as an example:

$$F_1 = \left\{ \frac{0}{1}, \frac{1}{1} \right\}; \quad F_2 = \left\{ \frac{0}{1}, \frac{1}{2}, \frac{1}{1} \right\}; \quad F_3 = \left\{ \frac{0}{1}, \frac{1}{3}, \frac{1}{2}, \frac{2}{3}, \frac{1}{1} \right\}. \tag{10}$$

Each ratio $\left(\frac{q}{p} \right)$ of the sequence is used in order to generate a corresponding projection (p, q) of the 2D MT.

In 3D, according to [5], the Haros-Farey sequence of order N, denoted by $\widehat{F}_N$, is the set of points $\left(\frac{y}{x}, \frac{z}{x} \right)$ such that $gcd(x, y, z) = 1$, between $[0, 0]$ and $[1, 1]$, and which denominator x does not exceed N. In other words, a point $\left(\frac{y}{x}, \frac{z}{x} \right) \in \widehat{F}_N$ if $x \leq N$, $0 \leq y \leq x$, $0 \leq z \leq x$ and $gcd(x, y, z) = 1$. Let $A_1 \left(\frac{y_1}{x_1}, \frac{z_1}{x_1} \right)$ and $A_2 \left(\frac{y_2}{x_2}, \frac{z_2}{x_2} \right)$, two points of $\widehat{F}_{N-1}$ such as $x_1 + x_2 = N$. The middle point between A_1 and A_2 has the coordinates $\left(\frac{y_1 + y_2}{x_1 + x_2}, \frac{z_1 + z_2}{x_1 + x_2} \right)$ [5]. Below, $\widehat{F}_1, \widehat{F}_3, \widehat{F}_3$ are given as an example, where each point $\left(\frac{y}{x}, \frac{z}{x} \right)$ is written as (x, y, z) [5]:

$$\widehat{F}_1 = \{(1, 0, 0), (1, 1, 0), (1, 1, 1)\};$$
$$\widehat{F}_2 = \{(1, 0, 0), (1, 1, 0), (1, 1, 1), (2, 1, 0), (2, 1, 1), (2, 2, 1)\};$$
$$\widehat{F}_3 = \widehat{F}_2 \bigcup \{(3, 1, 0), (3, 1, 1), (3, 2, 0), (3, 2, 1), (3, 2, 2),$$
$$(3, 3, 1), (3, 3, 1)\}. \tag{11}$$

The sequence is used in order to generate the (p, q, r) projections of the 3D MT, with (p, q, r) representing the point $\left(\frac{q}{p}, \frac{r}{p} \right)$.

3 Comparison

This section explains how the truncated lattices were constructed and how the projections were selected. The proposed criteria of comparison are also presented.

3.1 Methodology of Comparison

The goal is to compare the legacy MT using the Z^n lattice, with the MT using the densest lattice. Each lattice $(Z^n$ or $A^n)$ is truncated such that they have the same number of points N_{points}.

Construction of the Truncated Lattices. The first step is to create the truncated lattice. In order to do that, an iterative process is used, where from the 0 point, at each loop, we find the lattice points on successive embedded spheres. Locally, a basic pattern is used which gives for a lattice point its closed neighbours (see Figs. 2 and 4). The growing lattice process is stoped when the number of points N_{points} is reached (N_{points} is given by the user). Each point of the truncated lattices is set to a unitary value.

Selection of Projections. The minimal number of projections are chosen for the truncated lattice to be exactly reconstructible. Projection vectors are produced by sorting fractions of Haros-Farey series according to their squared Euclidean norms, i.e., respectively $x^2 + y^2$, $x^2 + y^2 + z^2$, $k^2 + l^2 - kl$ and $xx + yy + zz - yz - xz$ in lattices Z^2, Z^3, A^2 and A^3. For example, with F_5 sorted:

$$F_5 = \left(\frac{0}{1}, \frac{1}{5}, \frac{1}{4}, \frac{1}{3}, \frac{2}{5}, \frac{1}{2}, \frac{3}{5}, \frac{2}{3}, \frac{3}{4}, \frac{4}{5}, \frac{1}{1}\right);$$

$$\text{Sort}_{Z_2}(F_5) = \left(\frac{0}{1}, \frac{1}{1}, \frac{1}{2}, \frac{1}{3}, \frac{2}{3}, \frac{1}{4}, \frac{3}{4}, \frac{1}{5}, \frac{2}{5}, \frac{3}{5}, \frac{4}{5}\right); \quad (12)$$

$$\text{Sort}_{A_2}(F_5) = \left(\frac{0}{1}, \frac{1}{1}, \frac{1}{2}, \frac{1}{3}, \frac{2}{3}, \frac{1}{4}, \frac{3}{4}, \frac{2}{5}, \frac{3}{5}, \frac{1}{5}, \frac{4}{5}\right).$$

Before choosing another projection in the Haros-Farey, all equivalent projections by rotation are generated. The number of equivalent projections by rotation depends on the lattice (for instance, 3 other projections for the Z^2 lattice, and 5 other projections for the A^2 lattice).

In order to know if the truncated lattice is exactly reconstructible using the set of selected projections, the shape of successive dilatations of the projections directions is generated, as explained in Sect. 2.1. The truncated lattice is exactly reconstructible only when its radius is inferior or equal to the radius of the generated figure.

3.2   Comparison Criteria

Global criteria were used to compare the information distribution on the MT projections.

Redundancy. Redundancy is given by the following equation [1, Chap. 3]:

$$Red = \frac{nb_{bins}}{nb_{points}} - 1. \quad (13)$$

If redundancy is positive, it represents the percentage of extra bins compared to the number of points. If redundancy is negative, then there is no reconstructability of the truncated lattice. Here, by construction, Red is positive but small.

Number of Bins. B_i is the number of bins on the i-th projection (i.e. the i-th projection length). The mean $\bar{B}$, and the variance $Var(B)$, can be then calculated as following:

$$\bar{B} = \frac{1}{n}\sum_{i=1}^{n} B_i, \quad Var(B) = \frac{1}{n}\sum_{i=1}^{n}(B_i - \bar{B})^2, \quad (14)$$

where n is the number of projections in the set. This criteria measures the difference of the number of bins in the projections. The smaller $Var(B)$ for different projections, the higher those projections carry the same amount of information.

Number of Points per Bin. For the test case, each lattice point is set to 1, so each bin value (on every projections) equals to the number of points that contribute to the bin. The mean (considering all projections) is computed as:

$$Mean(\text{points per bin}) = \bar{b} = \frac{1}{m}\sum_{i=1}^{m} b_i, \tag{15}$$

where m is the total number of bins (considering all projections) and b_i is the value of the i-th bin. The higher the mean, the higher bins represent more points. This is directly related to the density of the lattice.

4 Experimental Results

In this section, the main results are presented and discussed. All the experiments were done in 2D and 3D, but we shall concentrate here onto the 3D case because it generalises the 2D case.

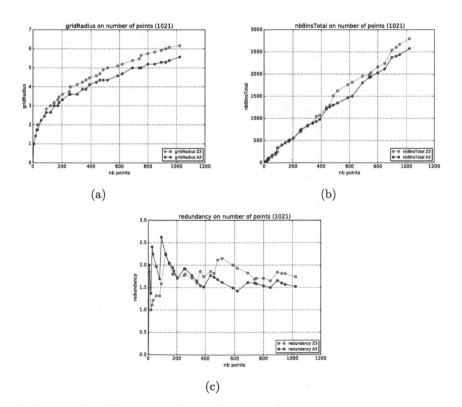

(a)

(b)

(c)

Fig. 5. Truncated lattice radius (a), Total number of bins (b), and Redundancy (c). The blue (resp. red) curves characterise the A^3 (resp. Z^3) lattice. (Color figure online)

The first feature displayed on Fig. 5(a) is the truncated lattice radius value computed from N_{points}, the number of generated points. For $N_{points} > 400$, the curves show the higher compacity of A^3 on Z^3.

The higher total number of bins of Z^3 (see Fig. 5(b)) explains its higher redundancy (see Fig. 5(c)).

The next features are the number of bins Mean (Fig. 6(a)) and Variance (Fig. 6(b)) according to the number of generated points. Concerning these 2 features, it seems that the two lattices perform almost equally, but the Fig. 6(c) shows that the projections number is different for the 2 lattices, a finer analysis at the projections level is then necessary.

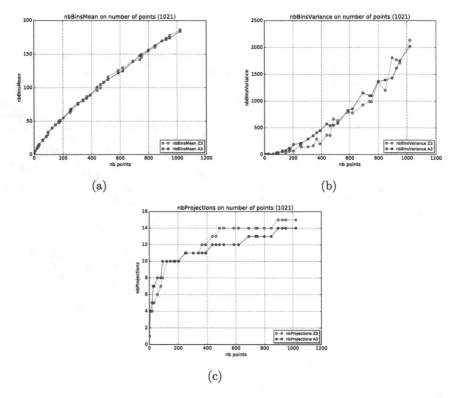

Fig. 6. Number of bins: Mean (a) and Variance (b), and Number of projections (c). The blue (resp. red) curves characterise the A^3 (resp. Z^3) lattice. (Color figure online)

We then use histograms. The Fig. 7 compares the projections densities considering the number of points Mean per projection. And the Fig. 8 compares the projections densities considering their lengths. The histograms with A^3 are slightly more uniform than the ones with Z^3, it shows the higher regularity of the projections when using A^3, these results are due to the high compacity of this lattice.

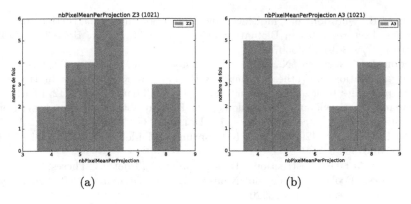

Fig. 7. Histograms of the number of points Mean per projection for Z^3 (a), and A^3 (b).

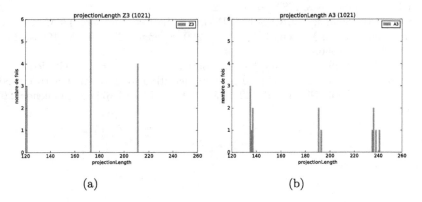

Fig. 8. Histograms of the projections length for Z^3 (a), and A^3 (b).

5   Conclusion and Perspectives

In this paper, we examined the behaviour of densest lattices in 2D and 3D from their discrete Mojette projections point of view. Exactly, the study focused on the 3D case because it generalises the 2D one. The analysis of Mojette Transform projections, when comparing the legacy MT with Z^3 versus the MT with A^3, shows some interesting differences both in terms of dimensions and in terms of projections regularity. Since the software has been developed to manage any dimension and lattice, future work will focus on higher dimensions. Indeed it seems interesting to try higher dimensions in order to see if the gap between A^n and Z^n lattices still increases.

References

1. Guedon, J.-P.: The Mojette transform: theory and applications. Wiley-ISTE (2009). ISBN 978-1-84821-080-6. https://hal.archivesouvertes.fr/hal-00367681

2. Katz, M.B.: Questions of Uniqueness and Resolution in Reconstruction from Projections. Lecture Notes in Biomathematics, vol. 26. Springer, Heidelberg (1978). doi:10.1007/978-3-642-45507-0

3. Guedon, J.-P., Normand, N., Lecoq, S.: Transformation Mojette en 3D: Mise en oeuvre et application en synthéses d image. GRETSI, Groupe d'Etudes du Traitement du Signal et des Images, September 1999. http://hdl.handle.net/2042/13101

4. Normand, N., Servières, M., Guédon, J.P.: How to obtain a lattice basis from a discrete projected space. In: Andres, E., Damiand, G., Lienhardt, P. (eds.) DGCI 2005. LNCS, vol. 3429, pp. 153–160. Springer, Heidelberg (2005). doi:10.1007/978-3-540-31965-8_15

5. Servieres, M.: Reconstruction Tomographique Mojette. Theses, Université de Nantes; Ecole Centrale de Nantes (ECN), December 2005. https://tel.archives-ouvertes.fr/tel-00426920

6. Normand, N., Guedon, J.-P.: La transformee mojette: Une representation redondante pour l'image. In: Comptes Rendus de l'Academie des Sciences - Series I - Mathematics, vol. 326, no. 1, pp. 123–126 (1998). http://dx.doi.org/10.1016/S0764-4442(97)82724-3, ISSN 0764-4442

7. Conway, J.H., Sloane, N.J.A.: Sphere Packings, Lattices and Groups. Grundlehren der mathematischen Wissenschaften, vol. 290. Springer, New York (1993). doi:10.1007/978-1-4757-6568-7

8. Rashid, M.A., Iqbal, S., Khatib, F., Hoque, M.T., Sattar, A.: Guided macromutation in a graded energy based genetic algorithm for protein structure prediction. Comput. Biol. Chem. **61**, 162–177 (2016). doi:10.1016/j.compbiolchem.2016.01.008

Fuzzy Directional Enlacement Landscapes

Michaël Clément$^{(\boxtimes)}$, Camille Kurtz, and Laurent Wendling

Université Paris Descartes, Sorbonne Paris Cité, LIPADE (EA 2517),
45 Rue des Saints-Péres, 75006 Paris, France
michael.clement@parisdescartes.fr

Abstract. Spatial relations between objects represented in images are of high importance in various application domains related to pattern recognition and computer vision. By definition, most relations are vague, ambiguous and difficult to formalize precisely by humans. The issue of describing complex spatial configurations, where objects can be imbricated in each other, is addressed in this article. A novel spatial relation, called enlacement, is presented and designed using a directional fuzzy landscape approach. We propose a generic fuzzy model that allows to visualize and evaluate complex enlacement configurations between crisp objects, with directional granularity. The interest and the behavior of this approach is highlighted on several characteristic examples.

1 Introduction

The spatial organization of objects is fundamental to increase the understanding of the perception of similarity between scenes or situations. Despite the fact that humans seem capable of apprehending spatial configurations, in many cases it is exceedingly difficult to quantitatively define these relations, mainly because they are highly prone to subjectivity. Standard all-or-nothing mathematical relations are clearly not suitable, and the interest of fuzzy relations was initially suggested by Freeman in the 70s [9], since they allow to take imprecision into account. Over the last few decades, numerous works were proposed on the analysis of spatial relationships in various domains, ranging from shape recognition to computer vision, with the main purpose of describing the relative positioning of objects in images [3]. These approaches provide a set of interesting features able to describe efficiently most of spatial situations. However, some configurations remain challenging to describe without ambiguities, especially when the objects are imbricated, or composed of multiple connected components. In this context, we propose to study new relations dedicated to the imbrication of objects.

This article is organized as follows. Section 2 presents related works to our approach. Section 3 recalls the model of directional enlacement, proposed in [6] for the description of complex spatial configurations. From the latter, we propose in Sect. 4 a generic model relying on fuzzy landscapes that allows to evaluate relative enlacement configurations between crisp objects, with directional granularity. Section 5 presents experimental results on different illustrative examples that allow to highlight the behavior and the interest of this model. Section 6 provides conclusions and perspectives.

© Springer International Publishing AG 2017
W.G. Kropatsch et al. (Eds.): DGCI 2017, LNCS 10502, pp. 171–182, 2017.
DOI: 10.1007/978-3-319-66272-5_15

2   Related Work

In the domain of spatial relations, two major research axes can be distinguished in the literature, based on two dual concepts: the one of *spatial relationship* and that of *relative position*. On the one hand, it is possible to formulate a fuzzy evaluation of a spatial relation (for example *"to the left of"*) for two objects, in order to describe their relative position. The fuzzy landscape model is a widely used method for providing this type of assessments [2]. This approach relies on the fuzzy modeling of a given spatial relation, directly in the image space, using morphological operators. Applications of this model can be found in various domains such as spatial reasoning in medical images [7] or the recognition of handwriting [8]. On the other hand, the location of an object with regards to another can be modeled by a quantitative representation, in the form of a relative position descriptor. Different spatial relations can be assessed from this representation and the associated descriptors can be integrated in pattern recognition systems to match similar spatial configurations. Among the various relative position descriptors, the histograms of forces [14] are widely used due to their ability to process pairwise information following a set of directions. They are applied in different works, such as the linguistic description of spatial relations [11] or image retrieval [5]. To summarize, fuzzy landscapes consist in determining the region of space matching a specific spatial relation, and relative position descriptors consist in characterizing the position of an object with regards to another, by combining different spatial features into a standalone descriptor.

Although these two types of approaches allow to interpret many spatial relations between objects, they usually fail at properly describing more complex spatial configurations, in particular when objects are concave, or composed of multiple connected components [4]. A typical complex spatial relation is the *"surrounded by"* relation, which was first studied by Rosenfeld [15] and deepened by Vanegas [16] with a dedicated approach based on fuzzy landscapes. Another specific spatial relation is *"between"*. This relation has been studied in details in [4], involving definitions based on convex hulls and specific morphological operators. Applications of this spatial configuration for the analysis of histological images have been proposed by [10]. Work has also been done to characterize the *"alignment"* and *"parallelism"* of objects in satellite images [17]. Recent works introduced the ϕ-descriptor [12,13], which is a powerful generic framework to assess any spatial relation from a set of specific operators, inspired by Allen intervals [1]. This descriptor can determine if two objects are imbricated or not, but it is not able to measure the depth of imbrication (such as, for instance, when two spirals are interlaced).

In this context, recent works [6] introduced both enlacement and interlacement descriptors, from the relative position point of view, in order to obtain a robust modeling of the imbricated parts of objects. Based on this model, in this article we propose to tackle the dual point of view, by considering fuzzy enlacement landscapes instead of enlacement descriptors. The goal of fuzzy enlacement landscapes is to visualize and evaluate these spatial configurations directly in the image space, by considering the concavities of the objects in a directional fashion.

3   Directional Enlacement Model

In this section, we present the model used to describe the relative enlacement of objects. This model was initially introduced in [6], mostly from the point of view of the relative position descriptors. Here, we recall the intuitive idea behind what is intended with the term *enlacement*, and we provide some useful definitions and notations for this model.

A two-dimensional object A of the Euclidean space is defined by its characteristic function $f_A : \mathbb{R}^2 \to \mathbb{R}$. This generic definition allows to handle both crisp and fuzzy objects. Let $\theta \in \mathbb{R}$ be an orientation angle, and $\rho \in \mathbb{R}$ a distance from the origin. We define the oriented line of angle θ at the altitude ρ by the non-finite set $\Delta^{(\theta,\rho)} = \{e^{i\theta}(t + i\rho), t \in \mathbb{R}\}$. The subset $A \cap \Delta^{(\theta,\rho)}$ represents a one-dimensional slice of the object A, also called a *longitudinal cut*. In the case of crisp objects, such a longitudinal cut of A is either empty (the oriented line does not cross the object) or composed of a finite number of segments. In the general case, a longitudinal cut of A along the line $\Delta^{(\theta,\rho)}$ can be defined as:

$$
\begin{aligned}
f_A^{(\theta,\rho)} &: \mathbb{R} \longrightarrow \mathbb{R} \\
t &\longmapsto f_A(e^{i\theta}(t + i\rho)).
\end{aligned}
\tag{1}
$$

Let (A, B) be a couple of objects. The goal is to describe how A is enlaced by B. The intuitive idea is therefore to capture the occurrences of points of A being *between* points of B. In order to determine such occurrences, objects are handled in a one-dimensional case, using longitudinal cuts along oriented lines. For a given oriented line $\Delta^{(\theta,\rho)}$, the idea is to combine the quantity of object A (represented by $f_A^{(\theta,\rho)}$) located simultaneously *before* and *after* object B (represented by $f_B^{(\theta,\rho)}$). Let f and g be two bounded measurable functions with compact support from $\mathbb{R}$ to $\mathbb{R}$. The enlacement of f with regards to g is defined as:

$$
E(f,g) = \int_{-\infty}^{+\infty} g(x) \int_x^{+\infty} f(y) \int_y^{+\infty} g(z) \, \mathrm{d}z \, \mathrm{d}y \, \mathrm{d}x.
\tag{2}
$$

The scalar value $E(f_A^{(\theta,\rho)}, f_B^{(\theta,\rho)})$ represents the enlacement of A by B along the oriented line $\Delta^{(\theta,\rho)}$. For crisp objects (*i.e.*, each point is either 0 or 1), it corresponds to the total number of ordered triplets of points on the oriented line, which can be seen as arguments to put in favor of the proposition "*A is enlaced by B*" in the direction θ. Algorithmically, this value can be derived by an appropriate distribution of segments lengths along the longitudinal cuts of both objects (see [6] for more details).

The set of all parallel lines $\{\Delta^{(\theta,\rho)}, \rho \in \mathbb{R}\}$ in the direction θ slices the objects into sets of longitudinal cut functions. To measure the global enlacement of an object with regards to another in this direction, we aggregate the one-dimensional enlacement values obtained for each of these longitudinal cuts. The enlacement of A by B in direction θ is defined by:

$$
\mathcal{E}_{AB}(\theta) = \frac{1}{\|A\|_1 \|B\|_1} \int_{-\infty}^{+\infty} E(f_A^{(\theta,\rho)}, f_B^{(\theta,\rho)}) \, \mathrm{d}\rho,
\tag{3}
$$

where $\|A\|_1$ and $\|B\|_1$ denote the areas of A and B. This normalization allows to achieve scale invariance. In the binary case, this definition corresponds to a number of triplets of points to put in favor of "A *is enlaced by B*" along the longitudinal cuts in this direction. Intuitively, it can be interpreted as the quantity of B traversed while sliding the object A in the direction θ, with regards to the quantity of B located on the opposite direction.

In [6] the enlacement model $\mathcal{E}_{AB}$ was considered from the point of view of the relative position descriptors by building a directional enlacement histogram, allowing to characterize how an object A is enlaced by another object B. In the next section, we involve this model in a novel evaluation point of view based on a fuzzy approach that allows to evaluate enlacement configurations directly in the image space, with directional granularity.

4   Fuzzy Enlacement Landscapes

We present here how to extend the directional enlacement model to evaluate the enlacement of objects in the image space from a local point of view, inspired by the works of Bloch [2] on fuzzy landscapes for classical spatial relations.

4.1   Definition

A fuzzy enlacement landscape of an object A should be a representation of the region of space that is enlaced by A. Since the initial enlacement model is essentially directional, we also propose to define directional enlacement landscapes. Let A be a crisp object (*i.e.*, represented as $f_A : \mathbb{R}^2 \to \{0, 1\}$). In a given direction θ, for a point outside of A located at (ρ, t) coordinates in the rotated frame, its local enlacement value can be defined as:

$$\mathcal{E}_A(\theta)(\rho, t) = \frac{1}{\|A\|_1} \int_t^{+\infty} f_A^{(\theta,\rho)}(x)\,\mathrm{d}x \int_{-\infty}^t f_A^{(\theta,\rho)}(x)\,\mathrm{d}x. \tag{4}$$

Therefore, $\mathcal{E}_A(\theta)$ can be seen as a landscape representing the local enlacement values of the points outside of the object A. This image can be normalized into the $[0,1]$ range of values in order to be interpreted as a fuzzy set, which we call a *Fuzzy Directional Enlacement Landscape (Fuzz-DEL)* of the object:

$$\mu_{\mathcal{E}}^A(\theta)(\rho, t) = \frac{\mathcal{E}_A(\theta)(\rho, t)}{\max\limits_{\rho,t} \mathcal{E}_A(\theta)(\rho, t)}. \tag{5}$$

Such a landscape allows to assess and visualize to which degree each point is enlaced by the object A in a fixed direction θ. It is interesting to note that the non-zero values of this landscape are necessarily located inside the object's concavities. This is particularly interesting from an algorithmic point of view, since it allows to restrict the computation to points located in the convex hull of A (and outside of A). Another point to highlight is that enlacement landscapes are symmetric, with period π (*i.e.*, $\mu_{\mathcal{E}}^A(\theta + \pi) = \mu_{\mathcal{E}}^A(\theta)$).

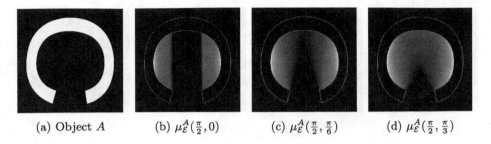

(a) Object A (b) $\mu_{\mathcal{E}}^{A}(\frac{\pi}{2},0)$ (c) $\mu_{\mathcal{E}}^{A}(\frac{\pi}{2},\frac{\pi}{6})$ (d) $\mu_{\mathcal{E}}^{A}(\frac{\pi}{2},\frac{\pi}{3})$

Fig. 1. Fuzzy directional enlacement landscapes of a crisp object A, with a fixed direction θ and an increasing width ω. In (b, c and d), A is outlined in white.

Since $\mu_{\mathcal{E}}^{A}(\theta)$ is focused on a single direction, we propose to aggregate such fuzzy landscapes across multiple orientation angles. Let $\theta \in [0, \pi]$ be an orientation angle and $\omega \in [0, \pi]$ a width parameter. The *Fuzz-DEL* on the interval $[\theta - \frac{\omega}{2}, \theta + \frac{\omega}{2}]$ is defined as follows:

$$\mu_{\mathcal{E}}^{A}(\theta,\omega)(\rho,t) = \frac{1}{\omega} \int_{\theta-\frac{\omega}{2}}^{\theta+\frac{\omega}{2}} \mu_{\mathcal{E}}^{A}(\alpha)(\rho,t)\,\mathrm{d}\alpha, \tag{6}$$

where θ represents the direction on which the fuzzy landscape is focused, while ω controls the width of the interval, allowing to measure either a narrow direction or a more global one. In particular, the landscape that aggregates all directions is denoted by $\widetilde{\mu}_{\mathcal{E}}^{A} = \mu_{\mathcal{E}}^{A}(\frac{\pi}{2},\pi)$.

In order to illustrate such definitions, Figs. 1 and 2 show the *Fuzz-DELs* obtained for two different objects. On the one hand, Fig. 1 illustrates the impact of the width parameter ω for a given vertical direction ($\theta = \frac{\pi}{2}$). Note that the landscape would be identical for the opposite vertical direction ($\theta = \frac{3\pi}{2}$) because of symmetry. We can observe the zero-valued points in the center of the object for $\omega = 0$, representing the fact that these points are not enlaced vertically by A. This can be interpreted by the idea that if another object was located here, it would be able to move in the vertical direction without crossing the other object (*i.e.*, the object could slide downwards). We also observe that when ω increases, the fuzzy landscape progressively gets smoother, taking into account a wider range of directions. On the other hand, Fig. 2 shows enlacement landscapes on another object for different directions θ (with a fixed width $\omega = \frac{\pi}{3}$). From these examples, one can note how a *Fuzz-DEL* allows to capture the object directional concavities. In the horizontal direction ($\theta = 0$), the local enlacement values are relatively high, and the values are higher the deeper we get inside the "snaked" shape. In the vertical direction ($\theta = \frac{\pi}{2}$), the *Fuzz-DEL* is mostly empty, except on some small concavities.

4.2   Fuzzy Evaluation

In the previous definitions, a reference object A is considered, and different *Fuzz-DELs* can be derived from it. These fuzzy landscapes allow to visualize the

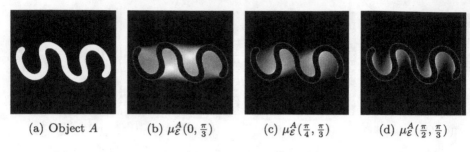

(a) Object A (b) $\mu_{\mathcal{E}}^{A}(0, \frac{\pi}{3})$ (c) $\mu_{\mathcal{E}}^{A}(\frac{\pi}{4}, \frac{\pi}{3})$ (d) $\mu_{\mathcal{E}}^{A}(\frac{\pi}{2}, \frac{\pi}{3})$

Fig. 2. Fuzzy directional enlacement landscapes of a crisp object A, with a fixed width ω and for different directions θ. In (b, c and d), A is outlined in white.

interaction area of A. In the following, we show how to exploit these landscapes to evaluate to which degree a target object B is enlaced by the reference object A, using classical fuzzy operators.

Let μ_A and μ_B be two fuzzy sets over $\mathbb{R}^2$. A typical way to evaluate how μ_B matches with μ_A is the necessity-possibility measure. The necessity N and possibility Π can be respectively defined as follows:

$$\Pi(\mu_A, \mu_B) = \sup_{x,y} t(\mu_A(x,y), \mu_B(x,y)), \tag{7}$$

$$N(\mu_A, \mu_B) = \inf_{x,y} T(\mu_A(x,y), 1 - \mu_B(x,y)), \tag{8}$$

where t is a fuzzy intersection (t-norm) and T is a fuzzy union (t-conorm). For the rest of this article, the min and max operators are chosen for t-norm and t-conorm respectively, but other fuzzy operators could be considered.

In our context, this fuzzy matching measure can be applied to evaluate how a target object B (represented by its membership function μ_B) matches with a *Fuzz-DEL* $\mu_{\mathcal{E}}^{A}(\theta, \omega)$ of a reference object A. The necessity-possibility interval $[N(\mu_{\mathcal{E}}^{A}(\theta, \omega), \mu_B), \Pi(\mu_{\mathcal{E}}^{A}(\theta, \omega), \mu_B)]$ constitutes a fuzzy evaluation of how B is enlaced by A in direction θ, with the necessity being a pessimist point of view, while the possibility represents an optimist point of view. The mean value $M(\mu_{\mathcal{E}}^{A}(\theta, \omega), \mu_B)$ can also be considered. This evaluation strategy will be further studied in the upcoming experiments.

5   Experimental Results

We present different illustrative examples to highlight the interest of our approach. These experiments are organized around two main applications. The first one is to evaluate the specific relation *"surrounded by"*. As mentioned previously, this relation can be considered as a particular case that can be derived from the directional enlacement model. The second application is to evaluate the spatial relation *"enlaced by"* in a more generic sense, in particular when the reference object has multiple degrees of concavities. We also propose some preliminary results on *interlacement* landscapes.

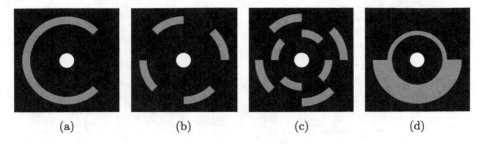

(a)   (b)   (c)   (d)

Fig. 3. Examples of typical surrounding configurations (gray: reference object A; white: target object B).

5.1   Surrounding

The *"surrounded by"* relation is easily apprehended by human perception, but is particularly challenging to evaluate quantitatively. It is usually modeled by the "all directions" point of view, *i.e.*, an object surrounds another object if it is located in all directions. In the following, we adopt the same insight, but we adapt it to the enlacement model: an object is surrounded if it is enlaced by the other object in all directions.

Figure 3 presents characteristic examples of surrounding configurations that we assessed using the proposed fuzzy evaluation strategy. In each image, the reference object A is in gray and the target object B is the white circle.

For this application, we propose a specific way to apply our approach. The target object B is projected into a *Fuzz-DEL* of A, and further normalized as a fuzzy set. Such a projection is defined as:

$$\mu_{\mathcal{E}}^{AB}(\theta, \omega) = \frac{\min_{\rho, t} \left(\mu_{\mathcal{E}}^{A}(\theta, \omega)(\rho, t), \mu_B(\rho, t)\right)}{\max_{\rho, t} \mu_{\mathcal{E}}^{A}(\theta, \omega)(\rho, t)}. \tag{9}$$

Then, the necessity $N(\mu_{\mathcal{E}}^{AB}(\theta, \omega), \mu_B)$ and possibility $\Pi(\mu_{\mathcal{E}}^{AB}(\theta, \omega), \mu_B)$ evaluations are performed for different values of $\theta \in [0, \pi]$. This results in informative directional necessity and possibility profiles, which can be also then exploited to derive a global evaluation of how B is surrounded by A. For the rest of this study, we fixed ω to a low value of $\frac{\pi}{36}$ (5°) to take into account different directions individually, while smoothing out some discretization issues. Following the "all directions" point of view, this global evaluation can be obtained with the following:

$$N_{\mathcal{S}}^{AB} = \frac{1}{\pi} \int_0^{\pi} N(\mu_{\mathcal{E}}^{AB}(\theta, \omega), \mu_B) \, \mathrm{d}\theta, \tag{10}$$

$$\Pi_{\mathcal{S}}^{AB} = \frac{1}{\pi} \int_0^{\pi} \Pi(\mu_{\mathcal{E}}^{AB}(\theta, \omega), \mu_B) \, \mathrm{d}\theta. \tag{11}$$

Figure 4 shows the directional necessity and possibility profiles obtained for the configurations of Fig. 3. In situation (a), the object is only partially surrounded. Both the pessimist and possibility evaluations agree that the reference

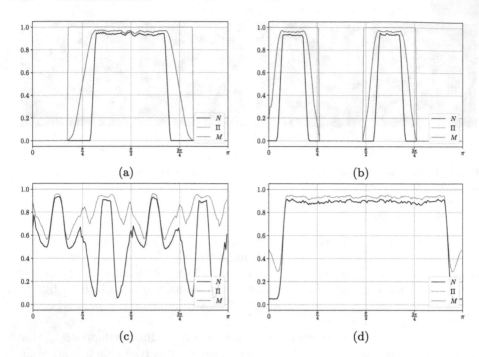

Fig. 4. Directional necessity, possibility and mean profiles measuring the surrounding configurations of Fig. 3.

object A is surrounded in the vertical directions, but not in the horizontal directions. The gradual transition of the situation is captured along the diagonal directions. The object is also partially surrounded in situation (b), where half of the surrounding circle has been cut out. In situation (c), small parts were added preventing the object to leave without crossing the surrounding object, and therefore the optimist point of view is 1 while the pessimist one oscillates but is never zero. Finally, evaluations tend to agree that the object is surrounded in situation (d). The optimist point of view is always 1, yet it takes into account that the object could escape by crossing a small portion of the surrounding object, resulting in low pessimist evaluations for the vertical directions.

Table 1. Fuzzy surrounding evaluations (necessity-possibility intervals and mean values) obtained for the configurations of Figs. 3 and 5.

	Vanegas et al. [16]	Enlacement $\mathcal{E}_{AB}$ [6]	$[N_S^{AB}, \Pi_S^{AB}]$
(a)	[0.70, 0.79], 0.76	[0.50, 0.63], 0.55	[0.36, 0.64], 0.48
(b)	[0.50, 0.54], 0.52	[0.40, 0.49], 0.45	[0.25, 0.53], 0.39
(c)	[0.93, 1.00], 0.97	[0.75, 1.00], 0.95	[0.54, 1.00], 0.79
(d)	[0.94, 1.00], 0.99	[0.48, 1.00], 0.82	[0.77, 1.00], 0.85
(arcachon)	[0.68, 0.85], 0.79	[0.35, 1.00], 0.62	[0.63, 1.00], 0.80

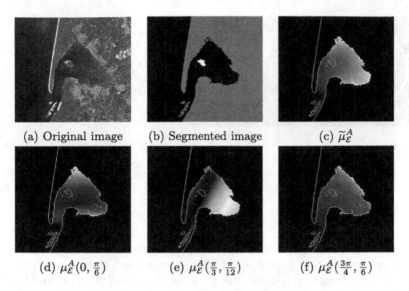

(a) Original image   (b) Segmented image   (c) $\widetilde{\mu}_{\mathcal{E}}^{A}$

(d) $\mu_{\mathcal{E}}^{A}(0, \frac{\pi}{6})$ (e) $\mu_{\mathcal{E}}^{A}(\frac{\pi}{3}, \frac{\pi}{12})$ (f) $\mu_{\mathcal{E}}^{A}(\frac{3\pi}{4}, \frac{\pi}{6})$

Fig. 5. Applicative example of a complex surrounding configuration. The satellite image represents the *Bassin d'Arcachon* (France). (b) Object A is gray and object B is white. (c–f) A is outlined in white and B is outlined in red. (Color figure online)

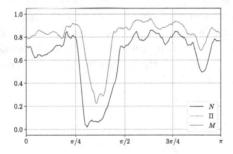

Fig. 6. Directional necessity, possibility and mean profiles measuring the surrounding of the island by the bay in Fig. 5.

To complement these results, Table 1 presents the surrounding necessity-possibility intervals $[N_{\mathcal{S}}^{AB}, \Pi_{\mathcal{S}}^{AB}]$, to evaluate the global surrounding of the target objects of Fig. 3. For comparison purposes, we also present the results of the approach of [16], which is also based on a fuzzy landscape framework, but dedicated to surrounding relation. It is based on a specific fuzzy landscape, considering only the visible concavities of the reference object. We also present the results obtained by [6] with the initial enlacement descriptors. Considering the fact that surrounding evaluations are highly subjective, our goal here is not to argue that an approach is better than another, but to illustrate that the proposed *Fuzz-DELs* can provide interesting point of views regarding this surrounding spatial relation.

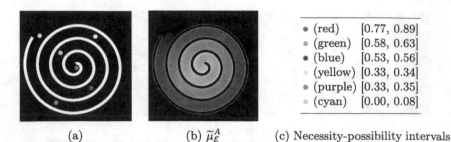

• (red)	[0.77, 0.89]
• (green)	[0.58, 0.63]
• (blue)	[0.53, 0.56]
• (yellow)	[0.33, 0.34]
• (purple)	[0.33, 0.35]
• (cyan)	[0.00, 0.08]

(a) (b) $\widetilde{\mu}_{\mathcal{E}}^{A}$ (c) Necessity-possibility intervals

Fig. 7. Fuzzy enlacement landscape of a spiral (reference object A) and evaluation for different target objects inside the spiral (represented in different colors). (Color figure online)

To show the potential of our approach on real data, we evaluated the *"surrounded by"* relation on geographical objects extracted from a satellite image (Fig. 5(a)). This image[1] represents the *Bassin d'Arcachon* (France) and has been acquired by the FORMOSAT-2 satellite. The image was segmented to produce a 3-class image (Fig. 5(b)) composed of an island enclosed into the bay (reference object A) and the land coast (target object B). For illustrative purposes, Fig. 5(c–f) present the *Fuzz-DELs* of the bay object for different directions θ and widths ω. In particular, (c) shows the overall landscape $\widetilde{\mu}_{\mathcal{E}}^{A}$ that aggregates all directions, and (e) shows the direction where the target object is the least enlaced (*i.e.*, for $\theta = \frac{\pi}{3}$). The related directional necessity and possibility profiles are shown in Fig. 6, and the respective fuzzy surrounding evaluations $[N_{\mathcal{S}}^{AB}, \Pi_{\mathcal{S}}^{AB}]$ are reported in Table 1.

5.2 Global Enlacement

To pursue our study and to go further the surrounding spatial relation, we consider in a more generic sense the spatial relation *"enlaced by"*, in particular when the reference object has multiple degrees of concavities. Figure 7 (a) presents a complex spatial configuration involving a spiral and different target objects enclosed into it, from the center of the spiral to its "tail". The spiral is the reference object A, and we consider here its *Fuzz-DEL* $\widetilde{\mu}_{\mathcal{E}}^{A}$ that aggregates all directions. From this landscape (Fig. 7 (b)), we can observe the decreasing pattern (from white pixels to dark gray pixels) as we shift away from the center of the spiral. To assess this behavior, Fig. 7 (c) presents the intervals $[N(\widetilde{\mu}_{\mathcal{E}}^{A}, \mu_B), \Pi(\widetilde{\mu}_{\mathcal{E}}^{A}, \mu_B)]$ measuring the global enlacement for the different target objects inside the spiral. Note that other surrounding approaches cannot take into account the depth within the spiral. For instance, the approach of [16] provides the same evaluations for the green, blue, yellow and pink objects (*i.e.*, around 0.50), because it does not consider the reference object as a whole, but only looks at the visible concavities from the target object.

---

[1] Thanks to the CNES agency and the Kalideos project (http://kalideos.cnes.fr/).

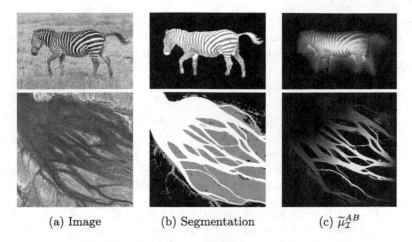

(a) Image	(b) Segmentation	(c) $\widetilde{\mu}_{\mathcal{I}}^{AB}$

Fig. 8. Examples of fuzzy interlacement landscapes (mapped into a "heat" color scale) obtained for different images (in (b), white: object A; gray: object B). (Color figure online)

5.3 Towards Fuzzy Interlacement Landscapes

We also propose some preliminary results on *interlacement* landscapes. The term interlacement is intended as a mutual enlacement of two objects. If we aggregate all directions, a fuzzy interlacement landscape between two objects A and B can be obtained by: $\widetilde{\mu}_{\mathcal{I}}^{AB} = \widetilde{\mu}_{\mathcal{E}}^{A} + \widetilde{\mu}_{\mathcal{E}}^{B}$. Fig. 8 shows the fuzzy interlacement landscapes obtained for two illustrative images, which have been respectively segmented into 3 classes. The first landscape is obtained from an image of a zebra whose coat features an alternating stripes pattern. We can observe the high interlacement values concentrated in the center of the animal's coat. The second landscape is obtained from a ASTER satellite image[2] covering a large delta river. Notice that the interlacement is mainly located around the ramifications between the river and the mangrove. Such interlacement visualization could be useful for instance for ecological landscape monitoring.

6 Conclusion

We introduced a generic fuzzy model for the evaluation of complex spatial configurations of binary objects represented in images. In particular, we focused on the enlacement spatial relation, which can be considered as a generalization of the notions of surrounding and imbrication of objects. Based on the directional enlacement model [6], our proposed evaluation approach exploits the concept of fuzzy landscapes to assess the enlacement of objects in the image space from a local point of view. An experimental study carried out on different illustrative examples highlighted the interest of this model to evaluate complex spatial

---

[2] U.S./Japan ASTER Science Team, NASA/GSFC/METI/ERSDAC/JAROS.

relations. In future works, we plan to further study how to exploit fuzzy interlacement landscapes, in particular with overlapping objects. We also plan to extend the model by integrating a measure of spacing in interlacement configurations, allowing to better take into account the distance between the objects.

References

1. Allen, J.F.: Maintaining knowledge about temporal intervals. Commun. ACM **26**(11), 832–843 (1983)
2. Bloch, I.: Fuzzy relative position between objects in image processing: a morphological approach. IEEE Trans. Pattern Anal. Mach. Intell. **21**(7), 657–664 (1999)
3. Bloch, I.: Fuzzy spatial relationships for image processing and interpretation: a review. Image Vis. Computing **23**(2), 89–110 (2005)
4. Bloch, I., Colliot, O., Cesar, R.M.: On the ternary spatial relation "Between". IEEE Trans. Syst. Man Cybern. B Cybern. **36**(2), 312–327 (2006)
5. Clément, M., Kurtz, C., Wendling, L.: Bags of spatial relations and shapes features for structural object description. In: Proceeding of ICPR (2016)
6. Clément, M., Poulenard, A., Kurtz, C., Wendling, L.: Directional enlacement histograms for the description of complex spatial configurations between objects. IEEE Trans. Pattern Anal. Mach. Intell. (2017, in press)
7. Colliot, O., Camara, O., Bloch, I.: Integration of fuzzy spatial relations in deformable models - application to brain MRI segmentation. Pattern Recogn. **39**(8), 1401–1414 (2006)
8. Delaye, A., Anquetil, E.: Learning of fuzzy spatial relations between handwritten patterns. Int. J. Data Min Model. Manage. **6**(2), 127–147 (2014)
9. Freeman, J.: The modelling of spatial relations. Comput. Graph. Image Process. **4**(2), 156–171 (1975)
10. Loménie, N., Racoceanu, D.: Point set morphological filtering and semantic spatial configuration modeling: application to microscopic image and bio-structure analysis. Pattern Recogn. **45**(8), 2894–2911 (2012)
11. Matsakis, P., Keller, J.M., Wendling, L., Marjamaa, J., Sjahputera, O.: Linguistic description of relative positions in images. IEEE Trans. Syst. Man Cybern. B Cybern. **31**(4), 573–88 (2001)
12. Matsakis, P., Naeem, M.: Fuzzy models of topological relationships based on the PHI-descriptor. In: Proceeding of FUZZ-IEEE, pp. 1096–1104 (2016)
13. Matsakis, P., Naeem, M., Rahbarnia, F.: Introducing the Φ-descriptor - a most versatile relative position descriptor. In: Proceeding of ICPRAM, pp. 87–98 (2015)
14. Matsakis, P., Wendling, L.: A new way to represent the relative position between areal objects. IEEE Trans. Pattern Anal. Mach. Intell. **21**(7), 634–643 (1999)
15. Rosenfeld, A., Klette, R.: Degree of adjacency or surroundedness. Pattern Recogn. **18**(2), 169–177 (1985)
16. Vanegas, M.C., Bloch, I., Inglada, J.: A fuzzy definition of the spatial relation "surround" - application to complex shapes. In: Proceeding of EUSFLAT, pp. 844–851 (2011)
17. Vanegas, M.C., Bloch, I., Inglada, J.: Alignment and parallelism for the description of high-resolution remote sensing images. IEEE Trans. Geosci. Remote Sens. **51**(6), 3542–3557 (2013)

Discrete Modelling and Visualization

An Introduction to Gamma-Convergence
for Spectral Clustering

Aditya Challa[1]([✉]), Sravan Danda[1], B.S. Daya Sagar[1], and Laurent Najman[2]

[1] Systems Science and Informatics Unit, Indian Statistical Institute, 8th Mile,
Mysore Road, Bangalore 560059, India
aditya.challa.20@gmail.com
[2] Université Paris-Est, LIGM, Equipe A3SI, ESIEE, Paris, France

Abstract. The problem of clustering is to partition the dataset into groups such that elements belonging to the same group are similar and elements belonging to the different groups are dissimilar. The unsupervised nature of the problem makes it widely applicable and also tough to solve objectively. Clustering in the context of image data is referred to as image segmentation. Distance based methods such as K-means fail to detect the non-globular clusters and hence spectral clustering was proposed to overcome this problem. This method detects the non globular structures by projecting the data set into a subspace, in which the usual clustering methods work well. Gamma convergence is the study of asymptotic behavior of minimizers of a family of minimization problems. Such a limit of minimizers is referred to as the gamma limit. Calculating the gamma limit for various variational problems has been proved useful - giving a different algorithm and insights into why existing methods work. In this article, we calculate the gamma limit of the spectral clustering methods, analyze its properties, and compare them with minimum spanning tree based clustering methods and spectral clustering methods.

1 Introduction

The problem of clustering is defined as - given a set of elements $\{x_i\}$, partition the set into non overlapping groups such that elements belonging to the same group are "similar", and elements belonging to different groups are "dissimilar". The importance of a solution to the problem of clustering is due to its wide range of applications [1,14,15]. Clustering in image data is also referred to as image segmentation. There exists several methods to solve the problem of clustering [2,10,11,18]. A comprehensive textbook on the subject is [1]. One of the most commonly used clustering methods is K-means [2,17]. However, it suffers from the problem of not being able to detect the non-globular structures. Spectral clustering methods were proposed to overcome this problem. Loosely speaking spectral clustering methods embed the data in a lower dimensional subspace, in which usual methods K-means clustering would be able to detect the non globular clusters as well.

Recently, in [6], seeded clustering/segmentation methods in [7,8,12,24] were extended by taking the limit of minimizers. This is referred to as the Γ-limit [19].

© Springer International Publishing AG 2017
W.G. Kropatsch et al. (Eds.): DGCI 2017, LNCS 10502, pp. 185–196, 2017.
DOI: 10.1007/978-3-319-66272-5_16

Γ-convergence is a tool to study the asymptotic behavior of families of minimum problems [3]. The aim of Γ-convergence is to replace a family of minimum problems with a single problem whose minima exhibits some interesting properties. For instance in [6] it has been shown that the Γ-limit revealed a new segmentation method which performs at least as well as graph cuts, random walker and shortest paths, if not better.

In this article our aim is to calculate the Γ-limit of the ratio cut spectral clustering. The Γ-limit of the ratio cut is referred to as *PRcut* and an algorithm to calculate the PRcut has been proposed. Due to various numerical precision errors and other constraints, a variant of the algorithm is implemented instead. Thus, during the exposition, we stick to the philosophy that theory is developed to be as general as possible, while the experiments are conducted with slightly modified theory based on practical considerations such as "small" clusters described later.

2   Background

Let $\{v_i\}$ be the given set of points in $\mathbb{R}^n$ which we would like to cluster. Taking each of these points as vertices, one can construct a similarity graph $\mathcal{G} = (V, E, W)$ with vertex set V, edge set E, and $W : E \to \mathbb{R}^+$ denotes weights assigned to each edge. Here $\mathbb{R}^+$ denotes the set of positive real numbers. With slight abuse of notation, we can write the weights as a $|V| \times |V|$ matrix, with w_{ij} denoting the edge weight between v_i and v_j. The degree of a vertex, d_i is given by

$$d_i = \sum_j w_{ij} \qquad (1)$$

Let D be the diagonal matrix $diag(d_1, d_2, \cdots, d_n)$. The *Laplacian* of a graph is then defined by

$$L = D - W \qquad (2)$$

We know that the Laplacian is a symmetric positive-semi definite matrix, and hence has non negative real eigenvalues, represented by $0 = \lambda_0 \le \lambda_1 \le \cdots \le \lambda_{n-1}$ [16]. The corresponding eigenvectors are denoted by $(e_0, e_1, \cdots, e_{n-1})$. Let $A \subseteq V$. Then the vector $\mathbf{1}_A(x)$ is given by

$$\mathbf{1}_A(x) = \begin{cases} 1 & if \ x \in A \\ 0 & otherwise \end{cases} \qquad (3)$$

Let p be a real parameter. Let $W^{(p)}$ be the matrix such that $W_{ij}^{(p)} = w_{ij}^p$. Let $D^{(p)}$ denote the matrix as constructed in (1) with weights w_{ij}^p. Let $L^{(p)} = D^{(p)} - W^{(p)}$.

2.1   Spectral Clustering

This section briefly reviews spectral clustering methods. For more details please refer to [20, 23, 25]. As noted before, spectral clustering methods work by embedding the data into a lower dimensional subspace. The three main steps are - (1)

Given a set of points $\{v_i\}$ (dataset), construct a similarity graph with each point as a vertex. (2) Construct the Laplacian for the obtained graph and calculate the first k eigenvectors. The value of k is fixed based on the number of clusters one would like to obtain. Let K be the matrix such that the ith column of K is the ith eigenvector e_{i-1}. (3) Using rows of the matrix K as new representation of the points v_i, use traditional clustering methods such as K-means to obtain the final clusters. Note that as a part of K-means step, the algorithm is run several times with random initialization of seeds.

Why does spectral clustering work? Although the definitive answer to this question still remains open, there exists several analyses which provide insights into this question [25]. One approach is to interpret the spectral clustering in an optimization framework. One of the measures to validate the appropriateness of the clusters is

$$cut(A_1, A_2, \cdots, A_k) = \frac{1}{2} \sum_{i=1}^{k} W(A_i, \overline{A_i}) \tag{4}$$

where $W(A, B) = \sum_{i \in A; j \in B} w_{ij}$, $\overline{A}$ denotes the complement of A in the vertex set V. $cut(.,.)$ measures how dissimilar the clusters are by taking the sum of the weights of the edges connecting distinct clusters. In practice minimizing the $cut(.,.)$ does not give good results, since it generally separates one vertex, and gives degenerate solutions. To solve this, it was proposed to use a slight modification of the above cost function. *Ratio-cut*, [25], is given by

$$Ratio - cut(A_1, A_2, \cdots, A_k) = \frac{1}{2} \sum_{i=1}^{k} \frac{W(A_i, \overline{A_i})}{|A_i|} \tag{5}$$

where $|A_i|$ is the cardinality of set A_i. It can be shown that minimizing the $Ratio-cut(.,.)$ for k clusters is approximately equivalent to solving the following optimization problem.

$$\underset{H \in \mathbb{R}^{n \times k}}{\text{minimize}} \quad Tr(H^t L H)$$
$$\text{subject to} \quad H^t H = I \tag{6}$$

Here I is the identity matrix and L is the laplacian as defined in (2). From the *Rayleigh-Ritz* theorem [16] we know that the solution to this optimization problem is obtained by considering the first k eigenvectors of L as columns of H.

2.2 Gamma Convergence

Let $\min\{F_p(x) : x \in X \subset \mathbb{R}^n\}$ be a family of minimum problems. Let x_p^* be a minimum of $F_p(x)$. We are interested in calculating the limit

$$x^* = \lim_{p \to \infty} \arg \min F_p \tag{7}$$

In other words, we are interested in the limit of a sequence of minimizers of the family $\{F_p(x)\}$. Note that there could be many such sequences. Now, consider a special case where

$$F_p(x) = \sum_{i=1}^{n} \alpha_i^p Q_i(x) \tag{8}$$

where $1 \geq \alpha_n > \alpha_{n-1} > \cdots > \alpha_1 > 0.$, and $Q_i(x)$ are smooth functions. We also assume that there exists a compact set C such that x_p^* belongs to C for all p. It turns out that, in this case, one can find a simple algorithm to calculate the limit of minimizers as described in the following Theorem 1 [19]. Define

$$M_n = \arg\min Q_n(x) \qquad x \in C \tag{9}$$

Recursively define for $k = n - 1, n - 2, \cdots, 1$

$$M_k = \arg\min Q_k(x) \qquad x \in M_{k+1} \tag{10}$$

Theorem 1. *Let $F_p(x)$ be as defined in (8) and x^* be the Γ-limit. Then $x^* \in M_1$.*

Refer to [19] for proof of Theorem 1. The main consequence of Theorem 1 is that it provides a method to calculate an "approximate" Γ-limit. One starts at the highest scale, optimizes the cost function at this scale, then moves on to the lower scale and repeats the process. Theorem 1 essentially states that one can obtain an approximate solution to the Γ-limit by this process. Approximate, in the sense that the Γ-limit belongs to the set M_1. The question of how good is the approximation, needs to be analyzed in the specific case.

3   Gamma Limit of Ratio - Cut

In this section, we calculate the approximate Gamma limit of the Ratio-cut. A few more notations are required. For a given graph $\mathcal{G}$, let $\mathcal{G}_k$ denote the graph (V, E_k, W_k). $E_k \subseteq E$ denotes the set of edges whose weight is w_k. According to the existence of the edge with weight w_k, W_k takes the values in $\{0, 1\}$. We refer to this graph as *scale graph/level graph* at level k. Just as for the original graph, one can construct a laplacian, L_k, for $\mathcal{G}_k$. A point to observe is that the entries in weight matrices of a level graph are either 1 or 0. We assume that the graph has distinct weights $w_1 < w_2 < \cdots < w_j$, where $j < |E|$. Given the notation as above, we have

$$Tr(H^t L H) = \sum_{k=1}^{j} w_k Tr(H^t L_k H) \tag{11}$$

Drawing a parallel with the Γ-convergence framework, we have that $Q_k(H) = Tr(H^t L_k H)$ and $\alpha_k = w_k$. We are thus interested in calculating the limit of minimizers of $\sum_{k=1}^{j} w_k^p Tr(H^t L_k H)$ as $p \to \infty$, subject to $H^t H = I$

Let $\mathcal{P}_k$ denote the following optimization problem.

$$\begin{aligned} \underset{H \in \mathbb{R}^{n \times m}}{\text{minimize}} \quad & Tr(H^t L_k H) \\ \text{subject to} \quad & H^t H = \mathbf{I} \end{aligned} \tag{12}$$

Thanks to Theorem 1, we have the following method to calculate the Γ-limit for spectral clustering.

1. Let $\mathcal{G}$ be a graph with distinct weights $w_1 < w_2 < \cdots < w_j$. Let M_{j+1} be the set of the all $n \times m$ matrices.
2. For each k going from j to 1, let M_k be the set of solutions for $\arg\min \mathcal{P}_k$ which belong to M_{k+1}.

The set M_1 is the output of the method. Note that the above steps are not implementable.

The main problem in finding an implementable version is to characterize all the solutions for the problem in (12). Let $\lambda_{(m)}$ denote the mth smallest eigenvalue. Let A_k be the matrix $[e_1, e_2, \cdots e_l]$, where e_i is the ith eigenvector, ordered in increasing order of the corresponding eigenvalue. The number of eigenvectors considered, l, are the number of eigenvalues less than or equal to $\lambda_{(m)}$. Let K be the matrix

$$K = \begin{bmatrix} I_{l_1 \times l_1} & 0 \\ 0 & k_{l_2 \times m - l_1} \end{bmatrix} \tag{13}$$

where $K^t K = I$. l_2 is the number of eigenvalues equal to $\lambda_{(m)}$, and $l = l_1 + l_2$. Let X be an orthogonal matrix such that $X^t X = I$. Then,

---

Algorithm 1. Efficient algorithm to compute Γ-limit for ratio-cut.

Input: A weighted graph, $\mathcal{G}$, with distinct weights $w_1 < w_2 < \cdots < w_j$. Number of clusters, m.
Output: N - A representation of the subspace spanned by the Γ-limit of the minimizers.
1: Set $k := j$.
2: **while** Number of connected components of $\mathcal{G}_{\geq k}$ is greater than or equal to m **do**
3:     Set $k := k - 1$ {We refer to this as an MST-Phase}
4: **end while**
5: Construct N by stacking the vectors $\mathbf{1}_{A_i}/\sqrt{|A_i|}$ in columns, where A_i is a connected component of $\mathcal{G}_{\geq k}$.
6: Set $l_1 := 0$ and $l_2 :=$ number of connected components in $G_{\geq k}$
7: Consider the graph $\mathcal{G}_k$ and let L_k be the corresponding laplacian.
8: Set $C = [N^t L_k N]_{l_2, l_2}$
9: Calculate the first eigenvectors of eigenvalue problem whose eigenvalue is less than or equal to $\lambda_{(m)}$.

$$Cx = \lambda x \tag{14}$$

10: Let A be the matrix obtained by stacking the eigenvectors as columns.
11: Construct $\hat{A}$ as

$$\hat{A} = \begin{bmatrix} I_{l_1 \times l_1} & 0 \\ 0 & A \end{bmatrix} \tag{15}$$

12: Update l_1 and l_2.
13: $N := N \times \hat{A}$
14: Set $k := k - 1$
15: **if** $k = 0$ **or** number of columns of N is equal to m **then**
16:     **return** N
17: **else**
18:     Goto Step (7)
19: **end if**

Proposition 1. *The set of all solutions to the optimization problem in* (12) *is of the form* $A_k K X$.

Proposition 1 results in an algorithm which is implementable. However, one requires to calculate all the eigenvectors at every stage, which is computationally expensive. Proposition 2 results in an efficient implementation of the algorithm as given in Algorithm 1. We refer the result obtained by this algorithm as *Power Rcut*. Also, $\mathcal{G}_{\geq k}$ denotes the graph with vertex set V and edge set $E_{\geq k} = \cup_{i \geq k} E_i$.

Proposition 2. *Given a graph* $\mathcal{G}$, *Let* $\mathcal{G}_{\geq k}$ *denote the graph, whose vertex set is* V *and edge set* E_k *containing all the edges whose weight is greater than or equal to* w_k. *At stage* k, *if* A *is a maximal connected component of* $\mathcal{G}_{\geq k}$, *then* $\mathbf{1}_A$ *is an eigenvector with eigenvalue* 0 *of the optimization problem* (12).

The proofs for the propositions and details of simplification to obtain the efficient Algorithm 1 will be discussed in detail in a complete version of the paper [4]. Observe that Algorithm 1 only gives an element of the set to which the Γ-limit actually belongs. The appropriateness of the solution thus obtained must be proved, which Proposition 3 gives.

Proposition 3. *Let* x *be the solution obtained by Algorithm 1 and let* x^* *be a* Γ-*limit. Then*

$$F_p(x^*) = F_p(x) \qquad \text{for all } p \tag{16}$$

Where F_p *is as given in* (8), *where* $Q_i(x) = Tr(x^t L_i x)$.

The above proposition implies that any Γ-limit, and the approximate one calculated have the same cost. In the context of spectral clustering, this implies that the solutions are equivalent and thus the solution obtained from Algorithm 1 is a good approximation. The proof of the above proposition will be detailed in later work.

4   Analysis with Experiments

In the rest of the article, we discuss how the algorithm works, its practical implementation and its similarities and dissimilarities with the MST based clustering and spectral clustering methods.

4.1   How the Algorithm Works?

Recall that the output of spectral clustering is a projection onto a subspace, and thus the algorithm produces a representation of the points in a subspace (denoted by N in Algorithm 1) which is a gamma limit. We assume that we need m clusters. In steps 2–4, we progressively add all the edges while there are at least m connected components in the threshold graph $\mathcal{G}_{\geq k}$. Thanks to Proposition 2 we know the first eigenvectors of the laplacian of the threshold graph are the indicator vectors as in (3) where A denotes each of the connected components. We construct the initial representation of the points N by taking the eigenvectors in step 5. Steps 7–14 update the representation N with respect to the lower weight edges. Note that, once the number of columns of the matrix N is equal to m, we need not update the representation anymore, since any other representation would just be an orthogonal transformation of the points and thus the clustering results would not change. This condition is checked in steps 15–19.

One issue with the gamma limit is that the property of non-trivial clusters for the Ratio-cut is not preserved in the limit. In the above algorithm, in practice, we get a lot of outliers and this results in small clusters. To avoid this, in steps 2–4, instead of calculating the number of connected components, we calculate the number of connected components whose size is greater than a given parameter *threshComp*. At the moment the algorithm jumps out of the while loop, all the components which have less than *threshComp* number of vertices are ignored. Recall that after the representation of the points is obtained, one has to perform K-means clustering to get the final partition. At this stage the ignored "small" clusters can either be returned as a different cluster, or combined with one of the larger clusters at random.

A simple application of the algorithm is illustrated in Fig. 1, where one can see that the algorithm correctly detects the object (flower in this case). Since small clusters are ignored, for the applications of image segmentation we do not get closed contours for the segments. Thus, for image segmentation the clusters are post-processed with an operator. Observe in Fig. 1(b) a few parts of the flower are missing. In this case, simple operators such as an opening works well, which gives the result as in Fig. 1(c). In general connected operators [22] preserve contours and are better for the post-processing of the image. Another important property of the Power Rcut clustering method is that, it results in smoother contours compared to the spectral clustering method. This is illustrated in Fig. 1(d)–(f).

4.2   Relation to MST-clustering

MST (Maximum Spanning Tree) based clustering is one of the earliest graph based clustering approach [5,13,21,26]. There exists several variations of the method. We consider here the simplest method - (a) Construct an MST (b) Iteratively remove the least weight edges until we get the required number of clusters. One of the most useful properties of spectral clustering is its ability to detect non-convex clusters in the data. This property is shared by the MST based clustering methods as well.

However, the problem with MST based clustering is breaking ties between edges of equal weight, which it does arbitrarily. Spectral clustering on the other hand ensures a single clustering (up to the arbitrariness of k-means step). In this sense, Power Rcut can be considered to be a method between these two clustering methods. Power Rcut follows a similar procedure as MST based clustering, and it breaks ties with spectral clustering on a subgraph. For example consider the graph in Fig. 2(a). Power Rcut segments the graph into two equal clusters, Fig. 2(b). The same behavior is also exhibited by spectral clustering. MST based

Fig. 1. (a) Original Image (Flower) (b) Power Rcut segment result (c) Power Rcut segment result post processing with opening (d) Original Image (Cameraman) (e) Contours obtained by Power R cut and (f) Contours obtained by Normalized spectral clustering.

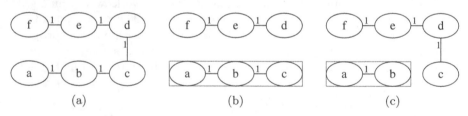

Fig. 2. (a) Basic Graph (b) Power Rcut Clustering (c) MST based clustering. Observe that Power Rcut splits the graph into two equal halves, where as MST based clustering does not.

clustering on the other hand segments the graph into non-equal parts since it breaks the ties arbitrarily, Fig. 2(c).

This is because, Power Rcut takes into consideration the sizes of the cluster while breaking ties. Consider another synthetic example in Fig. 3(a). An MST is highlighted in bold edges in Fig. 3(a). Since MST based clustering would break the ties randomly, it could result in clusters as in Fig. 3(b). Power Rcut clustering on the other would definitely not give clusters as in Fig. 3(b). An example Power Rcut clusters is given in Fig. 3(c).

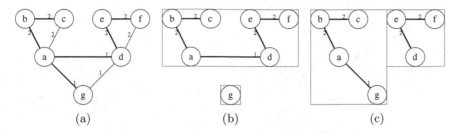

Fig. 3. (a) Basic Graph and MST edges highlighted in bold (b) Clustering with MST (c) Power Rcut Clustering. Observe that Power Rcut clustering tries to split the graph into equal parts.

Proposition 4. *If at a threshold t, we have that $\mathcal{G}_{\geq t}$ has exactly m clusters, then MST clustering and Power Rcut clustering results in the same clusters.*

Power Rcut clustering and MST based clustering are in fact closely related as suggested by the Proposition 4. The proof follows from the following observation - for any two vertices belonging to the same connected component has the same value in the embedded space, and hence belong to the same cluster. Since, we require m clusters, and there are m connected components, the k-means step results in each of these m components as a cluster.

4.3   Relation to Spectral Clustering

The Power Rcut solution can also be interpreted as being obtained by spectral clustering on every level graph $\mathcal{G}_k$. Since, the first few eigenvectors are the indicator of the connected components, this gives a heuristic explanation for steps 2–4 in Algorithm 1. This points out the similarity between Power Rcut solution and spectral clustering.

In low noise conditions, spectral clustering and Power Rcut clustering results are similar. However, as noise level increases, spectral clustering will not be able to identify the regions anymore. In Fig. 4(a),(b) and (c) data points are sampled from two concentric circles with noise. Figure 4(c) shows the results obtained by spectral clustering are shown. Notice that the structure of the two circles is not preserved. Figure 4(a) show the results obtained with Power Rcut. To generate the results of Power Rcut, as a post processing step, we assign each of the points the "small" clusters to the closest cluster, which results in Fig. 4(b). Although not perfect, Power Rcut results preserve some structure. For a quantitative view we calculate the Fowlkes-Mallows(FW) scores, given by the formula

$$Score = \frac{TP}{\sqrt{(TP + FP)}\sqrt{(TP + FN)}} \tag{17}$$

where TP is true positives, FP is false positives and FN is false negatives. The FW scores for various noise levels is plotted in Fig. 4(d). Note that Power Rcut performs better in high noise scenarios compared to spectral clustering.

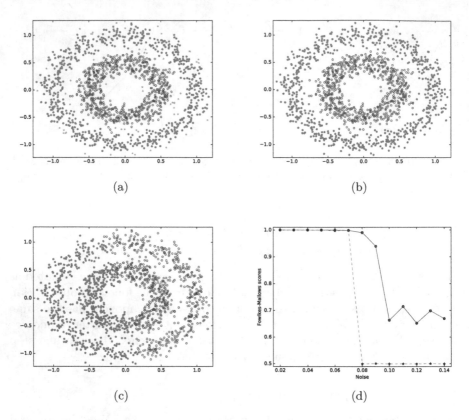

Fig. 4. Data is sampled from two concentric circles with noise. The clusters obtained are represented by red dots and blue circles. (a) Power Rcut clustering. The green stars indicate the "noise" clusters obtained. (b) Power Rcut clustering obtained by adjusting the noise clusters to the nearest cluster. (c) Spectral clustering (d) Fowlkes-Mallows scores for varying noise. The dashed line indicates spectral clustering score and continuous line indicate Power Rcut score. The higher the score, the better the procedure.

5   Conclusion and Future Work

In this article we outlined the basics of spectral clustering and discussed the concept of Γ-convergence. Important results to calculate the Γ-limit of Ratio-cut were outlined and the algorithm to calculate the gamma limit was obtained. The correctness of the algorithm was shown via a proposition. The similarities and dissimilarities between Power Rcut, MST clustering and Spectral clustering are analyzed stating the result that Power Rcut clustering is a specific kind of MST clustering. Power Rcut clustering was shown to be superior to MST clustering in dealing with ramp effects of the image, and superior to spectral clustering in noisy scenarios.

Note that none of the steps in spectral clustering methods indicate directly as to why spectral clustering methods could obtain non-convex clusters. This has been an open question for a long time. We believe that, the fact that gamma limit of spectral clustering is a method close to MST clustering provides an insight into this question. In particular, this provides a bridge between spectral clustering and MST clustering and allows us to dive into the question of why spectral methods work.

Note that the algorithm starts with combining the edges of highest weight until the number of clusters are obtained. This can be interepreted as a greedy method of "combining all the points which definitely belong to the same cluster", thus reducing the size of the dataset and allowing for faster computation. This in theory can also be done in parallel. This can allow spectral clustering to be applicable in the case of large datasets and also have efficient implementations. The error bounds and the exact algorithm to do this are a subject of future research.

In [9], the authors study in depth hierarchies and their equivalence with MST. Observe that the Algorithm 1 is inherently hierarchical. The question of how Algorithm 1 is related to the concepts in [9] is also a subject of future research.

Acknowledgements. AC and SD would like to thank Indian Statistical Institute. This work has been partly funded by ANR-15-CE40-0006 CoMeDiC and ANR-14-CE27-0001 GRAPHSIP research grants. BSDS would like to acknowledge the support received from the Science and Engineering Research Board (SERB) of the Department of Science and Technology (DST) with the grant number EMR/2015/000853, and the Indian Space Research Organization (ISRO) with the grant number ISRO/SSPO/Ch-1/2016-17.

References

1. Aggarwal, C.C., Reddy, C.K.: Data Clustering: Algorithms and Applications, 1st edn. Chapman & Hall/CRC, Boca Raton (2013)
2. Arthur, D., Vassilvitskii, S.: k-means++: the advantages of careful seeding. In: Proceedings of the Eighteenth Annual ACM-SIAM Symposium on Discrete Algorithms, pp. 1027–1035. Society for Industrial and Applied Mathematics (2007)
3. Braides, A.: Gamma-convergence for Beginners, vol. 22. Clarendon Press (2002)
4. Challa, A., Danda, S., Daya Sagar, B.S., Najman, L.: Power spectral clustering. Technical report, Université Paris-Est, LIGM, ESIEE Paris (working paper). https://hal.archives-ouvertes.fr/hal-01516649/
5. Chowdhury, N., Murthy, C.: Minimal spanning tree based clustering technique: relationship with bayes classifier. Pattern Recogn. **30**(11), 1919–1929 (1997)
6. Couprie, C., Grady, L., Najman, L., Talbot, H.: Power watershed: a unifying graph-based optimization framework. IEEE Trans. Pattern Anal. Mach. Intell. **33**(7), 1384–1399 (2011)
7. Cousty, J., Bertrand, G., Najman, L., Couprie, M.: Watershed cuts: minimum spanning forests and the drop of water principle. IEEE Trans. Pattern Anal. Mach. Intell. **31**(8), 1362–1374 (2009)

8. Cousty, J., Bertrand, G., Najman, L., Couprie, M.: Watershed cuts: thinnings, shortest path forests, and topological watersheds. IEEE Trans. Pattern Anal. Mach. Intell. **32**(5), 925–939 (2010)

9. Cousty, J., Najman, L., Kenmochi, Y., Guimares, S.: Hierarchical segmentations with graphs: quasi-flat zones, minimum spanning trees, and saliency maps. Research report, LIGM, July 2016. https://hal.archives-ouvertes.fr/hal-01344727

10. Dempster, A.P., Laird, N.M., Rubin, D.B.: Maximum likelihood from incomplete data via the EM algorithm. J. Roy. Stat. Soc, Ser. B (Methodol.) **39**(1), 1–38 (1977)

11. Ester, M., Kriegel, H.P., Sander, J., Xu, X., et al.: A density-based algorithm for discovering clusters in large spatial databases with noise. KDD **96**, 226–231 (1996)

12. Grady, L.: Random walks for image segmentation. IEEE Trans. Pattern Anal. Mach. Intell. **28**(11), 1768–1783 (2006)

13. Grygorash, O., Zhou, Y., Jorgensen, Z.: Minimum spanning tree based clustering algorithms. In: 2006 18th IEEE International Conference on Tools with Artificial Intelligence (ICTAI 2006), pp. 73–81. IEEE (2006)

14. Jain, A.K.: Data clustering: 50 years beyond k-means. Pattern Recogn. Lett. **31**(8), 651–666 (2010)

15. Jain, A.K., Murty, M.N., Flynn, P.J.: Data clustering: a review. ACM Comput. Surv. (CSUR) **31**(3), 264–323 (1999)

16. Lütkepohl, H.: Handbook of Matrices, 1st edn. Wiley, Chichester (1997)

17. MacQueen, J., et al.: Some methods for classification and analysis of multivariate observations. In: Proceedings of the Fifth Berkeley Symposium on Mathematical Statistics and Probability, Oakland, CA, USA, vol. 1, pp. 281–297 (1967)

18. McLachlan, G., Peel, D.: Finite Mixture Models. Wiley, New York (2004)

19. Najman, L.: Extending the powerwatershed framework thanks to Γ-convergence. Technical report, Université Paris-Est, LIGM, ESIEE Paris (to appear in SIAM Journal on Imaging Sciences). https://hal-upec-upem.archives-ouvertes.fr/hal-01428875

20. Ng, A.Y., Jordan, M.I., Weiss, Y., et al.: On spectral clustering: analysis and an algorithm. Adv. Neural Inform. Process. Syst. **2**, 849–856 (2002)

21. Peter, S.J.: Minimum spanning tree-based structural similarity clustering for image mining with local region outliers. Int. J. Comput. Appl. (0975–8887) Volume (2010)

22. Serra, J.: A lattice approach to image segmentation. J. Math. Imag. Vis. **24**(1), 83–130 (2006)

23. Shi, J., Malik, J.: Normalized cuts and image segmentation. IEEE Trans. Pattern Anal. Mach. Intell. **22**(8), 888–905 (2000)

24. Sinop, A.K., Grady, L.: A seeded image segmentation framework unifying graph cuts and random walker which yields a new algorithm. In: IEEE 11th International Conference on Computer Vision, ICCV 2007, pp. 1–8. IEEE (2007)

25. Von Luxburg, U.: A tutorial on spectral clustering. Stat. Comput. **17**(4), 395–416 (2007)

26. Zahn, C.T.: Graph-theoretical methods for detecting and describing gestalt clusters. IEEE Trans. Comput. **100**(1), 68–86 (1971)

Digital Surface Regularization by Normal Vector Field Alignment

David Coeurjolly[1]([✉]), Pierre Gueth[1], and Jacques-Olivier Lachaud[2]

[1] Université de Lyon, CNRS, LIRIS UMR 5205, 69622 Lyon, France
david.coeurjolly@liris.cnrs.fr
[2] Laboratoire de Mathématiques, CNRS, UMR 5127,
University Savoie Mont Blanc, Chambery, France

Abstract. Digital objects and digital surfaces are isothetic structures *per se*. Such surfaces are thus not adapted to direct visualization with isothetic quads, or to many geometry processing methods. We propose a new regularization technique to construct a piecewise smooth quadrangulated surface from a digital surface. More formally we propose a variational formulation which efficiently regularizes digital surface vertices while complying with a prescribed, eventually anisotropic, input normal vector field estimated on the digital structure. Beside visualization purposes, such regularized surface can then be used in any geometry processing tasks which operates on triangular or quadrangular meshes (e.g. compression, texturing, anisotropic smoothing, feature extraction).

1 Introduction

Objective. This paper addresses the problem of approximating the boundary of an object given as a 3d *binary* image (see Fig. 4 top for an example). Such input data is called a 3d digital object and its surface is called a digital surface. They arise as digitizations of continuous objects and segmentation of 3d images. Compared to gray-level volumetric images, such binary image model is much more complex to handle: the function takes only two values, it is not continuous and of course has no gradient, it is known only at regularly sampled places. For this data, direct use of former methods gives triangulated meshes with poor geometry (staircasing effects, few possible orientations).

Contribution. We propose an original method to construct a piecewise smooth approximation of the boundary of a digital object. It follows a convex variational principle that regularizes the digital surface according to three criteria: First, it must stay close to the input data; second, it should comply to a well-chosen normal vector field $\mathbf{u}$; third, its cells must be as regular as possible. Our method is not iterative and provides excellent piecewise smooth reconstructions when the normal vector field $\mathbf{u}$ is a good approximation of the normal vector field of the original continuous object. This is why we use recent digital normal vector estimators which offer multigrid convergence guarantees [5,9] and may also

© Springer International Publishing AG 2017
W.G. Kropatsch et al. (Eds.): DGCI 2017, LNCS 10502, pp. 197–209, 2017.
DOI: 10.1007/978-3-319-66272-5_17

detect normal discontinuities [4]. By this way, we are able to take into account the discrete and arithmetic nature of digital objects while using the powerful framework of convex optimization. Experiments show that the output surface is aligned with the prescribed normal vector field and is closer to the underlying continuous object than was the input digitized surface. Sharp features are nicely delineated and the quality of mesh faces is very good.

Related works. First of all, the numerous methods that extract isosurfaces from a volume image or function provide poor results with such input data, since they rely on the exact value of the image /function to determine the position of vertices. This includes marching cubes, dual marching cubes and the many variants designed to reconstruct meshes from implicit surfaces [12–14]. Standard mesh denoising methods could also be considered for removing staircasing effects of digital surfaces [8,15–17]. However, they tend either to consider all steps as features or to smooth everything out. Very few approaches take into account the peculiar nature of digital data. In 2d, we may quote early works for digital contour polygonalization, which use digital straightness properties to align digital points onto their estimated tangent line [2]. In 3d, reversible polyhedrization of digital surfaces can be achieved with greedy digital plane segmentation followed by Marching-Cubes sewing [3]. Although they are theoretically reversible, none of these techniques achieve similar visual quality compared to our proposal (see Fig. 3). Note that we experimentally show that our regularized surface is also very close to the original digital one.

Outline. Section. 2 gives the necessary notions and notations of digital geometry used throughout the paper. Section 3 presents the variational formulation of our digital surface regularization method. The convex optimization algorithm is described in Sect. 4. Experiments illustrating the qualities of the method are given in Sect. 5, as well as a quantitative asymptotic analysis of the surface reconstruction. Limitations and possible extensions of this method are discussed in Sect. 6.

2   Preliminaries

A 3d binary image is simply a function from $\mathbb{Z}^3$ toward $\{0, 1\}$. The set $Z \subset \mathbb{Z}^3$ of digital points where the function is not null represents the digital object. It is naturally embedded in $\mathbb{R}^3$ as the union of unit cubes centered on these digital points. Its topological boundary is called its digital boundary, and is a special case of digital surface. Since a digital surface is a set of unit squares sewn together, it has a cellular representation in cartesian cubical grid as a set $\mathbb{E}^0$ of $0-$cells (vertices), a set $\mathbb{E}^1$ of $1-$cells (edges), and a set $\mathbb{E}^2$ of $2-$cells (square faces). We do not detail the topological combinatorial structure, *á-la* Alexandrov, associated with subsets of the Cartesian cubic grid (and more specifically digital surfaces). Interested readers may refer to [10]. We just assume in the following that $(\mathbb{E}^0, \mathbb{E}^1, \mathbb{E}^2)$ defines a proper combinatorial 2-manifold without boundaries.

Our objective is to reconstruct a piecewise smooth surface from a digital boundary. Therefore we have in mind that the digital shape comes from some digitization or sampling of a continuous shape X, and our objective is to infer the original shape boundary ∂X with solely its digitization as input. The *multigrid convergence* framework is thus well adapted to evaluate objectively the qualities of our method (e.g. see [10,11]). For some continuous shape $X \subset \mathbb{R}^3$ and a digitization process parameterized by a sampling grid step h, we will evaluate if our regularized surface $\partial_h^* X$ (in red in Fig. 1) is closer to the continuous surface ∂X than the digitized boundary $\partial_h X$ (in orange). And of course, we wish that the finer is the resolution ($h \to 0$) the better is the approximation. Classical notations are recalled in Fig. 1. It is already known that, for the Gauss digitization (denoted $G_h(X)$ in Fig. 1), the digitized boundary $\partial_h X$ of a compact subset X of $\mathbb{R}^d$ with positive reach is Hausdorff close to ∂X, with a distance no greater than $\sqrt{d}h/2$ [11]. Hence our regularized surface $\partial_h^* X$ should also be Hausdorff close to ∂X. In Sect. 5, we show that this is indeed the case, and with a better constant than $\frac{\sqrt{3}}{2}h$.

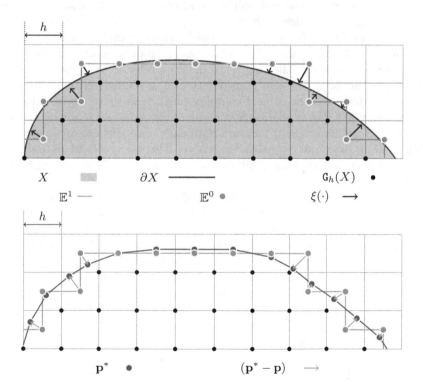

Fig. 1. Illustration of the digitization models and notations in dimension 2. $\xi(\mathbf{y})$ maps $\mathbf{y} \in \mathbb{R}^2 \setminus MA(X)$ to the closest point on ∂X ($MA(X)$ being the medial axis of X). (Color figure online)

Furthermore, the normals of the regularized surface $\partial_h^* X$ should also tend toward the normal of the original continuous surface ∂X. We achieve this property by adding as input to our regularization process a normal vector field associated to faces of the input digital surface. By choosing normal estimates given by some multigrid convergent normal estimator (like [5,9]), our variational formulation makes the normals of the regularized surface $\partial_h^* X$ align with the normals of ∂X.

3   Variational Formulation

Our variational model will simply move the vertices $\mathbb{E}^0$ of the input digital surface in order to regularize it. Although the cellular topology of the digital surface is used in the process, the output regularized surface has the same cellular topology as the input digital surface.

In the following, we denote $\mathbf{p} := \{\mathbf{p}_i\}$ the canonical coordinates in $\mathbb{R}^3$ of the $0-$cells $\{\sigma_i^0\}$ in the embedding Euclidean space. Similarly we introduce $\mathbf{p}^* := \{\mathbf{p}_i^*\} \in \mathbb{R}^3$ the set of regularized points coordinates in $\mathbb{R}^3$ associated with each $0-$cell. Furthermore, we denote by $\mathbf{u} := \{\mathbf{u}_k\}$ the input discrete normal vector field associated with the faces of the cubical complex (vector $\mathbf{u}_k$ is the estimated normal of the cell σ_k^2). In the following sections, we provide more details about the structure of these sets allowing us to define a reliable and efficient calculus on the digital surface. We also consider $\mathbf{S}$ as the embedded quadrangulated surface associated with the cubical complex (with vertices $\mathbf{p}$, edges induced by $\mathbb{E}^1$ and faces by $\mathbb{E}^2$). We denote by $\mathbf{S}^*$ the quadrangulated surface with the same structure as $\mathbf{S}$ with vertices $\mathbf{p}^*$. Please note that once vertices are regularized, the quads of $\mathbf{S}^*$ may not be planar anymore. The surface $\mathbf{S}$ is, by definition, $\partial_h X$ if we are considering the digitization of a shape X. The surface $\mathbf{S}^*$ then corresponds to $\partial_h^* X$ in this case.

Before formally defining $\mathbf{p}^*$, let us consider a given point set $\hat{\mathbf{p}} \subset \mathbb{R}^3$ and its associated quadrangulation $\hat{\mathbf{S}}$. We evaluate the energy associated with this point set. First, we want $\hat{\mathbf{p}}$ to be close to $\mathbf{p}$ (*data attachment*). Then, we want the discrete normal vector field associated with $\hat{\mathbf{p}}$ to comply with the input vector field $\mathbf{u}$. Since quads associated with $\hat{\mathbf{S}}$ may not be planar, we place this constraint onto the edges of each quad: we want each edge $\mathbf{d}\hat{\mathbf{p}}_j$ (where $\mathbf{d}\hat{\mathbf{p}}_j := \hat{\mathbf{p}}_{j_0} - \hat{\mathbf{p}}_{j_1}$ with $\partial\sigma_j^1 = \{\sigma_{j_0}^0, \sigma_{j_1}^0\}$, $\sigma_{j_0}^0$ being the tip of edge σ_j^1 and $\sigma_{j_1}^0$ being the origin of edge σ_j^1)[1] to be as orthogonal as possible to each neighboring face normal vector $\mathbf{u}_k$ (*normal vector alignment*). Those constraints are handled by the first and second terms of the following energy function :

$$\mathcal{E}(\hat{\mathbf{p}}) := \alpha \underbrace{\|\hat{\mathbf{p}} - \mathbf{p}\|^2}_{\mathcal{E}_{\text{data}}(\hat{\mathbf{p}})} + \beta \underbrace{\|\mathbf{d}\hat{\mathbf{p}} \cdot \mathbf{u}\|^2}_{\mathcal{E}_{\text{align}}(\hat{\mathbf{p}})} + \gamma \underbrace{\|\hat{\mathbf{p}} - \hat{\mathbf{b}}\|^2}_{\mathcal{E}_{\text{fairness}}(\hat{\mathbf{p}})} . \tag{1}$$

---

[1] For readers familiar with Discrete Exterior Calculus (DEC), $\mathbf{d}$ is similar to an exterior derivative operator on (triplets of) primal $0-$forms of the cubical complex.

where $\hat{\mathbf{b}}_i$ holds the coordinates of the barycenter of the neighboring vertices of $\mathbf{p}_i$. The last term (*fairness*) ensures that vertices of $\hat{\mathbf{S}}$ are well distributed along the surface: it moves the points onto their tangent planes so that the sampling is as regular as possible. In Sects. 3.1, 3.2 and 3.3, we detail the norms involved in each term of this functional.

From this formulation, we define the optimal regularized coordinates $\mathbf{p}^*$ as

$$\mathbf{p}^* := \underset{\hat{\mathbf{p}}}{\operatorname{argmin}} \ \mathcal{E}(\hat{\mathbf{p}}). \tag{2}$$

We detail now each energy term in a discrete calculus setting, which allows an efficient minimization of (2) (see Sect. 4).

3.1   Data Attachment Term

Since $\mathbf{p}$ contains point coordinates, it can be interpreted as a triplet of maps $\mathbf{p}_x$, $\mathbf{p}_y$ and $\mathbf{p}_z$ from $\mathbb{E}^0$ to $\mathbb{R}$ containing the vertex coordinates in embedding space. Thanks to a numbering of cells in $\mathbb{E}^0$, we use a vector representation of $\mathbf{p}$ as a single column vector concatenating vectors associated with $\mathbf{p}_x$, $\mathbf{p}_y$ and $\mathbf{p}_z$. In other words $\mathbf{p} = [\mathbf{p}_x^T, \mathbf{p}_y^T, \mathbf{p}_z^T]^T$. Data attachement term keeps $\mathbf{p}^*$ close to $\mathbf{p}$ and guarantees the convexity of the problem. It is defined as follows:

$$\mathcal{E}_{\text{data}}(\hat{\mathbf{p}}) := \|\hat{\mathbf{p}} - \mathbf{p}\|^2. \tag{3}$$

The norm in (3) is the norm of discrete $\mathbb{E}^0 \mapsto \mathbb{R}^3$ maps and is defined from the scalar products between discrete $\mathbb{E}^0 \mapsto \mathbb{R}$ maps $\langle \bullet, \bullet \rangle_0$:

$$\|\hat{\mathbf{p}} - \mathbf{p}\|^2 = \|\mathbf{a}\|^2 = \langle \mathbf{a}_x, \mathbf{a}_x \rangle_0 + \langle \mathbf{a}_y, \mathbf{a}_y \rangle_0 + \langle \mathbf{a}_z, \mathbf{a}_z \rangle_0. \tag{4}$$

To shorten notation in the previous equation, $\mathbf{a}$ is the difference map $(\hat{\mathbf{p}} - \mathbf{p})$.

One can simply consider classical Euclidean scalar products between vectors in $\mathbb{R}^{|\mathbb{E}^0|}$ to define (4). In Sect. 3.4, we propose an alternative definition which is more consistent with discrete calculus on combinatorial structures.

3.2   Normal Vector Alignment Term

The second term is the most complex one and tends to orthogonalize the direction of each edge with adjacent face normal vectors:

$$\mathcal{E}_{\text{align}}(\hat{\mathbf{p}}) := \|\mathbf{d}\hat{\mathbf{p}} \cdot \mathbf{u}\|^2. \tag{5}$$

First let us look at $\mathbf{d}\hat{\mathbf{p}} : \mathbb{E}^1 \mapsto \mathbb{R}^3$ (see definition in Sect. 3). There exists a linear operator $\mathbf{D} : (\mathbb{E}^0 \mapsto \mathbb{R}^3) \mapsto (\mathbb{E}^1 \mapsto \mathbb{R}^3)$ such that

$$\mathbf{d}\hat{\mathbf{p}} = \mathbf{D}\hat{\mathbf{p}}. \tag{6}$$

Thanks to the linearization of $\hat{\mathbf{p}}$ as a $3|\mathbb{E}^0|$ vector, such linear operator can be represented as a $3|\mathbb{E}^1| \times 3|\mathbb{E}^0|$ matrix. $\mathbf{d}\hat{\mathbf{p}} \cdot \mathbf{u} : \mathbb{E}^2 \mapsto \mathbb{R}^4$ holds the scalar products for each edge adjacent to all faces in embedding space:

$$(\mathbf{d}\hat{\mathbf{p}} \cdot \mathbf{u})_k = (\mathbf{d}\hat{\mathbf{p}}_{k_0} \cdot \mathbf{u}_k, \mathbf{d}\hat{\mathbf{p}}_{k_1} \cdot \mathbf{u}_k, \mathbf{d}\hat{\mathbf{p}}_{k_2} \cdot \mathbf{u}_k, \mathbf{d}\hat{\mathbf{p}}_{k_3} \cdot \mathbf{u}_k) \tag{7}$$

where $\partial \sigma_k^2 = \{\sigma_{k_0}^1, \sigma_{k_1}^1, \sigma_{k_2}^1, \sigma_{k_3}^1\}$ (always four edges in face border).

Furthermore, there exists a linear operator $\mathbf{U} : (\mathbb{E}^1 \mapsto \mathbb{R}^3) \mapsto (\mathbb{E}^2 \mapsto \mathbb{R}^4)$ such that :

$$d\hat{\mathbf{p}} \cdot \mathbf{u} = \mathbf{U} \mathbf{D} \hat{\mathbf{p}}. \tag{8}$$

Again, such operator can be represented as a $4|\mathbb{E}^2| \times 3|\mathbb{E}^1|$ matrix in our discrete calculus.

The norm in (5) is the norm of discrete $\mathbb{E}^2 \mapsto \mathbb{R}^4$ maps and is defined from the scalar products between discrete $\mathbb{E}^2 \mapsto \mathbb{R}$ maps $\langle \bullet, \bullet \rangle_2$:

$$\|d\hat{\mathbf{p}} \cdot \mathbf{u}\|^2 = \|\mathbf{b}\|^2 = \langle \mathbf{b}_{k_0}, \mathbf{b}_{k_0} \rangle_2 + \langle \mathbf{b}_{k_1}, \mathbf{b}_{k_1} \rangle_2 + \langle \mathbf{b}_{k_2}, \mathbf{b}_{k_2} \rangle_2 + \langle \mathbf{b}_{k_3}, \mathbf{b}_{k_3} \rangle_2. \tag{9}$$

In the previous equation, $\mathbf{b}$ is a shorthand for $d\hat{\mathbf{p}} \cdot \mathbf{u}$. Finally, the alignment term is thus simply expressed as

$$\mathcal{E}_{\text{align}}(\hat{\mathbf{p}}) = \|\mathbf{U} \mathbf{D} \hat{\mathbf{p}}\|^2. \tag{10}$$

3.3   Fairness Term

The last term is the fairness term which tends to flatten the regularized complex and to distribute the vertex positions with tangential displacements (see Fig. 2):

$$\mathcal{E}_{\text{fairness}}(\hat{\mathbf{p}}) := \|\hat{\mathbf{p}} - \hat{\mathbf{b}}\|^2. \tag{11}$$

The norm is the same as in (3) and relies on the same scalar product $\langle \bullet, \bullet \rangle_0$. $\hat{\mathbf{b}}_i$ is the barycenter of neighboring vertices to $\mathbf{p}_i$. More formally[2]:

$$\hat{\mathbf{b}}_i := \frac{1}{|\text{link}(\sigma_i^0)|} \sum_{\sigma_j^0 \in \text{link}(\sigma_i^0)} \hat{\mathbf{p}}_j. \tag{12}$$

This defines $\mathbf{B} : (\mathbb{E}^0 \mapsto \mathbb{R}^3) \mapsto (\mathbb{E}^0 \mapsto \mathbb{R}^3)$ a linear operator from and to $\mathbb{E}^0 \mapsto \mathbb{R}^3$ ($3|\mathbb{E}^0| \times 3|\mathbb{E}^0|$ matrix) allowing us to write $\hat{\mathbf{b}}$, expressed as a linearized column vector of positions, as the matrix-vector multiplication,

$$\hat{\mathbf{b}} := \mathbf{B} \hat{\mathbf{p}}. \tag{13}$$

Note that matrix $\mathbf{B}$ is sparse which leads to efficient factorization (see Sect. 4). The fairness term reduces to

$$\mathcal{E}_{\text{fairness}}(\hat{\mathbf{p}}) = \|(\mathbf{I} - \mathbf{B})\hat{\mathbf{p}}\|^2, \tag{14}$$

where $\mathbf{I}$ is the identity operator (identity matrix $3|\mathbb{E}^0| \times 3|\mathbb{E}^0|$).

---

[2] Link(σ_i^0) is the link operator on cubical complexes. As a consequence, $0-$cells in this set are *connected* to σ_i^0 by a $1-$cell in the complex.

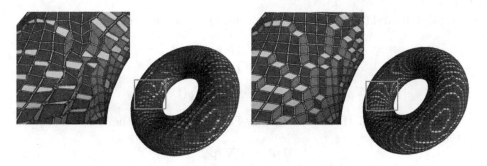

Fig. 2. Impact of the $\mathcal{E}_{\text{fairness}}$ term on the regularized quadrangulation (without on the left, with on the right).

3.4 Scalar Products

In previous definitions, scalar products $\langle \bullet, \bullet \rangle_0$ and $\langle \bullet, \bullet \rangle_2$ must be specified. More precisely, we need to specify the metric tensor (definite positive, therefore symmetric matrix) $\mathbf{W}_0$ (resp. $\mathbf{W}_2$) associated with maps $\mathbf{x}, \mathbf{y} : \mathbb{E}^0 \mapsto \mathbb{R}$ (resp. maps $\mathbf{u}, \mathbf{v} : \mathbb{E}^2 \mapsto \mathbb{R}$):

$$\langle \mathbf{x}, \mathbf{y} \rangle_0 := \mathbf{x}^T \mathbf{W}_0 \mathbf{y}, \tag{15}$$

$$\langle \mathbf{u}, \mathbf{v} \rangle_2 := \mathbf{u}^T \mathbf{W}_2 \mathbf{v}. \tag{16}$$

A simple choice consists in considering identity matrices for $\mathbf{W}_0$ and $\mathbf{W}_2$. However, specific weights can be set if embedding priors of the digital surface are known (see [6] for a complete discussion and Sect. 6).

Since we consider triplets or quadruple of respectively $0-$ and $2-$forms in previous formulations (due to the linearization of positions into a single vector and the linearization of the four dot products associated with each face edges), we need to extend scalar products of vector is $\mathbb{R}^{|\mathbb{E}^0|}$ to vectors in $\mathbb{R}^{3|\mathbb{E}^0|}$ (respectively to vectors in $\mathbb{R}^{4|\mathbb{E}^2|}$). We simply define

$$\bar{\mathbf{W}}_0 := \begin{bmatrix} \mathbf{W}_0 & 0 & 0 \\ 0 & \mathbf{W}_0 & 0 \\ 0 & 0 & \mathbf{W}_0 \end{bmatrix}, \bar{\mathbf{W}}_2 := \begin{bmatrix} \mathbf{W}_2 & 0 & 0 & 0 \\ 0 & \mathbf{W}_2 & 0 & 0 \\ 0 & 0 & \mathbf{W}_2 & 0 \\ 0 & 0 & 0 & \mathbf{W}_2 \end{bmatrix}. \tag{17}$$

4 Energy Minimization

Since (1) is convex, we can compute $\mathbf{p}^*$ as the unique solution to

$$\nabla_{\hat{\mathbf{p}}} \mathcal{E}(\mathbf{p}^*) = \alpha \, \nabla_{\hat{\mathbf{p}}} \mathcal{E}_{\text{data}}(\mathbf{p}^*) + \beta \, \nabla_{\hat{\mathbf{p}}} \mathcal{E}_{\text{align}}(\mathbf{p}^*) + \gamma \, \nabla_{\hat{\mathbf{p}}} \mathcal{E}_{\text{fairness}}(\mathbf{p}^*) = 0. \tag{18}$$

From (3), (10) and (14) gradients of each energy term can be expressed as follows:

$$\nabla_{\hat{\mathbf{p}}}\mathcal{E}_{\text{data}}(\hat{\mathbf{p}}) = 2\bar{\mathbf{W}}_0(\hat{\mathbf{p}} - \mathbf{p}), \tag{19}$$

$$\nabla_{\hat{\mathbf{p}}}\mathcal{E}_{\text{align}}(\hat{\mathbf{p}}) = 2\mathbf{D}^T\mathbf{U}^T\bar{\mathbf{W}}_2\mathbf{U}\mathbf{D}\hat{\mathbf{p}}, \tag{20}$$

$$\nabla_{\hat{\mathbf{p}}}\mathcal{E}_{\text{fairness}}(\hat{\mathbf{p}}) = 2(\mathbf{I} - \mathbf{B})^T\bar{\mathbf{W}}_0(\mathbf{I} - \mathbf{B})\hat{\mathbf{p}}. \tag{21}$$

Weighting up all these gradients in Eq. (18) leads to the following linear system in $\mathbf{p}^*$:

$$\mathbf{R}\mathbf{p}^* = \alpha\bar{\mathbf{W}}_0\mathbf{p}, \tag{22}$$

with

$$\mathbf{R} := \left(\alpha\bar{\mathbf{W}}_0 + \beta\mathbf{D}^T\mathbf{U}^T\bar{\mathbf{W}}_2\mathbf{U}\mathbf{D} + \gamma(\mathbf{I} - \mathbf{B})^T\bar{\mathbf{W}}_0(\mathbf{I} - \mathbf{B})\right). \tag{23}$$

As long as weights are strictly positive, the operator $\mathbf{R}$ is a $3|\mathbb{E}^0| \times 3|\mathbb{E}^0|$ matrix, which is symmetric definite-positive (see Appendix A). The linear system (22) is thus efficiently solved using classical linear algebra solvers [7]. In our experiments, we have used the following rules to balance the energy terms:

$$\beta = 1, 0 < \alpha \ll \gamma \ll \beta.$$

We choose parameters with different order of magnitudes to sequence the effect of the minimization of the three energy terms of Eq. (23). First, we wish to build a smooth surface leading to the highest value for β, the alignment term. Second, the α energy term is necessary to achieve uniqueness of the optimization problem. Its lowest order of magnitude ensures that the result is very close to the set of solutions of the alignment term alone. Finally, the user has freedom for the γ energy term, depending on the desired regularity of the output quadrangulation (see Fig. 2).

5   Experiments

In this section, we evaluate the quality of the regularization. We have considered two different normal vector fields as input: the first one is given by Integral Invariant approaches [5] and is known to be multigrid convergent. The second correspond to a piecewise smooth reconstruction of the normal vector field using [4]. This approach performs a normal vector field regularization while detecting and preserving sharp features. The discrete operators have been implemented using the DGTAL DEC package [1] with Eigen backend for the linear algebra solver [7].

In Fig. 4 we first illustrate the overall regularization on a Standford-bunny object using an Integral Invariant based normal vector field. In Fig. 5, we show that using an anisotropic, piecewise smooth normal vector field from [4], the regularized surface is able to capture sharp features. In Fig. 6, we demonstrate the robustness of the regularization in presence of noise. Note that beside the fact that the input normal vector field is robust for the alignment term, the

Fig. 3. Regularization example on a digital sphere ($r = 10$). From left to right the original Marching-Cubes surface, the simplified one using [3] and our regularization.

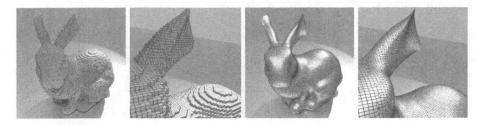

Fig. 4. Regularization example on a 128^3 `Stanford-bunny`. From left to right, input digital surface and regularized surface using the input normal vector field from [5] ($r = 6$).

fairness term allows us to obtain a smooth quadrangulation even in this case. If not specified, $\alpha = 10^{-3}$, $\beta = 1$ and $\gamma = 10^{-2}$ values have been used in these tests.

We also evaluate the asymptotic behavior of the regularization operator. As discussed in Sect. 2, we know that the Hausdorff distance between $\mathbf{S}$ and ∂X for some smooth shape X is in $O(h)$. We experimentally observe that $\mathbf{S}^*$ has the same asymptotic behavior and is even closer to the original surface than $\mathbf{S}$. We have considered asymptotic plane and sphere objects in Fig. 7 for various h (abscissa) tending to zero. In the first row, we show that the normal vector field $\mathbf{u}^*$ of the regularized surface $\mathbf{S}^*$ seems to converge to the input field $\mathbf{u}$ as h tends to zero[3]. The second row shows the proximity of $\mathbf{p}$ (i.e. the vertices of $\mathbf{S} = \partial_h X$) to ∂X (Theorem 1 of [11]). The $O(h)$ convergence speed is thus experimentally confirmed. In the third row, we evaluate the distance between $\mathbf{p}^*$ (the vertices of $\mathbf{S}^*$) and its closest point on ∂X. We also observe an experimental convergence speed in $O(h)$. Finally, the last row compares the two approximations. In this case, we see that $\mathbf{p}^*$ is experimentally closer to ∂X than $\mathbf{p}$.

---

[3] Since quads of $\mathbf{S}^*$ are not coplanar anymore, each vector of $\mathbf{u}^*$ is defined by averaging the two normal vectors of the quad triangles.

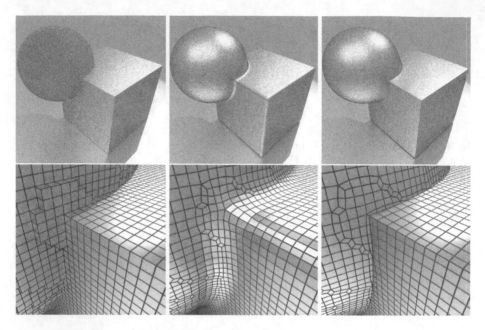

Fig. 5. Isotropic vs anistropic normal vector field as input data. *From left to right*: the original object, the regularization with Integral Invariant based normal vector field [5] ($r = 4$), the regularization after a piecewise smoothing of the same normal vector field using [4] ($\epsilon = [2, 0.25]$, $\lambda = 0.03$, $\alpha = 0.006$ and the Integral Invariant normal vector field with $r = 4$).

Fig. 6. Regularization examples without (*left*) or with noise *(right)* (Octaflower, 256^3). A highly specular surface has been used to highlight the smoothness of the reconstruction (normal vectors are given by the face geometry). The input normal vector field is given by [4].

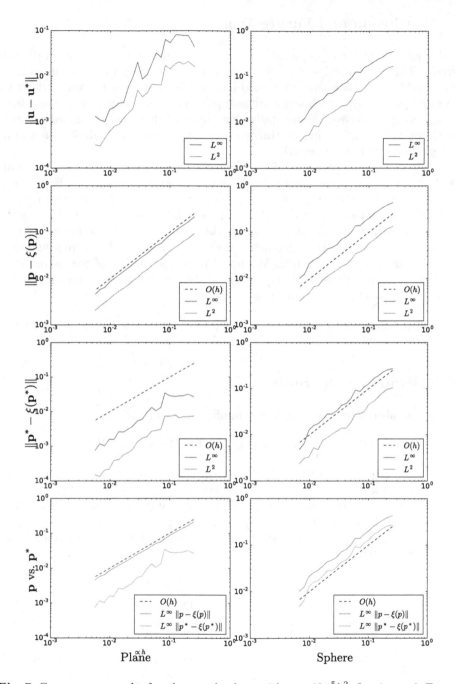

Fig. 7. Convergence results for plane and sphere with $\alpha = 10^{-5}h^2, \beta = 1, \gamma = 0$. Errors ($y$−axis) are given for both the L^2 and L^∞ norms. The grid resolution h (in abscissa) in the range $[6 \cdot 10^{-3}, 0.2]$. As h tends to zero, convergence graphs must be read from right to left.

6 Conclusion and Future Works

In this article, we have proposed a variational approach to regularize a digital surface. The regularized surface is consistent with respect to an input normal vector field, and has a smooth embedding. If the input normal vector field is piecewise smooth (*i.e.* with singularities), the regularization preserves these features. Finally, we have experimentally demonstrated that the regularized vertices are closer to the underlying continuous object than the Gauss digitization in a multigrid convergent framework.

Future works are twofold: First, similarly to Theorem 1 of [11], a formal proximity result is needed between $\hat{\mathbf{p}}$ and ∂X. At this point, we have been able to derived a formal proof for an L^2 proximity in $O(h)$ (average error). Further developments are required to obtain a L^∞ proximity in $O(h)$ (worst-case). Secondly, we would like to exploit the regularity and smoothness of the regularized surface in various geometry processing problems. Last, we would like to study the influence of metrics $\bar{\mathbf{W}}_0$ and $\bar{\mathbf{W}}_2$ in the quality of reconstruction. For now we have only used identity matrices, but we believe that metrics based on estimated local areas would improve results.

Acknowlegments. This work has been partly funded by the COMEDIC ANR-15-CE40-0006 research grant.

A   Details on Operator R

The $\mathbf{R}$ operator defined in (23) is a sum of three symmetric matrices,

$$\alpha\bar{\mathbf{W}}_0 + \beta\,\mathbf{D}^T\,\mathbf{U}^T\,\bar{\mathbf{W}}_2\,\mathbf{U}\,\mathbf{D} + \gamma(\mathbf{I} - \mathbf{B})^T\bar{\mathbf{W}}_0(\mathbf{I} - \mathbf{B})\,. \tag{24}$$

The first one is a positive definite matrix since it stands for a scalar product. From (5) and (11), one can clearly see that $\mathcal{E}_{\text{align}} \geq 0$ and $\mathcal{E}_{\text{fairness}} \geq 0$ as they are calculated as the sum of positive or null terms. Noting that $\mathcal{E}_{\text{align}}(\hat{\mathbf{p}}) = \hat{\mathbf{p}}^T\mathbf{D}^T\mathbf{U}^T\bar{\mathbf{W}}_2\mathbf{U}\mathbf{D}\hat{\mathbf{p}}$ and $\mathcal{E}_{\text{fairness}}(\hat{\mathbf{p}}) = \hat{\mathbf{p}}^T(\mathbf{I} - \mathbf{B})^T\bar{\mathbf{W}}_0(\mathbf{I} - \mathbf{B})\hat{\mathbf{p}}$, the second and third matrices are therefore positive semi-definite. Therefore we have $\forall\mathbf{x} \in \mathbb{R}^{3|\mathbb{E}^0|} \setminus \{0\}$:

$$\mathbf{x}^T\left(\alpha\bar{\mathbf{W}}_0\right)\mathbf{x} > 0\,, \tag{25}$$

$$\mathbf{x}^T\left(\beta\,\mathbf{D}^T\mathbf{U}^T\bar{\mathbf{W}}_2\mathbf{U}\,\mathbf{D}\right)\mathbf{x} \geq 0\,, \tag{26}$$

$$\mathbf{x}^T\left(\gamma(\mathbf{I} - \mathbf{B})^T\bar{\mathbf{W}}_0(\mathbf{I} - \mathbf{B})\right)\mathbf{x} \geq 0\,. \tag{27}$$

Assuming $\alpha > 0$, $\beta > 0$ and $\gamma > 0$, we have $\mathbf{x}^T\mathbf{R}\mathbf{x} > 0$ and $\mathbf{R}$ is positive definite. $\mathbf{R}$ is thus invertible and efficient inversion algorithms exist (*e.g.* using LDL^T Cholesky factorization [7]).

References

1. DGtal: Digital geometry tools and algorithms library. http://dgtal.org
2. Braquelaire, J., Vialard, A.: Euclidean paths: a new representation of boundary of discrete regions. Graph. Models Image Process. **61**(1), 16–43 (1999). http://dx.doi.org/10.1006/gmip.1999.0488
3. Coeurjolly, D., Dupont, F., Jospin, L., Sivignon, I.: Optimization schemes for the reversible discrete volume polyhedrization using marching cubes simplification. In: Kuba, A., Nyúl, L.G., Palágyi, K. (eds.) DGCI 2006. LNCS, vol. 4245, pp. 413–424. Springer, Heidelberg (2006). doi:10.1007/11907350_35
4. Coeurjolly, D., Foare, M., Gueth, P., Lachaud, J.: Piecewise smooth reconstruction of normal vector field on digital data. Comput. Graph. Forum **35**(7), 157–167 (2016). http://dx.doi.org/10.1111/cgf.13013
5. Coeurjolly, D., Lachaud, J., Levallois, J.: Multigrid convergent principal curvature estimators in digital geometry. Comput. Vis. Image Underst. **129**, 27–41 (2014). http://dx.doi.org/10.1016/j.cviu.2014.04.013
6. Grady, L.J., Polimeni, J.: Discrete Calculus: Applied Analysis on Graphs for Computational Science. Springer, London (2010)
7. Guennebaud, G., Jacob, B., et al.: Eigen v3. (2010). http://eigen.tuxfamily.org
8. He, L., Schaefer, S.: Mesh denoising via l_0 minimization. ACM Trans. Graph. (TOG) **32**(4), 64 (2013)
9. Jacques-Olivier Lachaud, D.C., Levallois, J.: Robust and convergent curvature and normal estimators with digital integral invariants. In: Modern Approaches to Discrete Curvature, vol. 2184. LNM, Springer International Publishing (to appear, 2017)
10. Klette, R., Rosenfeld, A.: Digital Geometry: Geometric Methods for Digital Picture Analysis. Series in Computer Graphics and Geometric Modelin. Morgan Kaufmann, San Francisco (2004)
11. Lachaud, J., Thibert, B.: Properties of gauss digitized shapes and digital surface integration. J. Math. Imaging Vis. **54**(2), 162–180 (2016). http://dx.doi.org/10.1007/s10851-015-0595-7
12. Lorensen, W.E., Cline, H.E.: Marching cubes: a high resolution 3d surface construction algorithm. In: ACM Siggraph computer Graphics, vol. 21, pp. 163–169. ACM (1987)
13. Ohtake, Y., Belyaev, A.G.: Dual/primal mesh optimization for polygonized implicit surfaces. In: Proceedings of the Seventh ACM Symposium on Solid Modeling and Application, pp. 171–178. ACM (2002)
14. Schaefer, S., Warren, J.: Dual marching cubes: primal contouring of dual grids. In: Proceedings of the 12th Pacific Conference on Computer Graphics and Applications, PG 2004, pp. 70–76. IEEE (2004)
15. Wang, R., Yang, Z., Liu, L., Deng, J., Chen, F.: Decoupling noise and features via weighted l_1-analysis compressed sensing. ACM Trans. Graph. (TOG) **33**(2), 18 (2014)
16. Wu, X., Zheng, J., Cai, Y., Fu, C.W.: Mesh denoising using extended rof model with l1 fidelity. Comput. Graph. Forum **34**(7), 35–45 (2015)
17. Zhang, H., Wu, C., Zhang, J., Deng, J.: Variational mesh denoising using total variation and piecewise constant function space. IEEE Trans. Visual Comput. Graphics **21**(7), 873–886 (2015)

Morphological Analysis

Opening Holes in Discrete Objects
with Digital Homotopy

Aldo Gonzalez-Lorenzo$^{(\boxtimes)}$, Alexandra Bac, and Jean-Luc Mari

Aix-Marseille Université, CNRS, LSIS UMR 7296, Marseille, France
aldo.gonzalez-lorenzo@univ-amu.fr

Abstract. Discrete objects are sets of pixels, voxels or their analog in higher dimension. A three-dimensional discrete object can contain holes such as tunnels, handles or cavities. Opening the holes of an object consists in erasing all its holes by removing some parts of it. The main idea is to take a point of the object and to dilate it inside the object without changing its homotopy type: the remaining points in the object are those which have to be removed. This process does not require the computation of the homology groups of the object and is only based on the identification of simple points.

In this experimental paper we propose two algorithms for opening the holes of a discrete object endowed with any adjacency relation in arbitrary dimension. Both algorithms are based on the distance transform of the object and differ in how the dilation is performed, favoring either time complexity or the quality of the output. Moreover, these algorithms contain a parameter that controls the thickness of the removed parts.

1 Introduction

Topological features such as connected components, tunnels, handles or cavities allow us to understand the essential structure of an object, which is invariant under a continuous deformation. They are defined through algebraic topology. Roughly speaking, homology theory defines a q-dimensional hole as a q-dimensional sphere which is not the boundary of a $(q + 1)$-dimensional ball. Hence, connected components are 0-dimensional holes; tunnels and handles are 1-dimensional holes and cavities are 2-dimensional holes. On the other hand, homotopy theory detects q-dimensional holes with q-dimensional spheres that cannot be continuously deformed into a point.

In order to erase a hole, one can add matter to the object so it disappears. This is called *closing* a hole. For instance, we can remove a connected component by adding a *bridge* to another connected component, a handle by adding a patch or a cavity by filling its interior. Intuitively, we can erase a q-dimensional hole by adding a $(q + 1)$-dimensional ball.

Another way to erase a hole is by removing matter from the object. We call this *opening* a hole. As an illustration, we can open a 0-dimensional hole by removing the whole connected component, a handle by cutting a slice or a

© Springer International Publishing AG 2017
W.G. Kropatsch et al. (Eds.): DGCI 2017, LNCS 10502, pp. 213–224, 2017.
DOI: 10.1007/978-3-319-66272-5_18

cavity by digging a well. In this case, if the object is embedded in the three-dimensional space, we say that we can open a q-dimensional hole by removing a $(3 - q)$-dimensional ball.

The problem of closing holes has been addressed in [1,7,8]. We recall that an object is *contractible* if it has the homotopy type of a point [6]. Hence, a contractible object has no holes at all. A trivial way of filling the holes of an object is to consider a contractible object that contains it: the difference between them is the matter that we have to add to close its holes. In order to optimize the quantity of matter that we add, we can shrink it while preserving its contractibility property and keeping the object inside.

In this experimental article we study the problem of opening the holes in an object. The main idea is to find a contractible subset of the object by taking a point in its interior and expanding it without changing its homotopy type. This has been shortly investigated in [10] for segmenting the brain cortex in magnetic resonance images. In this paper we develop farther this problem and design two algorithms for opening the holes of a discrete object using additional geometric conditions to obtain a visually pleasant output.

2 Preliminaries

2.1 Discrete Object

A d-dimensional *discrete object* is a (finite) subset of $\mathbb{Z}^d$. Its elements are called pixels when $d = 2$, voxels when $d = 3$ or points in general.

We endow a discrete object with an *adjacency relation*. Let us mention two of them: two points $x, y \in \mathbb{Z}^d$ are $(2d)$-adjacent (resp. $(3^d - 1)$-adjacent) if $\|x - y\|_1 \leq 1$ (resp. $\|x - y\|_\infty \leq 1$). Let $\alpha \in \{2d, 3^d - 1\}$, the α-neighborhood of a point x, denoted $N_\alpha(x)$, is the set of its α-adjacent points. We also denote $N_\alpha^*(x) := N_\alpha(x) - \{x\}$. The *outer α-boundary* of a discrete object X is $N_\alpha^*(X) := (\bigcup_{x \in X} N_\alpha(x)) - X$, the set of points in $\mathbb{Z}^d - X$ having an α-adjacent point in X.

2.2 Simple Points

Roughly speaking, a point is *simple* for a discrete object X if its addition or removal from X does not change the homotopy type of X. As a discrete object is just a set of isolated points, this only makes sense if we endow the discrete object with a topological space. This is usually done in terms of the adjacency relation considered for the object, so there are different characterizations of simple points. See [2,3,5,11] to cite a few.

We consider in this paper the $(3^d - 1)$-adjacency relation and its associated notion of simple point as described in [5], though any other definition of simple point can be used.

3   Homotopic Opening

A topological space is *contractible* if it has the homotopy type of a point. Intuitively, this means that we can shrink it to a point. We can translate this concept into the discrete context by using the notion of simple point. We recall that a point in a discrete object is simple if its removal does not change the homotopy type of the object. Consequently, a discrete object is contractible if we can reduce it to a point by a sequence of simple points deletions.

A contractible discrete object has no holes. Given a discrete object X, we say that $Y \subset X$ is a *homotopic opening* if $X - Y$ is contractible. Hence, the homotopic opening consists of a set of *cuts* that remove the holes from the object. Note that, if the object is not connected, then all the connected components except one will be in the homotopic opening, so it may be more interesting to consider the homotopic opening of each connected component separately.

The homotopic opening is clearly not unique. We could be interested in obtaining a minimal (in the number of points) homotopic opening, but this is not useful in the discrete context (see Fig. 1). Also, it can be useful to have thick cuts for better representing the holes. For this reason, we evaluate the homotopic opening by visual inspection and we do not try to define a unique or optimal homotopic opening.

Fig. 1. Two homotopic openings (black) for the same object (gray) with the same size. This example shows why we cannot judge a homotopic opening by its size.

4   Computing Homotopic Openings

A simple algorithm for computing a homotopic opening, which is the base for the more elaborate algorithms that we introduce later, is the following. Let X be a discrete object, choose some point $x \in X$ and set $C = \{x\}$. Then find a point in $X - C$ which is simple for C and add it to C. Repeat this operation while there are such points. At the end C is obviously contractible and thus $X - C$ is a homotopic opening for X. Note that, according to [10], C is a homotopic dilation of $\{x\}$ constrained by X.

Figure 2 illustrates the output of such algorithm. While the cuts are thin, they are far too long. A more elaborate algorithm consists in using the distance transform.

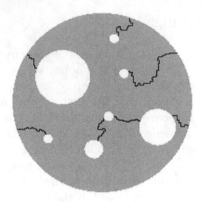

Fig. 2. A 2D discrete object (gray) and a homotopic opening (black) obtained by choosing simple points at random.

The (Euclidean) distance transform of a discrete object X is the map $dt_X : X \longrightarrow \mathbb{R}$ that assigns to each point of X its distance to $\mathbb{Z}^d - X$ or, in other words, its depth in the object. Hence, at the beginning we can set $C = \{x\}$ where x is a maximum for dt_X and then we give priority to points with higher dt_X value. By using the distance transform, we lead the propagation in order to obtain the cuts in the thinner parts of the object. In the following we describe two ways of doing this.

4.1   Random Propagation

A main concern when using the distance transform in the propagation of the contractible subset is how to deal with points with equal value. A simple idea for having an isotropic homotopic opening is to randomly pick a point among the points with equal distance transform value. Algorithm 1 describes this approach.

Algorithm 1. Homotopic opening with random propagation

Input: $X \subset \mathbb{Z}^d$
Output: Homotopic opening for X
$C \leftarrow \{x\}$ for a random $x \in X$ such that $dt_X(x)$ is maximal;
$S \leftarrow N_\alpha^*(x) \cap (X - C)$;
while $S \neq \emptyset$ **do**
$\quad$ $x \leftarrow$ random point in S such that $dt_X(x)$ is maximal;
$\quad$ $S \leftarrow S - \{x\}$;
$\quad$ **if** x *is simple for* C **then**
$\quad\quad$ $C \leftarrow C \cup \{x\}$;
$\quad\quad$ $S \leftarrow S \cup (N_\alpha^*(x) \cap (X - C))$;
return $X - C$;

Complexity of Algorithm 1. Let n denote the number of points in the bounding box of the discrete object. The distance transform of X can be computed in $\mathcal{O}(n)$ [4,9]. Since every point is inserted into S each time a neighbor is added to C, then it is added at most $(3^d - 1)$ times. Thus, the *while* loop is executed at most $3^d n$ times. The set S can be implemented as a priority queue using the distance transform as key. Hence, insertion and deletion can be performed in logarithmic time while finding the deepest point is done in constant time. Consequently, Algorithm 1 has complexity $\mathcal{O}(3^d n(\log n + f(d)))$, where $f(d)$ denotes the complexity of checking if a point is simple in $\mathbb{Z}^d$.

For dimension $d \leq 3$, we can use a look-up table for recognizing simple points, and hence the complexity of Algorithm 1 is $\mathcal{O}(n \log n)$.

Figure 3 illustrates some results of Algorithm 1. The holey disk (top-left) presents short cuts, though they do not seem *straight* segments, specially the bottom-left cut. This kind of issue is treated in Sect. 4.3. The hollow ball (top-center) and the sculpture (top-right) present well-shaped cuts. Regarding the torus (bottom), we can appreciate two differentiated cuts. One is around the central hole, which is well-shaped, while the other one is far from looking like a ring around the cavity.

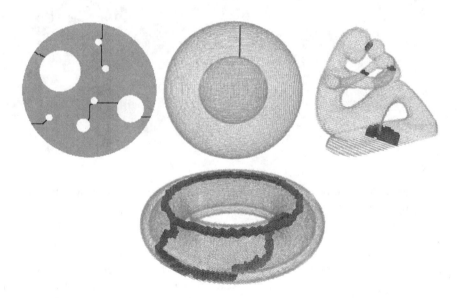

Fig. 3. Homotopic openings obtained with Algorithm 1.

This kind of problem is due to the propagation of the contractible subset. If we randomly propagate a contractible object inside a plane, we do not obtain a uniform disk, but a tree-shaped object. A 3D discrete object can contain a two-manifold-like set of points which are equidistant to the complement of the object, so the propagation has a similar behavior when it reaches their depth.

A concrete example is shown in Fig. 4 (top), which depicts a thickened torus—so its set of deepest points is a torus—and three steps in the propagation of the contractible subset. Hence, it is natural that Algorithm 1 produces strange homotopic openings on such objects. Note that this problem does not happen in 2D discrete objects.

4.2   Propagation by Layers

The previous example on the torus motivates the following algorithm. In this case, we treat the object by layers. The idea is that at each step we traverse the simple points in the outer boundary of the contractible subset to find the maximum distance transform value. Then we mark the points in the outer boundary whose distance transform value is close to that maximum and we try to add them to the object. This is described in Algorithm 2. This alternative strategy produces a more uniform propagation of the contractible subset. Figure 4 illustrates the difference between the propagations performed by Algorithms 1 and 2.

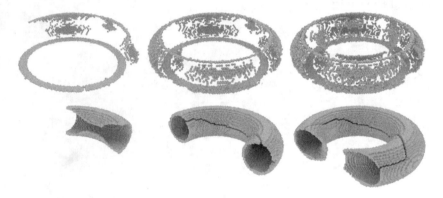

Fig. 4. Different steps of the propagation performed in Algorithms 1 (top) and 2 (bottom) for the same object.

Complexity of Algorithm 2. An upper bound for the number of executions of the *repeat* loop is n, since at each execution (except the last one), at least one point of $X - C$ is added to C. At each iteration we must find the outer boundary of the contractible subset and check if the points are simple, thus needing $\mathcal{O}(3^d n f(d))$ operations. The later traversal of the outer boundary does not affect the complexity of the iteration. Hence, the complexity of Algorithm 2 is $\mathcal{O}(3^d n^2 f(d))$.

Again, the complexity of Algorithm 2 is $\mathcal{O}(n^2)$ for dimension $d \leq 3$.

At each step, we put in the list L not only the deepest simple points in the outer boundary of the contractible subset, but all the points with distance transform value close to the maximum. Concretely, we consider the points whose

---

Algorithm 2. Homotopic opening with propagation by layers

Input: $X \subset \mathbb{Z}^d$
Output: Homotopic opening for X
$C \leftarrow \{x\}$ for a random $x \in X$ with highest dt_X value;
repeat
$\quad m \leftarrow \max \{dt_X(x) \mid x \in N_\alpha^*(C) \cap X, x \text{ simple for } C\};$
$\quad L \leftarrow dt_X^{-1}([m-1,m]) \cap N_\alpha^*(C);$
$\quad$**foreach** $x \in L$ **do**
$\quad\quad$**if** x *is simple for* C **then**
$\quad\quad\quad C \leftarrow C \cup \{x\};$

until *idempotency*;
return $X - C$;

---

distance transform value is in the interval $[m-a, m]$, for $a = 1$. We have chosen this value of a after visual inspection of several examples, but a different value can also be considered.

Some results of Algorithm 2 are depicted in Fig. 5. The cuts of the holey disk (top-left) look slightly better. There is no remarkable improvement for the hollow ball (top-center) nor the sculpture (top-right). However, the homotopic opening of the torus (bottom) is definitely better.

In conclusion, Algorithm 2 can be better than Algorithm 1, but it has worse complexity.

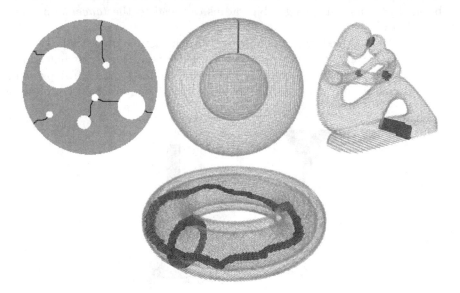

Fig. 5. Homotopic openings obtained with Algorithm 2.

4.3   More Than Simple Points

Both previous algorithms add as many simple points as possible to the contractible subset, which produces thin cuts. However, leaving thicker cuts can be interesting for two reasons:

1. Cuts usually look like polygonal lines instead of discretized straight lines. When two wavefronts collide, the next added simple points depend strongly on their neighborhood. This produces a chain reaction that makes the cut be far from what we would expect.
2. One-dimensional thin cuts in big objects can be very difficult to see.

We propose a simple method for obtaining thick cuts. Instead of adding simple points, we check if we can add a discrete ball without changing the homotopy type of the contractible subset, and then we just add its center point (which must be a simple point). Given $r \in \mathbb{N}$, we consider $B_r = \{x \in \mathbb{Z}^d \mid \|x\|_2 \leq r\}$, the discrete ball of radius r. Hence, given a discrete object X, a contractible subset C, a point p and a radius r, we check if we can reduce $(C \cup (p + B_r)) \cap X$ to C via a sequence of simple points deletions. A simple way of doing this is to put the $\mathcal{O}(r^d)$ points of $((p + B_r) \cap X) - C$ in a list and repeatedly traverse it and remove simple points until stability. Hence, the complexity of checking if a ball is simple is $\mathcal{O}(r^{2d}f(d))$.

By using *simple balls* instead of simple points, Algorithms 1 and 2 are enriched with a parameter r which is the radius of the discrete ball. Thus, their complexities are $\mathcal{O}(3^d n(\log n + r^{2d}f(d)))$ and $\mathcal{O}(3^d n^2 r^{2d}f(d))$ respectively.

Intuitively, using simple balls instead of simple points should be sufficient to obtain thick cuts, but using a big radius can lead to the *tangency problem* illustrated in Fig. 6.

The discrete ball (red) is not simple for the contractible object (black) since it creates a one-pixel hole. Consequently, the center point (green) is not added. This kind of phenomenon is quite frequent when we use a big ball. It prevents a correct propagation of the contractible subset which can eventually produce a

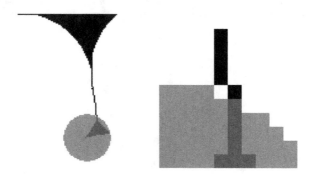

Fig. 6. Tangency problem. The discrete ball (red) is not simple for the object (black). (Color figure online)

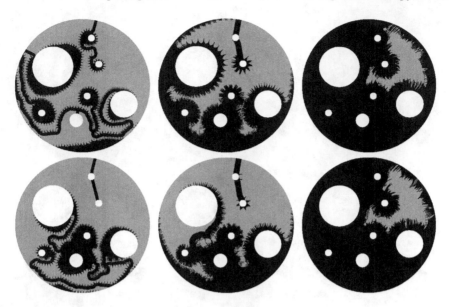

Fig. 7. Homotopic openings computed by Algorithms 1 (top) and 2 (bottom) with radiuses 10 (left), 20 (center) and 50 (right).

strange homotopic opening. This is illustrated in Fig. 7, which depicts the output of Algorithms 1 (top) and 2 (bottom) with radiuses 10, 20 and 50.

An attempt to avoid this kind of configurations is to center the ball not only in the point, but also in its $(2d)$-neighborhood. If one of the $2d + 1$ balls is simple for the contractible subset, then we add the point. We call this the *relaxed test for simple balls*. Figure 8 shows how the homotopic openings of Fig. 7 are when we use a relaxed test for simple balls. We can appreciate that, after using this approach, the outputs of Algorithms 1 (top) and 2 (bottom) are almost indistinguishable. Sadly, this idea does not seem to work for 3D objects. Figure 9 depicts the homotopic openings computed by Algorithms 1 (top) and 2 (bottom) using a ball of radius 5. The relaxed test for simple balls yet produces strange results, but we can appreciate that Algorithm 2 behaves notably better.

We observe that using a positive radius prevents fronts from approaching too much, so we obtain quite smooth cuts. We can then continue to propagate the contractible subset with balls of smaller radiuses in order to obtain thin and smooth cuts. We have chosen to divide the radius by two at each step. Figure 10 shows the homotopic openings computed by Algorithm 1 when using initial radiuses 0, 2, 4 and 8. We can appreciate how the cuts look more and more well-shaped as we increase the initial radius.

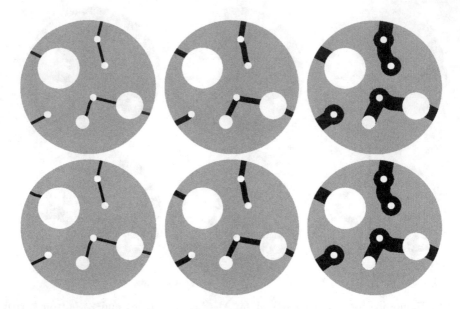

Fig. 8. Homotopic openings computed by Algorithms 1 (top) and 2 (bottom) with radiuses 10 (left), 20 (center) and 50 (right) and relaxed test for simple balls.

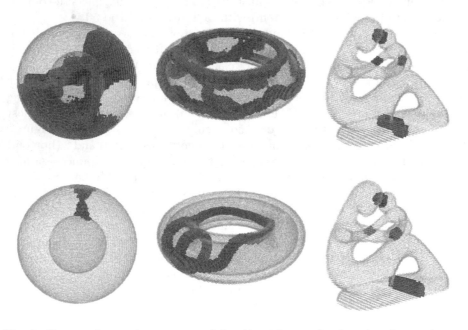

Fig. 9. Homotopic openings computed by Algorithms 1 (top) and 2 (bottom) with radius 5 and relaxed test for simple balls.

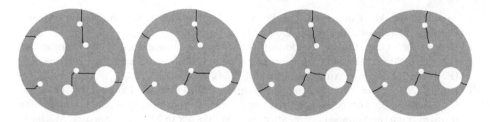

Fig. 10. Homotopic openings computed by Algorithm 1 with initial radiuses 0, 2, 4 and 8 and successive reductions. The cuts have been thickened for better visibility.

5   Conclusion and Future Work

This paper addresses the problem of erasing the holes of a discrete object by removing parts of it. We have defined the homotopic opening and we have presented two algorithms for computing it. The outputs of these algorithms are evaluated by visual inspection. While the second algorithm has worse complexity than the first one, it seems to produce better results. Then we have enriched these algorithms with a parameter corresponding to the radius of the balls used in the propagation of the contractible subset.

We do not obtain satisfactory results for big radiuses in 3D discrete objects, due to the tangency problem. We think that this problem deserves more attention. Our algorithms can also be generalized by considering other metrics and other structural elements for the propagation. Moreover, we suspect that the complexity of Algorithm 2 can be improved by smartly managing the layer update.

References

1. Aktouf, Z., Bertrand, G., Perroton, L.: A three-dimensional holes closing algorithm. Pattern Recogn. Lett. **23**(5), 523–531 (2002)
2. Bertrand, G.: New notions for discrete topology. In: Proceedings of the 8th International Conference Discrete Geometry for Computer Imagery, DCGI 1999, Marne-la-Vallée, France, 17–19 March 1999, pp. 218–228 (1999)
3. Bertrand, G., Malandain, G.: A new characterization of three-dimensional simple points. Pattern Recogn. Lett. **15**(2), 169–175 (1994)
4. Coeurjolly, D., Montanvert, A.: Optimal separable algorithms to compute the reverse Euclidean distance transformation and discrete medial axis in arbitrary dimension. IEEE Trans. Pattern Anal. Mach. Intell. **29**(3), 437–448 (2007)
5. Couprie, M., Bertrand, G.: New characterizations of simple points in 2D, 3D, and 4D discrete spaces. IEEE Trans. Pattern Anal. Mach. Intell. **31**(4), 637–648 (2009)
6. Hatcher, A.: Algebraic Topology. Cambridge University Press, Cambridge (2002)
7. Janaszewski, M., Couprie, M., Babout, L.: Geometric approach to hole segmentation and hole closing in 3D volumetric objects. In: Bayro-Corrochano, E., Eklundh, J.-O. (eds.) CIARP 2009. LNCS, vol. 5856, pp. 255–262. Springer, Heidelberg (2009). doi:10.1007/978-3-642-10268-4_30

8. Janaszewski, M., Couprie, M., Babout, L.: Hole filling in 3D volumetric objects. Pattern Recogn. **43**(10), 3548–3559 (2010)
9. Maurer Jr., C.R., Qi, R., Raghavan, V.: A linear time algorithm for computing exact Euclidean distance transforms of binary images in arbitrary dimensions. IEEE Trans. Pattern Anal. Mach. Intell. **25**(2), 265–270 (2003)
10. Rueda, A., Acosta, O., Couprie, M., Bourgeat, P., Fripp, J., Dowson, N., Romero, E., Salvado, O.: Topology-corrected segmentation and local intensity estimates for improved partial volume classification of brain cortex in MRI. J. Neurosci. Methods **188**(2), 305–315 (2010)
11. Saha, P.K., Chaudhuri, B.B., Chanda, B., Majumder, D.D.: Topology preservation in 3D digital space. Pattern Recogn. **27**(2), 295–300 (1994)

Well-Composedness in Alexandrov Spaces Implies Digital Well-Composedness in $\mathbb{Z}^n$

Nicolas Boutry[1,2(✉)], Laurent Najman[2], and Thierry Géraud[1]

[1] EPITA Research and Development Laboratory (LRDE),
Le Kremlin-Bicêtre, France
[2] Université Paris-Est, LIGM, Équipe A3SI, ESIEE, Champs-sur-Marne, France
nicolas.boutry@lrde.epita.fr

Abstract. In digital topology, it is well-known that, in 2D and in 3D, a digital set $X \subseteq \mathbb{Z}^n$ is *digitally well-composed (DWC)*, *i.e.*, does not contain any critical configuration, if its immersion in the Khalimsky grids $\mathbb{H}^n$ is *well-composed in the sense of Alexandrov (AWC)*, *i.e.*, its boundary is a disjoint union of discrete $(n - 1)$-surfaces. We show that this is still true in n-D, $n \geq 2$, which is of prime importance since today 4D signals are more and more frequent.

Keywords: Well-composed · Discrete surfaces · Alexandrov spaces · Critical configurations · Digital topology

1 Introduction

We recall that a subset of $\mathbb{Z}^n$, $n \geq 1$, is said to be *digital* if it is finite or if its complement in $\mathbb{Z}^n$ is finite. When $n \in \{2, 3\}$, the *immersion* $\mathcal{I}(X)$ into the *Khalimsky grid* $\mathbb{H}^n$ of a digital subset X of $\mathbb{Z}^n$ based on the *miss-strategy* is known to be *well-composed in the Alexandrov sense (AWC)* [13], *i.e.*, the connected components of the *boundary* of $\mathcal{I}(X)$ are *discrete $(n - 1)$-surfaces*, iff X is *digitally well-composed (DWC)* [5], *i.e.*, X does not contain any *critical configuration* (see Fig. 1). The aim of this paper is to show that AWCness (of the immersion of a set) implies DWCness (of the initial set) in n-D, $n \geq 2$.

In order to do so, we show that we can reformulate AWCness in a local way: a subset Y of $\mathbb{H}^n$ is AWC iff, for any z in the boundary $\mathfrak{N}$ of Y in $\mathbb{H}^n$, the subspace $|\beta_{\mathfrak{N}}^{\square}(z)|$ is a discrete $(n - 2 - \dim(z))$-surface. This property is of prime importance since we compare AWCness with DWCness, which is a local property. It is then sufficient to proceed by counterposition and to show that, if a digital subset X of $\mathbb{Z}^n$ contains a critical configuration in a block S such that $X \cap S = \{p, p'\}$ (primary case) or $S \setminus X = \{p, p'\}$ (secondary case) with p and p' two antagonists in S, then there exists some z^* in the boundary $\mathfrak{N}$ of $\mathcal{I}(X)$ satisfying that $|\beta_{\mathfrak{N}}^{\square}(z^*)|$ is not a $(n - 2 - \dim(z^*))$-surface. In fact, by choosing a particular z^* related to $\frac{p+p'}{2}$, we obtain that $|\beta_{\mathfrak{N}}^{\square}(z^*)|$ is the union of two disjoint $(n - 2 - \dim(z^*))$-surfaces, the first is a function of p and z^* and the second is

© Springer International Publishing AG 2017
W.G. Kropatsch et al. (Eds.): DGCI 2017, LNCS 10502, pp. 225–237, 2017.
DOI: 10.1007/978-3-319-66272-5_19

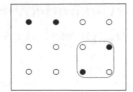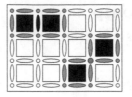

Fig. 1. From a digital set $X \subseteq \mathbb{Z}^2$ to its immersion $\mathcal{I}(X)$ in $\mathbb{H}^2$; X and $\mathcal{I}(X)$ are depicted in black, and the boundary of $\mathcal{I}(X)$ in $\mathbb{H}^2$ is depicted in gray and red. (Color figure online)

a function of p' and z^*. This way, $|\beta_{\mathfrak{N}}^{\square}(z^*)|$ is not a $(n-2-\dim(z^*))$-surface, which concludes the proof.

In Sect. 2, we recall some needed basics principles to formalize DWCness and AWCness. In Sect. 3, we present some new mathematical tools supporting our proof. In Sect. 4, after having detailed some properties of $\mathbb{Z}^n$ and $\mathbb{H}^n$, we show how they are related. In Sect. 5, we outline the proof of the paper's main result, while we conclude our work in Sect. 6.

2   Basic Concepts of Digital Topology

Let us reintroduce the notions of DWCness and AWCness.

2.1   Digital Topology and DWCness

Let $\mathbb{B} = \{e^1, \ldots, e^n\}$ be the canonical basis of $\mathbb{Z}^n$. We use the notation x_i, where i belongs to $[\![1, n]\!]$, to determine the ith coordinate of $x \in \mathbb{Z}^n$. We recall that the L^1-*norm* of a point $x \in \mathbb{Z}^n$ is denoted by $\|.\|_1$ and is equal to $\sum_{i \in [\![1,n]\!]} |x_i|$ where $|.|$ is the *absolute value*. Also, the L^∞-*norm* is denoted by $\|.\|_\infty$ and is equal to $\max_{i \in [\![1,n]\!]} |x_i|$. For a given point $x \in \mathbb{Z}^n$, an element of the set $\mathcal{N}_{2n}^*(x) = \{y \in \mathbb{Z}^n \; ; \; \|x - y\|_1 = 1\}$ (resp. of the set $\mathcal{N}^*(x) = \{y \in \mathbb{Z}^n \; ; \; \|x - y\|_\infty = 1\}$) is a $2n$-*neighbor* (resp. a $(3^n - 1)$-*neighbor*) of x. For any $z \in \mathbb{Z}^n$ and any $\mathcal{F} = (f^1, \ldots, f^k) \subseteq \mathbb{B}$, we denote by $S(z, \mathcal{F})$ the set $\left\{z + \sum_{i \in [\![1,k]\!]} \lambda_i f^i \mid \lambda_i \in \{0, 1\}, \forall i \in [\![1, k]\!]\right\}$. We call this set the *block* associated with the pair $(z, \mathcal{F})$; its *center* is $z + \sum_{f \in \mathcal{F}} \frac{f}{2}$, and its *dimension*, denoted by $\dim(S)$, is equal to k. More generally, a set $S \subset \mathbb{Z}^n$ is said to be a *block* iff there exists a pair $(z, \mathcal{F}) \in \mathbb{Z}^n \times \mathcal{P}(\mathbb{B})$ such that $S = S(z, \mathcal{F})$. Then, we say that two points $p, q \in \mathbb{Z}^n$ belonging to a block S are *antagonists* in S iff the distance between them equals the maximal distance using the L^1 norm between two points in S; in this case we write $p = \text{antag}_S(q)$. Note that the antagonist of a point p in a block S containing p exists and is unique. Two points that are antagonists in a block of dimension $k \geq 0$ are said to be k-*antagonists*; k is then called the *order of antagonism* between these two points. We say that a digital subset X of $\mathbb{Z}^n$ contains a *critical configuration* in a block S of dimension

$k \in [\![2, n]\!]$ iff there exists two points $\{p, p'\} \in \mathbb{Z}^n$ that are antagonists in S s.t. $X \cap S = \{p, p'\}$ (*primary case*) or s.t. $S \setminus X = \{p, p'\}$ (*secondary case*). Then, a digital set $X \subset \mathbb{Z}^n$ is said to be *digitally well-composed (DWC)* [5] iff it does not contain any critical configuration.

2.2   Axiomatic Digital Topology and AWCness

Let X be any set, and let $\mathcal{U}$ be a set of subsets of X satisfying that $X, \emptyset$ belong to $\mathcal{U}$, any union of any family of elements of $\mathcal{U}$ belongs to $\mathcal{U}$, and any finite intersection of any family of elements of $\mathcal{U}$ belongs to $\mathcal{U}$. Then $\mathcal{U}$ is a *topology* [1,10], and the pair $(X, \mathcal{U})$ is called a *topological space*. We abusively say that X is a topological space, assuming it is supplied with a topology $\mathcal{U}$. The elements of $\mathcal{U}$ are called the *open sets* of $(X, \mathcal{U})$, and the complement of an open set is said to be a *closed set* [1]. A set N containing an element p of a topological space X s.t. there exists $U \in \mathcal{U}$ satisfying $p \in U \subseteq N$ is said to be a *neighborhood* of p in X. We say that a topological space $(X, \mathcal{U})$ satisfies the T_0 *axiom of separation* [1,3,10] iff for any two different elements in X, for at least one of them there is an open neighborhood not containing the other element. A topological space which satisfies the T_0 axiom of separation is said to be a T_0-*space*, a topological space X is called *discrete* [2] iff the intersection of any family of open sets of X is open in X, and a discrete T_0-space is said to be an *Alexandrov space* [8].

Let Λ be an arbitrary set. A *binary relation* [4] R on Λ is a subset of $\Lambda \times \Lambda$, and for any $x, y \in \Lambda$, we denote by $x \mathrel{R} y$ the fact that $(x, y) \in R$, or equivalently $x \in R(y)$. A binary relation R is called *reflexive* iff, $\forall x \in \Lambda$, $x \mathrel{R} x$, is called *antisymmetric* iff, $\forall x, y \in \Lambda$, $x \mathrel{R} y$ and $y \mathrel{R} x$ imply $x = y$, and is called *transitive* iff, $\forall x, y, z \in \Lambda$, $x \mathrel{R} y$ and $y \mathrel{R} z$ imply $x \mathrel{R} z$. Also, we denote by $R^{\square}$ the binary relation defined such that, $\forall x, y \in \Lambda$, $\{x \mathrel{R^{\square}} y\} \Leftrightarrow \{x \mathrel{R} y \text{ and } x \neq y\}$. An *order relation* [4] on Λ is a binary relation which is reflexive, antisymmetric, and transitive; a set Λ of arbitrary elements supplied with an order relation R on Λ is denoted (Λ, R) or $|\Lambda|$ and is called a *poset* [4]; Λ is called the *domain* of $|\Lambda|$. According to Alexandrov (Theorem 6.52, p. 28 of [1]), we can identify any poset $|X| = (X, R)$ with the Alexandrov space *induced* by the order relation R. Let (X, α_X) be a poset and p an element of X, the *combinatorial closure* $\alpha_X(p)$ of p in $|X|$ is the set $\{q \in X ; (q, p) \in \alpha_X\}$, the *combinatorial opening* $\beta_X(p)$ of p in $|X|$ is the set $\{q \in X ; (p, q) \in \alpha_X\}$, and $\theta_X(p) := \alpha_X(p) \cup \beta_X(p)$; $\alpha_X(p)$ (resp. $\beta_X(p)$) is then the smallest closed (resp. open) set containing $\{p\}$ in X. Also, $\forall S \subseteq X$, $\alpha_X(S) := \cup_{p \in S} \alpha_X(p)$, $\beta_X(S) := \cup_{p \in S} \beta_X(p)$, and $\theta_X(S) := \cup_{p \in S} \theta_X(p)$. Assuming that $|X|$ is a poset and S is a subset of X, the *suborder* [4] of $|X|$ relative to S is the poset $|S| = (S, \alpha_S)$ with $\alpha_S := \alpha_X \cap S \times S$; we have then, for any $x \in S$, $\alpha_S(x) = \alpha_X(x) \cap S$, $\beta_S(x) = \beta_X(x) \cap S$, and $\theta_S(x) = \theta_X(x) \cap S$. For any suborder $|S|$ of $|X|$, we denote by $\mathrm{Int}_X(S)$ the open set $\{h \in X ; \beta_X(h) \subseteq S\}$. A set $S \subseteq X$ is said to be a *regular open set* (resp. a *regular closed set*) iff $S = \mathrm{Int}_X(\alpha_X(S))$ (resp. $S = \alpha_X(\mathrm{Int}_X(S))$). We call *relative topology* [8] induced in S by $\mathcal{U}$ the set of all the sets of the form $U \cap S$ where $U \in \mathcal{U}$. A set which is open in the relative topology of S is said to be a *relatively open set* [8]. A set $S \subseteq X$ is then said to be *connected* iff it is not the disjoint union of two non-empty relatively open subsets w.r.t. S. The largest connected set in $(X,\mathcal{U})$ containing $p \in X$ is called the *component* [1] of the point p in $(X,\mathcal{U})$ and we denote it $\mathcal{CC}(X,p)$. When $(X,\mathcal{U})$ is non-empty, the set of maximal components of X in the inclusion sense is denoted by $\mathcal{CC}(X)$ and is called the set of *connected components* of X. We call *path* [4] into $S \subseteq X$ a finite sequence $(p^0,\ldots,p^k)$ such that for all $i \in [\![1,k]\!]$, $p^i \in \theta_X^\square(p^{i-1})$, and we say that a set $S \subseteq X$ is *path-connected* [4] iff for any points p,q in S, there exists a path into S joining them. When $|X|$ is an Alexandrov space, any subset S of X is connected iff it is path-connected [4,8].

The *Khalimsky grid* [11] of dimension n is denoted $|\mathbb{H}^n| = (\mathbb{H}^n, \subseteq)$ and is the poset defined such that $\mathbb{H}_0^1 = \{\{a\} \, ; \, a \in \mathbb{Z}\}$, $\mathbb{H}_1^1 = \{\{a, a+1\} \, ; \, a \in \mathbb{Z}\}$, $\mathbb{H}^1 = \mathbb{H}_0^1 \cup \mathbb{H}_1^1$, and $\mathbb{H}^n = \{h_1 \times \cdots \times h_n \, ; \, \forall i \in [\![1,n]\!], h_i \in \mathbb{H}^1\}$. For any $h \in \mathbb{H}^n$, we have the following equalities: $\alpha(h) := \alpha_{\mathbb{H}^n}(h) = \{h' \in \mathbb{H}^n \, ; \, h' \subseteq h\}$, $\beta(h) := \beta_{\mathbb{H}^n}(h) = \{h' \in \mathbb{H}^n \, ; \, h \subseteq h'\}$, and $\theta(h) := \theta_{\mathbb{H}^n}(h) = \{h' \in \mathbb{H}^n \, ; \, h' \subseteq h$ or $h \subseteq h'\}$. For any suborder $|X|$ of $|\mathbb{H}^n|$, we obtain that $\alpha_X(h) = \{h' \in X \, ; \, h' \subseteq h\}$, $\beta_X(h) = \{h' \in X \, ; \, h \subseteq h'\}$, and $\theta_X(h) = \{h' \in X \, ; \, h' \subseteq h$ or $h \subseteq h'\}$. Any element h of $\mathbb{H}^n$ which is the Cartesian product of k elements, with $k \in [\![0,n]\!]$, of $\mathbb{H}_1^1$ and of $(n-k)$ elements of $\mathbb{H}_0^1$ is said to be of *dimension k* [12], which is denoted by $\dim(h) = k$, and the set of all the elements of $\mathbb{H}^n$ which are of dimension k is denoted by $\mathbb{H}_k^n$. Furthermore, for any $n \geq 1$, $|\mathbb{H}^n|$ is an Alexandrov space [4]. Finally, let A, B be two subsets of $\mathbb{H}^n$; we say that A and B are *separated* iff $(A \cap (\beta(B)) \cup (\beta(A) \cap B) = \emptyset$, or equivalently iff $A \cap \theta(B) = \emptyset$. The *rank* $\rho(x, |X|)$ of an element x in $|X|$ is 0 if $\alpha_X^\square(x) = \emptyset$ and is equal to $\max_{y \in \alpha_X^\square(x)}(\rho(y, |X|)) + 1$ otherwise. The *rank* of a poset $|X|$ is denoted by $\rho(|X|)$ and is equal to the maximal rank of its elements. An element x of X such that $\rho(x, |X|) = k$ is called *k-face* [4] of X. In Khalimsky grids, the dimension is equal to the rank.

Let $|X| = (X, \alpha_X)$ be a poset. $|X|$ is said to be *countable* iff its domain X is countable. Also, $|X|$ is called *locally finite* iff for any element $x \in X$, the set $\theta_X(x)$ is finite. A poset which is countable and locally finite is said to be a *CF-order* [4]. Now let us recall the definition of *n-surfaces* [9]. Let $|X| = (X, \alpha_X)$ be a CF-order; the poset $|X|$ is said to be a *(-1)-surface* iff $X = \emptyset$, or a *0-surface* iff X is made of two different elements $x, y \in X$ such that $x \notin \theta_X^\square(y)$, or an *$n$-surface*, $n \geq 1$, iff $|X|$ is connected and for any $x \in X$, $|\theta_X^\square(x)|$ is a $(n-1)$-surface. According to Evako *et al.* [9], $|\mathbb{H}^n|$ is an n-surface. Also, any n-surface $|X|$ is *homogeneous* [6], i.e., $\forall x \in X$, $\beta_X(x)$ contains an n-face. The *boundary* [13] of a digital subset S in an Alexandrov space X is defined as $\alpha_X(S) \cap \alpha_X(X \setminus S)$, and S is said to be *well-composed in the sense of Alexandrov (AWC)* iff the connected components of its boundary are discrete $(n-1)$-surfaces where $n \geq 0$ is the rank of X. Also, let us recall some properties about n-surfaces that will be useful in the sequel. Let $|X|, |Y|$ be two posets; it is said that $|X|$ and $|Y|$ can be *joined* [4] if $X \cap Y = \emptyset$. If $|X|$ and $|Y|$ can be joined, the *join* of $|X|$ and $|Y|$ is denoted $|X|*|Y|$ and is equal to $(X \cup Y, \alpha_X \cup \alpha_Y \cup X \times Y)$.

Property 1 ([7]). *Let $|X|$ and $|Y|$ be two orders that can be joined, and let n be an integer. The poset $|X|*|Y|$ is a $(n+1)$-surface iff there exists some $p \in [\![-1, n+1]\!]$ such that $|X|$ is a p-surface and $|Y|$ is a $(n-p)$-surface.*

Property 2 (Property 10 in [6]). *Let* $|X| = (X, \alpha_X)$ *be a poset. Then* $|X|$ *is an n-surface iff for any* $x \in X$, $|\alpha_X^\square(x)|$ *is a* $(k-1)$*-surface and* $|\beta_X^\square(x)|$ *is a* $(n-k-1)$*-surface, with* $k = \rho(x, |X|)$.

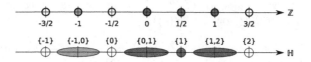

Fig. 2. Bijection between $(\mathbb{Z}/2)$ and $\mathbb{H}^1$.

2.3 A Bijection Between $(\mathbb{Z}/2)^n$ and $\mathbb{H}^n$

For A, B two arbitrary families of sets, we set $A \otimes B := \{a \times b \; ; \; a \in A, b \in B\}$, where $\times$ is the *Cartesian product*. For any $a \in \mathbb{H}^n$ and any $i \in [\![1, n]\!]$, we denote by a_i the ith coordinate of a into $\mathbb{H}^n$. Then, as a consequence of the Cartesian product, we obtain that $\forall a \in \mathbb{H}^n$, $\alpha(a) = \otimes_{m \in [\![1,n]\!]} \alpha(a_m)$ (resp. $\beta(a) = \otimes_{m \in [\![1,n]\!]} \beta(a_m)$). Also, we define the bijection $\mathcal{H} : (\mathbb{Z}/2) \to \mathbb{H}^1$ s.t. $\forall z \in (\mathbb{Z}/2)$, $\mathcal{H}(z) = \{z, z+1\}$ if $z \in \mathbb{Z}$ and $\mathcal{H}(z) = \{z + 1/2\}$ otherwise (see Fig. 2). Its inverse is denoted by $\mathcal{Z}$. Finally, we define the bijection $\mathcal{H}_n : \left(\frac{\mathbb{Z}}{2}\right)^n \to \mathbb{H}^n$ as the n-ary Cartesian product of $\mathcal{H}$ and $\mathcal{Z}_n : \mathbb{H}^n \to \left(\frac{\mathbb{Z}}{2}\right)^n$ its inverse.

3 Introducing a New Mathematical Background

Let us introduce new mathematical properties which show how $\mathbb{Z}^n$ and $\mathbb{H}^n$ are related to each other.

3.1 Complements About Antagonism in $\mathbb{Z}^n$

Lemma 1. *Let* x, y *be two elements of* $\mathbb{Z}^n$. *Then,* x *and* y *are antagonists in a block of* $\mathbb{Z}^n$ *of dimension* $k \in [\![0, n]\!]$ *iff:*

$$\begin{cases} \text{Card} \{m \in [\![1, n]\!] \; ; \; x_m = y_m\} = n - k, & (1) \\ \text{Card} \{m \in [\![1, n]\!] \; ; \; |x_m - y_m| = 1\} = k. & (2) \end{cases}$$

Proof: Let x, y be two elements of $\mathbb{Z}^n$ satisfying (1) and (2) with $k \in [\![0, n]\!]$. Now, let us take $c \in \mathbb{Z}^n$ such that $\forall i \in [\![1, n]\!]$, $c_i := \min(x_i, y_i)$, $\mathcal{I}_x := \{i \in [\![1, n]\!] \; ; \; c_i \neq x_i\}$ and $\mathcal{I}_y := \{i \in [\![1, n]\!] \; ; \; c_i \neq y_i\}$. Obviously, $\mathcal{I}_x \cap \mathcal{I}_y = \emptyset$, and then by (1), $\text{Card}(\mathcal{I}_x \cup \mathcal{I}_y) = k$. Since by (2) we have $x = c + \sum_{i \in \mathcal{I}_x} e^i$ and $y = c + \sum_{i \in \mathcal{I}_y} e^i$, then x and y belong to $S(c, \mathcal{F})$ where $\mathcal{F} := \{e^i \in \mathbb{B} \; ; \; i \in \mathcal{I}_x \cup \mathcal{I}_y\}$ is of cardinality k. Furthermore, the L^1 norm of $x - y$ is equal to k, and thus x and y maximize the L^1-distance between two points into $S(c, \mathcal{F})$. So, x and y are antagonists in $S(c, \mathcal{F})$. Conversely, let us assume that $x, y \in \mathbb{Z}^n$ are antagonists in a block $S(c, \mathcal{F})$ of dimension $k \in [\![0, n]\!]$. For any $i \in [\![1, n]\!]$, e^i belongs to $\mathcal{F}$ and hence $|x_i - y_i| = 1$, or it does not belong to $\mathcal{F}$ and hence $x_i = y_i$. Since $\text{Card}(\mathcal{F}) = k$ by hypothesis, this concludes the proof. $\qquad\square$

3.2   General Facts Between $\mathbb{Z}^n$ and $\mathbb{H}^n$

Let us present some properties relating $\left(\frac{\mathbb{Z}}{2}\right)^n$ and $\mathbb{H}^n$ that are induced by our bijection $\mathcal{H}_n$.

Lemma 2. *Let c be a value in $(\mathbb{Z}/2) \setminus \mathbb{Z}$, and let y be a value in $\mathbb{Z}$. Then, $y \in \{c - \frac{1}{2}, c + \frac{1}{2}\}$ iff $\beta(\mathcal{H}(y)) \subseteq \beta(\mathcal{H}(c))$.*

Proof: When c belongs to $(\mathbb{Z}/2) \setminus \mathbb{Z}$, $\mathcal{H}(c) = \{c + \frac{1}{2}\} \in \mathbb{H}_0^1$, and $\beta(\mathcal{H}(c)) = \{\{c - 1/2, c + 1/2\}, \{c + 1/2\}, \{c + 1/2, c + 3/2\}\}$. Also, when $y \in \mathbb{Z}$, $\mathcal{H}(y) = \{y, y + 1\} \in \mathbb{H}_1^1$, and $\beta(\mathcal{H}(y)) = \{\{y, y + 1\}\}$. If y belongs to $\{c - \frac{1}{2}, c + \frac{1}{2}\}$, we obtain that $\beta(\mathcal{H}(y)) \subseteq \beta(\mathcal{H}(c))$. Conversely, if $\{\{y, y + 1\}\} \subseteq \{\{c - 1/2, c + 1/2\}, \{c + 1/2\}, \{c + 1/2, c + 3/2\}\}$, it means that $y \in \{c - 1/2, c + 1/2\}$. □

Proposition 1. *Let S be a block in $\mathbb{Z}^n$, and let c be its center in $\left(\frac{\mathbb{Z}}{2}\right)^n$. Then $S = \mathcal{Z}_n(\beta(\mathcal{H}_n(c)) \cap \mathbb{H}_n^n)$.*

Proof: Let us remark that $S = \left\{ c + \sum_{i \in \frac{1}{2}(c)} \lambda_i e^i \; ; \; \forall i \in \frac{1}{2}(c), \lambda_i \in \{-\frac{1}{2}, \frac{1}{2}\} \right\}$ where $\frac{1}{2}(c)$ denotes the set of indices of the coordinates $i \in [\![1, n]\!]$ satisfying $c_i \in (\mathbb{Z}/2) \setminus \mathbb{Z}$. Then, for any $y \in S$, if $i \in [\![1, n]\!] \setminus \frac{1}{2}(c)$, then $y_i = c_i$, if $i \in \frac{1}{2}(c)$ such that $\lambda_i = 1/2$, then $y_i = c_i + 1/2$ with $c_i \in (\mathbb{Z}/2) \setminus \mathbb{Z}$, and if $i \in \frac{1}{2}(c)$ such that $\lambda_i = -1/2$, hence $y_i = c_i - 1/2$ with $c_i \in (\mathbb{Z}/2) \setminus \mathbb{Z}$. Then, for any $i \in [\![1, n]\!]$, by Lemma 2, $\mathcal{H}(y_i) \in \beta(\mathcal{H}(c_i))$, and then $\mathcal{H}_n(y) \in \beta(\mathcal{H}_n(c))$. Because $y \in \mathbb{Z}^n$, $\mathcal{H}_n(y) \in \mathbb{H}_n^n$, and then $\mathcal{H}_n(y) \in \beta(\mathcal{H}_n(c)) \cap \mathbb{H}_n^n$, which leads to $y \in \mathcal{Z}_n(\beta(\mathcal{H}_n(c)) \cap \mathbb{H}_n^n)$. Conversely, let us assume that $y \in \mathcal{Z}_n(\beta(\mathcal{H}_n(c)) \cap \mathbb{H}_n^n)$. Then, $\mathcal{H}_n(y) \in \beta(\mathcal{H}_n(c)) \cap \mathbb{H}_n^n$, which means that $y \in \mathbb{Z}^n$, and $\mathcal{H}_n(y) \in \beta(\mathcal{H}_n(c))$. In other words, for any $i \in [\![1, n]\!]$, $\mathcal{H}(y_i) \in \beta(\mathcal{H}(c_i))$. Two cases are then possible: $c_i \in \mathbb{Z}$, hence $y_i = c_i$, or $c_i \in (\mathbb{Z}/2) \setminus \mathbb{Z}$ and thus by Lemma 2, $y_i \in \{c_i - \frac{1}{2}, c_i + \frac{1}{2}\}$. This way, $y \in S$. □

3.3   Infimum of Two Faces in $\mathbb{H}^n$

Let X be a subset of $\mathbb{H}^n$. If there exists one element $x \in X$ such that for any $y \in X$, $y \subseteq x$, we say that x is the *supremum* of X, and we denote it $\sup(X)$. Now, let a, b be two elements of $\mathbb{H}^n$. When $\sup(\alpha(a) \cap \alpha(b))$ is well-defined, we denote it $a \wedge b$ and we call it the *infimum* between a and b.

Lemma 3. *Let a, b be two elements of $\mathbb{H}^n$. Then, $\alpha(a) \cap \alpha(b) \neq \emptyset$ iff $a \wedge b$ is well-defined. Furthermore, when $a \wedge b$ is well-defined, $a \wedge b = \times_{i \in [\![1, n]\!]} (a_i \wedge b_i)$, and $\alpha(a \wedge b) = \alpha(a) \cap \alpha(b)$.*

Proof: Let a_1, b_1 be two elements of $\mathbb{H}^1$, then it is easy to show by a case-by-case study that $\alpha(a_1) \cap \alpha(b_1) \neq \emptyset$ is equivalent to saying that $a_1 \wedge b_1$ is well-defined, and that $\alpha(a_1) \cap \alpha(b_1) = \alpha(a_1 \wedge b_1)$ when $a_1 \wedge b_1$ is well-defined. Then, when a, b belong to $\mathbb{H}^n$, $n \geq 1$ with $\alpha(a) \cap \alpha(b) \neq \emptyset$, we obtain that $\alpha(a) \cap \alpha(b)$ is equal to $\otimes_{i \in [\![1, n]\!]} (\alpha(a_i) \cap \alpha(b_i))$ which is non-empty, which means that for any $i \in [\![1, n]\!]$, $\alpha(a_i) \cap \alpha(b_i)$ is not empty, and then $a_i \wedge b_i$ is well-defined and

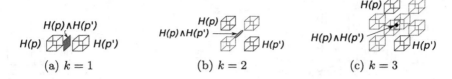

Fig. 3. Infima between images by $\mathcal{H}_n$ of the k-antagonists p and p' in 3D.

$\alpha(a_i) \cap \alpha(b_i) = \alpha(a_i \wedge b_i)$. This way, $\alpha(a) \cap \alpha(b)$ is equal to $\otimes_{i \in [\![1,n]\!]} \alpha(a_i \wedge b_i)$, and then is equal to $\alpha(\times_{i \in [\![1,n]\!]} (a_i \wedge b_i))$, and then the supremum of $\alpha(a) \cap \alpha(b)$ is $\times_{i \in [\![1,n]\!]} (a_i \wedge b_i)$, *i.e.*, exists and is unique, and so can be denoted by $a \wedge b$. Furthermore, it satisfies $\alpha(a \wedge b) = \alpha(a) \cap \alpha(b)$. Conversely, when $a \wedge b$ is well-defined, the supremum of $\alpha(a) \cap \alpha(b)$ exists and thus $\alpha(a) \cap \alpha(b) \neq \emptyset$. □

Lemma 4. $\forall p, p' \in \mathbb{Z}^n$, p and p' are k-antagonists, $k \in [\![0,n]\!]$, iff $\mathcal{H}_n(p) \wedge \mathcal{H}_n(p')$ is well-defined and belongs to $\mathbb{H}^n_{n-k}$.

Proof: The intuition of the proof is depicted on Fig. 3. Let p, p' be defined in $\mathbb{Z}^n$ and $k \in [\![0,n]\!]$ such that p and p' are antagonists in a block of dimension $k \in [\![0,n]\!]$. By Lemma 1, there exists $\mathfrak{I} \subseteq [\![1,n]\!]$ or cardinality k, and s.t. $\forall i \in \mathfrak{I}$, $|p_i - p'_i| = 1$, and $\forall i \in [\![1,n]\!] \setminus \mathfrak{I}$, $p_i = p'_i$. Since for each $i \in [\![1,n]\!]$, we have $p_i, p'_i \in \mathbb{Z}$, then $\mathcal{H}(p_i) = \{p_i, p_i + 1\}$, and $\mathcal{H}(p'_i) = \{p'_i, p'_i + 1\}$. Let us denote $z_i = \mathcal{H}(p_i)$, and $z'_i = \mathcal{H}(p'_i)$, then $z_i, z'_i \in \mathbb{H}^1_1$. When i is in $\mathfrak{I}$, $p'_i = p_i - 1$, and $\alpha(z_i) \cap \alpha(z'_i) = \{\{p_i\}\}$, and then $z_i \wedge z'_i = \{p_i\} \in \mathbb{H}^1_0$, or $p'_i = p_i + 1$, and $\alpha(z_i) \cap \alpha(z'_i) = \{\{p'_i\}\}$ and then $z_i \wedge z'_i = \{p'_i\} \in \mathbb{H}^1_0$. When i belongs to $[\![1,n]\!] \setminus \mathfrak{I}$, $z_i = z'_i$ and $\alpha(z_i) \cap \alpha(z'_i) = \alpha(z_i)$ and then $z_i \wedge z'_i = z_i \in \mathbb{H}^1_1$. It follows then that $\times_{i \in [\![1,n]\!]} (z_i \wedge z'_i)$ belongs to $\mathbb{H}^n_{n-k}$. Also, since $\alpha(z_i) \cap \alpha(z'_i) \neq \emptyset$ for any $i \in [\![1,n]\!]$, $\alpha(\mathcal{H}_n(p)) \cap \alpha(\mathcal{H}_n(p'))$ is equal to $\otimes_{i \in [\![1,n]\!]} (\alpha(z_i) \cap \alpha(z'_i))$ which is non-empty, and then, by Lemma 3, $\mathcal{H}_n(p) \wedge \mathcal{H}_n(p')$ exists and is equal to $\times_{i \in [\![1,n]\!]} (z_i \wedge z'_i)$, which belongs to $\mathbb{H}^n_{n-k}$. Let us now proceed to the converse implication. Let p, p' be two points of $\mathbb{Z}^n$, and $z = \mathcal{H}_n(p), z' = \mathcal{H}_n(p')$ such that $z \wedge z'$ is well-defined and belongs to $\mathbb{H}^n_{n-k}$. Then, we define $\mathfrak{I} = \{i \in [\![1,n]\!] ; z_i \wedge z'_i \in \mathbb{H}^1_0\}$, whose cardinality is equal to k thanks to Lemma 3. Now, let us observe that, for any $i \in [\![1,n]\!]$, $p_i \in \{p'_i - 1, p'_i + 1\}$ iff $z_i \wedge z'_i \in \mathbb{H}^1_0$, then p and p' have exactly k different coordinates, and they differ from one. Then, p and p' are antagonists in a block of dimension k by Lemma 1. □

Lemma 5. Let a, b be two elements of $\mathbb{Z}^n$ such that a and b are $(3^n - 1)$-neighbors in $\mathbb{Z}^n$ or equal. Then, $\mathcal{H}_n((a + b)/2) = \mathcal{H}_n(a) \wedge \mathcal{H}_n(b)$.

Proof: Since a and b are $(3^n - 1)$-neighbors in $\mathbb{Z}^n$, they are antagonists in a block of dimension $k \in [\![0,n]\!]$, and then by Lemma 4, $\mathcal{H}_n(a) \wedge \mathcal{H}_n(b)$ is well-defined. Now, let us prove that $(a + b)/2 = \mathcal{Z}_n(\mathcal{H}_n(a) \wedge \mathcal{H}_n(b))$. This is equivalent to say that for any $i \in [\![1,n]\!]$, we have $(a_i + b_i)/2 = \mathcal{Z}(\mathcal{H}(a_i) \wedge \mathcal{H}(b_i))$ by Lemma 3. Starting from the equality $\mathcal{H}(a_i) \wedge \mathcal{H}(b_i) = \{a_i, a_i + 1\} \wedge \{b_i, b_i + 1\}$ and observing that, since a and b are $(3^n - 1)$-neighbors in $\mathbb{Z}^n$ or equal, they satisfy for any

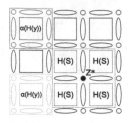

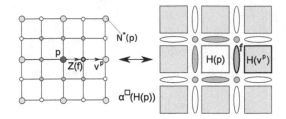

Fig. 4. When $y \notin S$ centered at $\mathcal{Z}_n(z^*)$, $\alpha(\mathcal{H}_n(y)) \cap \beta(z^*) = \emptyset$.

Fig. 5. $\alpha^\square(\mathcal{H}_n(p))$ is composed of the faces $\mathcal{H}_n(p) \wedge \mathcal{H}_n(v^p)$ such that $v^p \in \mathcal{N}^*(p)$.

$i \in [\![1, n]\!]$ that $a_i \in \{b_i - 1, b_i, b_i + 1\}$, we have 3 possible cases: $a_i = b_i - 1$, and then $\mathcal{H}(a_i) \wedge \mathcal{H}(b_i) = \{b_i - 1, b_i\} \wedge \{b_i, b_i + 1\} = \{b_i\}$, whose image by $\mathcal{Z}$ is equal to $b_i - \frac{1}{2} = (a_i + b_i)/2$, or we have $b_i = a_i - 1$, and then a symmetrical reasoning leads to the same result, or $b_i = a_i$, and then the result is immediate. $\square$

Proposition 2. *Let S be a block and let $p, p' \in S$ be any two antagonists in S. Then the center of the block S is equal to $\frac{p+p'}{2}$. Furthermore, its image by $\mathcal{H}_n$ into $\mathbb{H}^n$ is equal to $\mathcal{H}_n(p) \wedge \mathcal{H}_n(p')$.*

Proof: Starting from the two antagonists p, p' in S, we can compute $z \in \mathbb{Z}^n$ and $\mathcal{F} \subseteq \mathbb{B}$ such that $S = S(z, \mathcal{F})$. In fact, for all $i \in [\![1, n]\!]$, $z_i = \min(p_i, p'_i)$, and $\mathcal{F} = \{e^i \; ; \; i \in [\![1, n]\!], \; p_i \neq p'_i\}$. Then, it is clear that $p = (p-z)+z = z+\sum_{p_i \neq z_i} e^i$, and that $p' = (p' - z) + z = z + \sum_{p'_i \neq z_i} e^i$. Then, $p + p' = 2z + \sum_{f \in \mathcal{F}} f$, which shows that $\frac{p+p'}{2}$ is the center of S in $(\mathbb{Z}/2)^n$. The second part of the proposition follows from Lemma 5. $\square$

Lemma 6. *Let S be a block of $\mathbb{Z}^n$, and let $z^* \in \mathbb{H}^n$ be the image by $\mathcal{H}_n$ of the center of S. For all $y \in \mathbb{Z}^n$, $y \notin S$ implies that $\alpha(\mathcal{H}_n(y)) \cap \beta(z^*)$ is empty.*

Proof: This proof can be followed on Fig. 4. Let y be an element of $\mathbb{Z}^n$ s.t. $\alpha(\mathcal{H}_n(y)) \cap \beta(z^*)$ is not empty. Then, for all $i \in [\![1, n]\!]$, $\alpha(\mathcal{H}(y_i)) \cap \beta(z_i^*)$ is not empty. Now, let us show that y belongs to S. Since there exists $p_i \in \alpha(\mathcal{H}(y_i)) \cap \beta(z_i^*)$, then $\mathcal{H}(y_i) \in \beta(p_i)$ and $p_i \in \beta(z_i^*)$, which leads to $\mathcal{H}(y_i) \in \beta(z_i^*)$, and then $\mathcal{H}_n(y) \in \beta(z^*)$. Since $y \in \mathbb{Z}^n$, $\mathcal{H}_n(y) \in \mathbb{H}_n^n$, and then $\mathcal{H}(y) \in \beta(z^*) \cap \mathbb{H}_n^n$, which is equivalent to $y \in \mathcal{Z}_n(\beta(z^*) \cap \mathbb{H}_n^n)$, which is the reformulation of a block centered at z^* by Lemma 1. $\square$

Lemma 7. $\forall p \in \mathbb{Z}^n$, $\alpha^\square(\mathcal{H}_n(p)) = \bigcup_{v \in \mathcal{N}^*(p)} \alpha(\mathcal{H}_n(p) \wedge \mathcal{H}_n(v))$.

Proof: This proof is depicted on Fig. 5. Since $p \in \mathbb{Z}^n$, it can be easily proved that $\alpha^\square(\mathcal{H}_n(p))$ is equal to the set of elements f of $\mathbb{H}^n$ satisfying $\|\mathcal{Z}_n(f) - p\|_\infty = \frac{1}{2}$, i.e., $\|v^p - p\|_\infty = 1$ with $v^p := 2\mathcal{Z}_n(f) - p$. Then, $\alpha^\square(\mathcal{H}_n(p))$ is equal to the set of elements $f \in \mathbb{H}^n$ satisfying $v^p \in \mathcal{N}^*(p)$ and $f = \mathcal{H}_n((v^p + p)/2)$. By Lemma 5, we obtain that $\alpha^\square(\mathcal{H}_n(p))$ is equal to $\{\mathcal{H}_n(v^p) \wedge \mathcal{H}_n(p) \in \mathbb{H}^n \; ; \; v^p \in \mathcal{N}^*(p)\}$, which leads to the required formula by applying the α operator. $\square$

Lemma 8. *Let S be a block in $\mathbb{Z}^n$ of dimension $k \geq 2$. Now, let p, p' be two antagonists in S, and v be a $2n$-neighbor of p in S. Then, we have the following relation: $\mathcal{H}_n(p) \wedge \mathcal{H}_n(p') \in \alpha(\mathcal{H}_n(p) \wedge \mathcal{H}_n(v))$.*

Proof: By Lemma 4, $\mathcal{H}_n(p) \wedge \mathcal{H}_n(p')$ and $\mathcal{H}_n(p) \wedge \mathcal{H}_n(v)$ are well-defined (p and v are antagonists in a block of dimension 1). By Lemma 3, the first term of the relation is equal to $\times_{i \in [\![1,n]\!]} (\mathcal{H}(p_i) \wedge \mathcal{H}(p'_i))$. Likewise, the second term is equal to $\otimes_{i \in [\![1,n]\!]} (\alpha(\mathcal{H}(p_i)) \cap \alpha(\mathcal{H}(v_i)))$. Then we want to show that for all $i \in [\![1,n]\!]$, $\mathcal{H}(p_i) \wedge \mathcal{H}(p'_i)$ belongs to $\alpha(\mathcal{H}(p_i)) \cap \alpha(\mathcal{H}(v_i))$. Let $\mathfrak{J}$ be the family of indices $\{i \in [\![1,n]\!] \; ; \; p_i \neq p'_i\}$. Since v is a $2n$-neighbor of p into S, there exists an index i^* in $\mathfrak{J}$ such that $v_{i^*} \neq p_{i^*}$, i.e., $v_{i^*} = p'_{i^*}$, and $\forall i \in [\![1,n]\!] \setminus \{i^*\}, v_i = p_i$. When $i \in [\![1,n]\!] \setminus \mathfrak{J}$ or when $i = i^*$, the property is obviously true. When $i \in \mathfrak{J} \setminus \{i^*\}$, then $v_i = p_i$, which implies $\alpha(\mathcal{H}(p_i)) \cap \alpha(\mathcal{H}(v_i)) = \alpha(\mathcal{H}(p_i)) = \{\{p_i\}, \{p_i + 1\}, \{p_i, p_i + 1\}\}$. However, either $\mathcal{H}(p_i) \wedge \mathcal{H}(p'_i) = \{p_i\}$ (if $p'_i = p_i - 1$) or $\mathcal{H}(p_i) \wedge \mathcal{H}(p'_i) = \{p_i + 1\}$ (if $p'_i = p_i + 1$), then $\mathcal{H}(p_i) \wedge \mathcal{H}(p'_i) \in \alpha(\mathcal{H}(p_i)) \cap \alpha(\mathcal{H}(v_i))$. □

3.4 Some Additional Background Concerning n-surfaces

The following proposition results from the proof of Property 11 (p. 55) in [6].

Proposition 3. *Let $|X| = (X, \alpha_X)$ and $|Y| = (Y, \alpha_Y)$ be two n-surfaces, $n \geq 0$. Then, if $|X|$ is a suborder of $|Y|$, then $|X| = |Y|$.*

Proof: Let us proceed by induction. Initialization ($n = 0$): when $|X|$ and $|Y|$ are two 0-surfaces, the inclusion $X \subseteq Y$ implies directly that $X = Y$ since they have the same cardinality, and then $|X| = |Y|$. Heredity ($n \geq 1$): we assume that when two $(n-1)$-surfaces satisfy an inclusion relationship, they are equal. Now, let $|X|$ and $|Y|$ be two n-surfaces, $n \geq 1$, such that $|X|$ is a suborder of $|Y|$. Then, for all $x \in X$, $x \in Y$ and so we can write $\theta_X^{\square}(x) \subseteq \theta_Y^{\square}(x)$ since $X \subseteq Y$. However, $|\theta_X^{\square}(x)|$ and $|\theta_Y^{\square}(x)|$ are $(n-1)$-surfaces and $|\theta_X^{\square}(x)|$ is a suborder of $|\theta_Y^{\square}(x)|$, then $|\theta_X^{\square}(x)| = |\theta_Y^{\square}(x)|$. Now, let us assume that we have $X \subsetneq Y$. Then let x be a point of X and y a point of $Y \setminus X$. Since $|Y|$ is connected as an n-surface with $n \geq 1$, it is connected by path, and so $x, y \in Y$ implies that there exists a path π joining them into Y. This way, there exist $x' \in X$ and $y' \in Y \setminus X$ s.t. $y' \in \theta^{\square}(x')$. In other words, $y' \in \theta_Y^{\square}(x') = \theta_X^{\square}(x')$ since $x' \in X$. This leads to $y' \in X$. We obtain a contradiction. Thus we have $X = Y$, and consequently $|X| = |Y|$. □

Corollary 1. *Let $|X_1|$ and $|X_2|$ be two k-surfaces, $k \geq 0$, with $X_1 \cap X_2 = \emptyset$. Then $|X_1 \cup X_2|$ is not a k-surface.*

Proposition 4. *Let a, b be two elements of $\mathbb{H}^n$ with $a \in \beta^{\square}(b)$. Then $|\alpha^{\square}(a) \cap \beta^{\square}(b)|$ is a $(\dim(a) - \dim(b) - 2)$-surface.*

Proof: Since $|\mathbb{H}^n|$ is an n-surface, then $|\alpha^{\square}(a)|$ is a $(\rho(a, |\mathbb{H}^n|) - 1)$-surface by Property 2, and then is a $(\dim(a) - 1)$-surface. Now, we can remark that because b belongs to $\alpha^{\square}(a)$, we can write that $\alpha^{\square}(a) \cap \beta^{\square}(b) = \beta^{\square}_{\alpha^{\square}(a)}(b)$, and then, again by Property 2, $|\alpha^{\square}(a) \cap \beta^{\square}(b)|$ is a $((\dim(a) - 1) - \rho(b, |\alpha^{\square}(a)|) - 1)$-surface. Since $\rho(b, |\alpha^{\square}(a)|) = \rho(b, |\mathbb{H}^n|) = \dim(b)$, the proof is done.   □

4   Properties Specific to the Proof

From now on, we suppose n is an integer greater than or equal to 2, that X is a digital subset of $\mathbb{Z}^n$, that Y is the complement of X into $\mathbb{Z}^n$; also, we define the sets $\mathcal{X} := \mathcal{H}_n(X)$ and $\mathcal{Y} := \mathcal{H}_n(Y)$, and the *immersion* of X into $\mathbb{H}^n$ using the *miss strategy*: $\mathcal{I}(X) := \mathrm{Int}(\alpha(\mathcal{X}))$; its boundary is $\mathfrak{N} := \alpha(\mathcal{I}(X)) \cap \alpha(\mathbb{H}^n \setminus \mathcal{I}(X))$.

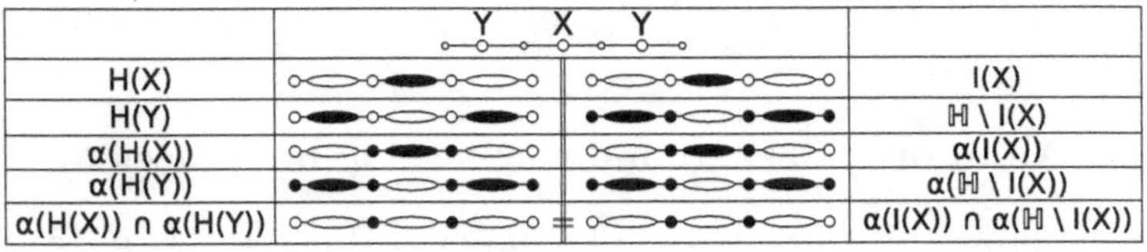

Fig. 6. $\alpha(\mathcal{H}_n(X)) \cap \alpha(\mathcal{H}_n(Y))$ vs. $\alpha(\mathcal{I}(X)) \cap \alpha(\mathbb{H}^n \setminus \mathcal{I}(X))$ in 1D.

Proposition 5. $\mathfrak{N}$ *is equal to* $\alpha(\mathcal{X}) \cap \alpha(\mathcal{Y})$.

Proof: An intuition of the proof is given in Fig. 6. Let us first remark that $\alpha(\mathcal{X})$ is a regular closed set. Effectively, $\mathrm{Int}(\alpha(\mathcal{X})) \subseteq \alpha(\mathcal{X})$ implies that $\alpha(\mathrm{Int}(\alpha(\mathcal{X}))) \subseteq \alpha(\mathcal{X})$ by monotonicity of α. Conversely, any element $x \in \mathcal{X}$ satisfies $\beta(x) = \{x\} \subseteq \mathcal{X}$, and so $\mathrm{Int}(\alpha(\mathcal{X}))$, which is equal to $\{h \in \alpha(\mathcal{X}) \; ; \; \beta(h) \subseteq \alpha(\mathcal{X})\}$, contains $\mathcal{X}$. This implies that $\alpha(\mathrm{Int}(\alpha(\mathcal{X}))) \supseteq \alpha(\mathcal{X})$. Thus $\alpha(\mathcal{X})$ is a regular closed set. We can then simplify the formula of $\mathfrak{N}$; by definition, $\mathfrak{N}$ is equal to $\alpha(\mathcal{I}(X)) \cap \alpha(\mathbb{H}^n \setminus \mathcal{I}(X))$, which is then equal to $\alpha(\mathcal{X}) \cap \alpha(\mathbb{H}^n \setminus \mathcal{I}(X))$. Since $\mathcal{I}(X)$ is open, $\alpha(\mathbb{H}^n \setminus \mathcal{I}(X)) = \mathbb{H}^n \setminus \mathcal{I}(X)$. Thus, $\mathfrak{N} = \alpha(\mathcal{X}) \setminus \mathcal{I}(X)$, which is equal to $\alpha(\mathcal{X}) \cap (\mathrm{Int}(\alpha(\mathcal{X})))^c$, and so $\mathfrak{N} = \alpha(\mathcal{X}) \cap \alpha(\mathrm{Int}(\mathcal{X}^c))$. Let us show that $\alpha(\mathcal{X}) \cap \alpha(\mathrm{Int}(\mathcal{X}^c))$ is equal to $\alpha(\mathcal{X}) \cap \alpha(\mathcal{Y})$. Since $\mathcal{Y} = \mathbb{H}^n_n \setminus \mathcal{X} \subseteq \mathbb{H}^n \setminus \mathcal{X}$, it is clear that $\mathrm{Int}(\mathcal{Y}) \subseteq \mathrm{Int}(\mathbb{H}^n \setminus \mathcal{X})$. Since $\mathcal{Y}$ is open as a set of n-faces, we obtain $\mathcal{Y} \subseteq \mathrm{Int}(\mathbb{H}^n \setminus \mathcal{X})$, and thus $\alpha(\mathcal{Y}) \subseteq \alpha(\mathrm{Int}(\mathbb{H}^n \setminus \mathcal{X})) = \alpha(\mathrm{Int}(\mathcal{X}^c))$. This way, $\alpha(\mathcal{X}) \cap \alpha(\mathcal{Y}) \subseteq \alpha(\mathcal{X}) \cap \alpha(\mathrm{Int}(\mathcal{X}^c))$. Now, let z be an element of $\alpha(\mathcal{X}) \cap \alpha(\mathrm{Int}(\mathcal{X}^c))$, then $\beta(z) \cap \mathbb{H}^n_n \subseteq \mathcal{X}$ (1), or $\beta(z) \cap \mathbb{H}^n_n \subseteq \mathcal{Y}$ (2), or $\beta(z) \cap \mathbb{H}^n_n \cap \mathcal{X} \neq \emptyset \neq \beta(z) \cap \mathbb{H}^n_n \cap \mathcal{Y}$ (3). Before treating the first case, let us prove that $\alpha(\beta(z)) = \alpha(\beta(z) \cap \mathbb{H}^n_n)$ (P). The converse inclusion is obvious. Concerning the direct inclusion, let a be an element of $\alpha(\beta(z))$. There exists $p \in \beta(z)$ such that $a \in \alpha(p)$. Also, $\mathbb{H}^n$ is an n-surface, and so is homogeneous. This implies that there exists $p^n \in \beta(p)$ s.t. $p^n \in \mathbb{H}^n_n$. Since $p^n \in \beta(p)$ and $p \in \beta(z)$, $p^n \in \beta(z) \cap \mathbb{H}^n_n$, and the fact that a belongs to $\alpha(p)$ implies that $a \in \alpha(\beta(z) \cap \mathbb{H}^n_n)$. This way, (P) is true. Now, we can treat the first case: $\beta(z) \cap \mathbb{H}_n^n \subseteq \mathcal{X}$ implies that $\mathrm{Int}(\alpha(\beta(z) \cap \mathbb{H}_n^n)) \subseteq \mathrm{Int}(\alpha(\mathcal{X}))$. Using (P), we obtain $\mathrm{Int}(\alpha(\beta(z))) \subseteq \mathrm{Int}(\alpha(\mathcal{X}))$. Since $\beta(z)$ is an open regular set, we obtain $\beta(z) \subseteq \mathrm{Int}(\alpha(\mathcal{X}))$. Yet, $\beta(z) \subseteq \alpha(\beta(z)) \subseteq \alpha(\mathcal{X})$, since $\alpha(\mathcal{X})$ is a regular closed set. However, this imples that $\beta(z) = \mathrm{Int}(\beta(z)) \subseteq \mathrm{Int}(\alpha(\mathcal{X}))$, and so $z \notin \alpha(\mathrm{Int}(\mathcal{X}^c))$, which is a contradiction. In the second case, $\beta(z) \cap \mathbb{H}_n^n \subseteq \mathcal{Y}$, which means that no $x \in \mathcal{X}$ exists such that $x \in \beta(z)$, which means that $z \notin \alpha(\mathcal{X})$, which leads once more to a contradiction. In the third case, $\beta(z) \cap \mathbb{H}_n^n \cap \mathcal{X} \neq \emptyset$ and $\beta(z) \cap \mathbb{H}_n^n \cap \mathcal{Y} \neq \emptyset$ implies that there exists some $x \in \mathcal{X}$ and $y \in \mathcal{Y}$ such that $z \in \alpha(x) \cap \alpha(y)$, and so $z \in \alpha(\mathcal{X}) \cap \alpha(\mathcal{Y})$. $\square$

Proposition 6. *For any $z \in \mathfrak{N}$, $|\alpha_{\mathfrak{N}}^{\square}(z)|$ is a $(\dim(z) - 1)$-surface.*

Proof: Since $\mathfrak{N}$ is closed, $\forall z \in \mathfrak{N}$, $|\alpha_{\mathfrak{N}}^{\square}(z)| = |\alpha^{\square}(z)|$, which is a $(\rho(z, |\mathbb{H}^n|) - 1)$-surface by Property 2 since $\mathbb{H}^n$ is an n-surface. Since $\rho(z, |\mathbb{H}^n|) = \dim(z)$, $|\alpha_{\mathfrak{N}}^{\square}(z)|$ is a $(\dim(z) - 1)$-surface. $\square$

Lemma 9. *$\mathcal{I}(X)$ is AWC iff $\forall z \in \mathfrak{N}$, $|\beta_{\mathfrak{N}}^{\square}(z)|$ is a $(n - 2 - \dim(z))$-surface.*

Proof: Let us recall that two disjoint components C_1 and C_2 of $\mathfrak{N}$ are separated: $C_1 \cap \theta(C_2) = \emptyset$. For this reason, for any $z \in \mathfrak{N}$, $|\theta_{\mathfrak{N}}^{\square}(z)| = |\theta^{\square}(z) \cap \bigcup_{C \in CC(\mathfrak{N})} C|$ $= |\bigcup_{C \in CC(\mathfrak{N})}(\theta^{\square}(z) \cap C)| = |\theta_{CC(\mathfrak{N},z)}^{\square}(z)|$. Since $n \geq 2$, $\mathcal{I}(X)$ is AWC iff $\forall C \in CC(\mathfrak{N})$, C is a $(n - 1)$-surface, *i.e.*, $\forall C \in CC(\mathfrak{N})$, $\forall z \in C$, $|\theta_C^{\square}(z)|$ is a $(n - 2)$-surface, which means that $\forall C \in CC(\mathfrak{N})$, $\forall z \in C$, $|\theta_{\mathfrak{N}}^{\square}(z)|$ is a $(n - 2)$-surface, or, in other words, by Property 1 and Proposition 6, $\forall z \in \mathfrak{N}$, $|\beta_{\mathfrak{N}}^{\square}(z)|$ is a $(n - 2 - \dim(z))$-surface. $\square$

Proposition 7. *Let S be a block of dimension $k \in [\![2, n]\!]$ s.t. $X \cap S = \{p, p'\}$ (resp. $Y \cap S = \{p, p'\}$) and $p' = \mathrm{antag}_S(p)$, then $\mathcal{H}_n\left(\frac{p+p'}{2}\right) \in \mathfrak{N}$.*

Proof: Let v be a $2n$-neighbor of p in S, then, by Lemma 4, $\mathcal{H}_n(p) \wedge \mathcal{H}_n(v)$ is well-defined, and by Lemma 3, $\alpha(\mathcal{H}_n(p) \wedge \mathcal{H}_n(v)) = \alpha(\mathcal{H}_n(p)) \cap \alpha(\mathcal{H}_n(v))$. Since $\dim(S) \geq 2$, $v \in Y$, and so $\alpha(\mathcal{H}_n(p)) \cap \alpha(\mathcal{H}_n(v)) \subseteq \mathfrak{N}$ by Proposition 5. Now, using Proposition 2, $\mathcal{H}_n\left(\frac{p+p'}{2}\right)$ is equal to $\mathcal{H}_n(p) \wedge \mathcal{H}_n(p')$, which belongs to $\alpha(\mathcal{H}_n(p) \wedge \mathcal{H}_n(v))$ by Lemma 8, and thus to $\mathfrak{N}$. $\square$

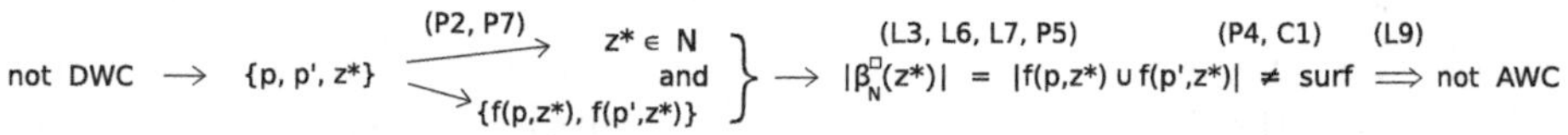

$$\text{not DWC} \rightarrow \{p, p', z^*\} \xrightarrow[\{f(p,z^*), f(p',z^*)\}]{\text{(P2, P7)}} \left.\begin{array}{c} z^* \in N \\ \text{and} \end{array}\right\} \rightarrow \overset{\text{(L3, L6, L7, P5)}}{|\beta_N^{\square}(z^*)|} = \overset{\text{(P4, C1)}}{|f(p,z^*) \cup f(p',z^*)|} \overset{\text{(L9)}}{\neq \text{surf}} \Rightarrow \text{not AWC}$$

Fig. 7. Summary of the proof; $f(q, z^*) := \alpha^{\square}(\mathcal{H}_n(q)) \cap \beta^{\square}(z^*)$, with $q \in \{p, p'\}$.

5 Proof of the Main Result

We want to prove that AWCness implies DWCness in n-D (see Fig. 7 for the summary of the proof). To this aim, we will show that if X is not DWC, then $\mathcal{I}(X)$ is not AWC. Since by Lemma 9, $\mathcal{I}(X)$ is AWC iff $\forall z \in \mathfrak{N}$, $|\beta_{\mathfrak{N}}^{\square}(z)|$ is a $(n - 2 - \dim(z))$-surface, it is sufficient to prove that when X is not DWC, then there exists an element z^* of $\mathfrak{N}$ such that $|\beta_{\mathfrak{N}}^{\square}(z^*)|$ is not a $(n-2-\dim(z^*))$-surface. So, let us assume that X is not DWC.

Fig. 8. $\beta_{\mathfrak{N}}^{\square}(\mathcal{H}_n(c))$ (in blue) when X admits a 2D/3D critical configuration in the block of center c (whose image by $\mathcal{H}_n$ is depicted in black) when $n = 3$. (Color figure online)

Now, X admits a critical configuration. Let us treat the primary case, since the reasoning for the secondary case is similar: let us assume that there exists a block S of dimension $k \in [\![2, n]\!]$ such that $X \cap S = \{p, p'\}$ with $p' = \mathrm{antag}_S(p)$. This way, we can compute the image z^* by $\mathcal{H}_n$ into $\mathbb{H}^n$ of the center of S. By Proposition 2, $z^* = \mathcal{H}_n(p) \wedge \mathcal{H}_n(p')$. Let us show that $|\beta_{\mathfrak{N}}^{\square}(z^*)|$ is not a $(n - 2 - \dim(z))$-surface. By Proposition 7, $z^* \in \mathfrak{N}$, so the expression $\beta_{\mathfrak{N}}^{\square}(z^*)$ is well-defined. Now, let us compute $|\beta_{\mathfrak{N}}^{\square}(z^*)|$. Using Lemmas 7 and 6, we obtain that $\mathsf{f}(p, z^*) := \alpha^{\square}(\mathcal{H}_n(p)) \cap \beta^{\square}(z^*)$ is equal to $\bigcup_{y \in S \setminus \{p\}} \alpha(\mathcal{H}_n(p) \wedge \mathcal{H}_n(y)) \cap \beta^{\square}(z^*)$. Then, since we know that $\alpha(\mathcal{H}_n(p) \wedge \mathcal{H}_n(p')) \cap \beta^{\square}(z^*) = \emptyset$, thus $\mathsf{f}(p, z^*)$ is equal to $\bigcup_{y \in S \setminus \{p, p'\}} \alpha(\mathcal{H}_n(p) \wedge \mathcal{H}_n(y)) \cap \beta^{\square}(z^*)$. By Lemma 3, $\mathsf{f}(p, z^*) = \alpha(\mathcal{H}_n(p)) \cap \alpha(\mathcal{H}_n(Y \cap S)) \cap \beta^{\square}(z^*)$. With a similar calculation based on p', we obtain that $\mathsf{f}(p', z^*) := \alpha^{\square}(\mathcal{H}_n(p')) \cap \beta^{\square}(z^*)$ is equal to $\alpha(\mathcal{H}_n(p')) \cap \alpha(\mathcal{H}_n(Y \cap S)) \cap \beta^{\square}(z^*)$. Next, $\mathsf{f}(p, z^*) \cup \mathsf{f}(p', z^*)$ is equal to $\alpha(\mathcal{H}_n(X \cap S)) \cap \alpha(\mathcal{H}_n(Y \cap S)) \cap \beta^{\square}(z^*)$, which is equal by Lemma 6 to $\alpha(\mathcal{X}) \cap \alpha(\mathcal{Y}) \cap \beta^{\square}(z^*)$, and then to $\beta_{\mathfrak{N}}^{\square}(z^*)$ by Proposition 5. Finally, we have that $|\beta_{\mathfrak{N}}^{\square}(z^*)|$ is equal to $|\mathsf{f}(p, z^*) \cup \mathsf{f}(p', z^*)|$. Figure 8 depicts examples of $\beta_{\mathfrak{N}}^{\square}(z^*)$ in the case $n = 3$. Let us now remark that $|\beta_{\mathfrak{N}}^{\square}(z^*)|$ is the disjoint union of $|\mathsf{f}(p, z^*)|$ and of $|\mathsf{f}(p', z^*)|$: $\alpha^{\square}(\mathcal{H}_n(p)) \cap \alpha^{\square}(\mathcal{H}_n(p')) \cap \beta^{\square}(z^*) = \emptyset$. However, by Proposition 4, $|\mathsf{f}(p, z^*)|$ and $|\mathsf{f}(p', z^*)|$ are both $(n - \dim(z^*) - 2)$-surfaces. Finally, by Corollary 1, $|\beta_{\mathfrak{N}}^{\square}(z^*)|$ is not a $(n - \dim(z^*) - 2)$-surface, and then $\mathcal{I}(X)$ is not AWC. □

6 Conclusion

Now that we have proved that AWCness implies DWCness for sets, we naturally conclude that, thanks to *cross-section topology*, this implication is also true for gray-level images: a gray-level image $u : \mathbb{Z}^n \to \mathbb{Z}$ will be DWC if the *span-based immersion* of u is AWC. As future work, we propose to study the converse implication, *i.e.*, if DWCness implies AWCness in n-D, $n \geq 2$.

References

1. Alexandrov, P.: Combinatorial topology, vol. 1–3. Graylock (1956)
2. Alexandrov, P.: Diskrete Räume. Matematicheskii Sbornik **2**(44), 501–519 (1937)
3. Alexandrov, P., Hopf, H.: Topologie I. Springer-Verlag, Heidelberg (2013)
4. Bertrand, G.: New notions for discrete topology. In: Bertrand, G., Couprie, M., Perroton, L. (eds.) DGCI 1999. LNCS, vol. 1568, pp. 218–228. Springer, Heidelberg (1999). doi:10.1007/3-540-49126-0_17
5. Boutry, N., Géraud, T., Najman, L.: How to make nD functions digitally well-composed in a self-dual way. In: Benediktsson, J.A., Chanussot, J., Najman, L., Talbot, H. (eds.) ISMM 2015. LNCS, vol. 9082, pp. 561–572. Springer, Cham (2015). doi:10.1007/978-3-319-18720-4_47
6. Daragon, X.: Surfaces discrètes et frontières d'objets dans les ordres. Ph.D. thesis, Université de Marne-la-Vallée, France (2005)
7. Daragon, X., Couprie, M., Bertrand, G.: Discrete frontiers. In: Nyström, I., Sanniti di Baja, G., Svensson, S. (eds.) DGCI 2003. LNCS, vol. 2886, pp. 236–245. Springer, Heidelberg (2003). doi:10.1007/978-3-540-39966-7_22
8. Eckhardt, U., Latecki, L.: Digital topology. Technical report, Institut für Angewandte Mathematik (1994)
9. Evako, A.V., Kopperman, R., Mukhin, Y.V.: Dimensional properties of graphs and digital spaces. J. Math. Imaging Vis. **6**(2–3), 109–119 (1996)
10. Kelley, J.L.: General Topology, Graduate Texts in Mathematics, vol. 27. Springer, Heidelberg (1955)
11. Khalimsky, E., Kopperman, R., Meyer, P.R.: Computer graphics and connected topologies on finite ordered sets. Topol. Appl. **36**(1), 1–17 (1990)
12. Kovalevsky, V.: Axiomatic digital topology. J. Math. Imaging Vis. **26**(1), 41–58 (2006)
13. Najman, L., Géraud, T.: Discrete set-valued continuity and interpolation. In: Hendriks, C.L.L., Borgefors, G., Strand, R. (eds.) ISMM 2013. LNCS, vol. 7883, pp. 37–48. Springer, Heidelberg (2013). doi:10.1007/978-3-642-38294-9_4

Discrete Shape Representation,
Recognition and Analysis

Heat Kernel Laplace-Beltrami Operator on Digital Surfaces

Thomas Caissard[1(✉)], David Coeurjolly[1], Jacques-Olivier Lachaud[2], and Tristan Roussillon[1]

[1] Univ Lyon, CNRS, INSA-Lyon, LIRIS, UMR 5205, 69621 Lyon, France
{thomas.caissard,david.coeurjolly,tristan.roussillon}@liris.cnrs.fr
[2] Université de Savoie, CNRS, LAMA UMR 5127, 73776 Chambéry, France
jacques-olivier.lachaud@univ-smb.fr

Abstract. Many problems in image analysis, digital processing and shape optimization can be expressed as variational problems involving the discretization of the Laplace-Beltrami operator. Such discretizations have been widely studied for meshes or polyhedral surfaces. On digital surfaces, direct applications of classical operators are usually not satisfactory (lack of multigrid convergence, lack of precision...). In this paper, we first evaluate previous alternatives and propose a new digital Laplace-Beltrami operator showing interesting properties. This new operator adapts Belkin *et al.* [2] to digital surfaces embedded in 3D. The core of the method relies on an accurate estimation of measures associated to digital surface elements. We experimentally evaluate the interest of this operator for digital geometry processing tasks.

1 Introduction

Objectives. In geometry processing, Partial Differential Equations (PDEs) containing Laplace-Beltrami operator arise in many applications such as surface fairing, mesh smoothing, mesh parametrization, remeshing, mesh compression, feature extraction or shape matching (see [16] for an extensive survey). On digital surfaces, few digital Laplace-Beltrami operators have been proposed and none has been evaluated in terms of multigrid convergence (convergence of the operator toward the continuous one in digitization of smooth manifolds on grid with decreasing gridstep).

Contributions. In this article, we propose a discrete Laplace-Beltrami operator on digital surfaces (boundaries of subsets of $\mathbb{Z}^2$ embedded in 3D). This new operator adapts Belkin *et al.* [2] on our specific data. The method uses an accurate estimation of areas associated with digital surface elements. This estimation is achieved through a convergent digital normal estimator described in [5]. We show experimental convergence of our operator but also that none of the existing approaches adapted to digital surfaces achieves such convergence. Finally, we illustrate the interest of the discretized Laplacian on digital surface geometry processing.

This work has been partly funded by CoMeDiC ANR-15-CE40-0006 research grant.

W.G. Kropatsch et al. (Eds.): DGCI 2017, LNCS 10502, pp. 241–253, 2017.
DOI: 10.1007/978-3-319-66272-5_20

Related works. The Laplace-Beltrami operator being a second order differential operator (divergence of the function gradient, see Sect. 2) a discrete calculus framework is required to define such operator on embedded combinatorial structures such as meshes or digital surfaces. First works on discrete calculus may be found in the Regge Calculus [21] for quantum physics, where tetrahedra in combination with edge lengths are used. Works on geometric acquisition devices and models drove studies toward calculus working on meshes and more generally on simplicial complexes. Early works include a definition of the Laplace-Beltrami operator using the classical cotangent formula [19] for solving the problem of minimal surfaces, which is an analog of the standard finite element method [16]. Exact calculus generalizing the cotangent discretization in 2D based on finite elements [20] emerged from the *German school* but with a restriction to triangular complexes.

In a more generic discrete calculus perspective, the Discrete Exterior Calculus (DEC) framework was then developed in the computational mathematics and geometry processing community. Another more recent formulation of the DEC comes from Hirani's thesis [11] and later by the monograph [8]. On triangular meshes, DEC based Laplace-Beltrami operator and the cotangent based one coincide.

In [10], authors show that under some strong assumptions, the cotangent Laplacian on a triangular mesh converges to the continuous one when the mesh interpolates a smooth manifold with increasing precision (with a continuous one-to-one map between the mesh and the manifold, which would not be the case on digital surfaces). The operator converges in the sense of distributions and the authors show that pointwise convergence in the l_2 sense does not generally hold. On triangular meshes, Belkin *et al.* [2] have proposed a first Laplace-Beltrami operator that converges in the l_∞ uniform case. The digital Laplacian operator we propose is an extension to digital surfaces of such operator.

In digital geometry, many estimators of differential quantities have been proposed and there exists multigrid convergent estimators for many quantities such as length, tangent and curvature in 2D (see [6] for a complete survey), surface area [14], normal vectors and curvature tensor in dimension 3 [5]. A preliminary approach can be found in [4,17]. It focuses on the conformal map computation of a digital surface, which is a related problem involving the definition of a Laplace-Beltrami operator. However, their definition is based on the cotangent formula, which lacks pointwise convergence. When designing a discrete version of the Laplacian, not all properties of the continuous one can be expected at the same time. In [24], entitled *"Discrete Laplace operators: No free lunch."*, the authors have proposed a formal evaluation of such properties. We position our new digital Laplace-Beltrami operator with respect to this analysis.

Outline. After introducing mathematical definitions, we review the classical approaches to define a discrete Laplacian and compare their properties in Sect. 2. We then formalize our operator in Sect. 3. In Sect. 4, we experimentally evaluate our proposal in terms of multigrid convergence and geometry processing applications.

2 Discretizations of the Laplace-Beltrami Operator and Their Properties

We first describe various discretizations of the Laplace-Beltrami operator on triangular meshes. We then check several desired properties of Laplacian using [24] as a baseline for comparisons.

2.1 Preliminaries and Classical Discretizations on Triangular Meshes

Let M be a 2-smooth manifold, with or without boundary, embedded in $\mathbb{R}^3$. The intrinsic smooth Laplace-Beltrami operator [22] is defined as:

$$\Delta : C^2(M) \to C^2(M)$$
$$u \mapsto \mathrm{div}(\nabla u), \tag{1}$$

where C^2 is the set of functions which are twice differentiable and with the second derivative continuous (in the literature, alternative definitions may consider "$-\mathrm{div}(\nabla u)$" for Δu).

Let Γ be a combinatorial structure (a triangular mesh for instance), $V(\Gamma)$ its set of vertices and $F(\Gamma)$ its faces. Let $u : M \to \mathbb{R}$ be a twice differentiable function. We suppose that $V(\Gamma)$ is a sampling of M (*i.e.* $V(\Gamma) \subset M$). In other words, $u(w)$ is perfectly defined for $w \in V(\Gamma)$.

A first simple discretization only considers the combinatorial structure of Γ. Such Laplacian is either called *graph Laplacian* or *combinatorial Laplacian* of Γ [25]:

$$(\mathcal{L}_{COMBI}\, u)(w) := -deg(w)u(w) + \sum_{p \in \mathrm{link}_0(w)} u(p), \tag{2}$$

for all $w \in V(\Gamma)$ where $\mathrm{link}_0(w)$ is the set of points $V(\Gamma)$ adjacent to w and $deg(w)$ is the degree of w in Γ.

A more complex approach can be defined using DEC operators [8,11]. Using an arbitrarily embedded dual structure of Γ, the Laplace operator can be shown to be a classical weighted double finite difference:

$$(\mathcal{L}_{DEC}\, u)(w) := \frac{1}{|\star w|} \sum_{p \in \mathrm{link}_0(w)} \frac{|\star e_{wp}|}{|e_{wp}|}(u(p) - u(w)), \tag{3}$$

where $\star$ is the Hodge-duality star operator acting on discrete forms (see [11]), and $|\cdot|$ the measure of a k-cell. As illustrated in Fig. 1, $|\star e_{wp}|$ would be the length of the segment orthogonal to e_{wp}. If we set all measures to one, $\mathcal{L}_{DEC}$ coincides with $\mathcal{L}_{COMBI}$.

By fixing the dual of Γ to be the Voronoi diagram of its vertices and by computing the measures as Euclidean lengths and areas of such dual complex, the DEC operator coincides exactly with the famous *cotan Laplacian* [19]:

$$(\mathcal{L}_{COT}\, u)(w) := \frac{1}{2A_w} \sum_{p\, \mathrm{link}_0(w)} (\cot(\alpha_{wp}) + \cot(\beta_{wp})) (u(p) - u(w)), \tag{4}$$

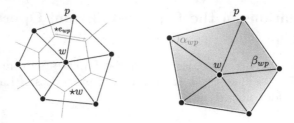

Fig. 1. Illustration of $\mathcal{L}_{DEC}$ (*left*), and $\mathcal{L}_{COT}$ (*right*) on triangular meshes. For $\mathcal{L}_{COT}$ the area of integration A_w is one third the area of all triangles incident on vertex w in green. For $\mathcal{L}_{DEC}$, the dual structure is in orange and the dual of the edge e_{wp} is in blue. (Color figure on line)

where A_w is one third of the area of all incident triangles to vertex w, α_{wp} and β_{wp} are the angles opposing the corresponding edge e_{wp} (see Fig. 1).

Finally, we detail the definition of the *mesh Laplacian* from [2]. Let $g : M \times (0, T) \to \mathbb{R}$ be a time-dependent function which solves the partial differential equation called the *heat equation*:

$$\Delta g(x, t) = \frac{\partial}{\partial t} g(x, t), \tag{5}$$

with initial condition $g_0 = g(\cdot, 0) : M \to \mathbb{R}$ which is the initial temperature distribution. An exact solution [22] is:

$$g(x, t) = \int_{y \in M} p(t, x, y) g_0(y) dy, \tag{6}$$

where $p \in C^{\infty}(R^+ \times M \times M)$ is called the heat kernel. The construction of the heat kernel involves complex technics [22]. Fortunately, there exists many approximations of p as t tends toward 0 (called small time asymptotics). Early work includes the famous Varadhan formula [23] and later extensions on a wider class of shapes [18]. Recently, an approximation using the ambient metric (*i.e.* the L_2 norm) have been proposed by Belkin *et al.* It is known to converge toward the real heat kernel for small t (see Lemma 5 of [1]):

$$g(x, t) \underset{t \to 0}{\sim} \frac{1}{4\pi t} \int_{y \in M} e^{-\frac{||x-y||^2}{4t}} g_0(y) dy. \tag{7}$$

Injecting Eq. (7) into Eq. (5), applying a finite time difference and knowing that the integral of e over M is one:

$$\Delta g(x, t) = \lim_{t \to 0} \frac{1}{4\pi t^2} \int_{y \in M} e^{-\frac{||x-y||^2}{4t}} (g_0(y) - g_0(x)) dy. \tag{8}$$

The previous equation can be seen as a convolution between differences of g and a time dependent Gaussian. Then, the mesh Laplace operator [2] on Γ is:

$$(\mathcal{L}_{MESH}\, u)(w) := \frac{1}{4\pi t^2} \sum_{f \in F(\Gamma)} \frac{A_f}{3} \sum_{p \in V(f)} e^{-\frac{\|p-w\|^2}{4t}} (u(p) - u(w)), \qquad (9)$$

where A_f is the area associated to the face f.

2.2   Desired Properties of a Discrete Laplacian

As discussed in [24], all properties of the continuous Laplace-Beltrami operator may not be preserved when discretizing it. We consider the discrete Laplacian as a linear operator acting on values $\mathbf{u} := \{u_p\}$ on $V(\Gamma)$ (represented as a vector in $\mathbb{R}^{|V(\Gamma)|}$). Such operator can thus be denoted as a matrix $\mathbf{L}$ with components l_{ij}. Hence, $\mathbf{v} := \mathbf{L}\mathbf{u}$ would be the resulting Laplacian of $\mathbf{u}$. Expected properties of the discrete Laplace-Beltrami operators are:

Symmetry (SYM). $\ell_{ij} = \ell_{ji}$ for $0 \leq i < |V(\Gamma)|$. This is very useful when it comes to solve linear systems as solvers are usually more performant when the matrix is symmetric.

Locality (LOC). $\ell_{ij} \neq 0$ if and only if vertices i and j share a common edge. The locality property gives very sparse matrices decreasing drastically memory consumption. It also opens a panel of very fast linear system solvers.

Linear Precision (LIN). $\mathbf{L}\mathbf{u} = 0$ whenever $\mathbf{u}$ is a linear function restricted to a plane.

Positive Weights (POS). $\ell_{ij} \geq 0$ for $i \neq j$. Furthermore, for each vertex i, there exists a vertex j such that $\ell_{ij} > 0$.

Positive Semi-Definiteness (PSD). The matrix is symmetric positive semi-definite regarding the standard inner product, and has a one-dimensional kernel. (SYM) and (POS) imply (PSD), but (PSD) does not imply (POS). This property ensures that the basis generated by the eigenvectors of L is orthogonal and that the eigenvalues are real.

Dirichlet Convergence (CON). When considering a sequence of meshes $\{\Gamma_i\}$ converging to M in a given sense as $i \to \infty$, we want the Laplacian sequence $\mathbf{L}_i$ to converge to Δ with respect to the discrete Dirichlet problem. The convergence is mandatory when we seek approximate solutions of partial differential equations.

The (CON) property requires a formal definition of the sequence $\{M_i\}$. In addition to [24], we add the following property

Pointwise Convergence (PCON). For a given sequence meshes $\{\Gamma_i\}$, we want the associated Laplacians $\{\mathbf{L}_i\}$ to converge to Δ in a pointwise l_2 or l_∞ sense. This notion of convergence is stronger than (CON) which is implied by (PCON).

For the Laplacian on digital surfaces (*e.g.* $L_h^\star$ we define in Sect. 3), we consider a sequence of combinatorial structures defined as the boundaries of the Gauss digitizations of M (see next section). We thus consider the (DPCON) property as the pointwise convergence of the operator for multigrid digital surfaces.

Table 1 compares classical Laplacian discretizations with respect to these properties. The (DPCON) property has been evaluated experimentally in Sect. 4. $\mathcal{L}_{COMBI}$ being purely combinatorial, the same operator can be used for both meshes and digital surfaces. For $\mathcal{L}_{COT}$ and $\mathcal{L}_{MESH}$, we have considered the marching-cubes representation of the digital surfaces (more precisely, a continuous analog of the dual surface [13]). Properties of $L_h^\star$ are discussed in the next section.

Table 1. Properties of various Laplacians. See text for the description of each property. ● means that the property is valid, whereas ○ means not. N.A. stands for Not Applicable and ? means that we do not know whether it is true or not. Finally exp. holds for experimental convergence with no known theoretical proof.

	Ref	SYM	LOC	LIN	POS	PSD	CON	PCON	DPCON
MEAN VALUE	[9]	○	●	●	●	○	○	?	N.A
INTRINSIC DEL	[3]	●	○	●	●	●	?	?	N.A
$\mathcal{L}_{COMBI}$	[25]	●	●	○	●	●	○	○	○
$\mathcal{L}_{COT}$	[8,10]	●	●	●	○	●	●	○	○
$\mathcal{L}_{MESH}$	[2]	○	○	?	●	●	●	●	○
$L_h^\star$	here	○	○	?	●	exp	?	N.A	exp

3 New Laplace-Beltrami Operator on Digital Surfaces

In the discretization schemes discussed above, the points of the combinatorial structure interpolate the underlying manifold. In order to define a discrete Laplace-Beltrami operator on digital surfaces, a challenge is to work with a combinatorial structure that only approximates the underlying manifold in a Hausdorff sense. First, we extend Eq. (9) to digital surfaces. Proper definitions of a digital surface can be found in [12,14]. Let us recall the Gauss digitization process:

Definition 1. (Gauss digitization). *Let $h > 0$ be the sampling grid step. The Gauss Digitization of an Euclidean shape $X \subset \mathbb{R}^d$ is defined as $\mathsf{D}_h(X) := X \cap (h\mathbb{Z})^d$ where d is the dimension.*

For smooth object X in dimension 3 with boundary $M := \partial X$, the digital surface is defined as the topological boundary of $\mathsf{D}_h(X)$, denoted $\partial_h X$ (see [14] for details). More precisely, the digital surface has a cellular representation in a Cartesian cubical grid and is composed of points of dimension 0 (*pointels*, $\mathbb{E}^0$), straight segments of dimension 1 (*linels*, $\mathbb{E}^1$) and squares of dimension 2 (*surfels*, $\mathbb{E}^2$).

The topological boundary $\partial_h X$ is an $O(h)$-Hausdorff approximation of M [14]. As a consequence, we need to map the smooth function u defined on M to $\partial_h X$:

Definition 2. (Extension of u to $\partial_h X$). *Given a smooth function u on M, we define the extension $\tilde{u}$ of u to $\partial_h X$ as*

$$\tilde{u}(\mathbf{s}) := u(\xi(\dot{\mathbf{s}})),$$

where $\dot{\mathbf{s}}$ is the centroid of the surfel $\mathbf{s} \in \mathbb{E}^2$, and ξ is the map that projects a point of $\partial_h X$ onto the closest point of M.

We show below how to adapt the definition of [2], recalled in Eq. (9), to digital surfaces. We chose this approach because $\mathcal{L}_{MESH}$ has an interesting pointwise convergence for triangular meshes. Our Laplace-Beltrami operator is thus defined as follows

Definition 3. (Digital Laplace-Beltrami operator). *The digital Laplace-Beltrami operator is defined on $\partial_h X$ as:*

$$(L_h^\star \tilde{u})(\mathbf{s}) := \frac{1}{4\pi t_h^2} \sum_{\mathbf{r} \in \mathbb{E}^2} e^{-\frac{||\dot{\mathbf{r}} - \dot{\mathbf{s}}||^2}{4t_h}} \mu(\mathbf{r})(\tilde{u}(\mathbf{r}) - \tilde{u}(\mathbf{s})), \tag{10}$$

where the sum is taken over all surfels of $\partial_h X$, $\dot{\mathbf{r}}$ is the centroid of the surfel $\mathbf{r}$, $\mu(\mathbf{s})$ is equal to the dot product between an estimated normal and the trivial normal orthogonal to the surfel $\mathbf{s}$ and t_h is a function of h tending to zero as h tends to zero.

The quantity $\mu(\mathbf{s})$ is called the measure of the surfel $\mathbf{s}$: it is the area of the projected surfel $\mathbf{s}$ onto the tangent plane induced by the estimated normal. Normal vectors are estimated using the estimator presented in [5,15] which has the multigrid convergence property. Note that summing μ for each surfel of the surface leads to an estimation of the global area of the shape boundary, which itself has a multigrid convergence property [14]. This surfel measure is a key ingredient of the digital formalization of the operator leading to an experimental multigrid convergence and isotropic properties when used for diffusion (see Sect. 4).

If we index surfels in $\mathbb{E}^2$, $L_h^\star$ has a $|\mathbb{E}^2| \times |\mathbb{E}^2|$ matrix representation $\boldsymbol{L}_h^\star$ defined as follows:

$$(\boldsymbol{L}_h^\star)_{ij} = \begin{cases} \dfrac{1}{4\pi t_h^2} e^{-\frac{||\dot{\mathbf{s}}_j - \dot{\mathbf{s}}_i||^2}{4t_h}} \mu(\mathbf{s}_j) & \text{if } i \neq j \\ -\displaystyle\sum_{k \neq i} (\boldsymbol{L}_h^\star)_{ik} & \text{if } i = j \end{cases} \tag{11}$$

In other words, if $\tilde{u}$ is constant and equal to 1, for a surfel $\mathbf{s} \in \mathbb{E}^2$ of index i, we have $(\boldsymbol{L}_h^\star \tilde{u})_i = (L_h^\star \tilde{u})(\dot{\mathbf{s}}_i)$.

For the properties listed in Table 1, we can first observe that Eq. (11) implies that we have property (POS) but not (SYM). As $L_h^\star$ performs a convolution on the complete surface with a Gaussian kernel, we do not have the (LOC) (similarly to $\mathcal{L}_{MESH}$). Although we do not provide a theoretical proof of (PSD) for $L_h^\star$, eigenvalues of $\boldsymbol{L}_h^\star$, have always been positive through all experiments. The (PCON) property is not applicable to our framework. The pointwise convergence (DPCON) is observed experimentally and discussed in Sect. 4.1.

4   Experiments

4.1   Experimental Convergence

We first evaluate the multigrid convergence of our Laplace-Beltrami operator. We consider a unit sphere $\mathbb{S}^2$ and three different smooth functions $u : \mathbb{S}^2 \to \mathbb{R}$, namely z, x^2 and e^x (see Fig. 2). Let θ be the azimutal angle, and ϕ the polar angle. The spherical Laplacian is then:

$$\Delta_{\mathbb{S}^2} u(\theta, \phi) = \frac{1}{\sin^2 \phi} \frac{\partial^2 u}{\partial \theta^2} + \frac{1}{\sin \phi} \frac{\partial}{\partial \phi} \left(\sin \phi \frac{\partial u}{\partial \phi} \right). \tag{12}$$

We compute the Gauss digitization $D_h(\mathbb{S}^2)$ of the sphere for decreasing grid steps h. Since the elements of $\partial_h \mathbb{S}^2$ do not interpolate the sphere, u is extended to $\tilde{u}$ as defined in Sect. 3. We compute $\mathcal{L}_{COMBI}$ and $L_h^\star$ directly on $\partial_h \mathbb{S}^2$, but $\mathcal{L}_{COT}$ and $\mathcal{L}_{MESH}$ on the associated marching-cubes triangulation. Since the vertices of this mesh coincide with the centroids of the surfels of $\partial_h \mathbb{S}^2$, all these operators $\mathcal{L}_{COMBI}$, $\mathcal{L}_{MESH}$ and $L_h^\star$ are evaluated at the same points.

For $L_h^\star$, we use the normal vector estimator described in [5] to estimate the measure of the surfels. In addition, for both $L_h^\star$ and $\mathcal{L}_{MESH}$, the parameter t_h is set to $0.1 \times h^{\frac{1}{3}}$. As the discretization becomes finer, the standard deviation $\sigma := \sqrt{2t_h}$ of the Gaussian function decreases and the number of points within the standard deviation σ increases. The constant factor 0.1 is a scale term derived from the unit sphere that sets the kernel to $1/10$ of the sphere.

For comparison, in order to mimic the setting of [2], we have also considered the Laplacian $\mathcal{L}_{MESH}^P$, which corresponds to $\mathcal{L}_{MESH}$ when the vertices of the marching-cubes are projected onto the sphere. In our framework, this operator is the gold standard, since the position errors are corrected by the projection, which is usually unknown.

For all the above operators, we plot in Fig. 2 the l_2 and l_∞ error between the computed Laplacians and the true spherical one against the grid step h. First, we observe that errors for $\mathcal{L}_{COT}$ (*in blue* ——) and $\mathcal{L}_{COMBI}$ (*in green* —○—) are constant: clearly both operators are non-convergent. Non-convergence is also observed for $\mathcal{L}_{MESH}$ (*in red* ——) but with lower errors. On the opposite, $\mathcal{L}_{MESH}^P$ (*in orange* ——) shows convergence behavior for both l_2 and l_∞ error, as expected in [2]. Concerning $L_h^\star$ (*in purple* ——), experimental convergence holds for the three functions. For the periodic function z, the convergence speed is slower than $\mathcal{L}_{MESH}^P$ whereas for the non-linear functions, we can see that convergence speed is the same. Moreover, the l_2 error for $L_h^\star$ tends toward $\mathcal{L}_{MESH}^P$, and its l_∞ error is close to that of $\mathcal{L}_{MESH}^P$.

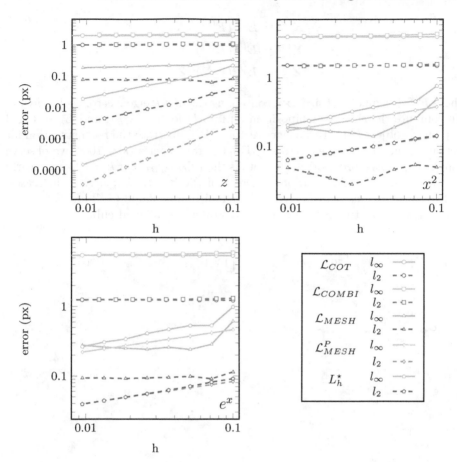

Fig. 2. Multigrid convergence graphs for various functions on $\mathbb{S}^2$, the unit sphere. Both l_2 error in plain line and l_∞ in dashed line are displayed for $\mathcal{L}_{COMBI}$, $\mathcal{L}_{MESH}$, $\mathcal{L}_{MESH}^P$, and $L_h^\star$.

4.2   Shape Approximation Using Eigenvectors Decomposition

In this section, we consider the *spectral analysis* framework to process shapes geometry [16]. Given a shape and its Laplace-Beltrami operator, we compute the eigenvalues and eigenvectors of the operator and project the geometry onto the eigenvector basis of the first k eigenvalues. More formally, given an operator L, we denote by $e_1, e_2, \ldots, e_n$ its normalized eigenvectors and the matrix E whose columns are those eigenvectors. By $\lambda_1, \lambda_2, \ldots, \lambda_n$ we denote the associated increasing eigenvalues where n is the number of rows of L. Given three input vectors (X, Y, Z) encoding the vertex positions in $\mathbb{R}^3$, we can approximate the input shape using a fixed number k of eigenvectors:

$$X^{(k)} = E^{(k)}(E^{(k)})^T X,$$
$$Y^{(k)} = E^{(k)}(E^{(k)})^T Y,$$
$$Z^{(k)} = E^{(k)}(E^{(k)})^T Z,$$

where $E^{(k)}$ is a matrix of size $n \times k$ containing the first k eigenvectors columnwise. We compute the eigen decomposition on $D_h(M)$ for $\mathcal{L}_{COMBI}$, and L_h^* in Fig. 3 on a bunny object (64^3, 13236 eigenvectors). We illustrate the reconstruction for increasing number of eigenvectors k. For low frequencies ($k \leq 100$), we observe that L_h^* captures more geometrical details than $\mathcal{L}_{COMBI}$. For $k = 100$, we clearly have a better approximation of both ears of the bunny shape. As k increases, both reconstructions converge to the original bunny shape ($E^{(n)}(E^{(n)})^T = I$). In Fig. 4, we show the first 20 eigenvectors on a axis aligned cube.

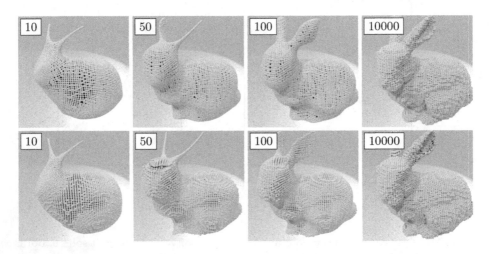

Fig. 3. Images of the reconstruction using an increasing number k of eigenvectors. (*First row*) using $\mathcal{L}_{COMBI}$, (*second row*) with L_h^* ($r = 6$ for [5] and $t_h = 3$).

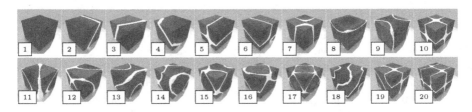

Fig. 4. Eigenfunctions are displayed on a simple cube with faces aligned with the grid axes (with a red to blue colormap and zero-crossing in white). (Color figure on line)

4.3   Heat Diffusion

In this section, we highlight an interesting isotropic property of $L_h^\star$ compared to the combinatorial one. We compute a heat diffusion when the source is a Dirac in the center of a rotated cube face (heat diffusion is a preliminary step of [7] to estimate geodesics on a manifold). We can derive an expression of $g(x,t)$ from Eq. (5):

$$(\boldsymbol{I} - t\boldsymbol{L})g(x,t) = g_0(x), \tag{13}$$

where $\boldsymbol{I}$ is the identity matrix, and $\boldsymbol{L}$ is the Laplacian matrix associated to any discrete laplace operator. This diffusion only makes sense for small t and around the Dirac $g_0(x)$. As the computed heat diffusion decreases exponentially, we display the absolute value of its log. For $\mathcal{L}_{COMBI}$ (first column in Fig. 5), both small and large values of t lead anisotropic estimations of the intrinsic metric due to the staircase effect of the face (concentric rhombi or concentric ellipses depending on t, see similar discussion in [7]). When using Eq. (13) with our matrix based operator $L_h^\star$ (second column), even if numerical instabilities occur far from the Dirac depending on t, the intrinsic metric is perfectly estimated (concentric circles). Note that both $\mathcal{L}_{MESH}$ and $L_h^\star$ already compute a diffusion without the approximation of Eq. (13). The third column shows the associated diffusion. Even if $\mathcal{L}_{MESH}$ does not have multigrid convergence properties, on this specific object, both methods provide good isotropic behaviors.

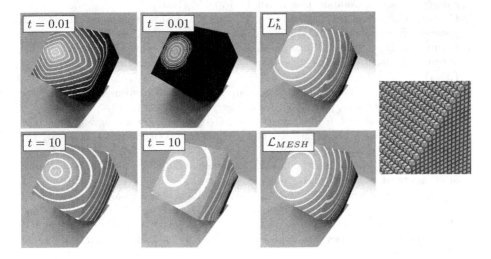

Fig. 5. Heat diffusion on a cube aligned with $\mathbb{R}^3$ axis. (*First column*) using $\mathcal{L}_{COMBI}$, (*second column*) using $L_h^\star$ and (*third column*, with $t_h = 4$) using the diffusion computed through the ambient heat kernel. The rightmost picture shows staircases on the rotated cube.

5   Conclusion and Future-Works

In this paper, we have investigated different discretization schemes of the Laplace-Beltrami operator on digital surfaces. The contribution is twofold: first, we have shown that classical schemes either do not asymptotically converge to the expected operator, or contain anisotropic artifacts when used for geometry processing tasks. Second, we have proposed a new Laplace-Beltrami operator that incorporates multigrid convergent surfel measures allowing us to have both an experimental multigrid convergence and isotropic properties on digital surfaces.

A natural future-work consists of focusing on the multigrid convergence proof of the operator. In dimension 2, a preliminary proof has been derived using digital integration results from [14] but it is still an open problem in dimension 3.

References

1. Belkin, M., Niyogi, P.: Towards a theoretical foundation for laplacian-based manifold methods. J. Comput. Syst. Sci. **74**(8), 1289–1308 (2008)
2. Belkin, M., Sun, J., Wang, Y.: Discrete laplace operator on meshed surfaces. In: Teillaud, M. (ed.) Proceedings of the 24th ACM Symposium on Computational Geometry, College Park, MD, USA, pp. 278–287. ACM, 9–11 June (2008)
3. Bobenko, A.I., Springborn, B.: A discrete laplace-beltrami operator for simplicial surfaces. Discrete Comput. Geom. **38**(4), 740–756 (2007)
4. Cartade, C., Mercat, C., Malgouyres, R., Samir, C.: Mesh parameterization with generalized discrete conformal maps. J. Math. Imaging. Vis. **46**(1), 1–11 (2013)
5. Coeurjolly, D., Lachaud, J.O., Levallois, J.: Multigrid convergent principal curvature estimators in digital geometry. Comput. Vis. Image Underst. **129**, 27–41 (2014)
6. Coeurjolly, D., Lachaud, J.O., Roussillon, T.: Multigrid Convergence of Discrete Geometric Estimators, pp. 395–424. Springer, Netherlands (2012)
7. Crane, K., Weischedel, C., Wardetzky, M.: Geodesics in heat: a new approach to computing distance based on heat flow. ACM Trans. Graph. (TOG) **32**(5), 152 (2013)
8. Desbrun, M., Hirani, A.N., Leok, M., Marsden, J.E.: Discrete exterior calculus. arXiv preprint math/0508341 (2005)
9. Dodgson, N.A., Floater, M.S., Sabin, M.A. (eds.): Advances in Multiresolution for Geometric Modelling. Springer, Heidelberg (2005)
10. Hildebrandt, K., Polthier, K., Wardetzky, M.: On the convergence of metric and geometric properties of polyhedral surfaces. Geom. Dedicata. **123**(1), 89–112 (2006)
11. Hirani, A.N.: Discrete exterior calculus. Ph.D. thesis, California Institute of Technology (2003)
12. Klette, R., Rosenfeld, A.: Digital Geometry: Geometric Methods for Digital Picture Analysis. The Morgan Kaufmann Series in Computer Graphics and Geometric Modeling. Elsevier, Amsterdam (2004)
13. Lachaud, J.O., Montanvert, A.: Continuous analogs of digital boundaries: a topological approach to iso-surfaces. Graph. Models Image Process. **62**, 129–164 (2000)

14. Lachaud, J.O., Thibert, B.: Properties of gauss digitized shapes and digital surface integration. J. Math. Imaging Vis. **54**(2), 162–180 (2016)
15. Levallois, J., Coeurjolly, D., Lachaud, J.-O.: Parameter-Free and Multigrid Convergent Digital Curvature Estimators. In: Barcucci, E., Frosini, A., Rinaldi, S. (eds.) DGCI 2014. LNCS, vol. 8668, pp. 162–175. Springer, Cham (2014). doi:10.1007/978-3-319-09955-2_14
16. Lévy, B., Zhang, H.: Spectral Mesh Processing. Technical. report, SIGGRAPH Asia 2009 Courses (2008)
17. Mercat, C.: Discrete Complex Structure on Surfel Surfaces. In: Coeurjolly, D., Sivignon, I., Tougne, L., Dupont, F. (eds.) DGCI 2008. LNCS, vol. 4992, pp. 153–164. Springer, Heidelberg (2008). doi:10.1007/978-3-540-79126-3_15
18. Molchanov, S.A.: Diffusion processes and riemannian geometry. Russ. Math. Surv. **30**(1), 1 (1975)
19. Pinkall, U., Polthier, K.: Computing discrete minimal surfaces and their conjugates. Exp. Math. **2**(1), 15–36 (1993)
20. Polthier, K., Preuss, E.: Identifying vector field singularities using a discrete Hodge decomposition. Vis. Math. **3**, 113–134 (2003)
21. Regge, T.: General relativity without coordinates. Il Nuovo Cimento Series 10 **19**(3), 558–571 (1961)
22. Rosenberg, S.: The Laplacian on a Riemannian Manifold. Cambridge University Press, Cambridge Books Online (1997)
23. Varadhan, S.: On the behavior of the fundamental solution of the heat equation with variable coefficients. Commun. Pure Appl. Math. **20**(2), 431–455 (1967)
24. Wardetzky, M., Mathur, S., Kaelberer, F., Grinspun, E.: Discrete Laplace operators: No free lunch. Eurographics Symposium on Geometry Processing, pp. 33–37 (2007)
25. Zhang, H.: Discrete combinatorial Laplacian operators for digital geometry processing. In: SIAM Conference on Geometric Design, pp. 575–592. (2004, press)

Efficiently Updating Feasible Regions for Fitting Discrete Polynomial Curve

Fumiki Sekiya[1(✉)] and Akihiro Sugimoto[2]

[1] Department of Informatics, SOKENDAI (The Graduate University
for Advanced Studies), Tokyo, Japan
[2] National Institute of Informatics, Tokyo, Japan
{sekiya,sugimoto}@nii.ac.jp

Abstract. We deal with the problem of fitting a discrete polynomial curve to 2D data in the presence of outliers. Finding a maximal inlier set from given data that describes a discrete polynomial curve is equivalent with finding the feasible region corresponding to the set in the parameter space. When iteratively adding a data point to the current inlier set, how to update its feasible region is a crucial issue. This work focuses on how to track vertices of feasible regions in accordance with newly coming inliers. When a new data point is added to the current inlier set, a new vertex is obtained as the intersection point of an edge (or a face) of the feasible region for the current inlier set and a facet (or two facets) of the feasible region for the data point being added. Evaluating all possible combinations of an edge (or a face) and a facet (or two facets) is, however, computationally expensive. We propose an efficient computation in this incremental evaluation that eliminates combinations producing no vertices of the updated feasible region. This computation facilitates collecting the vertices of the updated feasible region. Experimental results demonstrate our proposed computation efficiently reduces practical running time.

1 Introduction

Contour detection is unavoidable for many image processing and/or computer vision tasks such as object recognition, image segmentation and shape approximation. A contour is usually represented as a curve and, thus, contour detection is reduced to fitting a curve to noisy data. Since curves are discretized in the digital image, discrete curve fitting has been studied for decades for different classes of curves and different discretization models [1–6,9,11–13]. An important advantage of using a discrete curve over a continuous one, when used for fitting, is that it requires no empirical threshold in error to define an inlier that affects the output. We note that an underlying threshold that a discrete model uses to collect its points is usually designed only to achieve some properties such as connectivity (see [7,10] for example), and thus such a threshold is clearly justified.

This paper deals with the problem of fitting a discrete polynomial curve to 2D data in the presence of outliers, which is formulated as follows [8]: For a given

© Springer International Publishing AG 2017
W.G. Kropatsch et al. (Eds.): DGCI 2017, LNCS 10502, pp. 254–266, 2017.
DOI: 10.1007/978-3-319-66272-5_21

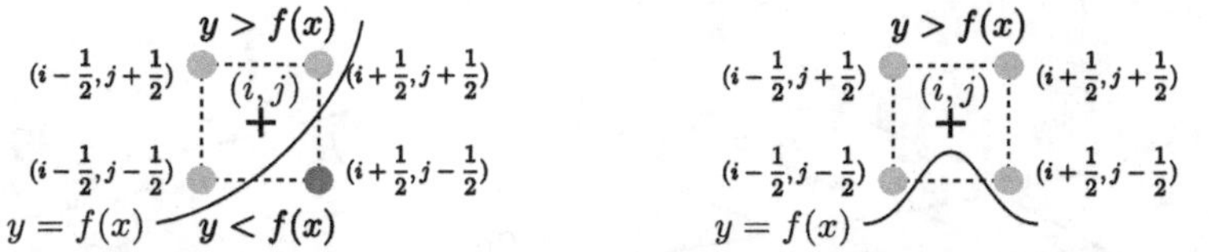

$$y > f(x)$$
$(i-\frac{1}{2}, j+\frac{1}{2})$ · · · · (i,j) · · · $(i+\frac{1}{2}, j+\frac{1}{2})$
$+$
$(i-\frac{1}{2}, j-\frac{1}{2})$ · · · · $(i+\frac{1}{2}, j-\frac{1}{2})$
$$y = f(x)\quad y < f(x)$$
(a) $(i,j) \in D(a)$.

$$y > f(x)$$
$(i-\frac{1}{2}, j+\frac{1}{2})$ · · · · (i,j) · · · $(i+\frac{1}{2}, j+\frac{1}{2})$
$+$
$(i-\frac{1}{2}, j-\frac{1}{2})$ · · · · $(i+\frac{1}{2}, j-\frac{1}{2})$
$$y = f(x)$$
(b) $(i,j) \notin D(a)$.

Fig. 1. Integer point in (not in) $D(a)$. The black curves depict the underlying continuous polynomial curve $y = f(x) = \sum_{l=0}^{d} a_l x^l$.

data set $P = \{(i_p, j_p) \in \mathbb{Z}^2 \mid p = 1, \ldots, n\}$ ($n < \infty$) and a degree d, the discrete polynomial curve fitting is to find the discrete polynomial curve $D(a)$ that has the maximum number of *inliers*, i.e., data points in $D(a)$. Here $D(a)$ is defined [10] by, with coefficients $a = (a_0, \ldots, a_d) \in \mathbb{R}^{d+1}$,

$$D(a) = \left\{ (i,j) \in \mathbb{Z}^2 \;\middle|\; \begin{array}{c} \min\limits_{s \in \{1,\ldots,4\}} \left[(j + y_s) - \sum\limits_{l=0}^{d} a_l (i + x_s)^l\right] \\ \leq 0 \leq \\ \max\limits_{s \in \{1,\ldots,4\}} \left[(j + y_s) - \sum\limits_{l=0}^{d} a_l (i + x_s)^l\right] \end{array} \right\}, \quad (1)$$

where $(x_1, y_1) = (-\frac{1}{2}, -\frac{1}{2})$, $(x_2, y_2) = (\frac{1}{2}, -\frac{1}{2})$, $(x_3, y_3) = (\frac{1}{2}, \frac{1}{2})$, $(x_4, y_4) = (-\frac{1}{2}, \frac{1}{2})$. Equation (1) collects $(i,j) \in \mathbb{Z}^2$ iff the four points $(i + x_s, j + y_s)$ for $s \in \{1, \ldots, 4\}$ lie on the both sides ($y > \sum_{l=0}^{d} a_l x^l$ and $y < \sum_{l=0}^{d} a_l x^l$) of the underlying continuous polynomial curve $y = \sum_{l=0}^{d} a_l x^l$, or at least one of them is on $y = \sum_{l=0}^{d} a_l x^l$ (Fig. 1). We employ this discretization model because the connectivity is guaranteed [7].

This problem can be discussed in the parameter (coefficient) space. For the data point (i_p, j_p) ($p \in \{1, \ldots, n\}$), the *feasible region* $R_p \subseteq \mathbb{R}^{d+1}$ is defined by

$$R_p = \left\{ a \in \mathbb{R}^{d+1} \;\middle|\; \min\limits_{s \in \{1,\ldots,4\}} h_{(p,s)}(a) \leq 0 \leq \max\limits_{s \in \{1,\ldots,4\}} h_{(p,s)}(a) \right\},$$

where $h_{(p,s)}(a) = (j_p + y_s) - \sum_{l=0}^{d} (i_p + x_s)^l a_l$ (Fig. 2(a)). We remark that $(i_p, j_p) \in D(a)$ iff $a \in R_p$. R_p is an unbounded concave polytope, which is the union of two unbounded convex polytopes defined by $h_{(p,1)}(a) \leq 0 \leq h_{(p,3)}(a)$ and $h_{(p,2)}(a) \leq 0 \leq h_{(p,4)}(a)$ (cf. Fig. 3(a)). For $\Pi \subseteq \{1, \ldots, n\}$, the *feasible region* R_Π is defined as the intersection of the feasible regions each of which is for a data point (i_p, j_p) where $p \in \Pi$: $R_\Pi = \bigcap_{p \in \Pi} R_p$ (Fig. 2(b)). Since $R_\Pi = \emptyset$ if there exists no $a \in \mathbb{R}^{d+1}$ that satisfies $\{(i_p, j_p) \mid p \in \Pi\} \subseteq D(a)$, we may assume $R_\Pi \neq \emptyset$ below. Accordingly, our fitting problem is formulated in the parameter

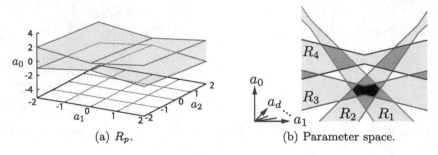

(a) R_p.   (b) Parameter space.

Fig. 2. Feasible region. (a) shows R_p for $d = 2$ and $(i_p, j_p) = (0,0)$. (b) shows inter-sections among the feasible regions for four data points indexed from 1 to 4. A darker region has a larger number of inliers.

space as follows: Given P and d, find $\Pi \subseteq \{1, \ldots, n\}$ with the maximum $|\Pi|$ and $a \in R_\Pi$ for that Π.

This problem requires evaluating R_Π for all $\Pi \subset \{1, \ldots, n\}$, which is reduced to classify each data point into an inlier or an outlier (we have 2^n instances). To this end, a heuristic based incremental approach [8] was proposed where it iteratively evaluates whether a data point can be added to the current inlier set until the inlier set does not have its superset. In this approach, the feasible region for the current inlier set is tracked by its vertices: when a new data point is added to the current inlier set, a vertex of the new feasible region can be obtained from the intersection points of an edge (or a face) of the current feasible region and a facet (or two facets) of the feasible region for the data point being added. Evaluating such possible combinations all is, however, computationally expensive. The contribution of this paper is to facilitate this incremental evaluation by introducing an efficient computation of the vertices of the new feasible region. Our introduced computation eliminates combinations producing no vertex of the new feasible region, based on the property that an edge or face of a bounded feasible region is inside the convex hull of its vertices. Though the computational complexity is not reduced, our introduced computation efficiently reduces running time in practice, as shown in experiments.

2   Brief Review of Incremental Approach

The incremental approach [8] starts with computing the feasible region for an initialized inlier set. It then evaluates each data point one by one to update the feasible region. If the updated feasible region is not empty, the data point is added to the inlier set; it is regarded as an outlier otherwise. How to represent and update the feasible region is a key issue there, which is briefly explained below.

2.1   Representing a Feasible Region Using Its Vertices

A vertex of a feasible region is defined as an intersection point of its facets. For $p = 1, \ldots, n$, and $s = 1, \ldots, 4$, a *facet* $F(p, s)$ *of* R_p is defined by

$$F(p, s) = \left\{ \boldsymbol{a} \in \mathbb{R}^{d+1} \;\middle|\; \begin{array}{l} h_{(p,s)}(\boldsymbol{a}) = 0 \text{ and} \\ s \in \underset{s' \in \{1,\ldots,4\}}{\arg \min}\, h_{(p,s')}(\boldsymbol{a}) \cup \underset{s' \in \{1,\ldots,4\}}{\arg \max}\, h_{(p,s')}(\boldsymbol{a}) \end{array} \right\}.$$

$F(p, s)$ is a part of the hyperplane $h_{(p,s)}(\boldsymbol{a}) = 0$ supporting R_p (Fig. 3(a)). Similarly, a *facet* $F_\Pi(p, s)$ *of* R_Π is defined by $F_\Pi(p, s) = F(p, s) \cap R_\Pi$ for $\Pi \subseteq \{1, \ldots, n\}$ and $(p, s) \in \Pi \times \{1, \ldots, 4\}$ (Fig. 3(b)). Note that $F_\Pi(p, s)$ may be empty for some (p, s).

A vertex of R_Π is given as the intersection of $d+1$ facets. The set V_Π of the vertices of R_Π is defined by

$$V_\Pi = \left\{ \boldsymbol{a} \in \mathbb{R}^{d+1} \;\middle|\; \begin{array}{l} \boldsymbol{a} \in \bigcap_{\lambda=1}^{d+1} F_\Pi(p_\lambda, s_\lambda) \text{ for some} \\ (p_1, s_1), \ldots, (p_{d+1}, s_{d+1}) \subseteq \Pi \times \{1, \ldots, 4\} \\ \text{such that } h_{(p_\lambda, s_\lambda)}(\boldsymbol{a}) = 0 \text{ for } \lambda = 1, \ldots, d+1 \\ \text{are linearly independent} \end{array} \right\}. \tag{2}$$

See Fig. 4 for an illustration of V_Π. Each combination of $d+1$ facets determining an element in V_Π is indicated by an element in Ψ_Π, which is defined by

$$\Psi_\Pi = \left\{ \begin{array}{l} \{(p_1, s_1), \ldots, (p_{d+1}, s_{d+1})\} \\ \subseteq \Pi \times \{1, \ldots, 4\} \end{array} \;\middle|\; \begin{array}{l} h_{(p_\lambda, s_\lambda)}(\boldsymbol{a}) = 0 \text{ for } \lambda = 1, \ldots, d+1 \\ \text{are linearly independent and} \\ \text{their solution is in } \bigcap_{\lambda=1}^{d+1} F_\Pi(p_\lambda, s_\lambda) \end{array} \right\}. \tag{3}$$

In this way, the feasible region of the inlier set Π can be represented by its vertices V_Π with the help of Ψ_Π. Note that different elements in Ψ_Π may determine the same element of V_Π.

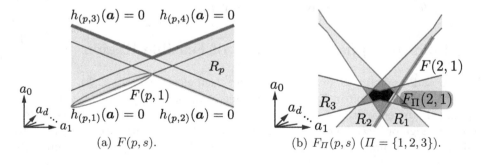

(a) $F(p, s)$.   (b) $F_\Pi(p, s)$ ($\Pi = \{1, 2, 3\}$).

Fig. 3. Facets of a feasible region. (a): $h_{(p,s)}(\boldsymbol{a}) = 0$, $s = 1, \ldots 4$ are depicted in blue lines. $\min_{s \in \{1,\ldots,4\}} h_{(p,s)}(\boldsymbol{a}) = 0$ is depicted in yellow, while $\max_{s \in \{1,\ldots,4\}} h_{(p,s)}(\boldsymbol{a}) = 0$ is depicted in pink. (Color figure online)

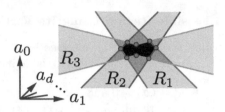

Fig. 4. Vertices V_Π (yellow) of R_Π ($\Pi = \{1, 2, 3\}$). (Color figure online)

2.2   Tracking the Vertices of the Feasible Region

Theorem 1 indicates that Ψ_Π plays an important role in tracking the vertices of the updated feasible region when a new inlier (represented by p^*) comes in. Note that R_Π is almost always bounded (see [8]).

Theorem 1 (Sekiya and Sugimoto[8]). *For $\Pi \subsetneq \{1, \ldots, n\}$ such that R_Π is bounded and $p^* \in \{1, \ldots, n\} \backslash \Pi$, $\Psi_{\Pi \cup \{p^*\}} \subseteq \Psi_\Pi \cup \Phi^1_{\Pi, p^*} \cup \Phi^2_{\Pi, p^*}$, where*

$$\Phi^1_{\Pi, p^*} = \left\{ \begin{array}{c} \{(p_1, s_1), \ldots, (p_d, s_d)\} \\ \cup \{(p^*, s^*)\} \end{array} \;\middle|\; \begin{array}{c} \{(p_1, s_1), \ldots, (p_d, s_d)\} \text{ is a subset of} \\ \text{an element in } \Psi_\Pi \text{ and } s^* = 1, \ldots, 4 \end{array} \right\},$$

$$\Phi^2_{\Pi, p^*} = \left\{ \begin{array}{c} \{(p_1, s_1), \ldots, (p_{d-1}, s_{d-1})\} \\ \cup \{(p^*, s_1^*), (p^*, s_2^*)\} \end{array} \;\middle|\; \begin{array}{c} \{(p_1, s_1), \ldots, (p_{d-1}, s_{d-1})\} \text{ is} \\ \text{a subset of an element in } \Psi_\Pi \\ \text{and } (s_1^*, s_2^*) \in \{(1, 2), (3, 4)\} \end{array} \right\}.$$

$\{(p_1, s_1), \ldots, (p_\omega, s_\omega)\}$ where $\omega = d$ and $d - 1$ respectively corresponds to an *edge* and a (2-dimensional) *face* of R_Π: $\bigcap_{\lambda=1}^\omega F_\Pi(p_\lambda, s_\lambda)$ (the intersection of ω facets). An element in Φ^1_{Π, p^*} (resp. Φ^2_{Π, p^*}) is thus considered to be the combination of an edge (resp. a face) of R_Π and a facet (resp. two facets) of R_{p^*}. Theorem 1 therefore indicates that a vertex of $R_{\Pi \cup \{p^*\}}$ is a vertex of R_Π or otherwise obtained as the intersection point of an edge (resp. a 2-dimensional face) of R_Π and a facet (resp. two facets) of R_{p^*}.

3   Efficient Update of Vertices of Feasible Region

Sekiya and Sugimoto [8] evaluates whether each element in $\Psi_\Pi \cup \Phi^1_{\Pi, p^*} \cup \Phi^2_{\Pi, p^*}$ satisfies the condition in Eq. (3) to extract elements in $\Psi_{\Pi \cup \{p^*\}}$. When R_Π is bounded, any edge or face of R_Π is inside the convex hull of the vertices of R_Π on that edge or face (Lemma 1). Based on this, we introduce a computation to eliminate elements in $\Phi^1_{\Pi, p^*} \cup \Phi^2_{\Pi, p^*}$ that cannot be in $\Psi_{\Pi \cup \{p^*\}}$. This enables us to compute $\Psi_{\Pi \cup \{p^*\}}$ efficiently.

We define a set of edges (or faces) of R_Π. Namely, for $\omega = d$ (edge), $d - 1$ (face), we define

$$\Psi_\Pi^{(\omega)} = \left\{ \begin{array}{c} \{(p_1, s_1), \ldots, (p_\omega, s_\omega)\} \\ \subseteq \Pi \times \{1, \ldots, 4\} \end{array} \;\middle|\; \{(p_1, s_1), \ldots, (p_\omega, s_\omega)\} \subseteq \xi \text{ such that } \xi \in \Psi_\Pi \right\}.$$

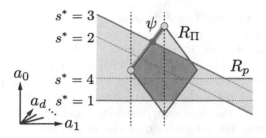

Fig. 5. Illustration of $A_{\Pi,p^*}(\psi)$. For $\psi \in \Psi_\Pi^{(\omega)}$, $\bigcap_{(p,s)\in\psi} F_\Pi(p,s)$ is depicted in green and $V_\Pi(\psi)$ in yellow. For $s^* = 1,\ldots,4$, $h_{(p^*,s^*)}(\boldsymbol{a}) = 0$ is depicted in a solid and dotted red line where the solid part depicts $F(p^*, s^*)$. In this example, s^* satisfies (i) in Eq. (4) if $h_{(p^*,s^*)}(\boldsymbol{a}) = 0$ runs between the two yellow vertices, and (ii) if $h_{(p^*,s^*)}(\boldsymbol{a}) = 0$ is depicted in a solid line on either of the dotted black lines which are parallel to the a_0 axis and passes through yellow vertices. $A_{\Pi,p^*}(\psi) = \{3\}$, accordingly. (Color figure online)

We note that an edge (or face) is determined as the intersection of d (or $d-1$) facets of R_Π. For an edge (or face) $\psi \in \Psi_\Pi^{(\omega)}$ ($\omega = d, d-1$), we denote by $V_\Pi(\psi)$ the vertices on ψ:

$$V_\Pi(\psi) = \{\boldsymbol{a} \in V_\Pi \mid \boldsymbol{a} \text{ is determined by } \xi \in \Psi_\Pi \text{ such that } \psi \subseteq \xi\}.$$

Let $\mathrm{Conv}(V_\Pi(\psi))$ denote the convex hull of $V_\Pi(\psi)$. Then we have Lemma 1 whose proof is provided in Appendix A.

Lemma 1. *For $\Pi \subseteq \{1,\ldots,n\}$ for which R_Π is bounded and $\psi \in \Psi_\Pi^{(\omega)}$ ($\omega = d, d-1$), $\bigcap_{(p,s)\in\psi} F_\Pi(p,s) \subseteq \mathrm{Conv}(V_\Pi(\psi))$.*

Suppose we are adding an new inlier p^* to the current inlier set Π. For each $\psi \in \Psi_\Pi^{(d)}$, Φ_{Π,p^*}^1 has four elements $\psi \cup \{(p^*, s^*)\}$, $s^* = 1,\ldots,4$. Lemma 1 allows to identify $\tilde{s}^* \in \{1,\ldots,4\}$ such that $\bigcap_{(p,s)\in\psi} F_\Pi(p,s) \cap F(p^*, \tilde{s}^*) = \emptyset$ (the facet of R_{p^*} corresponding to $\tilde{s}^*$ does not intersect with the edge ψ). We can thus define the subset of Φ_{Π,p^*}^1 by eliminating $\psi \cup \{(p^*, \tilde{s}^*)\}$ and use the subset instead of Φ_{Π,p^*}^1 itself in computing $\Psi_{\Pi \cup \{p^*\}}$. The same argument can be applied to Φ_{Π,p^*}^2.

To exclude $\tilde{s}^*$ above, we define $A_{\Pi,p^*}(\psi)$ for $\psi \in \Psi_\Pi^{(\omega)}$ ($\omega = d, d-1$), by

$$A_{\Pi,p^*}(\psi) =$$
$$\left\{ s^* \in \{1,\ldots,4\} \;\middle|\; \begin{array}{l} \text{(i)} \quad \min_{\boldsymbol{a}\in V_\Pi(\psi)} h_{(p^*,s^*)}(\boldsymbol{a}) \leq 0 \leq \max_{\boldsymbol{a}\in V_\Pi(\psi)} h_{(p^*,s^*)}(\boldsymbol{a}) \text{ and} \\ \text{(ii)} \; s^* \in \arg\min_{s'\in\{1,\ldots,4\}} h_{(p^*,s')}(\boldsymbol{a}) \cup \arg\max_{s'\in\{1,\ldots,4\}} h_{(p^*,s')}(\boldsymbol{a}) \\ \qquad \text{for some } \boldsymbol{a} \in V_\Pi(\psi) \end{array} \right\} \tag{4}$$

With $A_{\Pi,p^*}(\psi)$, we can identify the facets of R_{p^*} that potentially intersect with the edge/face ψ (see Fig. 5). (i) in Eq. (4) means that the hyperplane

$h_{(p^*,s^*)}(\boldsymbol{a}) = 0$ runs between two vertices of ψ, or passes through one of its vertices. Since the edge or face is inside the convex hull of its vertices (Lemma 1), it follows that the hyperplane intersects with ψ. This however does not indicate that the facet determined by s^* intersects with ψ (the facet is only a part of the hyperplane). We therefore have to evaluate whether the facet indeed intersects with ψ, for which (ii) plays the role. (ii) means that, for at least one vertex $\boldsymbol{a} \in V_\Pi(\psi)$, s^* achieves the maximum or minimum of $h_{(p^*,s')}(\boldsymbol{a}), s' \in \{1, \dots, 4\}$; unless (ii) is satisfied, the intersection of the hyperplane with ψ is out of the facet.

Using only $s^* \in \{1, \dots, 4\}$ in $A_{\Pi,p^*}(\psi)$ for each ψ, we can define the subsets of Φ^1_{Π,p^*} and Φ^2_{Π,p^*} that can be used in computing $\Psi_{\Pi \cup \{p^*\}}$.

$$X^1_{\Pi,p^*} = \left\{ \psi \cup \{(p^*, s^*)\} \;\middle|\; \psi \in \Psi^{(d)}_\Pi \text{ and } s^* \in A_{\Pi,p^*}(\psi) \right\},$$

$$X^2_{\Pi,p^*} = \left\{ \psi \cup \{(p^*, s^*_1), (p^*, s^*_2)\} \;\middle|\; \begin{matrix} \psi \in \Psi^{(d-1)}_\Pi \text{ and } s^*_1, s^*_2 \in A_{\Pi,p^*}(\psi) \\ \text{where } (s^*_1, s^*_2) = (1,2) \text{ or } (3,4) \end{matrix} \right\}.$$

Now we formally prove that no element in $\Phi^1_{\Pi,p^*} \backslash X^1_{\Pi,p^*}$ or $\Phi^2_{\Pi,p^*} \backslash X^2_{\Pi,p^*}$ is in $\Psi_{\Pi \cup \{p^*\}}$.

Theorem 2. *For* $\Pi \subsetneq \{1, \dots, n\}$ *such that* R_Π *is bounded and* $p^* \in \{1, \dots, n\} \backslash \Pi$, $\Psi_{\Pi \cup \{p^*\}} \subseteq \Psi_\Pi \cup X^1_{\Pi,p^*} \cup X^2_{\Pi,p^*}$.

Proof. Consider $\psi \in \Psi^{(\omega)}_\Pi$ ($\omega = d, d-1$). We show $\bigcap_{(p,s) \in \psi} F_\Pi(p,s) \cap F(p^*, s^*) = \emptyset$ for any $s^* \in \{1, \dots, 4\} \backslash A_{\Pi,p^*}(\psi)$. Note that this means $\psi' \notin \Psi_{\Pi \cup \{p^*\}}$ for any $\psi' \in \Phi^{d+1-\omega}_{\Pi,p^*} \backslash X^{d+1-\omega}_{\Pi,p^*}$.

We first assume that s^* does not satisfy (i) in Eq. (4): $h_{(p^*,s^*)}(\boldsymbol{a}) < 0$ for $\forall \boldsymbol{a} \in V_\Pi(\psi)$ or $h_{(p^*,s^*)}(\boldsymbol{a}) > 0$ for $\forall \boldsymbol{a} \in V_\Pi(\psi)$. From Lemma 1 and the linearity of $h_{(p^*,s^*)}$, we have $h_{(p^*,s^*)}(\boldsymbol{a}) < 0$ for $\forall \boldsymbol{a} \in \bigcap_{(p,s) \in \psi} F_\Pi(p,s)$ or $h_{(p^*,s^*)}(\boldsymbol{a}) > 0$ for $\forall \boldsymbol{a} \in \bigcap_{(p,s) \in \psi} F_\Pi(p,s)$. It follows that $\bigcap_{(p,s) \in \psi} F_\Pi(p,s) \cap F(p^*, s^*) = \emptyset$.

We next assume that s^* does not satisfy (ii): $s^* \notin \arg\min_{s' \in \{1,\dots,4\}} h_{(p^*,s')}(\boldsymbol{a}) \cup \arg\max_{s' \in \{1,\dots,4\}} h_{(p^*,s')}(\boldsymbol{a})$ for $\forall \boldsymbol{a} \in V_\Pi(\psi)$. Lemma 1 and the linearity of $h_{(p^*,s^*)}$ imply $s^* \notin \arg\min_{s' \in \{1,\dots,4\}} h_{(p^*,s')}(\boldsymbol{a}) \cup \arg\max_{s' \in \{1,\dots,4\}} h_{(p^*,s')}(\boldsymbol{a})$ for $\forall \boldsymbol{a} \in \bigcap_{(p,s) \in \psi} F_\Pi(p,s)$. It follows that $\bigcap_{(p,s) \in \psi} F_\Pi(p,s) \cap F(p^*, s^*) = \emptyset$. $\square$

4   Algorithm

Algorithm 1 [8][1] is the incremental approach to solve the discrete polynomial curve fitting for a given data point set P and a given degree d. It classifies each data index into two classes: Π (inlier) and Π^C (outlier). Π is first initialized to be a set of $d + 1$ data indices for which V_Π and Ψ_Π are computed at low cost

---

[1] Because the initial inlier selection is not the scope of this paper, Algorithm 1 is presented without any initial inlier set.

Algorithm 1. Incremental algorithm (Sekiya+ [8]).

Require: P, d.
Ensure: $\Pi \subseteq \{1, \ldots, n\}$ and V_Π.
1: Initialize $\Pi :=$ any $d + 1$ data indices in $\{1, \ldots, n\}$ for which R_Π is bounded.
2: Initialize $\Pi^{\complement} := \emptyset$.
3: Compute V_Π and Ψ_Π using Eqs. (2) and (3).
4: **while** $\{1, \ldots, n\} \setminus (\Pi \cup \Pi^{\complement}) \neq \emptyset$ **do**
5:     $p^* :=$ any data index in $\{1, \ldots, n\} \setminus (\Pi \cup \Pi^{\complement})$.
6:     Compute $V_{\Pi \cup \{p^*\}}$ and $\Psi_{\Pi \cup \{p^*\}}$
7:     **if** $V_{\Pi \cup \{p^*\}} \neq \emptyset$ **then**
8:         $\Pi := \Pi \cup \{p^*\}$ and update V_Π and Ψ_Π.
9:     **else**
10:         $\Pi^{\complement} := \Pi^{\complement} \cup \{p^*\}$.
11:     **end if**
12: **end while**
13: **return** Π and V_Π.

(see [8] for the sufficient condition that R_Π is bounded). In the following loop (Steps 4–12), we add new data indices to either Π or $\Pi^{\complement}$ one by one. When Π is updated, V_Π and Ψ_Π are also updated. Since $\Phi_{\Pi \cup \{p^*\}} \neq \emptyset$ if $R_{\Pi \cup \{p^*\}} \neq \emptyset$ (see [8]), an inlier set obtained by Algorithm 1 is guaranteed to have no superset.

The purpose of Algorithm 2 is to efficiently compute $V_{\Pi \cup \{p^*\}}$ and $\Psi_{\Pi \cup \{p^*\}}$ in Step 6 of Algorithm 1: efficient computation of the vertices of the feasible region updated by an additional data point. Some of the vertices are inherited from the current feasible region. So, the first loop (Steps 2–7) evaluates each vertex of the current feasible region to check if it serves as a vertex of the updated feasible region, where it suffices only to verify that the vertex is inside the feasible region for the additional data point (Step 5), since $F_\Pi(p, s) \cap R_{p^*} = F_{\Pi \cup \{p^*\}}(p, s)$ for any $(p, s) \in \Pi \times \{1, \ldots, 4\}$. The vertices appearing only in the updated feasible region are obtained from $X^1_{\Pi, p^*} \cup X^2_{\Pi, p^*}$ in the second loop (Steps 9–16). For each element in $X^1_{\Pi, p^*} \cup X^2_{\Pi, p^*}$, we first check if the hyperplanes corresponding to the element intersect at a unique point (Step 10), and if so, we then check if that intersection point serves as a vertex of the updated feasible region (Step 12). Note that the condition in Step 12 is equivalent with $\boldsymbol{a} \in \bigcap_{(p,s) \in \psi} F_{\Pi \cup \{p^*\}}(p, s)$; $\boldsymbol{a} \in \bigcap_{(p,s) \in \psi} F(p, s)$ implies $\boldsymbol{a} \in R_{p^*}$ since any $\psi \in X^1_{\Pi, p^*} \cup X^2_{\Pi, p^*}$ contains (p, s) such that $p = p^*$. The most efficiently working part is Step 8, which reduces the number of iterations in the second loop.

The computational complexity for this method is the same with [8]: $\mathcal{O}(n^{d+2})$ for a variable number n of data and a fixed degree d, because $\mathcal{O}(|X^\alpha_{\Pi, p^*}|) = \mathcal{O}(|\Phi^\alpha_{\Pi, p^*}|) = \mathcal{O}(|\Psi^{(\omega)}_\Pi|)$ for $\alpha = 1, 2$ where $\omega = d + 1 - \alpha$. The practical efficiency of the method is evaluated in the next section.

Algorithm 2. Efficient update of V_Π and Ψ_Π for an additional inlier.

Require: $P, d, \Pi \subsetneq \{1, \ldots, n\}, p^* \in \{1, \ldots, n\} \setminus \Pi, V_\Pi$ and Ψ_Π.
Ensure: $V_{\Pi \cup \{p^*\}}$ and $\Psi_{\Pi \cup \{p^*\}}$.
 1: Initialize $V := \emptyset$ and $\Psi := \emptyset$.
 2: **for all** $\psi \in \Psi_\Pi$ **do**
 3:     $a :=$ vertex determined by ψ.
 4:     **if** $a \in R_{p^*}$ **then**
 5:         $V := V \cup \{a\}$ and $\Psi := \Psi \cup \{\psi\}$.
 6:     **end if**
 7: **end for**
 8: Compute $A_{\Pi, p^*}(\psi)$ for all $\psi \in \Psi_\Pi^{(\omega)}$ ($\omega = d, d-1$) to have X^1_{Π, p^*} and X^2_{Π, p^*}.
 9: **for all** $\psi \in X^1_{\Pi, p^*} \cup X^2_{\Pi, p^*}$ **do**
10:     **if** $\{h_{(p,s)}(a) = 0 \mid (p, s) \in \psi\}$ has a unique solution a **then**
11:         $a :=$ the unique solution to $\{h_{(p,s)}(a) = 0 \mid (p, s) \in \psi\}$
12:         **if** $a \in \bigcap_{(p,s) \in \psi} F(p, s) \cap R_\Pi$ **then**
13:             $V := V \cup \{a\}$ and $\Psi := \Psi \cup \{\psi\}$.
14:         **end if**
15:     **end if**
16: **end for**
17: **return** $V = V_{\Pi \cup \{p^*\}}$ and $\Psi = \Psi_{\Pi \cup \{p^*\}}$.

---

5   Experiments

For $d = 2$, we generated input data sets P for $n = 200, 400, 600, 800, 1000$ as follows: setting $(a_0, a_1, a_2) = (450, -3.2, 0.0064)$, we randomly generated n integer points within $[0, 499] \times [0, 499]$ so that 80% of the points, called *ground-truth inliers*, are in $D(a_0, a_1, a_2)$ while the other 20% points, called *ground-truth outliers*, are not in $D(a_0, a_1, a_2)$, where each ground-truth outlier was generated so that its Euclidean distance from its closest point in $D(a_0, a_1, a_2)$ is in $[1, 4]$. In the same way, we generated data sets for $d = 3$ where we used $(a_0, a_1, a_2, a_3) = (250, 5, -0.03, 4.0 \times 10^{-5})$ to generate their ground-truth inliers and outliers. P is shown in Fig. 6 for $n = 200, 600, 1000$. In the experiments, we used a PC with an Intel Xeon 3.7 GHz processor with 64 GB memory.

We applied our proposed method (Algorithm 1 together with Algorithm 2) 100 times to each P to see the efficiency of our introduced computation. At each trial, we randomly initialized Π (Step 1) and selected p^* (Step 5), where the $d + 1$ data indices in the initial Π are chosen only from the ground-truth inliers. For comparison, we also applied Algorithm 1 alone (Sekiya+ [8]) 100 times to each P using the same initialization and data point selection. We then evaluated the recall (the ratio of ground-truth inliers in the output against the whole ground-truth inliers) and the computational time (i.e., processing time).

Tables 1 and 2 show the average of recalls over 100 trials for each P and the average of computational times over 100 trials for the two methods. We remark that the outputs by our proposed method are exactly the same as those by Algorithm 1 alone.

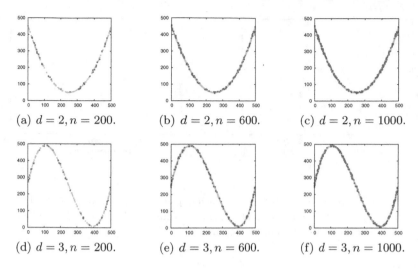

(a) $d = 2, n = 200.$ (b) $d = 2, n = 600.$ (c) $d = 2, n = 1000.$

(d) $d = 3, n = 200.$ (e) $d = 3, n = 600.$ (f) $d = 3, n = 1000.$

Fig. 6. Examples of input data sets. Ground-truth inliers are depicted in green while ground-truth outliers in red. (Color figure online)

We see in Table 2 that our proposed method achieves the same results much faster than Algorithm 1 alone. We see that Algorithm 2 indeed efficiently updates feasible regions in computational time thanks to $|X^1_{\Pi,p^*} \cup X^2_{\Pi,p^*}| < |\Phi^1_{\Pi,p^*} \cup \Phi^2_{\Pi,p^*}|$. To visualize this efficiency, we compared $|X^1_{\Pi,p^*} \cup X^2_{\Pi,p^*}|$ and $|\Phi^1_{\Pi,p^*} \cup \Phi^2_{\Pi,p^*}|$ in each iteration in a trial in the case of $n = 400$ (Fig. 7). We observe that (1) $|X^1_{\Pi,p^*} \cup X^2_{\Pi,p^*}|$ is significantly smaller than $|\Phi^1_{\Pi,p^*} \cup \Phi^2_{\Pi,p^*}|$ and that (2) $|X^1_{\Pi,p^*} \cup X^2_{\Pi,p^*}|$ is almost constant independent of the size of the inlier set. We remark that $X^1_{\Pi,p^*} \cup X^2_{\Pi,p^*} = \emptyset$ indicates $A_{\Pi,p^*}(\psi) = \emptyset$ for all $\psi \in \Psi^{(\omega)}_{\Pi}$ $(\omega = d, d - 1)$.

Table 1. Recall of ground-truth inliers (average over 100 trials).

n	200	400	600	800	1000
$d = 2$	0.870	0.860	0.810	0.859	0.855
$d = 3$	0.817	0.805	0.773	0.806	0.822

Table 2. Computational time (ms) (average over 100 trials).

	n	200	400	600	800	1000
$d = 2$	Sekiya+[8]	36.92	69.16	90.92	138.32	177.28
	Proposed	1.16	1.88	2.32	3.28	3.60
$d = 3$	Sekiya+[8]	154.16	267.32	366.16	488.36	627.16
	Proposed	8.68	9.12	10.36	9.76	11.12

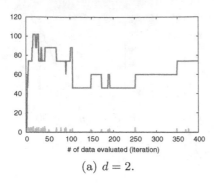
(a) $d = 2$.

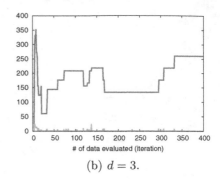
(b) $d = 3$.

Fig. 7. $|X^1_{\Pi,p^*} \cup X^2_{\Pi,p^*}|$ (red) and $|\Phi^1_{\Pi,p^*} \cup \Phi^2_{\Pi,p^*}|$ (green) in each iteration. The horizontal axis is the number of data points already evaluated (i.e., $|\Pi \cup \Pi^C|$). The results are from a trial of $n = 400$. (Color figure online)

6   Conclusion

We dealt with the problem of fitting a discrete polynomial curve to 2D data in the presence of outliers. We discussed how to efficiently compute the vertices of the feasible region for an incrementally updated inlier set in the parameter space. Based on the property that an edge or face of a bounded feasible region is inside the convex hull of its vertices, we introduced a computation to facilitate updating the vertices of the feasible region when a new data point is added to the current inlier set. The efficiency of our proposed computation was demonstrated by our experimental results.

Acknowledgements. This work is in part supported by Grant-in-Aid for Scientific Research (Grant No. 16H02851) of the Ministry of Education, Culture, Sports, Science and Technology of Japan.

A   Appendix: Proof of Lemma 1

Proof. For any $b \in \bigcap_{(p,s)\in\psi} F_\Pi(p, s)$, we prove that $b \in \text{Conv}(V_\Pi(\psi))$, i.e., b is represented as a convex combination of some vertices in $V_\Pi(\psi)$. In fact, it suffices to show that there exist $c_1, c_2 \in \bigcap_{(p,s)\in\psi} F_\Pi(p, s)$ and $(p_1, s_1), (p_2, s_2) \in \Pi \times \{1, \dots, 4\}$ that satisfy the following: $c_1 \in F_\Pi(p_1, s_1)$ and $c_2 \in F_\Pi(p_2, s_2)$, the linear systems $\{h_{(p,s)}(a) = 0 \mid (p, s) \in \psi \cup \{(p_1, s_1)\}\}$ and $\{h_{(p,s)}(a) = 0 \mid (p, s) \in \psi \cup \{(p_2, s_2)\}\}$ are respectively independent, and b is represented as a convex combination of c_1 and c_2. Why this proves $b \in \text{Conv}(V_\Pi(\psi))$ is explained as follows. For $\psi \in \Psi^{(d)}_\Pi$, it is obvious since we have $c_1, c_2 \in V_\Pi(\psi)$ immediately. For $\psi \in \Psi^{(d-1)}_\Pi$, next, each of c_1 and c_2 can be seen as b in the case of $\psi \in \Psi^{(d)}_\Pi$; $\psi \cup \{(p_1, s_1)\}, \psi \cup \{(p_2, s_2)\} \in \Psi^{(d)}_\Pi$ is proven by Lemma 1 in [8]. From the result already obtained for $\psi \in \Psi^{(d)}_\Pi$, therefore, c_1 and c_2 are respectively represented as a convex combination of two vertices in $V_\Pi(\psi \cup \{(p_1, s_1)\})$ and

$V_\Pi(\psi \cup \{(p_2, s_2)\})$. $\boldsymbol{b}$ is represented as a convex combination of the four vertices in $V_\Pi(\psi)$, accordingly.

We therefore prove for the existence of $\boldsymbol{c}_1, \boldsymbol{c}_2, (p_1, s_1), (p_2, s_2)$ as described above. If there exists $(p', s') \in (\Pi \times \{1, \ldots, 4\}) \backslash \psi$ such that $\boldsymbol{b} \in F_\Pi(p', s')$ and $\{h_{(p,s)}(\boldsymbol{a}) = 0 \mid (p, s) \in \psi \cup \{(p', s')\}\}$ is independent, then the required condition is immediately satisfied for $\boldsymbol{c}_1 = \boldsymbol{c}_2 = \boldsymbol{b}$ and $(p_1, s_1) = (p_2, s_2) = (p', s')$. We therefore give the proof for the other case. Let us first consider the $(d + 1 - \omega)$-dimensional flat $\{\boldsymbol{a} \in \mathbb{R}^{d+1} \mid h_{(p,s)}(\boldsymbol{a}) = 0 \text{ for } (p, s) \in \psi\}$, which includes $\bigcap_{(p,s) \in \psi} F_\Pi(p, s)$. We then consider an arbitrary half-line on the flat running from $\boldsymbol{b}$. Note that such a half-line necessarily exists since $d + 1 - \omega \geq 1$. A point in the half-line is represented by $\boldsymbol{c}(r) = \boldsymbol{b} + r\boldsymbol{v}$ for some vector $\boldsymbol{v} \in \mathbb{R}^{d+1} \backslash \{\boldsymbol{0}\}$ and a parameter $r \in \mathbb{R}_{\geq 0}$. For such a half-line, it has been already shown in the proof of Lemma 1 in [8] that, for some $r_1 > 0$ $(r_1 < \infty)$, $\boldsymbol{c}_1 = \boldsymbol{c}(r_1)$ satisfies $\boldsymbol{c}_1 \in F_\Pi(p_1, s_1)$ for $\exists(p_1, s_1) \in (\Pi \times \{1, \ldots, 4\}) \backslash \psi$ such that $\{h_{(p,s)}(\boldsymbol{a}) = 0 \mid (p, s) \in \psi \cup \{(p_1, s_1)\}\}$ is independent. $\boldsymbol{c}_2$, on the other hand, is found on the half-line running in the opposite direction from $\boldsymbol{b}$, whose point is represented by $\boldsymbol{c}'(r) = \boldsymbol{b} + r(-\boldsymbol{v})$. Since it is also a half-line on the same flat, for some $r_2 > 0$ $(r_2 < \infty)$, $\boldsymbol{c}_2 = \boldsymbol{c}'(r_2)$ satisfies $\boldsymbol{c}_2 \in F_\Pi(p_2, s_2)$ for $\exists(p_2, s_2) \in (\Pi \times \{1, \ldots, 4\}) \backslash \psi$ such that $\{h_{(p,s)}(\boldsymbol{a}) = 0 \mid (p, s) \in \psi \cup \{(p_2, s_2)\}\}$ is independent. Now, $\boldsymbol{b}$ is on the line segment connecting $\boldsymbol{c}_1$ and $\boldsymbol{c}_2$, which means that $\boldsymbol{b}$ is represented as a convex combination of $\boldsymbol{c}_1$ and $\boldsymbol{c}_2$. $\qquad\square$

References

1. Buzer, L.: An incremental linear time algorithm for digital line and plane recognition using a linear incremental feasibility problem. In: Braquelaire, A., Lachaud, J.-O., Vialard, A. (eds.) DGCI 2002. LNCS, vol. 2301, pp. 372–381. Springer, Heidelberg (2002). doi:10.1007/3-540-45986-3_33
2. Kenmochi, Y., Buzer, L., Talbot, H.: Efficiently computing optimal consensus of digital line fitting. In: International Conference on Pattern Recognition (ICPR 2010), pp. 1064–1067. IEEE (2010)
3. Largeteau-Skapin, G., Zrour, R., Andres, E.: $O(n^3\log n)$ time complexity for the optimal consensus set computation for 4-connected digital circles. In: Gonzalez-Diaz, R., Jimenez, M.-J., Medrano, B. (eds.) DGCI 2013. LNCS, vol. 7749, pp. 241–252. Springer, Heidelberg (2013). doi:10.1007/978-3-642-37067-0_21
4. Largeteau-Skapin, G., Zrour, R., Andres, E., Sugimoto, A., Kenmochi, Y.: Optimal consensus set and preimage of 4-connected circles in a noisy environment. In: Proceedings of the International Conference on Pattern Recognition (ICPR 2012), pp. 3774–3777. IEEE (2012)
5. Provot, L., Gerard, Y.: Recognition of digital hyperplanes and level layers with forbidden points. In: Aggarwal, J.K., Barneva, R.P., Brimkov, V.E., Koroutchev, K.N., Korutcheva, E.R. (eds.) IWCIA 2011. LNCS, vol. 6636, pp. 144–156. Springer, Heidelberg (2011). doi:10.1007/978-3-642-21073-0_15
6. Sekiya, F., Sugimoto, A.: Fitting discrete polynomial curve and surface to noisy data. Ann. Math. Artif. Intell. **75**(1–2), 135–162 (2015)
7. Sekiya, F., Sugimoto, A.: On connectivity of discretized 2D explicit curve. In: Ochiai, H., Anjyo, K. (eds.) Mathematical Progress in Expressive Image Synthesis II. MI, vol. 18, pp. 33–44. Springer, Tokyo (2015). doi:10.1007/978-4-431-55483-7_4

8. Sekiya, F., Sugimoto, A.: Discrete polynomial curve fitting guaranteeing inclusion-wise maximality of inlier set. In: Chen, C.-S., Lu, J., Ma, K.-K. (eds.) ACCV 2016. LNCS, vol. 10117, pp. 477–492. Springer, Cham (2017). doi:10.1007/978-3-319-54427-4_35

9. Sere, A., Sie, O., Andres, E.: Extended standard Hough transform for analytical line recognition. In: Proceedings of International Conference on Sciences of Electronics, Technologies of Information and Telecommunications (SETIT 2012), pp. 412–422. IEEE (2012)

10. Toutant, J.-L., Andres, E., Largeteau-Skapin, G., Zrour, R.: Implicit digital surfaces in arbitrary dimensions. In: Barcucci, E., Frosini, A., Rinaldi, S. (eds.) DGCI 2014. LNCS, vol. 8668, pp. 332–343. Springer, Cham (2014). doi:10.1007/978-3-319-09955-2_28

11. Zrour, R., Kenmochi, Y., Talbot, H., Buzer, L., Hamam, Y., Shimizu, I., Sugimoto, A.: Optimal consensus set for digital line and plane fitting. Int. J. Imaging Syst. Technol. 21(1), 45–57 (2011)

12. Zrour, R., Largeteau-Skapin, G., Andres, E.: Optimal consensus set for annulus fitting. In: Debled-Rennesson, I., Domenjoud, E., Kerautret, B., Even, P. (eds.) DGCI 2011. LNCS, vol. 6607, pp. 358–368. Springer, Heidelberg (2011). doi:10.1007/978-3-642-19867-0_30

13. Zrour, R., Largeteau-Skapin, G., Andres, E.: Optimal consensus set for nD fixed width annulus fitting. In: Barneva, R.P., Bhattacharya, B.B., Brimkov, V.E. (eds.) IWCIA 2015. LNCS, vol. 9448, pp. 101–114. Springer, Cham (2015). doi:10.1007/978-3-319-26145-4_8

A New Shape Descriptor Based on a Q-convexity Measure

Péter Balázs[1]([✉]) and Sara Brunetti[2]

[1] Department of Image Processing and Computer Graphics, University of Szeged, Árpád tér 2., Szeged 6720, Hungary
pbalazs@inf.u-szeged.hu
[2] Dipartimento di Ingegneria dell'Informazione e Scienze Matematiche, Via Roma, 56, 53100 Siena, Italy
sara.brunetti@unisi.it

Abstract. In this paper we define a new measure for shape descriptor. The measure is based on the concept of convexity by quadrant, called Q-convexity. Mostly studied in Discrete Tomography, this convexity generalizes hv-convexity to any two or more directions, and presents interesting connections with "total" convexity. The new measure generalizes that proposed by Balázs and Brunetti (*A measure of Q-convexity, LNCS 9647 (2016) 219–230*), and therefore it has the same desirable features: (1) its values range intrinsically from 0 to 1; (2) its values equal 1 if and only if the binary image is Q-convex; (3) its efficient computation can be easily implemented; (4) it is invariant under translation, reflection, and rotation by 90°. We test the new measure for assessing sensitivity using a set of synthetic polygons with rotation and translation of intrusions/protrusions and global skew, and for a ranking task using a variety of shapes. Based on the geometrical properties of Q-convexity, we also provide a characterization of any binary image by the matrix of its "generalized salient points", and we design a linear-time algorithm for the construction of the binary image from its associated matrix.

Keywords: Shape descriptor · Convexity measure · Q-convexity · Salient point

1 Introduction

Shape descriptors are widely used in image processing and computer vision for object detection, classification, and recognition [8,12]. One class of descriptors captures single geometrical or topological characteristics of shapes, like moments [7], orientation and elongation [15], circularity [10], just to mention a few. Among them, probably the most often studied descriptor is the measure of convexity. Depending on whether the interior or the boundary of the shape is investigated in order to determine the degree of convexity, these measures can be grouped into area-based [2,12,13] and boundary-based [14] categories.

© Springer International Publishing AG 2017
W.G. Kropatsch et al. (Eds.): DGCI 2017, LNCS 10502, pp. 267–278, 2017.
DOI: 10.1007/978-3-319-66272-5_22

In [1] we proposed a convexity measure which uses both boundary and area information, thus falls between the two above mentioned classes. It is based on the concept of Q-convexity [4,5], mostly studied in Discrete Tomography [9] for its good properties (it generalizes so-called hv-convexity to any two or more directions, and has interesting connections with "total" convexity). The notion of salient points of a Q-convex image has been introduced in [6] as the analogue of extremal points of a convex set. They have similar features, and in particular a Q-convex image is characterized by its salient points. Salient points can be generalized for any binary image, and they have been studied to model the "complexity" of a binary image which led to the convexity measure of [1,3].

The novel idea of this paper is to consider generalized salient points to have different weights – depending on "how" far they are from the boundary – when calculating the convexity measure. In this way we provide a flexible extension of the measure of [1]. For this purpose we introduce the matrix of generalized salient points of a binary image (shortly, GS matrix) and study its properties. We provide a linear-time algorithm for the construction of the binary image from its GS matrix and also describe how the measure can also be computed in linear time.

The structure of the paper is the following. In Sect. 2 we present the basic concepts and give the definition of the new measure of Q-convexity. In Sect. 3 we introduce the matrix of generalized salient points (GS matrix, for short) and describe its properties. We also design a linear-time algorithm for the construction of the binary image from its GS matrix. The aim of Sect. 4 is to briefly show how the introduced measure can be efficiently computed, in linear time in the size of the image. In Sect. 5 we present experimental results. Finally, Sect. 6 is for the conclusion.

2   New Q-convexity Measure

Any *binary image* $\mathcal{F}$ is a $m \times n$ binary matrix, and it can be represented by a set of black, *foreground pixels* denoted by F, and white, *background pixels* (unit squares) (see Fig. 1 left). Equivalently, foreground pixels can be regarded as points of $\mathbb{Z}^2$ contained in a lattice grid $\mathcal{G}$ (rectangle of size $m \times n$) up to a translation so that any binary image can be viewed as a subset of $\mathcal{G}$, also called *lattice set* (see Fig. 1 right). Throughout the paper, we will consider images different from the emptyset and use both representations as interchangeable, since notation for the latter one is more suitable to describe geometrical properties (even if the order of the points in the lattice and the order of the items in a matrix are different). For our convenience when not confusing, we use $\mathcal{F}$ for both the image and its representations, and we denote by $\mathcal{F}^c$ the *complement* of $\mathcal{F}$, i.e., the image obtained as the complement of its pixel values reversing foreground and background pixels. In the lattice representation, $\mathcal{F}^c$ corresponds to $\mathcal{G} \backslash F$.

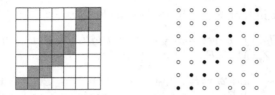

Fig. 1. A binary image represented as black and white pixels (left), and by a lattice set (right).

2.1 Background

Let us introduce the main definitions concerning Q-convexity [4,6]. In order to simplify our explanation, let us consider the horizontal and vertical directions, and denote the coordinate of any point M of the grid $\mathcal{G}$ by (x_M, y_M). Then, M and the directions determine the following four quadrants:

$$Z_0(M) = \{N \in \mathcal{G} : 0 \le x_N \le x_M, \, 0 \le y_N \le y_M\}$$
$$Z_1(M) = \{N \in \mathcal{G} : x_M \le x_N < m, \, 0 \le y_N \le y_M\}$$
$$Z_2(M) = \{N \in \mathcal{G} : x_M \le x_N < m, \, y_M \le y_N < n\}$$
$$Z_3(M) = \{N \in \mathcal{G} : 0 \le x_N \le x_M, \, y_M \le y_N < n\}.$$

Definition 1. *A lattice set F is Q-convex if $Z_p(M) \cap F \ne \emptyset$ for all $p = 0, \dots, 3$ implies $M \in F$.*

If $Z_p(M) \cap F = \emptyset$, we say that $Z_p(M)$ is a *background quadrant*. Thus, in other words, a binary image is Q-convex if there exists least a background quadrant $Z_p(M)$ for every pixel M in the background of $\mathcal{F}$. Figure 2 illustrates the above concepts.

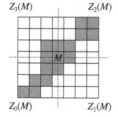 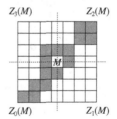

Fig. 2. Illustration of the concept of Q-convexity. A Q-convex (left) and a non-Q-convex (right) lattice set. Note that the image on the left is the Q-convex hull of the image on the right.

The Q-convex hull of $\mathcal{F}$ can be defined as follows:

Definition 2. *The Q-convex hull $\mathcal{Q}(F)$ of a lattice set F is the set of points $M \in \mathcal{G}$ such that $Z_p(M) \cap F \neq \emptyset$ for all $p = 0, \ldots, 3$.*

By Definitions 1 and 2, if F is Q-convex then $F = \mathcal{Q}(F)$. Differently, if F is not Q-convex, then $\mathcal{Q}(F) \backslash F \neq \emptyset$ (see Fig. 2, again, where for the lattice set F on the right, $\mathcal{Q}(F) \backslash F = \{M\}$).

We define a new measure in between region- and boundary-based measures exploiting some geometrical properties of the "shape".

Definition 3. *Let F be a lattice set. A point $M \in F$ is a salient point of F if $M \notin \mathcal{Q}(F \backslash \{M\})$.*

Denote the set of salient points of $\mathcal{F}$ by $\mathcal{S}(F)$. Of course $\mathcal{S}(F) = \emptyset$ if and only if $F = \emptyset$. In particular, Daurat proved in [6] that the salient points of $\mathcal{F}$ are the salient points of the Q-convex hull $\mathcal{Q}(F)$ of F, i.e. $\mathcal{S}(F) = \mathcal{S}(\mathcal{Q}(F))$. This means that if F is Q-convex, its salient points completely characterize F. If it is not, there are other points belonging to the Q-convex hull of F but not in F that "track" the non-Q-convexity of F. These points are called *generalized salient points* (abbreviated by g.s.p.). The set of generalized salient points $S_g(F)$ of F is obtained by iterating the definition of salient points on the sets obtained each time by discarding the points of the set from its Q-convex hull, i.e., using the set notation:

Definition 4. *If F is a lattice set, then the set of its generalized salient points (g.s.p.) $S_g(F)$ is defined by $S_g(F) = \bigcup_i S(F_i)$, where $F_1 = F$, $F_i = \mathcal{Q}(F_{i-1}) \backslash F_{i-1}$.*

With the obvious meaning we may denote the binary images related to $\mathcal{Q}(F)$, $\mathcal{S}(F)$, and F_i by $\mathcal{Q}(\mathcal{F})$, $\mathcal{S}(\mathcal{F})$, and $\mathcal{F}_i$, respectively. Figure 3 illustrates the definition in the lattice representation. We notice that $\mathcal{F}_i$ is contained in $\mathcal{F}_{i-1}^c$ (more precisely, in $\mathcal{Q}(F_{i-1}) \backslash F_{i-1}$), and if i is even, $\mathcal{F}_i$ is contained in $\mathcal{F}_1^c$, else if i is odd $\mathcal{F}_i$ is contained in $\mathcal{F}_1$. In the pixel-representation, this corresponds to say that foreground and background pixels in $\mathcal{F}_i$ correspond to white and black pixels for i even, and to black and white pixels for i odd, respectively. In this view,

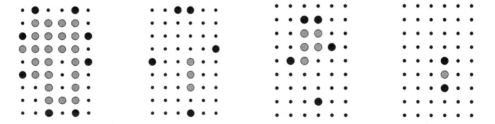

Fig. 3. Generalized salient points are in black. Leftmost: $\mathcal{F}_1$. Centre-left: $\mathcal{F}_2$. Centre-right: $\mathcal{F}_3$. Rightmost: $\mathcal{F}_4$.

the Q-convex hull of the foreground pixels of $\mathcal{F}_{i-1}$ contains the Q-convex hull of the foreground pixels of $\mathcal{F}_i$. Moreover if $\mathcal{F}$ and $\mathcal{F}'$ are two binary images, then $S_g(\mathcal{F})$ is different from $S_g(\mathcal{F}')$ (see Theorem 9 of [6]).

Let k be the index such that $F_{k+1} = \emptyset$. By definition, $\mathcal{S}(\mathcal{F}) = \mathcal{S}(\mathcal{F}_1) \subseteq S_g(\mathcal{F}) = \mathcal{S}(\mathcal{F}_1) \cup \mathcal{S}(\mathcal{F}_2) \cup \ldots \cup \mathcal{S}(\mathcal{F}_k)$ and the equality holds when $\mathcal{F}$ is Q-convex. Moreover, the points of $S_g(\mathcal{F})$ are chosen among the points of subsets of $\mathcal{Q}(\mathcal{F})$, thus $S_g(\mathcal{F}) \subseteq \mathcal{Q}(\mathcal{F})$.

2.2   The Generalized Shape Descriptor

In [1], we defined a shape measure in terms of proportion between salient points and generalized salient points. Denoting the cardinality of an arbitrary set $\mathcal{P}$ of points by $|\mathcal{P}|$, here we generalize the measure as follows:

Definition 5. *For a given binary image $\mathcal{F}$, its Q-convexity measure $\Psi_{(c_i)}(\mathcal{F})$ is defined by*

$$\Psi_{(c_i)}(\mathcal{F}) = \frac{|\mathcal{S}(\mathcal{F})|}{\sum_i c_i |\mathcal{S}(\mathcal{F}_i)|},$$

where $\mathcal{S}(\mathcal{F})$ and $\mathcal{S}(\mathcal{F}_i)$ are as in Definition 4, $c_1 = 1$, and each c_i is a non-negative real number.

Notice that $c_1 = 1$ must hold in order to get value 1 for Q-convex sets. Note also that the measure is purely qualitative because is independent from the size of the image. It coincides with the measure in [1] if $c_i = 1$, for all i, and, more generally, if the g.s.p. are many with respect to salient points, then $\mathcal{F}$ is far to be Q-convex. Besides, the dependence on successive $|\mathcal{S}(\mathcal{F}_i)|$ depends on the choice of c_i: if c_i is a decreasing function, then the measure scores heavily g.s.p. in the boundary with respect to the g.s.p. in the interior, and vice-versa in case of an increasing function. This approach provides, in fact, a family of shape descriptors (by setting the weights differently). Each member of the family could complement other ones, thus, giving a finely tuneable tool for solving pattern recognition issues, as different weightings can capture different aspects of the shapes (see the ranking examples in Sect. 5).

Since $\mathcal{S}(\mathcal{F}) \subseteq S_g(\mathcal{F}) \subseteq \mathcal{Q}(\mathcal{F})$, the Q-convexity measure satisfies the following properties:

- the Q-convexity measure ranges from 0 to 1;
- the Q-convexity measure equals 1 if and only if $\mathcal{F}$ is Q-convex.

In particular, $\mathcal{S}(\mathcal{F}) = S_g(\mathcal{F})$, and hence $\Psi_{(c_i)}(\mathcal{F}) = 1$, (for instance when $\mathcal{F}$ is a full rectangle as the rightmost image in Fig. 6) if and only if $\mathcal{F}$ is Q-convex. If $S_g(\mathcal{F}) = \mathcal{Q}(\mathcal{F})$ (for instance if $\mathcal{F}$ is a chessboard), $\Psi_{(c_i)}(\mathcal{F})$ decreases with the inverse of the size of $\mathcal{Q}(\mathcal{F})$ in the case where $c_i = 1$ for all i. So, if we consider the sequence of images starting from the full rectangle and ending with the chessboard, and having intermediate images obtained by deleting each time one suitable pixel iteratively row by row, they have measure values decreasing from 1 to $\frac{4}{mn}$ (circa $6 \cdot 10^{-5}$ for $m = n = 256$).

Moreover, since $\mathcal{S}(\mathcal{F})$, $\mathcal{S}(\mathcal{F}_i)$ and $\mathcal{Q}(\mathcal{F})$ are invariant under translation, reflection, and rotation by $90°$ for the horizontal and vertical directions, the measure is also invariant.

3   The *GS* Matrix

Let $\mathcal{F}$ be a $m \times n$ binary image, and $\mathcal{S}_g(\mathcal{F}) = \mathcal{S}(\mathcal{F}_1) \cup \mathcal{S}(\mathcal{F}_2) \cup \ldots \cup \mathcal{S}(\mathcal{F}_k)$. Consider the matrix representation of $\mathcal{F} = (f_{ij})$. We may associate $\mathcal{F}$ to the $m \times n$ integer matrix B of its generalized salient points defined as follows: $b_{ij} = h$, if and only if f_{ij} is a g.s.p. of $\mathcal{F}_h$; $b_{ij} = 0$ otherwise. Informally, items $0 < h(\leq k)$ of the integer matrix B correspond to g.s.p in $\mathcal{S}(\mathcal{F}_h)$; items 0 do not correspond to any g.s.p. of $\mathcal{F}$. For example, the rightmost matrix in Fig. 4 is the GS matrix associated to the leftmost binary matrix (which corresponds to $\mathcal{F}$ in Fig. 3). We call B, the GS matrix associated to $\mathcal{F}$, where GS stands for "generalized salient". The GS matrix is well-defined since by Definition 4, we have that $\cap_i \mathcal{S}(\mathcal{F}_i) = \emptyset$.

Theorem 1. *Any two binary images are equal if and only if their GS matrices are equal.*

Proof. The following construction permits to determine the binary image by its GS matrix: For all item i in the GS matrix considered in decreasing order (of i), compute the Q-convex hull of the pixels corresponding to the items i and fill the corresponding pixels not already considered with the foreground w.r.t. i.

It is easy to see that the construction is correct. Let k be the maximum value in the GS matrix; then, $\mathcal{F}_{k+1} = \emptyset$, and $\mathcal{F}_k$ is Q-convex. Therefore, the Q-convex hull of its g.s.p. is $\mathcal{F}_k$, and so the first step $i = k$ of the construction determines $\mathcal{F}_k = Q(\mathcal{S}(\mathcal{F}_k))$. In the second step $i = k - 1$, the construction determines $\mathcal{F}_{k-1}$: it computes the Q-convex hull of the pixels corresponding to the items $k - 1$ in the GS matrix, i.e. $Q(\mathcal{S}(\mathcal{F}_{k-1})) = Q(\mathcal{F}_{k-1})$, and fills the corresponding pixels not already considered with the foreground w.r.t. $k-1$, i.e. $Q(F_{k-1}) \backslash F_k$. By definition, $\mathcal{F}_k = Q(F_{k-1}) \backslash_{k-1}$, and since $Q(\mathcal{F}_{k-1}) = F_{k-1} \cup F_k$ and $F_{k-1} \cap F_k = \emptyset$, we have $\mathcal{F}_{k-1} = Q(\mathcal{F}_{k-1}) \backslash F_k$. By proceeding in this way, in the last step $i = 1$, the construction determines $\mathcal{F} = \mathcal{F}_1$ since $F_1 = Q(F_1) \backslash F_2$. Finally, since two different binary images have different GS matrices, there is a one-to-one correspondence between images and matrices. $\square$

In order to design an efficient algorithm based on the constructive proof of the theorem, we extend the definition of Q-convex hull as follows: The Q-convex hull $\mathcal{Q}(F_i)$ of the lattice set F_i is the set of points $M \in \mathcal{G}$ such that $Z_p(M) \cap F_i \neq \emptyset$ for all $p = 0, \ldots, 3$.

Therefore, pixel M belongs to the Q-convex hull of F_i if there is an item i in the GS matrix associated to $\mathcal{F}_i$ in each zone of M. Since the Q-convex hull of the foreground pixels of $\mathcal{F}_i$ contains the Q-convex hull of the foreground pixels of $\mathcal{F}_{i+1}$, pixel M belongs to $Q(F_i) \backslash Q(F_{i+1})$, if i is the minimum among the maximum items in the GS matrix in each zone in M. This ensures that every item is considered once. Let $Z_t = (z_{ij}^t)$ such that $z_{ij}^t = h$ iff h is the maximum item in the submatrix $Z_t(b_{ij})$, for $t = 0, 1, 2, 3$.

1: **procedure** (B) ▷ Construct the binary image $\mathcal{F}$ from its GS matrix B
2:     **for** each $b_{ij} = 0$ **do**
3:         find the maximum item of B in $Z_0(b_{ij})$ and store in z^0_{ij} of matrix Z^0
4:     **end for**
5:     **for** each $b_{ij} = 0$ **do**
6:         find the maximum item of B in $Z_1(b_{ij})$ and store in z^1_{ij} of matrix Z^1
7:     **end for**
8:     **for** each $b_{ij} = 0$ **do**
9:         find the maximum item of B in $Z_2(b_{ij})$ and store in z^2_{ij} of matrix Z^2
10:    **end for**
11:    **for** each $b_{ij} = 0$ **do**
12:        find the maximum item of B in $Z_3(b_{ij})$ and store in z^3_{ij} of matrix Z^3
13:    **end for**
14:    **for** each $b_{ij} = 0$ **do**
15:        $h \leftarrow \min(z^0_{ij}, z^1_{ij}, z^2_{ij}, z^3_{ij})$
16:        **if** h is odd **then** $f_{ij} \leftarrow 1$
17:        **else** $f_{ij} \leftarrow 0$
18:        **end if**
19:    **end for**
20:    **for** each $b_{ij} \neq 0$ **do**
21:        **if** b_{ij} is odd **then** $f_{ij} \leftarrow 1$
22:        **else** $f_{ij} \leftarrow 0$
23:        **end if**
24:    **end for**
25: **end procedure**

Starting from the GS matrix B in input, Algorithm 1 constructs $\mathcal{F}$ by using Z_0, Z_1, Z_2 and Z_3. For example:

$$B=\begin{bmatrix} 0&1&2&2&1&0 \\ 0&0&3&3&0&0 \\ 1&0&0&0&0&1 \\ 0&0&0&0&3&2 \\ 2&3&0&4&0&1 \\ 1&0&0&0&0&0 \\ 0&0&0&4&0&0 \\ 0&0&0&3&0&0 \\ 0&6&1&2&1&0 \end{bmatrix}, z^0=\begin{bmatrix} 2&0&0&0&2&4 \\ 2&3&0&0&4&4 \\ 0&2&3&4&4&0 \\ 2&3&3&4&0&0 \\ 0&0&3&0&4&0 \\ 0&1&1&4&4&4 \\ 0&0&1&0&4&0 \\ 0&0&1&0&3&0 \\ 0&0&0&0&0&0 \end{bmatrix}, z^1=\begin{bmatrix} 4&0&0&0&0&2 \\ 4&4&0&0&4&4 \\ 0&4&4&4&3&0 \\ 4&4&4&4&0&0 \\ 0&0&4&0&1&0 \\ 0&4&4&4&1&0 \\ 0&0&4&0&1&0 \\ 0&0&3&0&1&0 \\ 0&0&0&0&0&0 \end{bmatrix}, z^2=\begin{bmatrix} 2&0&0&0&0&0 \\ 3&3&0&0&1&0 \\ 0&3&3&3&1&0 \\ 3&3&3&3&0&0 \\ 0&0&4&0&3&0 \\ 0&4&4&4&3&2 \\ 0&0&4&0&3&0 \\ 0&0&4&0&3&0 \\ 0&0&0&0&0&0 \end{bmatrix}, z^3=\begin{bmatrix} 0&0&0&0&0&2 \\ 0&1&0&0&3&3 \\ 0&1&3&3&3&0 \\ 1&1&3&3&0&0 \\ 0&0&3&0&4&0 \\ 0&3&3&4&4&4 \\ 0&0&3&0&4&0 \\ 0&0&3&0&4&0 \\ 0&0&0&0&0&0 \end{bmatrix}.$$

If we consider for instance $b_{22} = 0$, since $\min\{z^0_{22} = 3, z^1_{22} = 4, z^2_{22} = 3, z^3_{22} = 3\} = 3$, then f_{22} is in $\mathcal{Q}(\mathcal{F}_3)$ and so $f_{22} = 1$. Note that Algorithm 1 reconstructs $\mathcal{F}$ of Fig. 4. The correctness of the algorithm derives by previous discussion.

Theorem 2. *Algorithm 1 computes the binary image from its GS matrix in linear time.*

Proof. Let $B = (b_{ij})$ be the GS matrix in input, and $\mathcal{F} = (f_{ij})$ be the binary matrix representation of the image associated to B. Initially, $f_{ij} = 0$, for all i, j.

The computation of the maximum in any zone for each item $b_{ij} = 0$ (statements 2–4) can be done in linear time in the size of the image. Indeed consider zone Z_0: for b_{ij}, by definition, $Z_0(b_{ij}) = Z_0(b_{i-1j}) \cup Z_0(b_{ij-1})$. Therefore the maximum in $Z_0(b_{ij})$ can be computed by previous computations for $Z_0(b_{i-1j})$ and $Z_0(b_{ij-1})$, and stored in a matrix Z^0. (Analogous, relations hold for Z_1, Z_2, Z_3.) For any item b_{ij}, the minimum among four corresponding values stored in the four matrices Z^0, Z^1, Z^2, Z^3 (statement 15), and the determination of the parity of the minimum cost $O(1)$ (statements 16–18, 21–23). Hence, the complexity of the algorithm is linear in the size of matrix B. $\square$

4   Computation of Ψ

The GS matrix and the shape measure $\Psi_{(c_1,\dots,c_k)}$ can be computed in linear time in the size of the binary image by the algorithm designed in [3] for the determination of generalized salient pixels.

Here we briefly describe the algorithm. The basic idea is that salient points and generalized salient points of a binary image $\mathcal{F}$ can be determined by implicit computation of the $Q(\mathcal{F})$. Indeed, the authors in [3] proved that $Q(\mathcal{F})$ is the complement of the union of maximal background quadrants. At each step i, the algorithm finds the foreground (generalized) salient pixels of $\mathcal{F}_i$ by computing the maximal background quadrants of $\mathcal{F}_i$. Pixels in the background quadrants are discarded and the remaining complemented image is considered in the next step being the Q-convex hull of $\mathcal{F}_i$ (recall that $\mathcal{F}_{i+1} = Q(\mathcal{F}_i)\backslash\mathcal{F}_i$). During the computation of generalized salient points, the algorithm constructs the GS matrix $B = (b_{ij})$. Indeed, $b_{ij} = h$, if f_{ij} is a g.s.p. in $\mathcal{F}_h$ and the algorithm finds it at step h (and $b_{ij} = 0$ for any item which is not a g.s.p.). Therefore, B is the matrix of the steps at which every g.s.p. is found. Figure 4 shows an example of the execution of the algorithm and the corresponding GS matrix.

```
   k=1               k=2               k=3               k=4                 GS

0 1 0 0 1 0       0 1 0 0 1 0       0 1 0 0 1 0       0 1 0 0 1 0       0 1 2 2 1 0
0 1 1 1 1 0       0 1 1 1 1 0       0 1 1 1 1 0       0 1 1 1 1 0       0 0 3 3 0 0
1 1 1 1 1 1       1 1 1 1 1 1       1 1 1 1 1 1       1 1 1 1 1 1       1 0 0 0 0 1
1 1 1 1 1 0       1 1 1 1 1 0       1 1 1 1 1 0       1 1 1 1 1 0       0 0 0 0 3 2
0 1 1 0 1 1       0 1 1 0 1 1       0 1 1 0 1 1       0 1 1 0 1 1       2 3 0 4 0 1
1 1 1 0 1 0       1 1 1 0 1 0       1 1 1 0 1 0       1 1 1 0 1 0       1 0 0 0 0 0
0 0 1 0 1 0       0 0 1 0 1 0       0 0 1 0 1 0       0 0 1 0 1 0       0 0 0 4 0 0
0 0 1 0 1 0       0 0 1 0 1 0       0 0 1 1 1 0       0 0 1 0 1 0       0 0 1 2 1 0
```

Fig. 4. Illustrative example of the algorithm for finding g.s.p. of image $\mathcal{F}$ (first matrix). In each step (first four matrices, from left to right) the identified g.s.p. are drawn bold and pixels of the background quadrants are grey. The positions inside the polygon constitute the Q-convex hull of the g.s.p. $Q(\mathcal{S}(\mathcal{F}_i))\backslash\mathcal{S}(\mathcal{F}_i)$ and will be investigated in the successive step. The rightmost matrix is the GS matrix of $\mathcal{F}$.

5   Experiments

Depending on the choice of c_i we obtain shape measures that score differently pixels closer to the boundary and those internal. In [3] we considered the case where $c_i = 1$ for all i, thus weighting all the g.s.p. in the same way. Here we investigate two pairs of opposite choices:

- $c_i = i$, and $c_i = i^2$ for $i = 1, \ldots, k$
- $c_i = 1/i$, and $c_i = (1/i)^2$ for $i = 1, \ldots, k$.

In the first experiment, we used the set of synthetic polygons in [11] to study the behavior of the measures in case of rotation, translation of intrusions/protrusions and global skew. In Fig. 5 are illustrated the results. We observe that $\Psi_{(c_i=i)}$ and $\Psi_{(c_i=i^2)}$ assign values lower than those assigned by $\Psi_{(c_i=1)}$, whereas $\Psi_{(c_i=1/i)}$ and $\Psi_{(c_i=1/i^2)}$ assign values greater than those assigned by $\Psi_{(c_i=1)}$. In this experiment, all the measures rank the images in the same order except $\Psi_{c_i=i^2}$ (which exchanges first with second image). Let us notice that measures are invariant under translation of intrusions and protrusions (see fourth and fifth shapes, for example), but are sensitive to rotations of angles different from $90°$ (see second shape).

$\Psi_{(c_i=1)}$	0.0175 0.0322 0.0645 0.0645 0.0645 0.0895 0.0895 0.1071 0.8638
$\Psi_{(c_i=i)}$	0.0079 0.0099 0.0211 0.0211 0.0211 0.0250 0.0250 0.0394 0.7245
$\Psi_{(c_i=i^2)}$	0.0033 0.0029 0.0057 0.0057 0.0057 0.0064 0.0064 0.0137 0.5119
$\Psi_{(c_i=1/i)}$	0.0368 0.0899 0.1576 0.1576 0.1576 0.2163 0.2163 0.2571 0.9361
$\Psi_{(c_i=1/i^2)}$	0.0737 0.2152 0.3091 0.3091 0.3091 0.3980 0.3980 0.4954 0.9703

Fig. 5. Synthetic shapes ranked into ascending order by shape measures. Values are rounded to four digits.

In the second experiment, we considered a variety of shapes, and we ranked them by each measure. In Fig. 6 the ranking in ascending order and the values for measure $\Psi_{(c_i=1)}$ are illustrated, whereas in Fig. 7 we report on the results for $\Psi_{(c_i=i)}$, $\Psi_{(c_i=i^2)}$, $\Psi_{(c_i=1/i)}$, and $\Psi_{(c_i=1/i^2)}$. Note that all the measures correctly assign value 1 to the "L" and rectangular shapes. Moreover, by definition, $\Psi_{(c_i=i)}$ and $\Psi_{(c_i=i^2)}$ assign lower values to shapes with the majority of g.s.p in the interior (thus, to images with many narrows but deep intrusions) than to shapes with the majority of g.s.p in the boundary, whereas $\Psi_{(c_i=1/i)}$ and $\Psi_{(c_i=1/i^2)}$ behave on the contrary. This is shown for example by the eagle and spiral images (fourth and eighth in Fig. 6, respectively). Indeed the spiral shape has most of its g.s.p. in the boundary so that it is in position four and two in the ranking for $\Psi_{(c_i=1/i)}$ and $\Psi_{(c_i=1/i^2)}$, respectively, whereas it is in position nine in the ranking for $\Psi_{(c_i=i)}$ and $\Psi_{(c_i=i^2)}$ in Fig. 7. The contrary happens for the eagle shape, where most of its g.s.p. are in the interior. This is in accordance of our expectations and shows that different weightings can be appropriate for different classification tasks.

In both experiments we used the images with original size (in both dimensions varying between 100 and 555 pixels), even if we illustrate them rescaled for better presentation quality. We also investigated scale invariance. This time we omitted the two fully convex images as their convexities are naturally scale-invariant. Taking the vectorized versions of the remaining 12 original images we digitized them on different scales (32×32, 64×64, 128×128, 256×256, and 512×512). Then, for each image we computed the convexity values and compared them to the convexity of the original sized image. Formally, we measured the normalized difference

$$\Delta\Psi = \frac{|\Psi(F_o) - \Psi(F_r)|}{\Psi(F_o)},$$

where $\Psi(F_o)$ and $\Psi(F_r)$ is the convexity of the original sized and the rescaled image, respectively. Table 1 shows the average of the measured convexity differences over the 12 pair of images. Of course, in lower resolutions the small details of the shapes disappear, therefore the shown difference values are higher. As expected, $\Psi_{(c_i=i)}$ and $\Psi_{(c_i=i^2)}$ are fairly intolerant to rescaling which has a high impact on the narrow and deep intrusions. On the other hand, the values for $\Psi_{(c_i=1/i)}$ and $\Psi_{(c_i=1/i^2)}$ in Table 1 are small (less than 0.1), from which we can deduce a reasonable scale-invariance of these convexity measures.

Table 1. Average difference of the convexity value of the original and rescaled images.

Size	$\Delta\psi_{(c_i=i)}$	$\Delta\psi_{(c_i=i^2)}$	$\Delta\psi_{(c_i=1/i)}$	$\Delta\psi_{(c_i=1/i^2)}$
32×32	8.9945	73.4749	0.0950	0.0601
64×64	2.2382	10.1241	0.0866	0.0674
128×128	0.7901	2.3950	0.0756	0.0459
256×256	0.1936	0.4378	0.0541	0.0227
512×512	0.1853	0.3452	0.0806	0.0312

0.0335 0.0353 0.0571 0.1171 0.1924 0.2040 0.2531 0.2766 0.3267 0.3366 0.5392 0.6516 1.0000 1.0000

Fig. 6. Shapes ranked into ascending order by $\Psi_{(c_i=1)}$. Values are rounded to four digits.

0.0010 0.0011 0.0060 0.0081 0.0157 0.0205 0.0546 0.0600 0.1046 0.1492 0.3452 0.3602 1.0000 1.0000

0.0000 0.0000 0.0003 0.0004 0.0007 0.0011 0.0048 0.0098 0.0304 0.0533 0.1146 0.2104 1.0000 1.0000

0.2927 0.3453 0.3529 0.5236 0.5746 0.5754 0.5821 0.6043 0.6819 0.7061 0.7118 0.8491 1.0000 1.0000

0.6390 0.7380 0.7543 0.7553 0.7705 0.8160 0.8303 0.8329 0.8818 0.8923 0.9113 0.9357 1.0000 1.0000

Fig. 7. Shapes ranked into ascending order by $\Psi_{(c_i=i)}$, $\Psi_{(c_i=i^2)}$, $\Psi_{(c_i=1/i)}$, and $\Psi_{(c_i=1/i^2)}$ measures, from top to bottom, respectively. Values are rounded to four digits.

6 Further Work

In this paper we presented a flexible extended version of the measure of Q-convexity defined in [1]. By introducing the matrix of generalized salient points we can give different weights for different groups of g.s.p. when calculating the degree of convexity.

Choosing the proper weightings depends, of course, always on the classification/recognition/image precessing problem and whether we are more interested in differentiating the images based on their boundaries or on their interiors. To find the proper weights, either trial-and-fail or more sophisticated methods, such as stochastic search or machine learning constructions can be used. This issue can be investigated in more detail in a further paper as well as the performance of the measures in real-life pattern recognition applications.

Acknowledgements. The authors thank P.L. Rosin for providing the dataset used in [11]. The collaboration of the authors was supported by the COST Action MP1207 "EXTREMA: Enhanced X-ray Tomographic Reconstruction: Experiment, Modeling, and Algorithms". The research of Péter Balázs was supported by the NKFIH OTKA [grant number K112998].

References

1. Balázs, P., Brunetti, S.: A measure of Q-convexity. In: Normand, N., Guédon, J., Autrusseau, F. (eds.) DGCI 2016. LNCS, vol. 9647, pp. 219–230. Springer, Cham (2016). doi:10.1007/978-3-319-32360-2_17

2. Boxter, L.: Computing deviations from convexity in polygons. Pattern Recogn. Lett. **14**, 163–167 (1993)
3. Brunetti, S., Balázs, P.: A measure of Q-convexity for shape analysis (submitted)
4. Brunetti, S., Daurat, A.: An algorithm reconstructing convex lattice sets. Theoret. Comput. Sci. **304**(1–3), 35–57 (2003)
5. Brunetti, S., Daurat, A.: Reconstruction of convex lattice sets from tomographic projections in quartic time. Theoret. Comput. Sci. **406**(1–2), 55–62 (2008)
6. Daurat, A.: Salient points of Q-convex sets. Int. J. Pattern Recognit. Artif. Intell. **15**, 1023–1030 (2001)
7. Flusser, J., Suk, T.: Pattern recognition by affine moment invariants. Patt. Rec. **26**, 167–174 (1993)
8. Gonzalez, R.C., Woods, R.E.: Digital Image Processing, 3rd edn. Prentice Hall, Upper Saddle River (2008)
9. Herman, G.T., Kuba, A. (eds.): Advances in Discrete Tomography and Its Applications. Birkhäuser Basel, Boston (2007). doi:10.1007/978-0-8176-4543-4
10. Proffitt, D.: The measurement of circularity and ellipticity on a digital grid. Patt. Rec. **15**, 383–387 (1982)
11. Rosin, P.L., Zunic, J.: Probabilistic convexity measure. IET Image Process. **1**(2), 182–188 (2007)
12. Sonka, M., Hlavac, V., Boyle, R.: Image Processing, Analysis, and Machine Vision, 3rd edn. Thomson Learning, Toronto (2008)
13. Stern, H.: Polygonal entropy: a convexity measure. Pattern Recogn. Lett. **10**, 229–235 (1998)
14. Zunic, J., Rosin, P.L.: A new convexity measure for polygons. IEEE T. Pattern Anal. **26**(7), 923–934 (2004)
15. Zunic, J., Rosin, P.L., Kopanja, L.: On the orientability of shapes. IEEE Trans. Image Process. **15**(11), 3478–3487 (2006)

Recognition of Digital Polyhedra with a Fixed Number of Faces Is Decidable in Dimension 3

Yan Gérard[(✉)]

LIMOS - UMR 6158 CNRS/Université Clermont Auvergne,
Clermont-ferrand, France
yan.gerard@uca.fr

Abstract. We consider a conjecture on lattice polytopes $Q \subset \mathbb{R}^d$ (the vertices are integer points) or equivalently on finite subsets $S \subset \mathbb{Z}^d$, Q and S being related by $Q \cap \mathbb{Z}^d = S$ or $Q = \text{conv}(S)$: given the vertices of Q or the list of points of S and an integer n, the problem to determine whether there exists a (rational) polyhedron $P \subset \mathbb{R}^d$ with at most n faces and verifying $P \cap \mathbb{Z}^d = S$ is decidable.

In terms of computational geometry, it's a problem of polyhedral separability of S and $\mathbb{Z}^d \setminus S$ but the infinite number of points of $\mathbb{Z}^d \setminus S$ makes it intractable for classical algorithms. This problem of digital geometry is however very natural since it is a kind of converse of Integer Linear Programming.

The conjecture is proved in dimension $d = 2$ and in arbitrary dimension for non hollow lattice polytopes Q [6]. The purpose of the paper is to extend the result to hollow polytopes in dimension $d = 3$. An important part of the work is already done in [5] but it remains three special cases for which the set of outliers can not be reduced to a finite set: planar sets, pyramids and *marquees*. Each case is solved with a particular method which proves the conjecture in dimension $d = 3$.

Keywords: Pattern recognition · Geometry of numbers · Polyhedral separation · Digital polyhedron · Hollow lattice polytopes

1 Introduction

The recognition of digital primitives is a classical task of pattern recognition and digital geometry. It is usually question of recognizing digital primitives such as digital straight segments, conics or more generally some families of shapes in several dimensions. These problems can be stated in the following terms:

Problem 1 (Recognition($d, \mathscr{F}, S$)). **Input:** Let $\mathscr{F}$ be a family of subsets F of $\mathbb{R}^d$ and S be a subset of $\mathbb{Z}^d$.
Output: Does there exists a set F of $\mathscr{F}$ verifying $F \cap \mathbb{Z}^d = S$?

We focus in this paper on the problem [Recognition($d, \mathscr{F}, S$)] where the family $\mathscr{F}$ is a set of polyhedra with a prescribed number of faces. By denoting

© Springer International Publishing AG 2017
W.G. Kropatsch et al. (Eds.): DGCI 2017, LNCS 10502, pp. 279–290, 2017.
DOI: 10.1007/978-3-319-66272-5_23

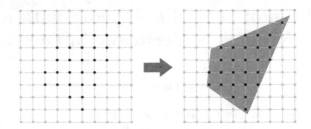

Fig. 1. An instance of [Recognition(d, $\mathscr{P}_n$, S)] is the given of dimension d (here $d = 2$), a number of faces n (here $n = 4$) and a finite set of integer points. The problem is to find a polyhedron $P \subset \mathscr{P}_n$ with the prescribed number of faces containing S but no other integer point: $P \cap \mathbb{Z}^d = S$. The problem can be restated as a problem of polyhedral separability [PolyhedralSeparability($d, n, S, \mathbb{Z}^d \setminus S$)] of S and its complementary $\mathbb{Z}^d \setminus S$ by a polyhedron of $\mathscr{P}_n$.

$\mathscr{P}_n$ the set of polyhedra of $\mathbb{R}^d$ defined as intersections of at most n linear half-spaces (by convention, a polyhedron of $\mathscr{P}_n$ is in $\mathscr{P}_{n+1}$), we investigate [Recognition(d, $\mathscr{P}_n$, S)] (Fig. 1).

This problem [Recognition(d, $\mathscr{P}_n$, S)] is mentioned in the review of open questions in digital geometry [1]. Until 2015, it has been only investigated in dimension 2 with specific polyhedra such as squares and rectangles [4,7,10]. The difficulty of [Recognition(d, $\mathscr{P}_n$, S)] is that even its decidability is not straightforward. By definition, [Recognition(d, $\mathscr{P}_n$, S)] enters in the class of problems of *polyhedral separability* investigated in Computational Geometry:

Problem 2 (PolyhedralSeparability(d, n, S, T)).
Input: A dimension d, an integer n, a set $S \subset \mathbb{R}^d$ of inliers and a set $T \subset \mathbb{R}^d$ of outliers.
Output: Find a polyhedron $P \subset \mathscr{P}_n$ such that all the points of S and none point of T are in P: $S \subset P \subset \mathbb{R}^d \setminus T$.

The problem [Recognition(d, $\mathscr{P}_n$, S)] can be stated as [PolyhedralSeparability ($d, n, S, \mathbb{Z}^d \setminus S$)] where the set of outliers T is the complementary of S in $\mathbb{Z}^d$. For finite sets S and T, polyhedral separability can be solved in linear time if $n = 1$ [8] or in $O((|S| + |T|)\, log(|S| + |T|))$ if $d = 2$ for any n [3]. It becomes NP-complete in arbitrary dimension even with $n = 2$ [9]. With any finite sets S and T, the decidability of [PolyhedralSeparability(d, n, S, T)] is completely straightforward but it is no more the case with the infinite set of outliers $T = \mathbb{Z}^d \setminus S$ considered for [Recognition(d, $\mathscr{P}_n$, S)]. The problem is intractable for classical algorithms of polyhedral separability.

This problem is however interesting because it is a kind of converse problem of Integer Linear Programming [11]. In ILP, the input is a set of n linear inequalities and the purpose is to provide the integer points which satisfy them. In [Recognition(d, $\mathscr{P}_n$, S)], we have the set of integer points and we want to recover a prescribed number of inequalities which characterize it. Although the geometry of numbers provides a powerful framework to work on lattice polytopes, the classical algorithms of this field do not allow to solve directly [Recognition(d, $\mathscr{P}_n$, S)].

We can however imagine a direct strategy by reducing the infinite set of outliers $T = \mathbb{Z}^d \setminus S$ to the subset of its minimal elements according to a partial order relation "is in the shadow of" [5,6]. The minimal elements of $\mathbb{Z}^d \setminus S$ are called the *lattice jewels* of S while the non minimal elements do not need to be taken into account. This approach allows to prove the decidability of the problem if the number of lattice jewels is finite. It holds in dimension $d = 2$ or in arbitrary dimension, if the polytope $Q = \operatorname{conv}(S)$ is not hollow (its interior contains at least an integer point) [6]. In dimension $d = 3$, the hollow 3-polytopes with a finite number of lattice jewels are characterized in [5]. For them, [Recognition(3, $\mathscr{P}_n$, S)] is also decidable. In this dimension, it just remains the cases of the hollow 3-polytopes with an infinite number of lattice jewels (Fig. 2), namely if

- S is coplanar,
- or S is a *pyramid* of lattice height 1, namely the lattice width of S is 1 and one of the two consecutive planar sections of S is reduced to a point,
- or S is a *marquee*: the lattice width of S is 1 and one of the two consecutive planar sections of S in the thin direction is reduced to a segment.

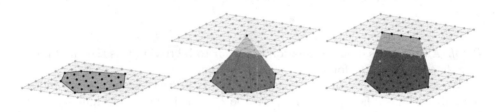

Fig. 2. The three types of 3-polytopes S with infinitely many lattice jewels (thus the decidability of [Recognition(3, $\mathscr{P}_n$, S)] in dimensions $d = 3$ is an open question). On the left, S is planar. In the middle, S is a *pyramid* of lattice height 1. On the right, the lattice width of S is again 1 but one of its consecutive planar sections in the thin direction is reduced to a segment. We refer to this case as a (circus) *marquee*.

In both cases, we provide an algorithm to decide [Recognition(3, $\mathscr{P}_n$, S)] in a finite time. It solves the last cases allowing to prove the following theorem:

Theorem 1. *The problem [Recognition(3, $\mathscr{P}_n$, S)] is decidable.*

Section 2 is devoted to the easiest cases of planar sets and pyramids while marquees are investigated in Sect. 3.

2 Decidability for Planar Sets and Pyramids

For planar set S of $\mathbb{Z}^3$, the strategy to solve [Recognition(3, $\mathscr{P}_n$, S)] is to consider the problem [Recognition(2, $\mathscr{P}_n$, S)] in the sublattice of $\mathbb{Z}^3$ containing S.

If [Recognition$(2, \mathscr{P}_n, S)$] admits no solution, neither [Recognition$(3, \mathscr{P}_n, S)$]. If [Recognition$(2, \mathscr{P}_n, S)$] has a solution P, then we have to expand P in a 3-polytope without adding any integer point and by preserving the number of faces.

2.1 Polytope's Expansion in Higher Dimensions

The following lemma is not easy to find in the state of the art may be while its dual is trivial.

Lemma 1. *Any polytope P of $\mathbb{R}^d$ with $n \geq d+2$ faces is the planar section of a polytope of $\mathbb{R}^{d+1}$ with the same number n of faces (Fig. 3).*

Fig. 3. According to Theorem 1, any polytope of $\mathbb{R}^d$ with at least $d+2$ faces is the planar section of a polytope of $\mathbb{R}^{d+1}$ with the same number of faces.

Proof. Although its dual formulation is not far to be trivial, this sketch of proof (Fig. 4) is necessary to follow the proof of Lemma 2.

We assume that P is of full dimension in $\mathbb{R}^d$. Otherwise, we proceed by induction. Since P has n faces, the polar polytope P^* of P is the convex hull of a finite set $A \subset \mathbb{R}^d$ of n points (we consider as origin O or pole an interior point of P) [12]. We can elevate the points of A in a set A' of full dimension in $\mathbb{R}^{d+1}$. The intersection of the convex hull of A' with the vertical line passing through the origin is not reduced to a point (since the interior of the projection of the convex hull of A contains the origin). It's a segment. If it does not contain the origin, we can translate A' to obtain this property. It provides a polytope $(P^+)^*$. Its polar P^+ is a polytope with n faces and P as planar section (Fig. 4).   □

There is n degrees of freedom which allow to build an expanded polyhedron with complementary constraints. We express it in the following lemma which holds in general dimension but is given here for $d = 2$.

Lemma 2. *Given a polytope $P \subset \mathbb{R}^2$ with n non parallel faces ($n \geq 4$), embedded in the plane $z = 0$ and a point X with $z_X \neq 0$, there exists a pyramid P^X with n faces, X as apex, a basis as close as we want from P and P as planar section of height $z = 0$.*

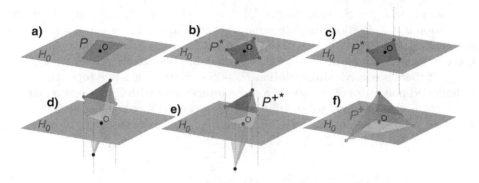

Fig. 4. In (a), we consider a d-polytope with n faces and $n \geq d+2$. In (b), we introduce its polar polytope P^* for instance with respect to the barycenter of P. In (c), we introduce the elevation lines where we can move the vertices of P^*. In (d), we provide a vertical expansion of P^* but the origin is not in the $d+1$-polytope. In (e), we translate it vertically so that the origin enters in its interior. In (f), we obtain a $d+1$-polytope $(P^+)^*$ whose polar polytope has n faces and P as planar section.

Proof. The construction follows from the proof of Lemma 1. The property that P has no parallel faces means that there are no collinear vertices in its polar polytope. We denote A the set of the vertices of P^*. According to Carathéodory Theorem, we can remove a vertex of A (it remains at least 3 vertices) and provide a reduced set of vertice A' with the origin in its convex hull. The condition of non collinearity of the points of A guarantees that the origin is in the interior of the convex hull of A'. Then we elevate the points of A' at level $+1$ and send the point of $A \setminus A'$ with a negative height h. This elevation puts the points in the position of the vertices of a pyramid $P^{*h} \subset \mathbb{R}^3$. There is a limit value h_0 for which the origin enters in the pyramid P^{*h}. Its polar polyhedron with $h < h_0$ is a pyramid P^h with the prescribed horizontal section. We can change the apex of P^h with linear transformations preserving the horizontal plane and send it to X. And as h is tending to $-\infty$, the lower face of the solution P^h -its basis- is becoming closer and closer and as close as we want from P. $\qquad\square$

2.2 Proof of Decidability for Planar Sets

We prove the decidability of [Recognition($3, \mathscr{P}_n, S$)] for planar subsets of $\mathbb{Z}^3$:

Lemma 3. *If S is a finite planar set of $\mathbb{Z}^3$ and $n \geq 4$, [Recognition($3, \mathscr{P}_n, S$)] is decidable.*

Notice that polyhedra with only $n = 2$ faces and a finite intersection with the lattice $\mathbb{Z}^3$ exist but they are a bit pathological (they are planes with irrational normal directions and an intersection with the lattice reduced to a point).

Proof. If S is reduced to a point or to a segment, the answer is positive for $n \geq 4$. Then we assume that the dimension of S is 2. We consider the sublattice of a support plane H_S of S. According to [5], [Recognition($d, \mathscr{P}_n, S$)] is decidable in dimension $d = 2$. It allows to determine whether there exists a two dimensional polyhedron $P \subset H_S$ with n faces separating S from the other integer points. If there is no polyhedron of $\mathbb{R}^2$ solution in the sublattice, there exists no solution in $\mathbb{R}^3$. If there is a polyhedron solution P, as S is finite, P is a polytope (the only unbounded polyhedra of $\mathbb{R}^2$ having a finite intersection with $\mathbb{Z}^2$ have necessarily a recession cone reduced to an irrational direction. It's only possible if S is a singleton). Then Lemma 1 allows to expand it in dimension 3. It provides a solution of [Recognition($3, \mathscr{P}_n, S$)]. □

2.3   Proof of Decidability for Pyramids

We prove the decidability of [Recognition($3, \mathscr{P}_n, S$)] for pyramids of $\mathbb{Z}^3$ of lattice height 1:

Lemma 4. *If S is an pyramid of $\mathbb{Z}^3$ of lattice height 1, [Recognition($3, \mathscr{P}_n, S$)] is decidable (we assume $n \geq 4$).*

Proof. Up to an unimodular affine isomorphism preserving $\mathbb{Z}^3$, we can assume that S is the union of a basis B in the horizontal plane $z = 0$ and the point $(0, 0, 1)$. We decompose the pyramid in its basis that we denote B and its vertex y. The basis B is a planar set embedded in a plane H_B. As for Lemma 3, we consider the problem [Recognition($2, \mathscr{P}_n, B$)] in the sublattice of the plane $z = 0$. If [Recognition($2, \mathscr{P}_n, B$)] is not feasible, neither [Recognition($3, \mathscr{P}_n, S$)]. Conversely if [Recognition($2, \mathscr{P}_n, B$)] has a solution P, we have to consider the parallelism of the faces of P in order to provide a three dimensional solution.

If P has no parallel faces, then Lemma 2 allows to expand with a finite number of computations P in a pyramid having $(0, 0, 1)$ as apex and a basis as close as we want from the plane $z = 0$. It allows to provide solutions whose only integer points are $(0, 0, 1)$ and the ones of B. In other words, the expansion does not introduce any new integer point in the polytope. It proves that [Recognition($3, \mathscr{P}_n, S$)] is feasible.

If P has parallel faces, we can perturb it to avoid the difficulty that it occurs. By definition, a solution P of [Recognition($2, \mathscr{P}_n, B$)] contains all the points of B and none point of $\mathbb{Z}^2 \setminus B$. As it is compact, its minimal distance to $\mathbb{Z}^2 \setminus B$ is strictly positive. It means that there exists an $\epsilon > 0$ for which P and $(1 + \epsilon)P$ are both solutions of [Recognition($2, \mathscr{P}_n, B$)]. The space between P and $(1+\epsilon)P$ allows to move vertices and break the parallelism of the faces (Fig. 5).

With a finite number of computations, we have reduced [Recognition($3, \mathscr{P}_n, S$)] to [Recognition($2, \mathscr{P}_n, B$)] which is a decidable problem [6]. □

3   Decidability for Marquees

The approach to prove the decidability of [Recognition($3, \mathscr{P}_n, S$)] for marquees can not be done just by considering its basis. Up to an affine isomorphism of $\mathbb{Z}^d$ sending S in a reference position, we can assume that the set S is covered by the

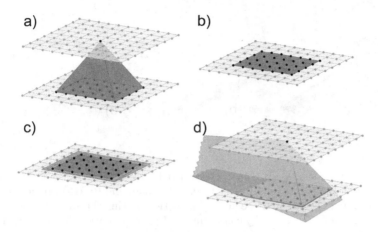

Fig. 5. In (a), we consider a pyramid as instance of [Recognition(3, $\mathscr{P}_4, S$)] of basis B. In (b), we provide a two-dimensional solution P of [Recognition(2, $\mathscr{P}_4, B$)] but its faces are parallel, which does not allow to use Lemma 2. In (c), we can pertub P in P' in order to break the faces parallelism by remaining between P and $(1+\epsilon)P$. Then Lemma 2 allows to expand P'. The only new integer point in the expanded polyhedron is the apex of the pyramid.

two consecutive planes $z = 0$ and $z = 1$ and that the section S_1 of S is a segment containing at least two integer points on the line passing through $(0,0,1)$ in the direction x. The section of S in the lower plane $z = 0$ is denoted S_0. We are interested in the width $\text{width}_y(S_0)$ of S_0 in the y direction. We decompose the proof of the decidability of [Recognition(3, $\mathscr{P}_n, S$)] for marquees according to the value of the width $\text{width}_y(S_0)$:

- if $\text{width}_y(S_0) = 0$, the marquee S is a planar set (previously solved).
- if $\text{width}_y(S_0) = 1$, the basis of the marquee is reduced to two consecutive segments in the x direction. It is a particular case to which we refer as a *prism* (Fig. 6).
- if $\text{width}_y(S_0) \geq 2$, we have a general case which requires some specific work.

3.1   Decidability for Prisms

The problem for prisms is particular because there are three lines of lattice jewels (the three lines in the x direction passing through the points $(0, -1, 1)$, $(0, 1, -1)$ and $(0, 1, 1)$) but it is easy to solve. They are intersections of a tetrahedron with the lattice:

Lemma 5. *If $S \subset \mathbb{Z}^3$ is a prism namely a finite set unimodularly equivalent to the union of three segments in the x direction passing through the three points $(0,0,0)$, $(0,1,0)$ and $(0,0,1)$, then [Recognition(3, $\mathscr{P}_n, S$)] is feasible for any $n \geq 4$.*

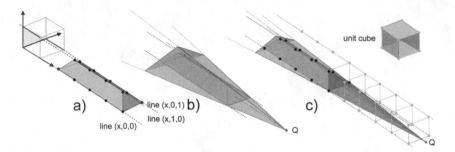

Fig. 6. In (a), we consider a prism S contained by three lines in the x direction and passing through the points $(0,0,0)$ (red), $(0,1,0)$ (blue) and $(0,1,0)$ (green). In (b) we introduce a vertex Q (red) on a line in the x direction passing through the interior of the prism. We build a tetrahedron T_Q containing S. In (c), we notice that by construction, the spike (the right part of the tetrahedron in red) can not contain any integer point. By choosing Q far enough, we can guarantee that the tetrahedron does not contain other integer points than the ones of the prism. (Color figure online)

Proof. Notice that the notion of prism has here a very precise meaning. We introduce a vertex Q on a line in the x direction crossing the interior of the prism. Then we consider the tetrahedron containing S, with Q as apex and the plane of the opposite face of Q in the prism as basis (Fig. 6). We choose a first position of Q_0. The corresponding tetrahedron T_{Q_0} might contain a finite number of unwanted integer points (the important point is here that it is finite). Then we push Q far enough to remove these integer points from the tetrahedron T_Q. The key-point is that by driving away the point Q in the x direction, the spike is increasing but by construction, it does not contain any integer point. The rear part of the tetrahedron is decreasing. It means that by choosing Q far enough, we can exclude all the unwanted integer points from T_Q.

3.2 Strategy for General Marquees

It remains to establish the decidability of [Recognition$(3, \mathscr{P}_n, S)$] for the general marquees.

Lemma 6. *If $S \subset \mathbb{Z}^3$ is a marquee and not a prism, then [Recognition$(3, \mathscr{P}_n, S)$] is decidable.*

Before sketching a proof of Lemma 6, let us consider the particular case where the basis S_0 is of dimension 1. The marquee is made of two segments whose convex hull is a tetrahedron. In this case, the convex hull of S is a solution of [Recognition$(3, \mathscr{P}_n, S)$] for any $n \geq 4$. In the remaining case, a first result provides a localization of an infinite set of the lattice jewels of S (Fig. 7).

Lemma 7. *For a finite marquee $S \subset \mathbb{Z}^3$ which is not a prism, with a non degenerated basis and placed in the reference position, we denote $T^- = \{(k, -1, 1) | k \in \mathbb{Z}\}$ and $T^+ = \{(k, 1, 1) | k \in \mathbb{Z}\}$ (Fig. 7). We have two properties:*

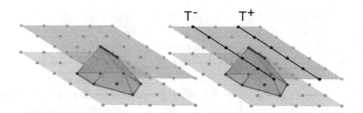

Fig. 7. On the left a marquee which is not a prism and on the right, the sets of points T^- and T^+ are the two main sets of lattice jewels (with a finite number of other integer points not colored here).

- For any $k \in K$, the points $(k, 1, 1)$ and $(k, -1, 1)$ are lattice jewels of S.
- The set of the other jewels $T^0 = \text{jewels}(S) \setminus (T^- \cup T^+)$ is finite.

The proof of Lemma 7 is based on the same arguments of compacity than the ones used in [5] but due to the lack of space, it is absent from the paper.

Let us prove now Lemma 6 in the case of a basis S_0 of dimension 2:

Proof. The approach provided in [5,6] allows to reduce [Recognition$(3, \mathscr{P}_n, S)$] to [PolyhedralSeparability$(3, n, S, \text{jewels}(S))$]. With Lemma 7, we rewrite it [PolyhedralSeparability$(3, n, S, T^0 \cup T^- \cup T^+)$]. Then the strategy is to process differently with the constraints coming from S and T^0 than for the ones excluding the points of T^- and T^+.

As T^0 is finite, the problem [PolyhedralSeparability$(3, n, S, T^0)$] is decidable. The first key point is to decompose [PolyhedralSeparability$(3, n, S, T^0)$] in n instances [PolyhedralSeparability$(3, n, S, T_i)$] with an index i going from 1 to n where the sets T_i define a partition of T. We notice that any solution P of [PolyhedralSeparability$(3, n, S, T^0)$] is the intersection of n half-planes H_i respectively solutions of some instances [PolyhedralSeparability$(3, n, S, T_i)$] where the sets T_i define a partition of T. Conversely, since any solution can be decomposed in this way, our strategy is to consider all the partitions of T^0 in sets T_i. Given such a partition, each one of the n half-space H_i has to be chosen in a set of half-spaces K_i defined by the linear constraints expressing [PolyhedralSeparability$(3, n, S, T_i)$]. By denoting $a_i x + b_i y + c_i z \leq h_i$ an equation of the half-space H_i, the set K_i is a convex cone defined by the linear inequalities $a_i u + b_i v + c_i w \leq h_i$ where (u, v, w) is in S and $a_i u' + b_i v' + c_i w' > h_i$ where (u', v', w') is in T_i. Choosing the n half-spaces H_i in K_i guarantees that their intersection contains S and no point of T^0. The separation from S and T^0 being already taken into account, it remains to add the constraints of exclusion of the points of T^- and T^+.

The restriction of H_i to the lines $z = 1$ and $y = \delta$ with $\delta = \pm 1$ (these two lines contain respectively T^- and T^+) is given by the linear inequalities $H_i : a_i x + b_i \delta + c_i \leq h_i$. Our task is to determine coefficients a_i, b_i, c_i and h_i in each K_i so that no integer $x \in \mathbb{Z}$ with $\delta = +1$ or -1 satisfies the n conditions.

The sets K_i are polyhedral cones in the space of dimension 4 of coordinates (a_i, b_i, c_i, h_i). They can be described by their three sections by the hyperplanes

$a_i = 1$, $a_i = 0$ or $a_i = -1$. The section of K_i with the hyperplane $a_i = \alpha$ is denoted K_i^α with $\alpha = -1$ or 0 or $+1$.

For $a_i = 1$, the linear inequality becomes $H_i : x \le h_i - b_i \delta - c_i$ with linear constraints on the coefficients h_i, b_i and c_i. For $a_i = 0$, we have $H_i : 0 \le h_i - b_i \delta - c_i$ and for $a_i = -1$, $x \ge b_i \delta + c_i - h_i$. Our problem is to decide if we can choose the coefficients a_i equal to -1, 0 or 1 and the coefficients h_i, b_i and c_i so that we can exclude all the points of T^- and T^+ namely all the integers x with $\delta \in \{-1, 1\}$. We can decompose again the problem in the following questions:

1. given K_i, does there exist an half-space in K_i^0 excluding all the points of T^+? of T^-? of both?
2. given K_i^α and $K_j^{-\alpha}$ with $i \ne j$, does there exist a pair of half-spaces in $K_i^\alpha \times K_j^{-\alpha}$ excluding all the points of T^+? of T^-? of both? (α and $-\alpha$ because their orientation in the direction x should not be the same)?
3. given K_i^α, $K_j^{-\alpha}$ and $K_{j'}^{-\alpha}$ with different indices, does there exist a triplet of half-spaces excluding T^- and T^+?
4. same questions with a pair excluding T^+ and a pair excluding T^-, but it can be reduced to the second question.

There is no reason to increase the size of the tuple considered in these questions, because if three intervals of the form $]\infty, \beta]$ and $[\gamma, +\infty[$ and $[\gamma', +\infty[$ are excluding the integers, one of them is redundant. Then if we cannot exclude T^+ (or T^-) with two half-spaces, we cannot exclude them at all.

The case 1 is solved by comparing $h_i - b_i \delta - c_i$ to 0 with $\delta = +1$ for T^+ and $\delta = -1$ for T^-.

Let us focus now on the case 2. The equation of H_i can be rewriten $x \le -b_i \delta + d_i$ with $d_i = h_i - c_i$ for $H_i \in K_i^{+1}$ and $x \ge b_j \delta - d_j$ for $H_j \in K_j^{-1}$ with $d_j = h_j - c_j$. By replacing the coordinates c and h by the coordinate $d = h - c$, we proceed to a projection of the convex sets K_i^α. Its image by this projection in the space of parameters (b, d) is denoted $K_i'^\alpha$. It is a two-dimensional convex set described by a finite number of inequalities which can be obtained from the inequalities characterizing K_i^α by Fourier-Motzkin elimination.

We notice now that two constraints issued from K_i^{+1} and K_j^{-1} exclude T^+ if and only if there is no integer x verifying $b_j \delta - d_j \le x \le -b_i \delta + d_i$ for $\delta = +1$. We can determine the existence of such pair of points $(b_i, d_i) \in K_i'^{+1}$ and $(b_j, d_j) \in K_j'^{-1}$ by computing the maximum $\max_{+i}'^{+1}$ and $\max_{+j}'^{-1}$ of $b - d$ for $(b_i, d_i) \in K_i'^{+1}$ and $(b_j, d_j) \in K_j'^{-1}$. It follows that T^+ can be excluded by a pair of constraints coming from K_i^{+1} and K_j^{-1} if and only if the interval $[\max_{+j}^+, -\max_{+i}^+]$ does not contain any integer. This last question can be solved by Linear Programming (a similar approach holds for T^- with $-b - d$ instead of $b - d$) (Fig. 8).

We end the proof with the case 3. In order to determine whether three constraints coming from $K_i'^{+1}$, $K_j'^{-1}$ and $K_{j'}'^{-1}$ can exclude T^- and T^+, we use Linear Programming in the same manner. We compute again the maximum $\max_{+j}'^{-1}$ of $b - d$ for $(b_j, d_j) \in K_j'^{-1}$ and the maximum $\max_{-j'}'^{-1}$ of $-b - d$ for $(b_j, d_j) \in K_j'^{-1}$. Then, we determine whether the set $-K_i'^{+1}$ has

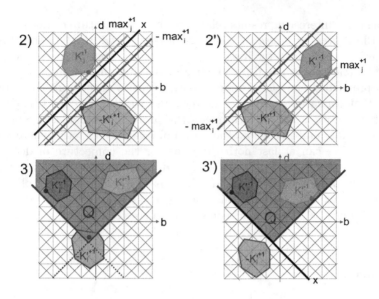

Fig. 8. In the cases (2) and (2'), excluding all the outliers of T^+ with two half-spaces in K_i^{+1} and K_j^{-1} is equivalent with finding a pair of points $(b_i, d_i) \in -K_i'^{+1}$ and $(b_j, d_j) \in K_j'^{-1}$ with no integer x verifying $b_j - d_j \leq x \leq -b_i + d_i$. It's not possible in (2) because there is a diagonal line $b - d = x \in \mathbb{Z}$ with $K_j'^{-1}$ above and $-K_i'^{+1}$ below. It is possible in (2). This possibility is determined by the extreme points in the diagonal direction namely the result of the comparison of $\lceil \max_{+j}^+ \rceil$ and $-\max_{+i}^+$. In the cases (3) and (3'), we deal with the possibility to exclude all the outliers of T^- and T^+ with only three constraints coming from the sets K_i^{+1}, K_j^{-1} and $K_{j'}^{-1}$. We can compute a quadrant Q determined by the extreme points of $K_j'^{-1}$ and $K_{j'}'^{-1}$ and determine its intersection with $K_i'^{+1}$. The outliers of $T^- \cup T^+$ can all be excluded in this manner iff the intersection is non empty (as in (3) and not in (3')).

a non empty intersection with the quadrant $Q = \{(b_i, d_i) \in \mathbb{R}^2 | -b_i + d_i < \lceil \max_{+j}^{-1} \rceil$ and $b_i + d_i < \lceil \max_{+j}^{-1} \rceil\}$. It can be done with a linear program. $\square$

3.3 Perspectives

We have proved the decidability of the recognition of digital polyhedra [Recognition$(3, \mathscr{P}_n, S)$] in dimension 3. This result is weak and the three cases considered in the paper can be considered as marginal. They are not because they require to deal with an infinite number of irreducible constraints but we have proved that by using their geometry, it is possible to decide in a finite time.

The reader can however believe that this standalone approach is not appropriate and that some known results coming from the lattice polytope's theory allow to prove stronger results with less work. The problem [Recognition$(d, \mathscr{P}_n, S)$] is a kind of converse of Integer Linear Programming but the idea that some kinds of ILP approaches could avoid the difficulty requires more than an intuition.

One of the most interesting results related with the conjecture could be the existence of the *finiteness threshold width* [2]: for larger width than the threshold denoted $w^\infty(d)$ (we have for instance $w^\infty(3) = 1$), there exists only a finite number of lattice polytopes (up to lattice preserving affine isomorphisms) containing a prescribed number of integer points. Such a deep result could be used to prove that for lattice polytopes Q verifying width$(Q) > w^\infty(d)$, the number of lattice jewels is finite which makes the problem decidable. It remains the mystery of what happens below the threshold namely exactly where infiniteness occures. We can at last notice that as for the conjecture of decidability of [Recognition$(d, \mathscr{P}_n, S)$] which is unsolved in dimension $d \geq 4$ for hollow polytopes, the infiniteness threshold width is also related with hollow polytopes... It can explain that for this specific class of objects, the decidability of [Recognition$(d, \mathscr{P}_n, S)$] which seems to be a so weak question remains challenging.

References

1. Asano, T., Brimkov, V.E., Barneva, R.P.: Some theoretical challenges in digital geometry: a perspective. Discrete Appl. Math. **157**(16), 3362–3371 (2009)
2. Blanco, M., Haase, C., Hofmann, J., Santos, F.: The finiteness threshold width of lattice polytopes (2016)
3. Edelsbrunner, H., Preparata, F.: Minimum polygonal separation. Inform. Comput. **77**(3), 218–232 (1988)
4. Forchhammer, S., Kim, C.: Digital squares, vol. 2, pp. 672–674. IEEE (1988)
5. Gerard, Y.: About the decidability of polyhedral separability in the lattice z^d. J. Math. Imaging Vis. (2017)
6. Gérard, Y.: Recognition of digital polyhedra with a fixed number of faces. In: Normand, N., Guédon, J., Autrusseau, F. (eds.) DGCI 2016. LNCS, vol. 9647, pp. 415–426. Springer, Cham (2016). doi:10.1007/978-3-319-32360-2_32
7. Krishnaswamy, R., Kim, C.E.: Digital parallelism, perpendicularity, and rectangles. IEEE Trans. Pattern Anal. Mach. Intell. **9**(2), 316–321 (1987)
8. Megiddo, N.: Linear-time algorithms for linear programming in r^3 and related problems. SIAM J. Comput. **12**(4), 759–776 (1983)
9. Megiddo, N.: On the complexity of polyhedral separability. Discrete Comput. Geom. **3**(4), 325–337 (1988)
10. Nakamura, A., Aizawa, K.: Digital squares. Comput. Vis. Graph. Image Process. **49**(3), 357–368 (1990)
11. Schrijver, A.: Theory of Linear and Integer Programming. Wiley, New York (1986)
12. Ziegler, G.: Lectures on Polytopes. Graduate Texts in Mathematics. Springer, New York (1995)

Reconstructions of Noisy Digital Contours with Maximal Primitives Based on Multi-scale/Irregular Geometric Representation and Generalized Linear Programming

Antoine Vacavant[1]([✉]), Bertrand Kerautret[2], Tristan Roussillon[3],
and Fabien Feschet[1]

[1] Université Clermont Auvergne, CNRS, SIGMA Clermont, Institut Pascal,
63000 Clermont-Ferrand, France
{antoine.vacavant,fabien.feschet}@uca.fr
[2] LORIA, UMR CNRS 7503, Université de Lorraine,
54506 Vandœuvre-lès-Nancy, France
bertrand.kerautret@univ-lorraine.fr
[3] Univ Lyon, INSA-LYON, LIRIS UMR 5205, 69622 Villeurbanne, France
tristan.roussillon@liris.cnrs.fr

Abstract. The reconstruction of noisy digital shapes is a complex question and a lot of contributions have been proposed to address this problem, including blurred segment decomposition or adaptive tangential covering for instance. In this article, we propose a novel approach combining multi-scale and irregular isothetic representations of the input contour, as an extension of a previous work [Vacavant et al., A Combined Multi-Scale/Irregular Algorithm for the Vectorization of Noisy Digital Contours, CVIU 2013]. Our new algorithm improves the representation of the contour by 1-D intervals, and achieves afterwards the decomposition of the contour into maximal arcs or segments. Our experiments with synthetic and real images show that our contribution can be employed as a relevant option for noisy shape reconstruction.

Keywords: Digital shape analysis · Irregular isothetic grids · Multi-scale analysis · Decomposition into maximal arcs · Decomposition into maximal segments

1 Introduction

The representation of digital contours is an important task in image analysis applications, since binary shapes obtained by image processing algorithms (pre-processing and segmentation) may be altered by noise. A lot of efforts have been made on these algorithms to produce *smooth contours*, by developing sophisticated deblurring and denoising algorithms [14], or by integrating regularization

© Springer International Publishing AG 2017
W.G. Kropatsch et al. (Eds.): DGCI 2017, LNCS 10502, pp. 291–303, 2017.
DOI: 10.1007/978-3-319-66272-5_24

terms in segmentation process for instance [26]. However, these approaches significantly raise the computational complexity of the complete image analysis pipeline, and include other input-noise-dependent parameters to be tediously set for any new specific applications.

Hence, another approach consists in obtaining a faithful geometrical representation directly from any noisy digital contours. A lot of research works have addressed this question by *fitting parametric curves* (*e.g.* B-splines, rational Gaussian curves) to the input points [4,8,10]. These approaches require a parameter depending on input noise scale, in order to fit the objective function at best. In general, they do not use the fact that digital points belong to $\mathbb{Z}^2$, as this is always the case in the image plane.

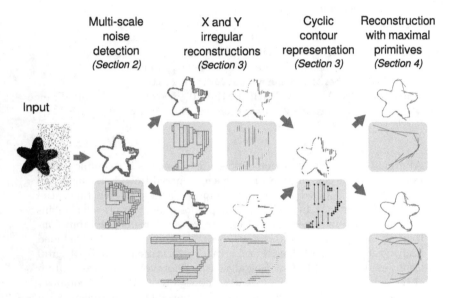

Fig. 1. Global pipeline of our approach. Input: a noisy contour. Output: maximal geometric primitives. Stages are: 1- Extraction of a multi-scale representation (unsupervised geometric noise detector); 2- Irregular isothetic representation (non overlapping cells) in X and Y directions; 3- 1-D intervals representation; 4- Fusion of the two directions to achieve a faithful geometric structure of the input contour.

In the digital geometry community, an important literature has been dedicated to this problem since the 80's, by representing contours with several kinds of primitives (segments, arcs of circle). Thanks to theoretical concepts designed in digital geometry, these approaches extend the scheme of *vectorization*, well-known in document analysis [3], consisting in converting pixels to line segments. In particular, some publications tackle the issue of fitting both straight segments and circular arcs to digital contours at the same time. The famous approach of Rosin and West [18] relies on least square fitting and is non parametric. Another parametric technique has been designed by Hilaire and Tombre [9], based on the notions of fuzzy digital segments and fuzzy digital arcs, and Faure and Feschet [5] by using α-thick decomposition and combinatorial optimization. All of them are robust and accurate whilst the former two suffer from a high time complexity and are restricted to one pixel wide digital curves. Since multi-primitives decomposition can be viewed as a competition between primitives, the complexity can be tackled with an efficient and unified representation of the multiple decompositions with all individuals primitives. Relying on the work of Feschet and Tougne [6], each decomposition can be represented by a circular arc-graph in linear-time. Decomposition into several primitives can be solved in $\mathcal{O}(qn)$ where q is the minimum number of intersecting primitives in the graph [2]. Other recent methods from state-of-the-art opt for different strategies, such as the adaptation of tangential cover [15] or the detection of dominant points [16].

In this article, we propose a novel *unsupervised approach* for reconstructing noisy digital contours by combining multi-scale and irregular isothetic representations as presented in Fig. 1. The complete pipeline of our approach works as follows. From an input (supposedly noisy) closed contour obtained from any image (first column), we first extract a multi-scale object, containing overlapping boxes, with an unsupervised geometric noise detector (second column). We then represent this structure by irregular isothetic objects that is with cells without any overlapping (third column). This is done by following two directions (X and Y axes) simultaneously. Then these two X and Y axis aligned boxes are represented as lists of 1-D intervals between irregular cells (fourth column). Then, we combine both intervals to achieve a faithful geometric structure of the input contour using both X and Y oriented segments (fifth column) in a unified representation of the contour. At the end (sixth column) we compute maximal primitives within this last representation using same Generalized Linear Programming approach for segments and arcs allowing us to produce decomposition of the contour into maximal straight line segments or circular arcs.

The article is organized as follows. In Sect. 2, we present the first step of our approach, aiming at representing the input contour as a multi-scale set of bounding boxes. These are then analyzed and converted into irregular isothetic structures, exposed in Sect. 3. Then, we describe the way to obtain decompositions into maximal primitives (Sect. 4), and experimental results with real and synthetic contours (Sect. 5) before concluding this article in Sect. 6.

2   Multi-scale Noise Detection

The noise level detection on a digital contour is an important problem, which can influence the quality of geometric estimators or contour representation algorithms. From the digital geometry domain, a method was proposed to automatically detect the amount of noise present on a digital structure [11]. This detection is based on the meaningful scale detection computed from asymptotic properties of the maximal segments. In particular, it is based on a theorem describing the evolution of the lengths of the maximal segments computed on the border of a shape on finer and finer grid sizes [13]. From such a multiscale analysis,

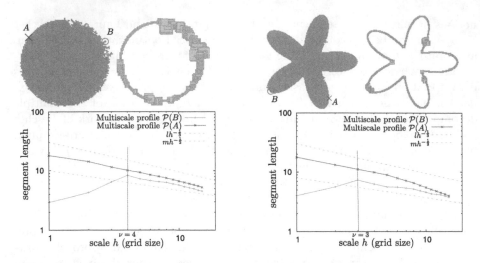

Fig. 2. Conversion of two noisy digital contours into meaningful scales (first row) and illustration of the multiscale profiles for two points A and B (second row)

the proposed algorithm consists in constructing, for each contour points C, a multiscale profile $(\mathcal{P}(C))$ defined by the segment length of all segments covering P for larger and larger grid sizes h (see graph of Fig. 2 second row). From each profile, the noise level is determined by the first scale (nu) minus one for which the slope of $\mathcal{P}$ is decreasing and follows the awaited theoretical bounds between $h^{-\frac{1}{2}}$ and $h^{-\frac{1}{3}}$ if C is on a non null curvature area and near h^{-1} on flat part. On the examples of Fig. 2, the noise levels of points A are 0 since $\mathcal{P}(A)$ is always decreasing and B for the circle (resp. flower) has a noise level of 3 (resp. 3) since $\mathcal{P}(B)$ is increasing until scale 4 (resp. 3). This uncertainty can be represented as boxes and as exposed in Fig. 2 first row, a high noise in the contour will lead to a large box, and *vice-versa*. The algorithm can be tested on-line from any digital contour given by a netizen [12].

3   Irregular Isothetic Cyclic Representation

In this section, we first recall the $\mathbb{I}$-grid (Irregular Isothetic grid) model [22]:

Definition 1 (2-D $\mathbb{I}$-grid). *Let $\mathcal{R}$ be a closed rectangular subset of $\mathbb{R}^2$. A 2-D $\mathbb{I}$-grid G is a tiling of $\mathcal{R}$ with closed rectangular cells whose edges are parallel to the X and Y axes, and whose interiors have a pairwise empty intersection. The position of each cell R is given by its center point $(x_R, y_R) \in \mathbb{R}^2$ and its length along X and Y axes by $(l_R^x, l_R^y) \in \mathbb{R}_+^{*\,2}$.*

This model permits to generalize many irregular image representations such as quad-trees, kd-trees, run-length encodings, and the geometry of frames encoded within video coding standards like MPEG, H.264, *etc.* For the rest of the article, we consider the following definitions for $\mathbb{I}$-grids.

Definition 2 (*ve-**adjacency and *e***-adjacency**). *Let R_1 and R_2 be two cells. R_1 and R_2 are ve-adjacent (vertex and edge adjacent) if:*

$$or \begin{cases} |x_{R_1} - x_{R_2}| = \frac{l^x_{R_1} + l^x_{R_2}}{2} \text{ and } |y_{R_1} - y_{R_2}| \leq \frac{l^y_{R_1} + l^y_{R_2}}{2} \\ |y_{R_1} - y_{R_2}| = \frac{l^y_{R_1} + l^y_{R_2}}{2} \text{ and } |x_{R_1} - x_{R_2}| \leq \frac{l^x_{R_1} + l^x_{R_2}}{2} \end{cases}$$

R_1 and R_2 are e-adjacent (edge adjacent) if we consider an exclusive "or" and strict inequalities in the above ve-adjacency definition. The letter k may be interpreted as e or ve in the following definitions.

A k-path from R to R' is a sequence of cells $(R_i)_{1 \leq i \leq n}$ with $R = R_1$ and $R' = R_n$ such that for any i, $2 \leq i < n$, R_i is k-adjacent to R_{i-1} and R_{i+1}.

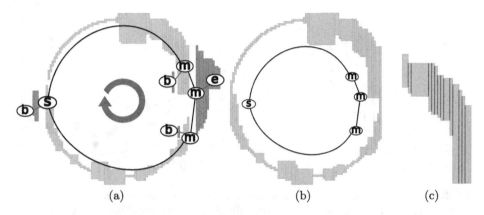

(a)   (b)   (c)

Fig. 3. From the meaningful scales of the *Circle* sample, we reconstruct a set of k-arcs converted to a k-curve thanks to the underlying graph (b). (c) presents a part of the cells obtained with intervals (red) (Color figure online)

Definition 3 (*k-curve*). *Let $A = (R_i)_{1 \leq i \leq n}$ be a k-path from R_1 to R_n. Then A is a k-curve iff each cell R_i has exactly two k-adjacent cells in A.*

As shown in Fig. 2, the meaningful boxes (denoted afterwards by the set $\mathcal{M}$) overlap and thus cannot be viewed as an irregular isothetic object directly (Definition 1). However each one contains a given number of pixels (at the initial resolution) so that the set of boxes $\mathcal{M}$ covers a subset of the input image. This subset $\mathcal{P}$, which is an irregular isothetic object, is transformed into k-arcs, *i.e.* open k-curves, and the respective adjacencies relations between arcs is represented by a Reeb graph structure [22], as illustrated in Fig. 3a [21,23]. In that graph, each edge is associated to an irregular k-arc reconstructed. This process is driven by considering a given order relation, along X or Y axis (in this figure, X axis has been chosen). With the support of the Reeb graph, we are then able to produce a cyclic representation of the contour by parsing k-arcs in a given order (*e.g.* clock-wise from the top-most element). We also remove extra branches of the graph, *i.e.* edges corresponding to k-arcs not belonging to the cycle (associated to red parts in Fig. 3a). In particular, graph edges comporting a node of degree 1 are removed. At the end of this process, we obtain a single k-curve (Fig. 3b), associated to a cyclic graph, and we consider the interface (Euclidean segment shared) between two consecutive cells in the k-curve, *i.e.* 1-D intervals.

By combining the intervals computed from both X and Y axes, we have thus two lists of segments representing the input contour, denoted by $\mathcal{S}_X$ and $\mathcal{S}_Y$, as shown in Fig. 4a. We set the internal and external points of these straight segments (black and white points in Fig. 4b) by considering the barycenter of the global shape as we did in [21].

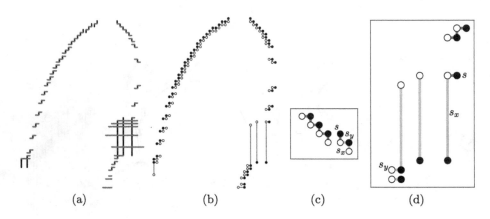

(a)   (b)   (c)   (d)

Fig. 4. A part of the two sets $\mathcal{S}_X$ (black) and $\mathcal{S}_Y$ (red) intervals from the *Flower* sample (a), and converted into a single set of intervals $\mathcal{S}_{XY}$ (b). Internal points are dotted in white, external ones in black. We also illustrate some cases of our process, with specific examples of segments s, s_x and s_y (see text) in (c) and (d). In the images, red intervals are already in place in $\mathcal{S}_{XY}$ and green ones are to be processed from $\mathcal{S}_X$, $\mathcal{S}_Y$ (Color figure online)

We then build a single list of intervals $\mathcal{S}_{XY} = \{\mathcal{S}_{XY}[i]\}_{1 \leq i \leq n}$ with X- and Y-aligned elements by first removing segments from $\mathcal{S}_Y$ intersecting one or several ones in $\mathcal{S}_X$ (see again Fig. 4a and b). The converse choice can done (*i.e.* removing segments from $\mathcal{S}_X$ overlapping some of $\mathcal{S}_Y$), nevertheless, our option leads to a faithful representation of the input contour, as exposed later in experiments.

Then, we simultaneously parse the sets $\mathcal{S}_X$ and $\mathcal{S}_Y$ to construct this list by a linear and incremental approach, according to the size of these two lists. At an iteration of this process, consider the last segment added in $\mathcal{S}_{XY}$, denoted by s, and the next segments to be added from $\mathcal{S}_X$ and $\mathcal{S}_Y$, denoted by s_x and s_y respectively. We add in $\mathcal{S}_{XY}$ the closest interval from s. As an illustration, in Fig. 4c, we add s_y, and in Fig. 4d, s_x.

During this process, adding segments of $\mathcal{S}_X$ and $\mathcal{S}_Y$ in $\mathcal{S}_{XY}$ is realized in in $\mathcal{O}(|\mathcal{S}_X| + |\mathcal{S}_Y|)$. We can observe that we build a valid list of segments in $\mathcal{S}_{XY}$, since two successive segments in the final list (not sharing the same end point) respect this condition:

$$\overrightarrow{\mathcal{S}_{XY}[i]}.\overrightarrow{\mathcal{S}_{XY}[(i+1)\bmod n]} \geq 0, \ \forall i = 1,\ldots,n, \tag{1}$$

wherein each segment is considered as a vector with the orientation given by internal and external endpoints. For instance, in Fig. 4c, $\vec{s}.\vec{s_y} = 0$ and are added successively in $\mathcal{S}_{XY}$, in Fig. 4d, $\vec{s}.\vec{s_x} = 0$, and any two consecutive parallel segments $\mathcal{S}_{XY}[i]$ and $\mathcal{S}_{XY}[i+1]$ will respect a strict positive dot product in Eq. 1. The validity of the list $\mathcal{S}_{XY}$ also means that the list of internal points are ordered in the clockwise order, and follow the curvilinear abscissa of the input contour (and this is the same for the list of external points).

4 Recognition of Straight Segments and Circular Arcs

Even if the arrangement of straight segments in $\mathcal{S}_{XY}$ is not completely random, it lacks regularity. For instance, the X-coordinates of the endpoints do not necessarily increase and for this reason, we cannot use the algorithm of O'Rourke [17] for the recognition of straight segments.

In this work, you use a general algorithm for the recognition of both straight segments and circular arcs, formulated as two instances of a *Generalized Linear Programming* (GLP) problem [1].

Our notations and definitions follow [1]. A GLP problem is a family H of constraints and an objective function ω from subfamilies of H to some totally ordered set S. In addition, H and ω must be such that:

(C1) *Monotonicity*: $\forall F \subseteq G \subseteq H, \ \omega(F) \leq \omega(G)$,

(C2) *Locality*: $\forall F \subseteq G \subseteq H$ s.t. $\omega(F) = \omega(G)$ and for each $h \in H$: $\omega(F \cup h) > \omega(F)$ iff $\omega(G \cup h) > \omega(G)$.

Note that the set S must contain a special maximal element Ω so that $G \subseteq H$ is *unfeasible* if $\omega(G) = \Omega$ and *feasible* otherwise.

In our framework, the constraint set H is given by the endpoints of the set of the n straight segments of $\mathcal{S}_{XY}$, with $n \geq 1$. Each straight segment has two endpoints: one with label "white", the other with label "black", as depicted in Fig. 4b. Let us denote the set of white (resp. black) endpoints by $P^\circ := \{p_i^\circ\}_{i=1\ldots n}$ (resp. $P^\bullet := \{p_i^\bullet\}_{i=1\ldots n}$). Let $\mathbb{P}$ (resp. $\mathbb{D}$) be the set of all possible half-planes (resp. disks). For a given $\mathbb{X} \in \{\mathbb{P}, \mathbb{D}\}$, we want to find a shape $X \in \mathbb{X}$ that contains one point set, *e.g.* P°, but not the other. In other words, we want to find $X \in \mathbb{X}$ under the constraint set $H := \{h_{2i-1}, h_{2i}\}_{i=1\ldots n}$, where

$$\forall i = 1,\ldots,n, \ h_{2i-1} := p_i^\bullet \in X, \ h_{2i} := p_i^\circ \notin X. \tag{2}$$

The problem is unfeasible if it does not exist such a X, but feasible otherwise. In the latter case, we search for X minimizing a given objective function.

For any $\mathbb{X} \in \{\mathbb{P}, \mathbb{D}\}$, there exists an objective function $\omega_\mathbb{X}$ so that (C1) and (C2) are true, which means that the above problem reduces to a GLP problem.

The objective function $\omega_{\mathbb{D}}$ is chosen to either return Ω if the problem is unfeasible or the radius of a smallest separating disk for H otherwise. By definition, the pair $(H, \omega_{\mathbb{D}})$ satisfies the monotonicity condition (C1), which coarsely says that the larger the constraint set is, the larger the smallest separating disk for this set is. In addition, since $n \geq 1$, the smallest separating disk, if it exists, is unique, which implies locality (C2).

The objective function $\omega_{\mathbb{P}}$ returns Ω if the problem is unfeasible. Otherwise, the convex hulls of the point set to enclose and the point set to not enclose are well-defined because $n \geq 1$ and do not intersect. In this case, $\omega_{\mathbb{P}}$ returns the inverse of the minimal distance between the two convex hulls. The inverse is taken so that adding a non-redundant constraint makes the objective function increase. Again, the pair $(H, \omega_{\mathbb{P}})$ satisfies conditions (C1) and (C2). As a result, depending on $\omega_{\mathbb{X}}$, we have to solve two different kinds of GLP problem.

There exists an easy-to-implement and randomized algorithm that solves these two kinds of GLP problem in expected linear-time [20]. It comes from the well-known randomized algorithm for the smallest enclosing circle problem [25]. It takes a pair $(H, \omega_{\mathbb{X}})$ and returns a basis, i.e. a minimal subfamily $B \in H$ such that $\omega_{\mathbb{X}}(B) = \omega_{\mathbb{X}}(H)$. The combinatorial dimension d of the problem is the maximum size of any basis for any feasible family. For instance, $d = 3$ for $\omega_{\mathbb{D}}$ (resp. $\omega_{\mathbb{P}}$) because at most three constraints uniquely define a disk (resp. the width between two convex polygons).

The algorithm is incremental and recursive. It may be coarsely described as follows. We iteratively add constraints. For each constraint, we check whether the new constraint violates the current basis or not. If yes, then we try to update the basis from the new constraint by recursively calling the same algorithm with all the previous constraints.

It is useful to have an on-line algorithm in order to compute the whole set of maximal segments [7] or arcs [19]. Since the original algorithm [20] is incremental, adding the constraints in order straightforwardly leads to an on-line algorithm. The drawback is that the random order can be used only in the recursive calls but not during the constraint discovery, which results to an increase of the expected time-complexity from linear to quadratic. However, we experimentally observe short running times. The next section shows results of our pipeline, employing this on-line algorithm for the reconstruction of maximal segments or arcs.

5   Experimental Results

We first present in Fig. 5 the whole set of maximal segments and arcs for the *Flower* image. Contrary to the previous work dedicated to pure vectorization [23], we do not calculate a unique polyline from a complex structure of k-arcs. Thanks to the cyclic irregular representation of the input contour, we are capable of reconstructing maximal primitives, bounded by 1-D intervals, whose lengths only depend on local input noise. Without any parameter, we obtain faithful representations of noisy shapes. Moreover, the results do not depend on any starting point, as it could be the case for other methodologies employing

(a) Maximal straight segments   (b) Maximal circular arcs

Fig. 5. Top: Decomposition into maximal segments (a) and maximal arcs (b) in a part of the *Flower* sample (green primitives), superimposed on input intervals (presented as in Fig. 4). Center and bottom: The maximal primitives are shown over original digital contour of *Flower* and *Circle* (Color figure online)

greedy algorithms. Even in the case of an high amount of local noise, our algorithm successfully reconstructs sets of primitives, as illustrated in Fig. 5 (bottom) wherein the input digital contour is significantly corrupted and contains large discontinuities and holes (top-right of the shape). As in [21], we can also obtain the circle passing through the contour, by choosing maximal arcs (Fig. 5b).

We finally present the meaningful representation, and sets of maximal straight segments and circular arcs obtained with our algorithm, for two real images, in Fig. 6 (one contour) and Fig. 7 (two contours). The *Char* image leads to a noisy contour (Fig. 6b), which is accurately represented thanks to our algorithm. Maximal primitives (Fig. 6e, f) represent the complete contour, while one of our previous contributions (d) [23], an accurate vectorization by MLP (or Minimum Length Polyline), misses a part of the object, and produces abrupt angles in round parts. The *Sign* image (350 × 350 pixels) allows us to test the scalability of our method. The contours we have extracted generate a high number of meaningful boxes (1,364 boxes for external part, 1,234 for internal part) that we have processed without any extra effort.

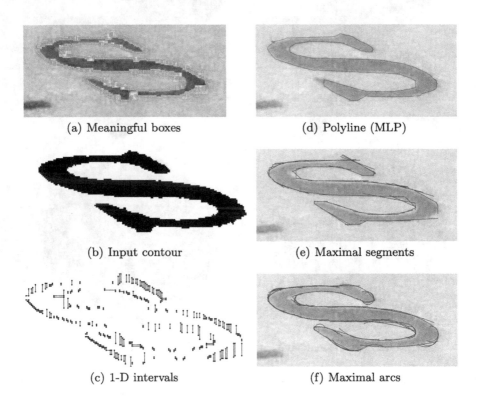

(a) Meaningful boxes   (d) Polyline (MLP)

(b) Input contour   (e) Maximal segments

(c) 1-D intervals   (f) Maximal arcs

Fig. 6. Results of our algorithm with the real image *Char* of size 185 × 85 pixels

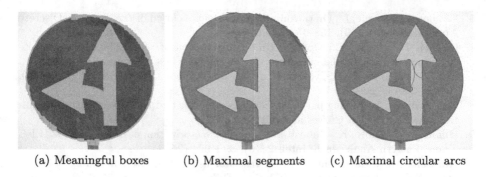

(a) Meaningful boxes   (b) Maximal segments   (c) Maximal circular arcs

Fig. 7. Results of our algorithm with the real image *Sign* of size 350 × 350 pixels

6   Conclusion and Future Works

In this article, we have proposed a novel approach combining multi-scale and irregular isothetic representations for the geometrical reconstruction of digital noisy contours. Our algorithm calculates a set of 1-D bounding intervals of the input shape, which permits to apply an on-line and incremental recognition algorithm. Our contribution has been successfully applied on synthetic and real images, encouraging us to exploit it in concrete image analysis contexts, and to investigate several lines of research.

Our first concern will consists in adapting the tangential cover approach [6] to our cyclic irregular representation. This will enable the calculation of a structure containing successive primitives, instead of overlapping maximal segments or arcs. Second, we would like to compare our contribution with other methods selected from state-of-the-art, *e.g.* [15,16], and to test their robustness [24] with challenging data-sets of binary shapes, such as KIMIA. As a longer term, we plan to investigate the more general question of reconstructing digital shapes with other geometrical primitives, like B-splines and other parametric curves, with a similar framework we have presented herein.

References

1. Amenta, N.: Helly-type theorems and generalized linear programming. Discrete Comput. Geom. **12**(3), 241–261 (1994)
2. Atallah, M.J., et al.: An optimal algorithm for shortest paths on weighted interval and circular-arc graphs. Appl. Algorithmica **14**(5), 429–441 (1995)
3. H.S., B.: Structured Document Image Analysis. Springer, Heidelberg (1992). doi:10.1007/978-3-642-77281-8
4. Bo, P., et al.: A graph-based method for fitting planar b-spline curves with intersections. J. Comput. Des. Eng. **3**(1), 14–23 (2016)
5. Faure, A., Feschet, F.: Multi-primitive analysis of digital curves. In: Wiederhold, P., Barneva, R.P. (eds.) IWCIA 2009. LNCS, vol. 5852, pp. 30–42. Springer, Heidelberg (2009). doi:10.1007/978-3-642-10210-3_3

6. Feschet, F., Tougne, L.: On the min DSS problem of closed discrete curves. Discrete Appl. Math. **151**(1–3), 138–153 (2005)
7. Feschet, F., Tougne, L.: Optimal time computation of the tangent of a discrete curve: application to the curvature. In: Bertrand, G., Couprie, M., Perroton, L. (eds.) DGCI 1999. LNCS, vol. 1568, pp. 31–40. Springer, Heidelberg (1999). doi:10. 1007/3-540-49126-0_3
8. Goshtasby, A.A.: Fitting parametric curves to dense and noisy points. In: International Conference on Curves and Surfaces (1999)
9. Hilaire, X., Tombre, K.: Robust and accurate vectorization of line drawings. IEEE Trans. Pattern Anal. Mach. Intell **28**(6), 890–904 (2006)
10. Karasalo, M., et al.: Contour reconstruction using recursive smoothing splines - algorithms and experimental validation. Robot. Auton. Syst. **57**(6–7), 617–628 (2009)
11. Kerautret, B., Lachaud, J.O.: Meaningful scales detection along digital contours for unsupervised local noise estimation. IEEE Trans. Pattern Anal. Mach. Intell. **34**(12), 2379–2392 (2012)
12. Kerautret, B., Lachaud, J.O.: Meaningful scales detection: an unsupervised noise detection algorithm for digital contours. Image Process. On Line **4**, 98–115 (2014)
13. Lachaud, J.O., Non-Euclidiens, E., d'Image, A.: Modèles Déformables Riemanniens et Discrets, Topologie et Géométrie Discrète. Habilitation à Diriger des Recherches, Université Bordeaux 1 (2006). (en francais)
14. Lebrun, M., et al.: Secrets of image denoising cuisine. Acta Numer. **21**, 475–576 (2012)
15. Ngo, P., Nasser, H., Debled-Rennesson, I., Kerautret, B.: Adaptive tangential cover for noisy digital contours. In: Normand, N., Guédon, J., Autrusseau, F. (eds.) DGCI 2016. LNCS, vol. 9647, pp. 439–451. Springer, Cham (2016). doi:10.1007/ 978-3-319-32360-2_34
16. Nguyen, T.P., Debled-Rennesson, I.: Decomposition of a curve into arcs and line segments based on dominant point detection. In: Heyden, A., Kahl, F. (eds.) SCIA 2011. LNCS, vol. 6688, pp. 794–805. Springer, Heidelberg (2011). doi:10.1007/ 978-3-642-21227-7_74
17. O'Rourke, J.: An on-line algorithm for fitting straight lines between data ranges. Commun. ACM **24**(9), 574–578 (1981)
18. Rosin, P.L., West, G.A.W.: Nonparametric segmentation of curves into various representations. IEEE Trans. Pattern Anal. Mach. Intell. **17**(12), 1140–1153 (1995)
19. Roussillon, T., Lachaud, J.-O.: Accurate curvature estimation along digital contours with maximal digital circular arcs. In: Aggarwal, J.K., Barneva, R.P., Brimkov, V.E., Koroutchev, K.N., Korutcheva, E.R. (eds.) IWCIA 2011. LNCS, vol. 6636, pp. 43–55. Springer, Heidelberg (2011). doi:10.1007/978-3-642-21073-0_7
20. Sharir, M., Welzl, E.: A combinatorial bound for linear programming and related problems. In: Finkel, A., Jantzen, M. (eds.) STACS 1992. LNCS, vol. 577, pp. 567–579. Springer, Heidelberg (1992). doi:10.1007/3-540-55210-3_213
21. Toutant, J.-L., Vacavant, A., Kerautret, B.: Arc recognition on irregular isothetic grids and its application to reconstruction of noisy digital contours. In: Gonzalez-Diaz, R., Jimenez, M.-J., Medrano, B. (eds.) DGCI 2013. LNCS, vol. 7749, pp. 265–276. Springer, Heidelberg (2013). doi:10.1007/978-3-642-37067-0_23
22. Vacavant, A., Coeurjolly, D., Tougne, L.: Topological and geometrical reconstruction of complex objects on irregular isothetic grids. In: Kuba, A., Nyúl, L.G., Palágyi, K. (eds.) DGCI 2006. LNCS, vol. 4245, pp. E1–E1. Springer, Heidelberg (2006). doi:10.1007/11907350_58

23. Vacavant, A., et al.: A combined multi-scale/irregular algorithm for the vectorization of noisy digital contours. Comput. Vis. Image Underst. **117**(4), 438–450 (2013)

24. Vacavant, A.: A novel definition of robustness for image processing algorithms. In: Kerautret, B., Colom, M., Monasse, P. (eds.) RRPR 2016. LNCS, vol. 10214, pp. 75–87. Springer, Cham (2017). doi:10.1007/978-3-319-56414-2_6

25. Welzl, E.: Smallest enclosing disks (balls and ellipsoids). In: Maurer, H. (ed.) New Results and New Trends in Computer Science. LNCS, vol. 555, pp. 359–370. Springer, Heidelberg (1991). doi:10.1007/BFb0038202

26. Wirjadi, O.: Survey of 3D image segmentation methods. Berichte des Fraunhofer ITWM, p. 23 (2007)

Discrete and Combinatorial Topology

Euclidean and Geodesic Distance Profiles

Ines Janusch$^{(\boxtimes)}$, Nicole M. Artner, and Walter G. Kropatsch

Pattern Recognition and Image Processing Group,
Institute of Computer Graphics and Algorithms, TU Wien, Vienna, Austria
{ines,artner,krw}@prip.tuwien.ac.at

Abstract. This paper presents a boundary-based, topological shape descriptor: the distance profile. It is inspired by the LBP (= local binary pattern) scale space – a topological shape descriptor computed by a filtration with concentric circles around a reference point. For rigid objects, the distance profile is computed by the Euclidean distance of each boundary pixel to a reference point. A geodesic distance profile is proposed for articulated or deformable shapes: the distance is measured by a combination of the Euclidean distance of each boundary pixel to the nearest pixel of the shape's medial axis and the geodesic distance along the shape's medial axis to the reference point. In contrast to the LBP scale space, it is invariant to deformations and articulations and the persistence of the extrema in the profiles allows pruning of spurious branches (i.c. robustness against noise on the boundary). The distance profiles are applicable to any shape, but the geodesic distance profile is especially well-suited for articulated or deformable objects (e.g.applications in biology).

Keywords: Distance profile · Shape description · Local topology · Persistence · Local binary patterns · LBP scale space · Medial axis

1 Introduction

Shape is a widely used feature to describe and distinguish objects for a variety of computer vision tasks such as image retrieval, object classification and recognition, segmentation or tracking. A shape descriptor should be invariant to transformations, distortions and occlusions of the shape. Furthermore, it is desirable for a shape descriptor to be independent of the application and to be low in computational complexity (especially for online image retrieval).

Zhang et al. [21] divide shape descriptors into two classes: region-based and boundary-based (contour-based) approaches. Each class can be further divided into structural and global approaches. Structural approaches describe shapes by segments while global approaches describe shapes as a whole.

The distance profiles presented in this paper are boundary-based descriptors for 2D shapes without holes. They allow a global description of the shape, but are also able to divide the shape into segments based on intervals of topological persistence (see Sects. 2.2 and 3.2). Two distance profiles are presented: (I) Euclidean distance profile (DP) and (II) geodesic distance profile (DP^*).

© Springer International Publishing AG 2017
W.G. Kropatsch et al. (Eds.): DGCI 2017, LNCS 10502, pp. 307–318, 2017.
DOI: 10.1007/978-3-319-66272-5_25

These profiles are based on the idea of the LBP scale space [11] (see Sect. 2), which describes a shape based on a filtration process from a chosen reference point. This filtration yields a translation- and rotation-invariant topological description of the shape.

In comparison to the LBP scale space, the distance profiles (Euclidean and geodesic) in this paper speed up the computation of the descriptor and are in addition invariant against scaling, articulation and deformation (in case of DP^*). Furthermore, they allow the pruning of spurious branches based on persistence.

1.1 State of the Art

The proposed distance profiles are mostly related to topological shape descriptors, shape signatures and spectral descriptors.

Verri et al. [20] use topological persistence for shape description in the form of size functions, which represent the persistence of the connected components. Carlsson et al. [4] propose persistence barcodes for shape description and classification. Barcodes visualize the lifetime for which features persist. For shape retrieval, shapes may be compared based on their persistence diagram using the matching distance as presented by Cerri et al. [5]. Another topological shape representation are Reeb graphs [1]. The simplest way to obtain a Reeb graph of a 2D or 3D shape is to use a height function as Morse function. In the same way persistence diagrams or barcodes can be determined using a filtration (see Sect. 2.2) based on a height function. Topological shape representations in general depend on the filtration. Height functions for example are not invariant to rotations and therefore lack in representational power.

Distance profiles are also related to shape signatures, which are one dimensional functions derived from the shape's boundary. There are different kinds of shape signatures [7,19,21]: centroidal profile, complex coordinates, centroid distance, tangent angle, cumulative angle, curvature, area and chord-length. Shape signatures are sensitive to noise on the boundary. Small changes in the boundary may cause large errors when matching the shapes (e.g. image retrieval). Considering the persistence of extrema, it is possible to filter out noise (i.e. spurious branches of the medial axis) and make distance profiles invariant to noise on the boundary.

Spectral descriptors, such as Fourier descriptors (FD) [6,9,22] and wavelet descriptors (WD) [18], are another kind of boundary-based shape descriptors, which are usually derived from a spectral transform on a shape signature. They overcome the problem of noisy boundaries by analyzing shapes in the spectral domain. Spectral descriptors are derived from spectral transforms on one dimensional shape signatures.

1.2 Overview of the Paper

Sect. 2 recalls the LBP scale space and its persistence. Section 3 presents the proposed distance profiles and their local extrema. Furthermore, Sect. 3 explains how persistence is defined on the distance profiles. First experiments on the distance profiles of discrete shapes are presented in Sect. 4. Conclusions are given in Sect. 5.

2   Recall: LBP Scale Space

Originally, LBPs were introduced for texture classification in 1996 [14]. A LBP describes the local texture around a pixel $p = (x, y)$ by a bit pattern BP. This bit pattern, results from a comparison of the grayvalue of the center pixel $g(p)$ and the grayvalues $g(q_i)$ along a circle sub-sampled with P points $q_i, i = 1, \ldots, P$:

$$BP(q_i) = \begin{cases} 1 & \text{if } g(p) - g(q_i) \geq 0 \\ 0, & \text{otherwise} \end{cases} \tag{1}$$

Two parameters determine the computation of a LBP: r defines the radius of the circular neighborhood around p, and $P = |BP|$ fixes the number of sampling points along the circle, i.e. the number of bits $BP = (s_1, s_2, \ldots, s_P)$ [15].

2.1   Describing Local Topology with LBPs

LBPs can be employed to describe the local topology of a binary image region around a reference point p. Based on the bit pattern *LBP types* can be defined by counting the number of transitions from 0 to 1 and vice versa. A transition equals an intersection b_i of the boundary B with the LBP circle of radius r at position p: $B \cap c(p, r)$. Following topological types can be derived from LBPs:

(local) maximum: no transition, bit pattern contains only 0 s and $g(p) = 1$;
(local) minimum: no transition, bit pattern contains only 1 s and $g(p) = 0$;
plateau: no transition, bit pattern contains only 1 s and $g(p) = 1$;
slope: two transitions (compare uniform patterns [15]);
saddle point: four or more transitions [10].

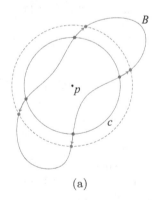

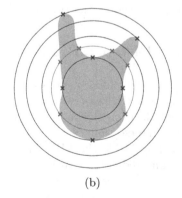

(a)   (b)

Fig. 1. (a) Increasing radius r shifts intersections along the boundary B (blue arrows). (b) LBP scale space in the continuous case – critical points marked. (Color figure online)

2.2 LBP Based Persistence

The persistence of a feature (e.g. LBP type) – its lifetime, is measured by the filtration of a space, which is "a nested sequence of subspaces that begins with the empty and ends with the complete space" [8, p. 5].

Filtration Based on LBPs: One possibility to perform a filtration using LBPs is by varying the radius r of the LBP computation for a fixed reference point p and a boundary B [11]. Varying r corresponds to a movement along B (blue arrows in Fig. 1a). This filtration process changes the intersection points of the circle c with the boundary B, i.e. the bit patterns BP (number of transitions) and thus influences the topological LBP types.

The persistence of an LBP type is measured by these $2n$ intersection points, which divide the shape in $n+1$ regions and the boundary in $2n$ segments (persistence diagram: "birth" for each region/segment). By increasing r the intersection points move along B. Once two intersection points coincide the LBP type changes (persistence diagram: "death" of the respective segment/region).

2.3 LBP Scale Space

The *LBP scale space* was proposed as shape descriptor in [11]. From a chosen reference point p inside a shape, a filtration based on LBPs is performed. The filtration may start with a radius $r = 0$, which is increased according to a predefined sampling scheme (for discrete case: r increased by 1 covers all integer radii). For the shape descriptor, the number of transitions observed for each of the LBP radii is stored, i.e. the changes in the local topology. Figure 1b illustrates the LBP scale space for sampling at critical points.

Besides applications in classification or recognition, the LBP scale space enables the reconstruction of a shape when extended by polar coordinates [11].

3 Distance Profiles

The brute force computation (see Fig. 1) of the LBP scale space starts with a small circle for which the intersections with the shape's boundary are computed. Then the radius of the LBP circle is increased and the computation of the intersections is done again. This process is repeated until the shape is completely inside the LBP circle and no more intersections can be observed for a connected shape. Such an implementation of the LBP scale space is computationally expensive. Hence, this paper proposes an alternative and more efficient way to compute the LBP scale space based on the definition of distance profiles.

First, the distances d of all points b_i along the shape's boundary B to the reference point p are computed:

$$\mathrm{DP}(b_i, p) = ||b_i - p||, \ b_i, p \in \Re^2 \tag{2}$$

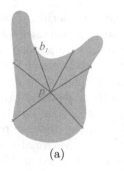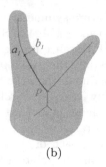

(a)   (b)

Fig. 2. LBP scale space computation using the (a) Euclidean distance profile DP and (b) geodesic distance profile DP^*. (Color figure online)

We call $DP : B \mapsto \Re^+$ the *Euclidean distance profile* for each $b_i \in B$ (see Eq. (2) and Fig. 2a). The LBP scale space [11] corresponds to this Euclidean distance profile DP. By changing the metric and using the medial axis of the shape S, a *geodesic distance profile* $DP^* : B \mapsto \Re^+$ can be defined in two parts: (I) the distance of a point $b_i \in B$ to the closest point a_l of the shape's medial axis $MA(S)$ plus (II) the geodesic distance along this axis from a_l to the reference point p:

$$DP^*(b_i, p) = ||b_i - a_l|| + \text{arclength}(a_l, p), p \in MA(S) \tag{3}$$

A visualization of the geodesic distance profile DP^* is given in Fig. 2b.

Note that a connected medial axis MA is essential for this approach. Therefore, we assume that a connected medial axis can be derived for a given binary shape and do not focus on the computation of the MA itself. Furthermore, the input of the proposed method is currently limited to binary shapes without holes.

3.1   Local Extrema Along a Distance Profile

Let $DT : S \subset \Re^2 \mapsto \Re^+$ denote the Euclidean distance transform [16] of the shape. It assigns to each point $p \in S$ inside the shape the radius of the largest inscribed circle with radius $DT(p)$, which touches the shape's boundary B at a boundary point $b_i : ||p - b_i|| = DT(p)$. Let p be the LBP scale space center.

Lemma 1. *If the reference point $p \in S$ is inside the shape S, then there is at least one local minimum along the distance profile.*

Since $p \in S$, $DT(p)$ is the radius of a circle touching B. Increasing the (maximal) circle at p would cross the boundary assigning larger distance values to neighbors of b_i. Hence, the touching point is a local minimum of DP.

For the geodesic distance profile DP^* we consider the medial axis $MA \subset S$: every circle with center on the MA touches the shape's boundary B at two or more points.

Leyton's curvature-symmetry duality relates each local curvature maximum with an end point $a_e \in MA$ [12]. Such a maximum in curvature also produces a maximum in both distance profiles for reference points p along the MA-branch of a_e. This is generally the case if p is located inside the shape farther away from the curvature maximum than the radius of the osculating circle.

The maximally inscribed circle at a branching point of MA touches the boundary of the shape in at least as many points as there are branches. If the branching point is taken as the reference point for the distance profile, it shows a local minimum at each of these touching points. Maximal circles touch the shape's boundary in two points, if their centers are located at MA of the shape, but not at an end or at a branching point of the MA.

The extrema of the distance profiles may be used for shape description and representation. At these extrema the topology changes: a connected component either starts or ends at the distance associated with the local extremum for this distance profile. The LBP scale space [11] similarly describes a shape based on a sequence of changes in the topology of the shape through a filtration for a certain LBP center. The Euclidean distance profile DP and the LBP scale space representations are identical, but the computation of the DP is more efficient. In contrast, the geodesic distance profile DP^* provides a similar representation based on a different metric, which is more robust against articulations and deformations.

The number of maxima in a distance profile is determined by the number of end points of MA (i.e. the number of positive local curvature maxima), whereas the number of minima must be equal to the number of maxima along B since a minimum has to be located between every pair of maxima. This is the smallest number of extrema for a shape and it depends exclusively on the number of MA-branches.

Consequently, spurious branches of MA generate extra extrema, which can be removed by the concept of persistence (see Sect. 3.2). The Euclidean DP may contain more extrema than the geodesic DP^* due to bent branches.

3.2   Persistence Defined on the Distance Profile

As in classical persistence [8], we consider the lifetime of connected components generated by thresholding the distance profile. The corresponding space is the boundary B of the shape S and the sub-level sets of the profile function give the filtration. This corresponds to the choice of a particular radius in the LBP scale space. The transitions from 0 to 1 and vice versa in the LBP code for a certain radius correspond to the transition between the different connected components.

The extrema $E = (b_1, b_2, \ldots, b_{2M})$, $b_i \in B$ derived from DP are alternating (i.e. maximum – minimum – maximum – ...), where M is the number of maxima. The persistence P of each of the extrema b_j, $j = 1 \ldots 2M$, is defined by the smallest difference to the adjacent two extrema:

$$P(b_j) = \min\{|DP(b_{j-1}, p) - DP(b_j, p)|, |DP(b_{j+1}, p) - DP(b_j, p)|\}. \qquad (4)$$

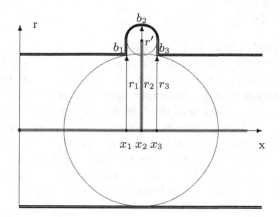

Fig. 3. A spurious branch generates non-persistent extrema. *MA* in yellow. (Color figure online)

If an extremum is close to an adjacent extremum, then this difference is small and a small modification of the threshold used for filtration would suffice to change the intersections between an LBP circle and the shape's boundary.

Figure 3 shows part of an elongated true branch (in direction x) and a typical spurious branch generated by a small bump (circle with small radius r') along the boundary of the actual branch. The bump itself (centered at position x_2) is a local maximum of the distance profile (we denote by b_i the boundary point corresponding to the axis point x_i)

$$DP^*(b_2) = x_2 + r_2 = (x_1 + r') + (r_1 + 2r') = (x_1 + r_1) + 3r' \qquad (5)$$
$$= DP^*(b_1) + 3r' > DP^*(b_1)$$

while the return to the main branch at x_3 is a local minimum with distance

$$DP^*(b_3) = x_3 + r_3 = (x_1 + 2r') + r_1 = (x_1 + r_1) + 2r' \qquad (6)$$
$$= DP^*(b_1) + 2r' < DP^*(b_2)$$

b_2 is clearly a local maximum and b_3 a local minimum since $DP^*(b(x)) > DP^*(b_3)$ for $x > x_3$ and the width of the branch is constant, but the geodesic distance to the reference point increases.

The difference between the two extrema b_2 and b_3 (their persistence) is related to the size of the bump, e.g. r', but independent of the width of the main branch. As in many pruning strategies for the *MA*, branches with a long axis are considered reliable while short branches often occur due to noise. Large differences between local minima and maxima of the distance profile also indicate long distances along the axis and thus long branches. A highly persistent reference point (LBP center) should induce a small number of extrema of the distance profile and favor the center of the diameter of the *MA* as primary locus.

4   Experiments

The DP and the DP^* defined in the continuous case in Sect. 3 have been implemented for a practical evaluation in the discrete case. Morphological thinning has been used as a pre-processing step to obtain a connected skeleton. The two distance profiles together with persistence applied to DP^* have been evaluated in experiments on binary shapes of the Kimi99 [17] and the Myth [2,3] dataset.

For the DP the Euclidean distance from every pixel along a shape's boundary to a fixed reference point on the skeleton of the shape is calculated. The DP^* is computed as the geodesic distance along the skeleton from the reference point to every skeleton pixel plus the medial axis radius (computed using a distance transformation) at each skeleton position. Figure 4 shows a binary input shape, the skeleton of the shape, as well as the DP and the DP^*.

Extrema of the distance profiles are determined as local minima and maxima in the ordered sequence of distances. For the DP, this sequence is the sequence of distances observed when tracing a shape's boundary in either clockwise or counter-clockwise direction. In the case of the DP^*, the ordered sequence of distances is retrieved by tracing the skeleton clockwise or counter-clockwise. Note that due to this difference in the computation the length of the sequences of distances of the two distance profiles may vary.

Figures 5a and b show the extrema marked along DP and DP^* for the binary hand shape shown in Fig. 4a. The DP shows a higher number of extrema, as it is prone to noise due to small variations along the boundary. Additional extrema in the DP can also be caused by the deformation of a shape through bending. This deformation however does not affect the extrema of DP^*. Note that no filtering based on persistence has been done for Fig. 5a and b.

These observations have further been evaluated on all shapes of the Kimia99 dataset. In total 4824 extrema were found along DP for the 99 shapes and 2162 extrema in total along DP^*. The minimum difference in the number of extrema per shape on the Kimia99 dataset between DP and DP^* is 0, however the maximum difference is 96. In average DP^* produces 27.8 less extrema than DP for every shape in the Kimia99 dataset.

This difference in number of extrema of the two distance profiles $\Delta = |DP - DP^*|$ has further been used to cluster the shapes in the Kimia99 dataset and

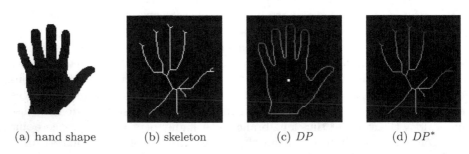

(a) hand shape   (b) skeleton   (c) DP (d) DP^*

Fig. 4. Binary input shape and obtained distances for DP and DP^*, reference point = white square, blue indicates small distances, red large distances. (Color figure online)

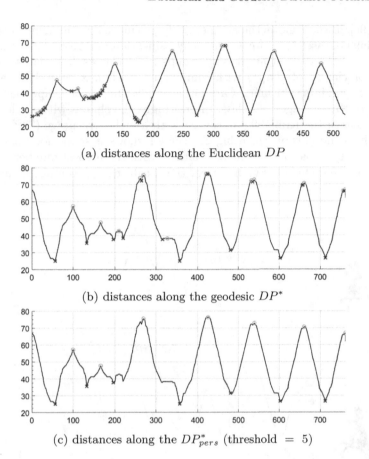

(a) distances along the Euclidean DP

(b) distances along the geodesic DP^*

(c) distances along the DP^*_{pers} (threshold $= 5$)

Fig. 5. Distance profiles (a) DP and (b) DP^* and (c) persistence (threshold $= 5$) applied to DP^*: DP^*_{pers} for the hand shape. ○ indicate maxima, × indicate minima. (Color figure online)

Table 1. Shapes in the Kimia99 dataset clustered according to the difference in number of extrema in the distance profiles $\Delta = |DP - DP^*|$.

$\Delta =$	$[0, 10]$	$[11, 30]$	$[31, 50]$	$[51, 96]$
# of images	25	32	33	10
representative image				
recall of class	100%	82%	73%	75%

to identify common features among shapes within a cluster. The dataset was partitioned into four cluster with $\Delta = [0, 10]$, $\Delta = [11, 30]$, $\Delta = [31, 50]$ or

$\Delta = [51, 96]$. Table 1 shows the number of shapes in each of these intervals. Since the Kimia99 dataset is meant for shape classification and retrieval experiments, the images of the dataset are mainly grouped into classes. The dataset consists of 6 major classes of shapes with minimum 7 and maximum 11 images per class. The remaining 40 images are either single shapes or shape classes with only two or three images per class. Interestingly, the shapes within one of these major classes of the Kimia99 dataset are with a high percentage (73% or more) also in the same cluster regarding Δ. Table 1 shows a shape for each cluster, representing the class of the Kimia99 dataset with the highest number of images in the respective cluster. Furthermore, the recall, the percentage of images of

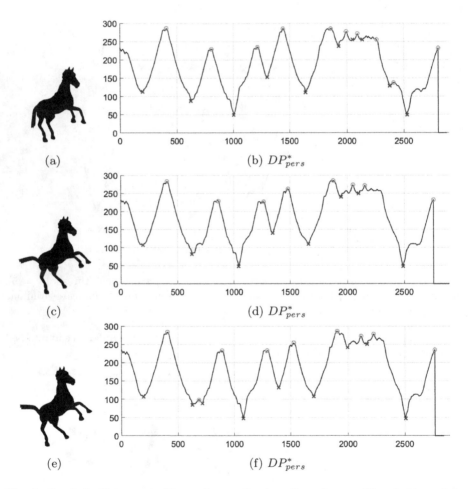

(a)   (b) DP^*_{pers}

(c)   (d) DP^*_{pers}

(e)   (f) DP^*_{pers}

Fig. 6. Geodesic distance profiles under application of persistence (threshold = 8 for the three shapes shown in the left column respectively. ○ indicate maxima, × indicate minima. (Color figure online)

each such class in the cluster to the total number of images in the class, is given in the last row of Table 1 (recall of class).

The persistence defined on the distance profile DP^* (see Sect. 3.2) has been subject to further experiments. If persistence is applied to the distance profile DP^* (DP^*_{pers}) with a very small distance threshold of 5, it further reduces the total number of extrema for all shapes in the Kimia99 dataset to 1174 DP^*_{pers} (compared to 2162 for DP^* and 4824 for DP). In average the number of extrema in DP^* is already reduced by 54% for this small threshold in DP^*_{pers}. Figure 5c shows DP^*_{pers} for the hand shape introduced in Fig. 4a.

The robustness of the geodesic distance profile with persistence based pruning DP^*_{pers} to articulated deformations is demonstrated on three shapes of the Myth dataset. The three shapes together with their respective distance profiles DP^*_{pers} are shown in Fig. 6. While legs and tail of the horse move from Fig. 6a to c and the horse is angling its torso upwards from Fig. 6c to e, the respective distance profiles given in the Figs. 6b, d and f show high similarity for all three shapes.

5   Conclusion and Future Work

This paper presented the Euclidean and geodesic distance profiles, which are consistent with Leyton's shape evolution as expressed in his *process-grammar* [13]. The distance profiles are shape descriptors based on the ideas of the LBP scale space. Both profiles are invariant against translation, rotation and scaling. The geodesic distance profile is also invariant against articulations and deformations. Local extrema along the distance profiles can be filtered with their persistence to prune spurious branches of the MA. Hence, the filtered distance profiles are robust against noise on the boundary. First experiments on the computation of the profiles result in the expected alternating sequence of local minima and maxima.

References

1. Biasotti, S., Giorgi, D., Spagnuolo, M., Falcidieno, B.: Reeb graphs for shape analysis and applications. Theor. Comput. Sci. **392**(1–3), 5–22 (2008)
2. Bronstein, A.M., Bronstein, M.M., Bruckstein, A.M., Kimmel, R.: Analysis of two-dimensional non-rigid shapes. Int. J. Comput. Vis. **78**(1), 67–88 (2008)
3. Bronstein, A.M., Bronstein, M.M., Kimmel, R.: Numerical Geometry of Non-rigid Shapes. Springer, New York (2008). doi:10.1007/978-0-387-73301-2
4. Carlsson, G., Zomorodian, A., Collins, A., Guibas, L.J.: Persistence barcodes for shapes. Int. J. Shape Model. **11**(02), 149–187 (2005)
5. Cerri, A., Fabio, B., Medri, F.: Multi-scale approximation of the matching distance for shape retrieval. In: Ferri, M., Frosini, P., Landi, C., Cerri, A., Fabio, B. (eds.) CTIC 2012. LNCS, vol. 7309, pp. 128–138. Springer, Heidelberg (2012). doi:10.1007/978-3-642-30238-1_14
6. Dalitz, C., Brandt, C., Goebbels, S., Kolanus, D.: Fourier descriptors for broken shapes. EURASIP J. Adv. Signal Process. **2013**(1), 161 (2013)

7. Davies, E.R.: Machine Vision: Theory, Algorithms, Practicalities. Elsevier, Amsterdam (2004)
8. Edelsbrunner, H.: Persistent homology: theory and practice (2014)
9. El-ghazal, A., Basir, O., Belkasim, S.: A new shape signature for Fourier descriptors. In: 2007 IEEE International Conference on Image Processing, vol. 1, pp. I-161–I-164, September 2007
10. Gonzalez-Diaz, R., Kropatsch, W.G., Cerman, M., Lamar, J.: Characterizing configurations of critical points through LBP. In: SYNASC 2014 Workshop on Computational Topology in Image Context (2014)
11. Janusch, I., Kropatsch, W.G.: Persistence based on LBP scale space. In: Bac, A., Mari, J.-L. (eds.) CTIC 2016. LNCS, vol. 9667, pp. 240–252. Springer, Cham (2016). doi:10.1007/978-3-319-39441-1_22
12. Leyton, M.: Symmetry-curvature duality. Comput. Vis. Graph. Image Process. **38**(3), 327–341 (1987)
13. Leyton, M.: A Generative Theory of Shape. LNCS, vol. 2145. Springer, Heidelberg (2001). doi:10.1007/3-540-45488-8
14. Ojala, T., Pietikäinen, M., Harwood, D.: A comparative study of texture measures with classification based on featured distributions. Pattern Recognit. **29**(1), 51–59 (1996)
15. Pietikäinen, M., Hadid, A., Zhao, G., Ahonen, T.: Computer Vision Using Local Binary Patterns. Computational Imaging and Vision, vol. 40. Springer, London (2011). doi:10.1007/978-0-85729-748-8
16. Rosenfeld, A., Pfaltz, J.L.: Sequential operations in digital picture processing. J. ACM (JACM) **13**(4), 471–494 (1966)
17. Sebastian, T., Klein, P., Kimia, B.: Recognition of shapes by editing shock graphs. In: IEEE International Conference on Computer Vision, vol. 1, p. 755. IEEE Computer Society (2001)
18. Tieng, Q.M., Boles, W.W.: Recognition of 2D object contours using the wavelet transform zero-crossing representation. IEEE TPAMI **19**(8), 910–916 (1997)
19. van Otterloo, P.J.: A Contour-Oriented Approach to Shape Analysis. Prentice Hall International Series in Acoustics, Speech & Si. Prentice Hall, Englewood (1991)
20. Verri, A., Uras, C., Frosini, P., Ferri, M.: On the use of size functions for shape analysis. Biol. Cybern. **70**(2), 99–107 (1993)
21. Zhang, D., Lu, G.: Review of shape representation and description techniques. Pattern Recognit. **37**(1), 1–19 (2004)
22. Zhang, D., Lu, G., et al.: A comparative study on shape retrieval using Fourier descriptors with different shape signatures. In: Proceedings of the International Conference on Intelligent Multimedia and Distance Education, pp. 1–9. Citeseer (2001)

Greyscale Image Vectorization from Geometric Digital Contour Representations

Bertrand Kerautret[1]([⊠]), Phuc Ngo[1], Yukiko Kenmochi[2],
and Antoine Vacavant[3]

[1] LORIA, UMR CNRS 7503, Université de Lorraine,
54506 Vandœuvre-lès-Nancy, France
Bertrand.Kerautret@univ-lorraine.fr
[2] LIGM, UMR CNRS 8049, Université Paris-Est, 77454 Marne-la-Vallée, France
[3] Institut Pascal, Université Clermont Auvergne, UMR 6602 CNRS/UCA/SIGMA,
63171 Aubière, France

Abstract. In the field of digital geometry, numerous advances have been recently made to efficiently represent a simple polygonal shape; from dominant points of a curvature-based representation, a binary shape is efficiently represented even in presence of noise. In this article, we exploit recent results of such digital contour representations and propose an image vectorization algorithm allowing a geometric quality control. All the results presented in this paper can also be reproduced online.

1 Introduction

Image vectorization is a classic problem with potential impacts and applications in computer graphic domains. Taking a raw bitmap image as input, this process allows us to recover geometrical primitives such as straight segments, arcs of circles and Bézier curved parts. Such a transformation is exploited in design softwares such as *Illustrator* or *Inkscape* in particular when users need to import bitmap images. This domain is large and also concerns document analyses and a variety of graphical objects such as engineering drawings [1], symbols [2], line drawings [3], or circular arcs [4]. Related to these applications, different theoretical advances are regularly proposed in the domains of pattern recognition and digital geometry. In particular, their focus on digital contour representations concerns various methodologies: dominant point detection [5,6], relaxed straightness properties [7], multi-scale analysis [8], irregular isothetic grids and meaningful scales [9] or curvature-based approach [10].

These last advances are limited to digital shapes (*i.e.* represented within a crisp binary image). Indeed, even if a digital contour can be extracted from a non-binary image, their adaptation to other types of images such as greyscale and color images is not straightforward. There are other methods, which are designed to handle greyscale images; for example, Swaminarayan and Prasad proposed a method to vectorize an image by using its contours (from simple Canny or Sobel filters) and by performing Delaunay triangulation [11]. Despite its interest, the

© Springer International Publishing AG 2017
W.G. Kropatsch et al. (Eds.): DGCI 2017, LNCS 10502, pp. 319–331, 2017.
DOI: 10.1007/978-3-319-66272-5_26

proposed approach does not take into account the geometrical characteristics of edge contours and other information obtained from low intensity variations. More oriented to segmentation methods, Lecot and Levy introduced an algorithm based on Mumford and Shah's minimization process to decompose an image into a set of vector primitives and gradients [12]. Inspired from this technique, which includes a triangulation process, other methods were proposed to represent a digital image by using a gradient mesh representation allowing to preserve the topological image structure [13]. We can also mention the method proposed by Demaret *et al.* [14], which is based on linear spline over adaptive Delaunay triangulation or the work of Xia *et al.* [15] exploiting a triangular mesh followed of Bézier curves patch based representation. From a more interactive process, the vectorization method of Price and Barrett [16] was designed for interactive image edition by exploiting graph cut segmentation and hierarchical object selections.

In this article, we propose to investigate the solutions exploiting the geometric nature of image contours, and apply these advanced representations to handle greyscale images. In particular, we will mainly focus on recent digital-geometry based approaches of different natures: (i) the approach based on dominant point detection by exploiting the maximal straight segment primitives [6,17], (ii) the one from the digital level layers with the algorithms proposed by Provot *et al.* [18,19], (iii) the one using the Fréchet distance [20,21], and (iv) the curvature based polygonalization [10].

In the next section (Sect. 2), we first present a strategy to reconstruct the image by considering the greyscale image intensities. Different strategies are presented with pros and cons. Afterwards, in Sect. 3, we recall different polygonalization methods. Then, we present the main results and comparisons obtained with different polygonalization techniques (Sect. 4).

2   Vectorizing Images from Level Set Contours

The proposed method is composed of two main steps. The first one is to extract the level set contours of the image. This step is parameterized by the choice of the intensity step (δ_I), which defines the intensity variations from two consecutive intensity levels. From these contours, the geometric information can then be extracted and selected by polygonalization algorithms (as described in Sect. 3). The second step concerns the vectorial image reconstruction from these contours.

Step 1: Extracting Level Set Contours and Geometrical Information
The aim of this step is to extract geometrical information from the image intensity levels. More precisely, we propose to extract all the level set contours by using different intensity thresholds defined from an intensity step δ_I. For this purpose, we apply a simple boundary extraction method defined from an image intensity predicate and from a connectivity definition. The method is based on the definition of the Khalimsky space and can work in N dimension [22]. This algorithm has the advantage to be implemented in the *DGtal* library [23] and can be tested online [24].

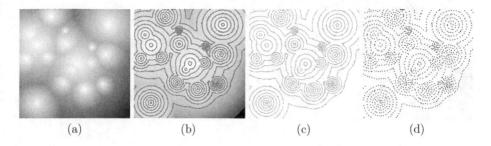

Fig. 1. Extraction of the level set contours (b, c) from the input image (a) and sampling resulting contours (d)

Figure 1 illustrates such a level set extraction (Fig. 1(b, c)) with a basic contour sampling defined from a point selection taken at the frequency of 20 pixels (Fig. 1(d)). Naturally, the more advanced contour polygonalization algorithms, as described in section Sect. 3, will be exploited to provide better contours, with relevant geometrical properties.

Step 2: Vectorial Image Reconstruction
We explore several choices for the reconstruction of the resulting vectorial images: the application of a triangulation based process, a reconstruction from sequence of intensity intervals, and a component tree representation.

Representation from Delaunay triangulation: A first solution was inspired from the method proposed by Swaminarayan and Prasad [11], which uses a Delaunay triangulation defined on the edges detected in the image. However the strategy is different since the triangulation is performed after an extraction of the main geometric features like dominant points, curvature or Fréchet distance. The first direct integration of the triangulation was performed by using the *Triangle* software [25] and by reconstructing the image with the mean intensity represented in the triangle barycenter. Though, even if such strategy appears promising, the quality is degraded if we have a closer view to the digital structure, as illustrated for the DGCI logo in Fig. 2. To obtain a better quality of image reconstruction, the level set intensity has to be integrated into the reconstruction structure in order to make a better decision for the triangle intensity value.

Representation by filling single intensity intervals: The filling of the polygonal region boundaries of the input image could be a possibility to obtain a better image reconstruction. For instance, the image in Fig. 4(a) can be reconstructed by filling inside the region $\mathcal{R}_0$ defined from its borders ($\delta^0_{\mathcal{R}_0}$, $\delta^1_{\mathcal{R}_0}$, and $\delta^2_{\mathcal{R}_0}$) and in the same way for the other regions $\mathcal{R}_1$, $\mathcal{R}_2$, $\mathcal{R}_3$, $\mathcal{R}_4$ and $\mathcal{R}_5$. However, such a strategy may depend on the quality of the polygonalization, which has no guarantee to generate the polygonal contours with correspondence between the polygon vertices (according to the border of the current region and the border of their adjacent region). For instance, the vertices of the polygonal representations

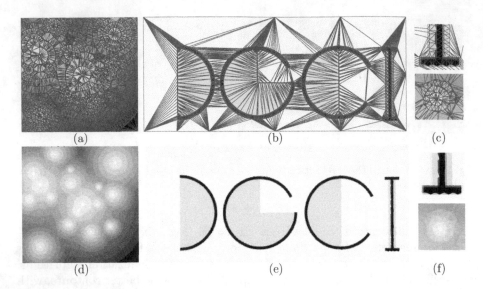

Fig. 2. Illustration of the Delaunay based reconstruction from sampled level set contours: resulting mesh representation (a–c) and final vectorial rendering (d–f)

of $\delta^2_{\mathcal{R}_0}$ and $\delta^0_{\mathcal{R}_2}$ do not necessarily share the same vertices. Such limitations are illustrated in the Fig. 3(a–e) where numerous defects are present with null-color areas which are not filled by any color.

Component tree based representation: To overcome the limitations of the previous reconstruction, we propose to exploit a mathematical morphology based representation defined on the component tree [26,27]. This method allows to represent an image by considering the successive intensity levels and by maintaining the tree representing inclusions of connected components. As an illustration, Fig. 4(d) shows such a representation with the root $\mathcal{R}_0$ associated to the first level; it contains all the pixels of the image (with values higher or equal to 0). The edge between the nodes $\mathcal{R}_4$ and $\mathcal{R}_2$ indicates that the region $\mathcal{R}_4$ is included in the region $\mathcal{R}_2$.

With such a representation, each region is -by definition- included in its anchors and as a consequence a reconstruction starting from the root region can be executed without any non-filled areas (even with poor-quality polygonalization algorithms). As shown in Fig. 3(f–j), the reconstruction results have no empty-color region.

Algorithm 1 describes the global process of the proposed method where we can use the different polygonalization methods that are detailed in the next section.

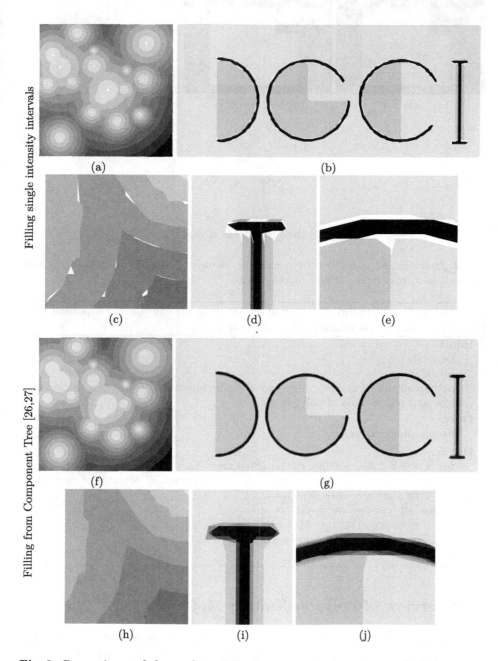

Fig. 3. Comparisons of the quality of the two reconstructions types: single intensity interval (a–e) and component tree based (f–j)

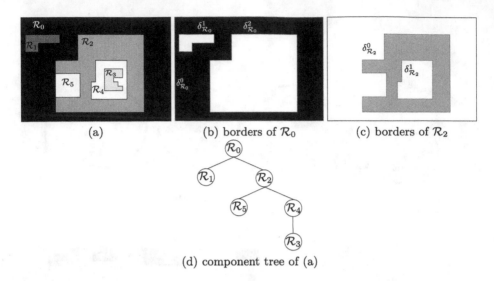

(a)   (b) borders of $\mathcal{R}_0$ (c) borders of $\mathcal{R}_2$

(d) component tree of (a)

Fig. 4. Image border extraction from a simple intensity predicate (images (a–c)) and illustration of its component tree representation

---

Algorithm 1. Algorithm to vectorize a bitmap image based on component tree representation and polygonalization algorithm.

Input

    Image2D im ▷ The input bitmap image.

    Int $iStep$ ▷ Intensity step.

    polygonalizationAlgo(cnt) ▷ Polygonalization algorithm, which takes as input a contour and return a list of polygon vertices.

Output

    VectorialImage $vectIm$ ▷ Resulting vectorial image

Begin

    for $i = 0; i < 255 - iStep; i = i + iStep$ **do**

        listCnt = EXTRACTALLCNT($0, iStep$)   ▷ Extract all contours of connected regions with pixel predicate: $I(p) >= i$

        for all contour cnt in listCnt **do**

            POLYGONALIZATIONALGO(cnt)

            ADDTOIMAGEVECTOR($vectIm, cnt$)

End

---

3   Overview of Polygonalization Methods

In the proposed vectorization method, the polygonalization step can play an important role in the resulting image quality. Before showing some results, we overview several methods with different principles.

3.1   Dominant Points Based Polygonalization

The polygonalization method proposed by Nguyen and Debled-Rennesson allows to find the characteristic points on a contour, called *dominant points*, to build a polygon representing a given contour. The method consists in using a discrete curve structure based on *width ν tangential cover* [28] (see Fig. 5(a)). This structure is widely used for studying geometrical characteristics of a discrete curve. More precisely, the width ν tangential cover of a contour is composed of a sequence of all maximal blurred segments of width ν [29] along the contour. Using this structure, a method proposed in [6,17] detects the dominant points of a contour. The key idea of the method is that the candidates to be dominant points are located in common zones of successive maximal blurred segments. By a simple measure of angle, the dominant point in each common zone is determined as the point having the smallest angle with respect to the left and right endpoints of the left and right maximal blurred segments constituting the common zone. Then, the polygon representing the contour is constructed from the detected dominant points (see Fig. 5(b)). In the next section the acronym DPP will refer to this method.

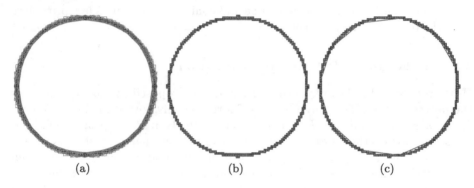

(a)   (b)   (c)

Fig. 5. (a) Width 2 tangential cover of a curve. (b) Dominant points detected on the curve in (a) and the associated polygonal representation. (c) Polygonalization result based on Fréchet distance

3.2   DLL Based Polygonalization [18, 19]

A method of curve decomposition using the notion of *digital level layers* (DLL) has been proposed in [18,19]. Roughly speaking, DLL is an analytical primitive which is modeled by a double inequality $-\omega \leq f(x) \leq \omega$. Thanks to this analytic model of DLL, the method allows a large possible classes of primitives such as lines, circles and conics.

Based on a DLL recognition algorithm, the method consists in decomposing a given curve into the consecutive DLL and, by this way, an analytical representation of the curve is obtained. Figure 6 illustrates the DLL decomposition using

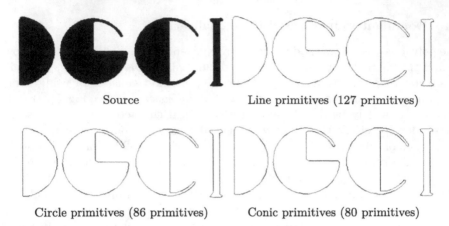

Source   Line primitives (127 primitives)

Circle primitives (86 primitives)   Conic primitives (80 primitives)

Fig. 6. Decomposing results of DLL based method using different primitives

line, circle and conic primitives. As the shape contains many line segments and arcs, the decompositions based on circle and conic primitives uses less primitives than the one with line primitives.

3.3 Fréchet Based Polygon Representation

Using the exact Fréchet distance, the algorithm proposed in [20,21] computes the polygonal simplification of a curve. More precisely, given a polygonal curve P, the algorithm simplifies P by finding the *shortcuts* in P such that the Fréchet distance between the shortcut and the associated original part of the curve is less than a given error e. As a consequence, this method guarantees a minimum number of vertices in the simplified curve according to the Fréchet distance and a maximal allowed error e (see Fig. 5(c)).

3.4 Visual Curvature Based Polygon Representation

The method was introduced by Liu, Latecki and Liu [10]. It is based on the measure of the number of extreme points presenting on a height function defined from several directions. In particular, the convex-hull points are directly visible as extreme points in the function and a scale parameter allows to retain only the extreme points being surrounded by *large enough* concave or convex parts. Thus, the non significant parts can be filtered from the scale parameter. The main advantage of this estimator is the feature detection that gives an interesting multi-scale contour representation even if the curvature estimation is only qualitative. Figure 7 illustrates some examples.

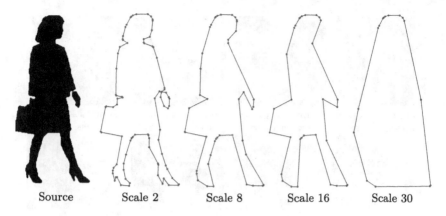

Source   Scale 2   Scale 8   Scale 16   Scale 30

Fig. 7. Illustration of the visual curvature obtained at different scales

4   Results and Comparisons

In this section, we present some results obtained with the proposed method using the previous polygonalization algorithms. Figure 8 shows the results with their execution times (t) in link to the number of polygon points ($\#$). The comparisons were obtained by applying some variations on the scale parameter (when exists: ν for DPP, e for Fréchet) and on the intensity interval size s ($s = 255/iStep$, with $iStep$ of Algorithm 1). From the presented results, we can observe that the Fréchet polygonalization methods provides a faster reconstruction but contains more irregular contours. By comparing DLL and DPP, the DLL looks, at low scale, slightly smoother than the others (Fig. 8(a) and (f)). The execution time mainly depends of the choice of the input parameter s and the choice of the method (see time measure given in Fig. 8). Due to page limitation the results obtained with the visual-curvature based polygonalization method are not presented but they can be obtained from an **Online demonstration** available at:

http://ipol-geometry.loria.fr/~kerautre/ipol_demo/DGIV_IPOLDemo/

To conclude this section we also performed comparisons with *Inkscape*, a free drawing software [30] (Fig. 9). We can observe that our representation provides fine results, with a lighter PDF file size.

5   Conclusion and Perspectives

Exploiting recent advances from the representation of digital contours, we have proposed a new framework to construct a vectorial representation from an input bitmap image. The proposed algorithm is simple, can integrate many other polygonalization algorithms and can be reproduced online. Moreover, we propose a vectorial representation of the image with a competitive efficiency, w.r.t. *Inkscape* [30].

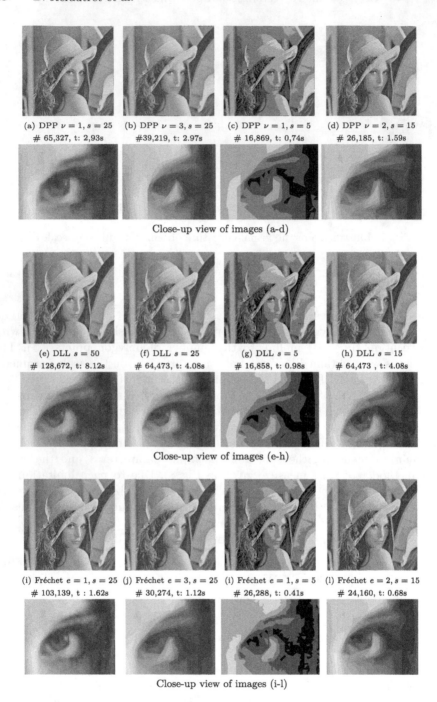

(a) DPP $\nu = 1, s = 25$ (b) DPP $\nu = 3, s = 25$ (c) DPP $\nu = 1, s = 5$ (d) DPP $\nu = 2, s = 15$
\# 65,327, t: 2,93s   \#39,219, t: 2.97s   \# 16,869, t: 0,74s   \# 26,185, t: 1.59s

Close-up view of images (a–d)

(e) DLL $s = 50$ (f) DLL $s = 25$ (g) DLL $s = 5$ (h) DLL $s = 15$
\# 128,672, t: 8.12s   \# 64,473, t: 4.08s   \# 16,858, t: 0.98s   \# 64,473 , t: 4.08s

Close-up view of images (e–h)

(i) Fréchet $e = 1, s = 25$ (j) Fréchet $e = 3, s = 25$ (i) Fréchet $e = 1, s = 5$ (l) Fréchet $e = 2, s = 15$
\# 103,139, t : 1.62s   \# 30,274, t: 1.12s   \# 26,288, t: 0.41s   \# 24,160, t: 0.68s

Close-up view of images (i–l)

Fig. 8. Experiments of the proposed level set based method on the Lena image with three polygonalization algorithms: (a–d) dominant points (DPP), (e–h) digital level layers with line primitive (DLL) and (i–l) Fréchet distance based

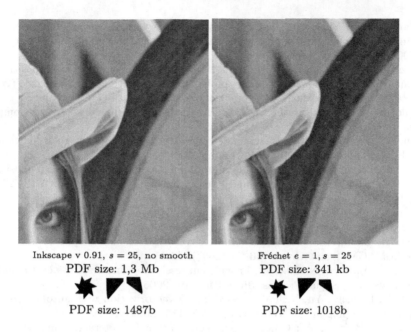

Inkscape v 0.91, $s = 25$, no smooth   Fréchet $e = 1, s = 25$
PDF size: 1,3 Mb   PDF size: 341 kb

PDF size: 1487b   PDF size: 1018b

Fig. 9. Comparisons of vectorization obtained with Inkscape and Fréchet (resp. DLL) based vectorization (first row, resp. second row).

This work opens new perspectives with for instance the integration of the meaningful scale detection to filter only the significant intensity levels [31]. Also, we would like to evaluate the robustness of our approach with different polygonalization algorithms, by testing images corrupted by noise.

References

1. Tombre, K.: Analysis of engineering drawings: state of the art and challenges. In: Tombre, K., Chhabra, A.K. (eds.) GREC 1997. LNCS, vol. 1389, pp. 257–264. Springer, Heidelberg (1998). doi:10.1007/3-540-64381-8_54
2. Cordella, L., Vento, M.: Symbol recognition in documents: a collection of techniques? Int. J. Doc. Anal. Recognit. **3**, 73–88 (2000)
3. Hilaire, X., Tombre, K.: Robust and accurate vectorization of line drawings. IEEE Trans. Pattern Anal. Mach. Intell. **28**, 890–904 (2006)
4. Nguyen, T.P., Debled-Rennesson, I.: Arc segmentation in linear time. In: Real, P., Diaz-Pernil, D., Molina-Abril, H., Berciano, A., Kropatsch, W. (eds.) CAIP 2011. LNCS, vol. 6854, pp. 84–92. Springer, Heidelberg (2011). doi:10.1007/978-3-642-23672-3_11
5. Marji, M., Siy, P.: Polygonal representation of digital planar curves through dominant point detection – a nonparametric algorithm. Pattern Recognit. **37**, 2113–2130 (2004)
6. Nguyen, T.P., Debled-Rennesson, I.: A discrete geometry approach for dominant point detection. Pattern Recognit. **44**, 32–44 (2011)

7. Bhowmick, P., Bhattacharya, B.B.: Fast polygonal approximation of digital curves using relaxed straightness properties. IEEE Trans. Pattern Anal. Mach. Intell. **29**, 1590–1602 (2007)
8. Feschet, F.: Multiscale analysis from 1D parametric geometric decomposition of shapes. In: International Conference on Pattern Recognition, pp. 2102–2105 (2010)
9. Vacavant, A., Roussillon, T., Kerautret, B., Lachaud, J.O.: A combined multi-scale/irregular algorithm for the vectorization of noisy digital contours. Comput. Vis. Image Underst. **117**, 438–450 (2013)
10. Liu, H., Latecki, L.J., Liu, W.: A unified curvature definition for regular, polygonal, and digital planar curves. Int. J. Comput. Vis. **80**, 104–124 (2008)
11. Swaminarayan, S., Prasad, L.: Rapid automated polygonal image decomposition. In: 35th IEEE Applied Imagery and Pattern Recognition Workshop (AIPR 2006), p. 28 (2006)
12. Lecot, G., Levy, B.: Ardeco: automatic region detection and conversion. In: 17th Eurographics Symposium on Rendering-EGSR 2006, pp. 349–360 (2006)
13. Sun, J., Liang, L., Shum, H.Y.: Image vectorization using optimized gradient meshes. ACM Trans. Graph. **26**, 11 (2007). ACM
14. Demaret, L., Dyn, N., Iske, A.: Image compression by linear splines over adaptive triangulations. Signal Process. **86**, 1604–1616 (2006)
15. Xia, T., Liao, B., Yu, Y.: Patch-based image vectorization with automatic curvilinear feature alignment. Trans. Graph. **28**, 115 (2009). ACM
16. Price, B., Barrett, W.: Object-based vectorization for interactive image editing. Vis. Comput. **22**, 661–670 (2006)
17. Ngo, P., Nasser, H., Debled-Rennesson, I.: Efficient dominant point detection based on discrete curve structure. In: Barneva, R.P., Bhattacharya, B.B., Brimkov, V.E. (eds.) IWCIA 2015. LNCS, vol. 9448, pp. 143–156. Springer, Cham (2015). doi:10.1007/978-3-319-26145-4_11
18. Gérard, Y., Provot, L., Feschet, F.: Introduction to digital level layers. In: Debled-Rennesson, I., Domenjoud, E., Kerautret, B., Even, P. (eds.) DGCI 2011. LNCS, vol. 6607, pp. 83–94. Springer, Heidelberg (2011). doi:10.1007/978-3-642-19867-0_7
19. Provot, L., Gerard, Y., Feschet, F.: Digital level layers for digital curve decomposition and vectorization. Image Process. On Line **4**, 169–186 (2014)
20. Sivignon, I.: A near-linear time guaranteed algorithm for digital curve simplification under the Fréchet distance. In: Debled-Rennesson, I., Domenjoud, E., Kerautret, B., Even, P. (eds.) DGCI 2011. LNCS, vol. 6607, pp. 333–345. Springer, Heidelberg (2011). doi:10.1007/978-3-642-19867-0_28
21. Sivignon, I.: A near-linear time guaranteed algorithm for digital curve simplification under the Fréchet distance. Image Process. On Line **4**, 116–127 (2014)
22. Lachaud, J.O.: Coding Cells of Digital Spaces: A Framework to Write Generic Digital Topology Algorithms, vol. 12, pp. 337–348. Elsevier (2003)
23. DGTal-Team: DGtal: digital geometry tools and algorithms library (2017). http://dgtal.org
24. Coeurjolly, D., Kerautret, B., Lachaud, J.O.: Extraction of connected region boundary in multidimensional images. Image Process. On Line **4**, 30–43 (2014)
25. Shewchuk, J.R.: Delaunay refinement algorithms for triangular mesh generation. Comput. Geom. **22**, 21–74 (2002)
26. Najman, L., Couprie, M.: Building the component tree in quasi-linear time. Trans. Image Process. **15**, 3531–3539 (2006)
27. Bertrand, G.: On the dynamics. Image Vis. Comput. **25**, 447–454 (2007)
28. Faure, A., Buzer, L., Feschet, F.: Tangential cover for thick digital curves. Pattern Recognit. **42**, 2279–2287 (2009)

29. Debled-Rennesson, I., Feschet, F., Rouyer-Degli, J.: Optimal blurred segments decomposition of noisy shapes in linear time. Comput. Graph. **30**, 30–36 (2006)
30. https://inkscape.org/fr/
31. Kerautret, B., Lachaud, J.O.: Meaningful scales detection along digital contours for unsupervised local noise estimation. IEEE Trans. Pattern Anal. Mach. Intell. **34**, 2379–2392 (2012)

Discrete Models and Tools

The Boolean Map Distance: Theory and Efficient Computation

Filip Malmberg[1]([✉]), Robin Strand[1], Jianming Zhang[2], and Stan Sclaroff[3]

[1] Department of Information Technology, Centre for Image Analysis,
Uppsala University, Uppsala, Sweden
`filip.malmberg@it.uu.se`
[2] Adobe Research, San Jose, USA
[3] Department of Computer Science, Boston University, Boston, USA

Abstract. We propose a novel distance function, the *boolean map distance* (BMD), that defines the distance between two elements in an image based on the probability that they belong to different components after thresholding the image by a randomly selected threshold value. This concept has been explored in a number of recent publications, and has been proposed as an approximation of another distance function, the *minimum barrier distance* (MBD). The purpose of this paper is to introduce the BMD as a useful distance function in its own right. As such it shares many of the favorable properties of the MBD, while offering some additional advantages such as more efficient distance transform computation and straightforward extension to multi-channel images.

1 Introduction

Distance functions and their transforms (DTs, where each pixel is assigned the distance to a set of *seed pixels*) are used extensively in many image processing applications. Here, we introduce a novel distance function, the *boolean map distance* (BMD), that defines the distance between two elements in an image based on the probability that they belong to different components after thresholding the image by a randomly selected threshold value. The idea of considering connectivity with respect to randomly selected thresholds was first introduced by Zhang and Sclaroff [5], who considered the probability that a pixel does not belong to a component that touches the image border after thresholding with a randomly selected value. This probability was used to identify salient objects in an image, and the concept was named *boolean map saliency*. Here, we consider the probability that *any* pair of pixels are not connected with respect to a random threshold, and thus use the term *boolean map distance* (BMD). Ideas related to this concept have further been explored in a number of recent publications [1,4,6], and have been used as an approximation of another distance function, the *minimum barrier distance* (MBD) first proposed by Strand et al. [4].

The purpose of this paper is to introduce the BMD as a useful distance function in its own right. As such the BMD shares many of the favorable properties

© Springer International Publishing AG 2017
W.G. Kropatsch et al. (Eds.): DGCI 2017, LNCS 10502, pp. 335–346, 2017.
DOI: 10.1007/978-3-319-66272-5_27

of the MBD, while offering some additional advantages such as more efficient distance transform computation and straightforward extension to multi-channel images. Specifically, our contributions are as follows:

- We provide a formal definition of the BMD, for both continuous and discrete images.
- We show that the (continuous and discrete domain) BMD is a pseudo-metric, thus motivating the name boolean map distance.
- We prove the equivalence between the (continuous and discrete domain) BMD and the φ mapping introduced by Strand et al. [4] as an approximation to the MBD. Thereby, we strengthen the connection between the BMD and the MBD. Previously, this equivalence was only established in the discrete case [6].
- We summarize available algorithms for computing distance transforms for the discrete BMD. Specifically, we note that the equivalence between the discrete BMD and φ functions allows the BMD to be expressed as the difference between two well-known distance functions, whose distance transforms can be computed using the *image foresting transform* [2], a generalization of Dijkstra's algorithm. We demonstrate empirically that the resulting algorithm is an order of magnitude faster than previously reported algorithms, while still producing exact results.

2   The Boolean Map Distance in $\mathbb{R}^n$

We define a n-dimensional gray-scale image I as a pair $I = (D, f)$, where $D \subset \mathbb{R}^n$ and $f : D \to [0, 1]$ is a continuous function. The restriction of the image values to the range $[0, 1]$ does, for the purposes considered here, not imply a loss of generality [6].

For $p, q \in D$, a *path* from p to q (in D) is any continuous function $\pi : [0, 1] \to D$ with $\pi(0) = p$ and $\pi(1) = q$. We use the symbol $\Pi_{p,q}^D$ to denote the family of all such paths. The *reverse* π^{-1} of a path π is defined as $\pi^{-1}(s) = \pi(1 - s)$ for all $s \in [0, 1]$. Recall that $D \subset \mathbb{R}^n$ is *path connected* if for every $p, q \in D$ there exists a path $\pi : [0, 1] \to D$ from p to q. For the remainder of this section, we assume that the set D is path connected.

Let $t \in [0, 1]$. Given an image $I = (D, f)$, we define the function $T_{I,t} : D \to \{0, 1\}$ by

$$T_{I,t}(p) = \begin{cases} 0, & \text{if } f(p) < t \\ 1, & \text{otherwise} \end{cases}, \tag{1}$$

for all $p \in D$. We refer to any function that maps image elements to the set $\{0, 1\}$ as a *boolean map*. The boolean map $T_{I,t}$ represents the *thresholding* of the image I by t. For any $p, q \in D$, we say that p and q belong to the same component of $T_{I,t}$ if there exists a path $\pi \in \Pi_{p,q}^D$ such that either $T_{I,t}(\pi(s)) = 0$ for all $s \in [0, 1]$ or $T_{I,t}(\pi(s)) = 1$ for all $s \in [0, 1]$. A path satisfying either of these criteria is called a *connecting path*. Otherwise, p and q belong to different components of $T_{I,t}$. We use the notation $p \underset{t}{\sim} q$ to indicate that p and q belong to the same component of $T_{I,t}$, while the notation $p \underset{t}{\nsim} q$ indicates that they belong to different components.

Definition 1. *Let t be a random value sampled from a uniform probability distribution over $[0,1]$. The continuous domain boolean map distance $BMD : D \times D \to [0,1]$ is defined as*

$$BMD(p,q) = P(p \underset{t}{\nsim} q) = 1 - P(p \underset{t}{\sim} q) \tag{2}$$

for all $p, q \in D$, where $P(A)$ denotes the probability of the event A.

Definition 2. *A function $d : D \times D \to [0,\infty)$ is a pseudo-metric on a set D if, for every $p, q, r \in D$,*

  (i) $d(p,p) = 0$ (identity)
  (ii) $d(p,q) \geq 0$ (non-negativity)
  (iii) $d(p,q) = d(p,q)$ (symmetry)
  (iv) $d(p,r) \leq d(p,q) + d(q,r)$ (triangle inequality)

If additionally it holds that $d(p,q) = 0 \Leftrightarrow p = q$ for all p, q, then d is a *metric*.

In the proof that BMD is a pseudo-metric, we will use the following notion. The *concatenation* $\pi_1 \cdot \pi_2$ of the paths π_1 and π_2 such that $\pi_1(1) = \pi_2(0)$ is

$$(\pi_1 \cdot \pi_2)(s) = \begin{cases} \pi_1(2s) & \text{if } s \in [0, 1/2] \\ \pi_2(2s) & \text{otherwise} \end{cases}, \tag{3}$$

Theorem 1. *BMD is a pseudo-metric.*

Proof. First, we show that BMD obeys property (i). Consider a path $\pi(s)$ such that $\pi(s) = p$ for any $s \in [0,1]$. This path is obviously connecting p to itself. Thus, $p \underset{t}{\sim} p$ for all t, and so $P(p \underset{t}{\nsim} q) = 0$.

Since BMD is defined as a probability, we have $BMD(p,q) \in [0,1]$ for all p,q. Thus, BMD clearly obeys property (ii).

Next, we show that BMD obeys property (iii). If, for a given threshold t, there exists a connecting path π from p to q then the reverse path π^{-1} is a connecting path from q to p. Thus, $p \underset{t}{\nsim} q \Leftrightarrow q \underset{t}{\nsim} p$, and so $P(p \underset{t}{\nsim} q) = P(q \underset{t}{\nsim} p)$.

Finally, we show that BMD obeys property (iv). If, for a given threshold t, there exists a connecting path π_1 from p to q and another connecting path π_2 from q to r, then the concatenation $\pi_1 \cdot \pi_2$ of these two paths is a connecting path from p to r. Thus, the set of thresholds t for which $q \underset{t}{\sim} p$ and $q \underset{t}{\sim} r$ is a subset of the set of thresholds for which $p \underset{t}{\sim} r$, and so $BMD(p,r) \leq BMD(p,q) + BMD(q,r)$. $\square$

Note that for a constant function f we have $P(p \underset{t}{\nsim} q) = 0$ for all $p,q \in D$, and thus BMD is not in general a metric.

2.1   Equivalence Between *BMD* and φ

In this section, we prove the equivalence between the proposed *BMD* mapping and the φ mapping defined by Strand et al. (Definition 3 in [4]). We recall the definition of φ:

Definition 3. *Let* $I = (D, f)$. *The mapping* $\varphi : D \times D \to [0, \infty)$ *is defined as*

$$\varphi(p, q) = \inf_{\pi_1 \in \Pi_{p,q}^D} \max_s f(\pi_1(s)) - \sup_{\pi_2 \in \Pi_{p,q}^D} \min_s f(\pi_2(s)) \tag{4}$$

for all $p, q \in D$.

As stated by Strand et al. [4] the minimum/maximum over the numbers $s \in [0, 1]$ are attained, while neither the infimum nor supremum operators can be replaced by maximum/minimum.

Theorem 2. *The mappings BMD and φ are equal, i.e.* $\varphi(p, q) = BMD(p, q)$ *for all* $p, q \in D$.

Proof. We begin our proof by observing that

$$BMD(p, q) = P(p \underset{t}{\not\sim} q) = 1 - P(p \underset{t}{\sim} q) = 1 - P(A \vee B) \tag{5}$$

where A is the event $(p \underset{t}{\sim} q \wedge f(p) < t \wedge f(q) < t)$ and B is the event $(p \underset{t}{\sim} q \wedge f(p) \geq t \wedge f(q) \geq t)$. Since the events A and B are mutually exclusive, it follows that

$$BMD(p, q) = 1 - P(A) - P(B). \tag{6}$$

First, we study the probability $P(A)$ that p and q belong to the same component of $T_{I,t}$, and $f(p)$ and $f(q)$ are both less than t. This is true if there exists a path $\pi \in \Pi_{p,q}^D$ such that

$$f(\pi(s)) < t \text{ for all } s \in [0, 1]. \tag{7}$$

Let $c = \inf_{\pi_1 \in \Pi_{p,q}^D} \max_s f(\pi_1(s))$. Then a path π satisfying the condition given in Eq. (7) exists if $t > c$, but does not exist if $t < c$. If $t = c$, then the existence of π is not possible to determine in the general case but depends on whether, for the specific f, p and q at hand, there exists a path whose maximum value $\max_s f(\pi(s))$ attains the infimum over all paths in $\Pi_{p,q}^D$. Let $\phi : [0, 1] \to \{0, 1\}$ be an indicator function for the event A, defined by

$$\phi(t) = \begin{cases} 1 & \text{if } A \\ 0 & \text{otherwise} \end{cases} \tag{8}$$

Depending on the specific values of f, p and q, we have either $\phi = \phi_1$ or $\phi = \phi_2$ where

$$\phi_1(t) = \begin{cases} 1 & \text{if } t \geq c \\ 0 & \text{otherwise} \end{cases} \qquad \phi_2(t) = \begin{cases} 1 & \text{if } t > c \\ 0 & \text{otherwise} \end{cases}. \tag{9}$$

We note that

$$P(A) = \int_0^1 \phi(t)dt = \int_0^1 \phi_1(t)dt = \int_0^1 \phi_2(t)dt = 1-c = 1- \inf_{\pi_1 \in \Pi_{p,q}^D} \max_s f(\pi_1(s)).$$
(10)

Thus, regardless of the existence of the path π in the case where $t = c$, the probability of the event A occurring is $P(A) = 1 - \inf_{\pi_1 \in \Pi_{p,q}^D} \max_s f(\pi_1(s))$.

Next, we study the probability $P(B)$ that p and q belong to the same component of $T_{I,t}$, and that $f(p)$ and $f(q)$ are both greater than or equal to t. This is true if there exists a path $\pi \in \Pi_{p,q}^D$ such that $f(\pi(s)) \geq t$ for all $s \in [0,1]$. Such a path exists iff $t < \sup_{\pi_2 \in \Pi_{p,q}^D} \min_s f(\pi_2(s))$. The probability of the event B occurring is $P(B) = \sup_{\pi_2 \in \Pi_{p,q}^D} \min_s f(\pi_2(s))$.

From Eq. (6), it thus follows that

$$BMD(p,q) = \inf_{\pi_1 \in \Pi_{p,q}^D} \max_s f(\pi_1(s)) - \sup_{\pi_2 \in \Pi_{p,q}^D} \min_s f(\pi_2(s)) = \varphi(p,q).$$
(11)

□

3 The Discrete Boolean Map Distance

In this section, we introduce a discrete formulation of the BMD.

We define a discrete gray-scale digital image $\hat{I}$ as a pair $\hat{I} = (\hat{D}, f)$ consisting of a finite set $\hat{D}$ of image elements and a mapping $f : \hat{D} \rightarrow [0,1]$. We will refer to elements of $\hat{D}$ as *pixels*, regardless of the dimensionality of the image. Additionally, we define a mapping $\mathcal{N} : \hat{D} \rightarrow \mathcal{P}(\hat{D})$ specifying an *adjacency* relation over the set of pixels $\hat{D}$. For any $p, q \in \hat{D}$, we refer to $\mathcal{N}(p)$ as the *neighborhood* of p and say that q is *adjacent* to p if $q \in \mathcal{N}(p)$. We require the adjacency relation to be symmetric, so that $q \in \mathcal{N}(p) \Leftrightarrow p \in \mathcal{N}(q)$ for all $p, q \in \hat{D}$.

A discrete path $\hat{\pi} = \langle \hat{\pi}(0), \hat{\pi}(1), \ldots \hat{\pi}(k) \rangle$ of length $|\hat{\pi}| = k + 1$ from $\hat{\pi}(0)$ to $\hat{\pi}(k)$ is an ordered sequence of pixels in $\hat{D}$ where each consecutive pair of pixels are *adjacent*. We use the symbol $\hat{\Pi}_{p,q}^{\hat{D}}$ to denote the set of all discrete paths from p to q. For a set of pixels $S \subseteq \hat{D}$, the symbol $\hat{\Pi}_{p,S}^{\hat{D}}$ denotes the set of all discrete paths from p to any $q \in S$. The *reverse* $\hat{\pi}^{-1}$ of a discrete path $\hat{\pi}$ is defined as $\hat{\pi}^{-1}(i) = \hat{\pi}(k - i)$ for all $i \in \{0, 1, \ldots, k\}$. Given two discrete paths $\hat{\pi}_1$ and $\hat{\pi}_2$ such that the endpoint of $\hat{\pi}_1$ equals the starting point of $\hat{\pi}_2$, we denote by $\hat{\pi}_1 \cdot \hat{\pi}_2$ the concatenation of the two paths.

Throughout, we assume that the combination of the set $\hat{D}$ and the adjacency relation $\mathcal{N}$ defines a connected graph, so that for every pair of pixels $p, q \in \hat{D}$ there exists a path between them.

Let $t \in [0,1]$. Given a discrete image $\hat{I} = (\hat{D}, f)$, we define the thresholding $T_{\hat{I},t} : \hat{D} \to \{0,1\}$ of $\hat{I}$ by t as

$$T_{\hat{I},t}(p) = \begin{cases} 0, & f(p) < t \\ 1, & \text{otherwise} \end{cases}, \tag{12}$$

For any $p, q \in \hat{D}$, we say that p and q belong to the same component of $T_{\hat{I},t}$ if there exists a path $\hat{\pi} \in \hat{\Pi}_{p,q}^{\hat{D}}$ such that either $T_{\hat{I},t}(\hat{\pi}(i)) = 0$ for all $i \in \{0,1,\dots,k\}$ or $T_{\hat{I},t}(\hat{\pi}(i)) = 1$ for all $i \in \{0,1,\dots,k\}$. A path satisfying either of these criteria is called a *connecting path*. As before, we use the notation $p \underset{t}{\sim} q$ to indicate that p and q belong to the same component of $T_{\hat{I},t}$, while the notation $p \underset{t}{\not\sim} q$ indicates that they belong to different components. Additionally, for a set of pixels S, the notation $p \underset{t}{\sim} S$ indicates that $p \underset{t}{\sim} q$ for at least one $q \in S$ while the notation $p \underset{t}{\not\sim} S$ indicates that $p \underset{t}{\not\sim} q$ for all $q \in S$.

Definition 4. *Let t be a random value sampled from a uniform probability distribution over $[0,1]$. For any set of pixels $S \subseteq \hat{D}$ and pixel $p \in \hat{D}$, the discrete boolean map distance $B\hat{M}D : \hat{D} \times \mathcal{P}(\hat{D}) \to [0,1]$ is defined as*

$$B\hat{M}D(p, S) = P(p \underset{t}{\not\sim} S) \tag{13}$$

In the above definition, $\mathcal{P}(\hat{D})$ denotes the power set of $\hat{D}$. If the set S consists of a single element q, we can consider $B\hat{M}D$ to be a mapping from $\hat{D} \times \hat{D}$ to $[0,1]$, and the definition can in this case be reduced to $B\hat{M}D(p,q) = P(p \underset{t}{\not\sim} q)$.

Theorem 3. *Let $S \subseteq \hat{D}$ consist of a single element q and let $p \in \hat{D}$. Then the discrete $B\hat{M}D$, viewed as a mapping from $\hat{D} \times \hat{D}$ to $[0,1]$, is a pseudo-metric.*

The proof of Theorem 3 is identical to that of Theorem 1, provided that the relevant continuous notions defined in Sect. 2 are swapped out for their discrete counterparts defined in this section.

3.1 Equivalence Between the Discrete $B\hat{M}D$ and $\hat{\varphi}$

In this section, we prove the equivalence between the discrete $B\hat{M}D$ mapping and the $\hat{\varphi}$ mapping defined by Strand et al. (Eq. (4) in [4]) as a discrete counterpart of the φ mapping. We provide a slightly extended definition of $\hat{\varphi}$:

Definition 5. *The mapping $\hat{\varphi} : \hat{D} \times \mathcal{P}(\hat{D}) \to [0,1]$ is defined as*

$$\hat{\varphi}(p, S) = \min_{\hat{\pi}_1 \in \hat{\Pi}_{p,S}^{\hat{D}}} \left(\max_{i \in \{0,1,\dots,k\}} I(\hat{\pi}_1(i)) \right) - \max_{\hat{\pi}_2 \in \hat{\Pi}_{p,S}^{\hat{D}}} \left(\min_{i \in \{0,1,\dots,k\}} I(\hat{\pi}_2(i)) \right) \tag{14}$$

for all $p \in \hat{D}$ and $S \subseteq \hat{D}$.

Note that if S consists of a single element q, the above definition of $\hat{\varphi}$ reduces to the definition given by Strand et al. [4].

Theorem 4. *The mappings $B\hat{M}D$ and $\hat{\varphi}$ are equal, i.e., $\hat{\varphi}(p, S) = B\hat{M}D(p, S)$ for all $p \in \hat{D}$ and $S \subseteq \hat{D}$.*

Proof. We start by observing that

$$B\hat{M}D(p, S) = P(p \underset{t}{\nsim} S) = 1 - P(p \underset{t}{\sim} S) = 1 - P(A \vee B) \tag{15}$$

where A is the event $(p \underset{t}{\sim} q_1 \wedge f(p) < t \wedge f(q_1) < t)$ for some $q_1 \in S$ and B is the event $(p \underset{t}{\sim} q_2 \wedge f(p) \geq t \wedge f(q_2) \geq t)$ for some $q_2 \in S$. Since the events A and B are mutually exclusive, it follows that

$$B\hat{M}D(p, S) = 1 - P(A) - P(B). \tag{16}$$

First, we study the probability $P(A)$ that p and q_1 belong to the same component of $T_{\hat{f},t}$, and $f(p)$ and $f(q_1)$ are both less than t. This is true for some p_1 if there exists a path $\hat{\pi} \in \hat{\Pi}_{p,S}^D$ such that $f(\hat{\pi}(i)) < t$ for all $i \in \{0, 1, \ldots, |\hat{\pi}| - 1\}$. Such a path exists iff $t \geq \min\limits_{\hat{\pi}_1 \in \hat{\Pi}_{p,S}^D} \max\limits_{i \in \{0,1,\ldots,k\}} f(\hat{\pi}_1(i))$. The probability of this event occurring is $P(A) = 1 - \min\limits_{\hat{\pi}_1 \in \hat{\Pi}_{p,S}^D} \max\limits_{i \in \{0,1,\ldots,k\}} f(\hat{\pi}_1(i))$.

Next, we study the probability $P(B)$ that p and q_2 belong to the same component of $T_{\hat{f},t}$, and that $f(p)$ and $f(q_2)$ are both greater than or equal to t. This is true for some q_2 if there exists a path $\hat{\pi} \in \hat{\Pi}_{p,S}^{\hat{D}}$ such that $f(\hat{\pi}(i)) \geq t$ for all $i \in \{0, 1, \ldots, |\hat{\pi}| - 1\}$. Such a path exists iff $t < \max\limits_{\hat{\pi}_2 \in \hat{\Pi}_{p,S}^{\hat{D}}} \min\limits_{i \in \{0,1,\ldots,k\}} f(\hat{\pi}_2(i))$. The probability of the event B occurring is $P(B) = \max\limits_{\hat{\pi}_2 \in \hat{\Pi}_{p,q}^{\hat{D}}} \min\limits_{i \in \{0,1,\ldots,k\}} f(\hat{\pi}_2(i))$.

From Eq. (16), it thus follows that

$$B\hat{M}D(p, q) = \min\limits_{\hat{\pi}_1 \in \hat{\Pi}_{p,S}^{\hat{D}}} \max\limits_{i \in \{0,1,\ldots,k\}} f(\hat{\pi}_1(i)) - \max\limits_{\hat{\pi}_2 \in \hat{\Pi}_{p,S}^{\hat{D}}} \min\limits_{i \in \{0,1,\ldots,k\}} f(\hat{\pi}_2(i)) = \hat{\varphi}(p, S).$$
$$\tag{17}$$
$\square$

A proof of Theorem 4 was previously provided by Zhang and Sclaroff [6].

4   Computing Distance Transforms for the Discrete BMD

Given a discrete image $\hat{I} = (\hat{D}, f)$ and a set of *seed pixels* $S \subseteq \hat{D}$, the *distance transform* for the discrete BMD is a map assigning to each pixel $p \in \hat{D}$ the value $BMD(p, S)$, i.e., each pixel is assigned the discrete boolean map distance to the set S. In this section, we study various methods for computing distance transforms for the discrete BMD.

4.1 Monte Carlo Approximation

From the definition of the BMD, it is straightforward to devise a Monte Carlo algorithm for approximating the BMD distance transform by iteratively selecting a random threshold, performing thresholding, and using a flood-fill operation to find the set of pixels connected to at least one seed-point. As the number of iterations increases, the relative frequency with which each pixel belongs to the complement of this set approaches the correct distance transform value.

4.2 The Zhang-Sclaroff Algorithm

Assume that all intensities present in a given image can be written as i/k, for some fixed integer k and some i in the set $\{1, 2, \ldots, k\}$. This situation occurs in practice if we, e.g., remap an image with integer intensity values to the range $[0, 1]$. If gray levels are stored as 8-bit integers, for example, we can take $k = 256$. Then the algorithm proposed by Zhang and Sclaroff for calculating Boolean Map Saliency [6] can be used directly for computing the exact BMD distance transform from any set of seed-points. Pseudo-code for this algorithm is listed in Algorithm 1.

In this algorithm each iteration of the **foreach**-loop requires $\mathcal{O}(n)$ operations, where n is the number of image pixels. Thus, the entire algorithm terminates in $\mathcal{O}(nk)$ operations which, since k can be considered a constant, equals $\mathcal{O}(n)$ operations.

Algorithm 1. The Zhang-Sclaroff algorithm for computing the discrete BMD distance transform.

    Input: An image I, a set of seed-points S, an integer k
    Output: Distance transform D
1 Set $D(p) = 0$ for all pixels p in I;
2 **foreach** $i \in \{1, 2, \ldots, k\}$ **do**
3   Set $B \leftarrow T_{I, i/k}$;
4   Perform a flood-fill operation to identify the set of pixels belonging to the same component as at least one seed-point in B;
5   Increase D by $1/(k)$ for all pixels not in this set;
6 **end**

4.3 Dijkstra's Algorithm

As shown in Sect. 3.1, the discrete BMD can be written as the difference between two functions:

$$\min_{\hat{\pi} \in \hat{\Pi}_{p,S}^{\hat{D}}} \left(\max_{i \in \{0, 1, \ldots, k\}]} I(\hat{\pi}(i)) \right) \tag{18}$$

and

$$\max_{\hat{\pi} \in \hat{\Pi}_{p,S}^{\hat{D}}} \left(\min_{i \in \{0, 1, \ldots, k\}} I(\hat{\pi}(i)) \right). \tag{19}$$

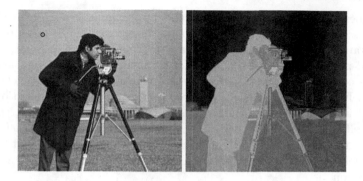

Fig. 1. Left: "Cameraman" image used in the experiments. The location of the single seed-point is indicated in green. Right: The corresponding BMD distance transform. (Color figure online)

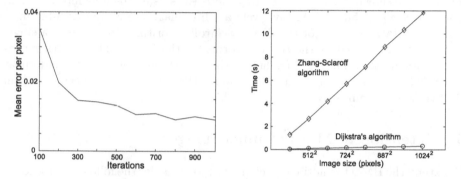

Fig. 2. Left: Approximation error of Monte Carlo estimation as a function of the number of samples. Right: Empirical comparison of running time for the Zhang-Sclaroff algorithm and Dijkstra's algorithm.

Both of these functions are *path based distance functions*, and are *smooth* in the sense defined by Falcão et al. [2]. Therefore, distance transforms for each term can be computed in $\mathcal{O}(n \log n)$ operations using the *image foresting transform* [2], a generalization of Dijkstra's algorithm. In the case where the magnitude of the set of all intensities present in the image is bounded by a fixed integer, this can further be reduced to $\mathcal{O}(n)$ operations [2].

4.4 Empirical Comparison of Running Time

In this section, we perform an empirical comparison of the algorithms described above.

First, we consider the Monte Carlo approximation method. This algorithm differs from the others in that it only produces approximate results. The error of the approximation decreases as the number of iterations is increased, but this also increases the computation time. We are thus interested in investigating the trade-off between number of iterations and approximation error. To this end, we use the Monte Carlo approximation method to compute an approximate BMD distance transform of the "Cameraman" image shown in Fig. 1, from the single seed-point indicated in the figure for a varying number of iterations. Each result was compared to the true distance transform, computed using Dijkstra's algorithm. Figure 2 (left) shows the average error per pixel as a function of the number of iterations. Due to the stochastic nature of the algorithm, the mean error is itself noisy. To increase clarity the figure therefore shows, for each number of samples, the average error obtained when repeating the experiment 20 times. As Fig. 2 (left) shows, a large number of samples is required to obtain an accurate approximation.

The Zhang-Sclaroff algorithm and the Dijkstra based method both have linear time complexity, but the constants involved differ substantially. To compare the algorithms empirically we calculated distance transforms for the "Cameraman" image shown in Fig. 1, scaled to various sizes using bi-cubic interpolation, using both algorithms. The gray-levels in this image are stored as 8-bit integers, so we take $k = 256$ for the Zhang-Sclaroff algorithm. In all cases, a single seed-point was placed at the top left corner of the image. The resulting running times are shown in Fig. 2 (right). Both algorithms show the expected linear dependence between image size and running time, but the approach based on Dijkstra's algorithm is faster by about a factor 30–40.

5   Extension to Multi-channel Images

To extend the BMD to multi-channel images, we consider the following procedure for creating a boolean map:

1. Randomly select one of the image channels according to some probability distribution over the set of image channels.
2. Randomly select a threshold from a uniform distribution over $[0, 1]$ and threshold the selected image channel at this value.

The multi-channel BMD between two pixels in an image with m channels is then defined as the probability that they belong to different components of the boolean map obtained by the above procedure. This probability is given by

$$BMD(p, q) = w_1 BMD_1(p, q) + w_2 BMD_2(p, q) + \ldots + w_m BMD_m(p, q), \quad (20)$$

where w_i denotes the probability of selecting channel i and BMD_i denotes the single channel BMD defined on channel i. For example, we may chose $w_i = 1/m$ for all $i \in \{1, 2, ..., m\}$. Note that the above result applies to both the continuous and discrete BMD. In the discrete case, this means that we can compute BMD distance transforms for multi-channel images by computing the single channel BMD on each channel, and forming a weighted average of the results.

An illustration of computing a multi-channel BMD distance transform on a color image is shown in Fig. 3.

Fig. 3. Top left: "Flower" image (from Rhemann et al. [3]) with seedpoints overlaid in white. Top right: Gray-scale image. Bottom left: BMD distance transform of the color image, after transformation to the CIEL*a*b* color space. Bottom right: BMD distance transform of the gray-scale image. The values of both distance transforms have been scaled for display purposes. Compared to the single-channel distance transform, the multi-channel BMD distance transform better captures the contrast between the flower and the background.

6   Conclusion

We have introduced the *Boolean map distance* (BMD), a pseudo-metric that measures the distance between elements in an image based on the probability that they belong to different components after thresholding the image by a randomly selected value. Formal definitions of the BMD have been given in both the continuous and discrete settings. The equivalence between the BMD and the φ mapping proposed by Strand et al. was previously shown in the discrete case [6]. We have extended this proof to also cover the continuous case, thereby further strengthening the connection between the BMD and the MBD.

We have summarized available algorithms for computing distance transforms for the discrete BMD. From the empirical comparison, we conclude that the Monte Carlo approximation method is not suitable for practical applications, given the existence of efficient exact algorithms. In the comparison between exact algorithms, we found that the approach based on Dijkstra's algorithm was faster than the Zhang-Sclaroff algorithm by an order of magnitude for calculating distance transforms on gray-scale images stored using 8-bit pixels. With increased color depth, the difference in computation time will increase.

Various aspects of the ideas presented have been explored in previous publications [1,4–6]. The BMD, however, has not to our knowledge previously been proposed as a distance function in its own right. By compiling and extending ideas previously scattered across multiple publications, we hope to highlight the BMD as a valuable distance function for image processing tasks.

References

1. Strand, R., Malmberg, F., Saha, P.K.: Efficient algorithm for finding the exact minimum barrier distance. Comput. Vis. Image Underst. **123**, 53–64 (2014)
2. Falcão, A.X., Stolfi, J., de Alencar Lotufo, R.: The image foresting transform: theory, algorithms, and applications. IEEE Trans. Pattern Anal. Mach. Intell. **26**(1), 19–29 (2004)
3. Rhemann, C., Rother, C., Wang, J., Gelautz, M., Kohli, P., Rott, P.: A perceptually motivated online benchmark for image matting. In: IEEE Conference on Computer Vision and Pattern Recognition (CVPR 2009), pp. 1826–1833. IEEE (2009)
4. Strand, R., Ciesielski, K.C., Malmberg, F., Saha, P.K.: The minimum barrier distance. Comput. Vis. Image Underst. **117**(4), 429–437 (2013)
5. Zhang, J., Sclaroff, S.: Saliency detection: a boolean map approach. In: Proceedings of the IEEE International Conference on Computer Vision, pp. 153–160 (2013)
6. Zhang, J., Sclaroff, S.: Exploiting surroundedness for saliency detection: a boolean map approach. IEEE Trans. Pattern Anal. Mach. Intell. **38**(5), 889–902 (2016)

Fast and Efficient Incremental Algorithms for Circular and Spherical Propagation in Integer Space

Shivam Dwivedi[1], Aniket Gupta[1], Siddhant Roy[1], Ranita Biswas[1(✉)], and Partha Bhowmick[2]

[1] Department of Computer Science and Engineering,
Indian Institute of Technology, Roorkee, India
shivamdw22@gmail.com, aniketguptaknp@gmail.com,
raj.siddhant.rohit@gmail.com, biswas.ranita@gmail.com
[2] Department of Computer Science and Engineering,
Indian Institute of Technology, Kharagpur, India
bhowmick@gmail.com

Abstract. Space filling circles and spheres have various applications in mathematical imaging and physical modeling. In this paper, we first show how the thinnest (i.e., 2-minimal) model of digital sphere can be augmented to a space filling model by fixing certain "simple voxels" and "filler voxels" associated with it. Based on elementary number-theoretic properties of such voxels, we design an efficient incremental algorithm for generation of these space filling spheres with successively increasing radius. The novelty of the proposed technique is established further through circular space filling on 3D digital plane. As evident from a preliminary set of experimental result, this can particularly be useful for parallel computing of 3D Voronoi diagrams in the digital space.

Keywords: Digital circle · Digital sphere · Space filling curve · Space filling surface · Spherical propagation · Voronoi diagram

1 Introduction

Space filling curves and surfaces find numerous applications in scientific computing, especially when the space is discretized by a well-defined grid or lattice. In digital geometry, the discretization usually refers to a collection of isotropic pixels in 2D and isotropic voxels in 3D. Our domain of interest is discrete 3D space or a discrete 3D plane, which eventually needs to be filled by voxels in a spherical or in a circular fashion. There exist few techniques in the literature on this; see, for example, [17,18], where a marching sphere approach to wall distance calculation is proposed. It uses a modified mid-point circle algorithm to produce spheres at each level while ensuring that no gap is produced in between two consecutive concentric digital spheres. However, the adopted model of digital sphere conforms to only 16-symmetry and hence is not uniform with respect

© Springer International Publishing AG 2017
W.G. Kropatsch et al. (Eds.): DGCI 2017, LNCS 10502, pp. 347–359, 2017.
DOI: 10.1007/978-3-319-66272-5_28

Table 1. List of symbols used and their meaning

Symbol	Meaning
$\mathcal{S}(r)$	Real sphere of radius r
$\mathsf{S}(r)$	Naive sphere of radius r
$\mathbb{S}(r)$	(isothetic) $\frac{1}{2}$-offset digital sphere of r
$\mathbf{S}(r)$	Space filling digital sphere of radius r
$\mathbf{S}^*(r)$	Solid digital sphere of radius r
$\chi(r)$	Set of simple voxels corresponding to $\mathsf{S}(r)$
$\mathsf{F}(r)$	Set of filler voxels between $\mathsf{S}(r)$ and $\mathsf{S}(r+1)$
$\mathsf{B}(r)$	Set of inter-octant boundary voxels of $\mathsf{S}(r)$

to different coordinates. Nonetheless, that work suggests the usability of sphere propagation in a real-life situation where distance from a moving or deforming body needs be estimated with reasonable efficiency in computation.

During spherical propagation, gaps or absentee voxels is a crucial issue, which needs to be tackled with guarantee and efficiency. Work on absentee characterization between concentric circles and spheres in digital space can be seen in [4–6]. Gap-free solid sphere generation using layering approach is also reported in [7], which is however suitable for voxel-based rapid prototyping but not for spherical propagation. In this paper, we show how space filling digital sphere can be propagated with efficient computation, which can subsequently be used for successful creation of Voronoi diagrams in 3D digital space.

For the basic terms and definitions useful for our work, we refer to [16]. For more specific terms and concepts related to metrics and topology in the voxel space, we refer to [8,9,13–15]. For formal definitions and details on paths, tunnels, and gaps in discrete objects, we refer to [11].

For a discrete object $A \subset \mathbb{Z}^3$, a coordinate plane, say, xy, is functional for A, if for every voxel $v(x_0, y_0, z_0) \in A$ there is no other voxel in A with the same first two coordinates. For example, let $A = \{(2,5,3),(2,6,3),(3,5,3)\}$; here the functional plane of A is the xy-plane, since a bijection exists between A and its projection on the xy-plane.

In Table 1, we have shown the list of important symbols used in this paper. Some related images for easy understanding are shown in Fig. 1.

Our contribution. A naive sphere of integer radius and integer center is 2-minimal and unique by composition. It has 48 symmetric parts (termed *quadraginta octants* or q-octants in short), each with a definite functional plane, and hence can be constructed very efficiently [8,9]. However, being not space filling, as we show in this paper, it cannot readily be used for spherical or circular propagations in the integer space. To circumvent this, we show how to augment the 2-minimal naive sphere to a space filling digital sphere. We exploit certain number-theoretic properties of naive sphere to characterize the voxels lying in between two consecutive naive spheres. Once this is fixed, the space

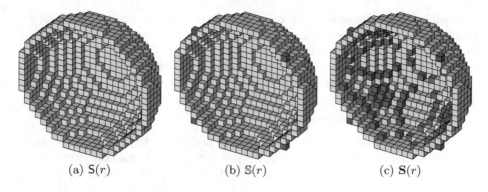

| (a) S(r) | (b) 𝕊(r) | (c) **S**(r) |

Fig. 1. Hemispheres of radius 11 (voxels: naive = white, simple = blue, filler = red). (Color figure online)

filling digital spheres are generated in succession very efficiently. We design an incremental algorithm that uses the current solid sphere to build the next one with simple integer operations. This, in turn, makes out algorithms for spherical propagation in 3D space and circular propagation on any 3D digital plane. We show several results to demonstrate its usefulness.

2   Space Filling Digital Sphere

There are different discretization models for representing geometric objects in digital space, each with its own advantages and disadvantages. Sphere is one of the most important primitive objects and computationally interesting due to its non-linearity. Its usual discretization models are naive, standard, supercover, arithmetic, and τ-offset. Among these, standard and supercover models produce overlapping spheres for consecutive integer radii and hence not computationally attractive for space filling. The τ-offset model gives rise to different models with varying values of τ taken in the Euclidean metric space, as recently shown in [10]. An arithmetic sphere is identical with $\frac{1}{2}$-offset sphere and produces distinct voxel sets for consecutive radii. However, the algorithm for its construction is computationally heavy due to its generality to the real domain [1]. As tested by us, the integer algorithm for arithmetic sphere in [1] takes around thrice as much time as the algorithm proposed in this paper (Sect. 3). For example, with radius 95, it takes 2.2 ms against 0.8 ms needed by our algorithm on the same computing platform.

As shown in [8,9], a naive sphere is computationally very efficient, as its construction is based on simple integer operations like addition, increment, and comparison, but devoid of any multiplication or division. This owes to the minimality of naive sphere. Due to its very minimality, though space filling is not possible by successive naive spheres, they can be augmented by requisite filler voxels to serve the purpose. These filler voxels admit interesting characterization, as we show in this section.

Each q-octant has one of the coordinate planes as the functional plane [9]. This allows us to represent the q-octant merely as a 2D array, which results in its quick computation. We show in this paper that the filler voxels can efficiently be found from few circular arcs in 3D, which are effectively elliptical arcs in 2D projection of those circular arcs on the functional plane.

2.1   Properties of Naive Sphere

We recall in particular some basic properties and characterization of the naive (i.e., 2-minimal or thinnest) model of digital sphere [8]. These concepts help us formalize the notion of filler voxels and eventually propose the concept of space filling spheres and spherical propagation while generating a higher-radius sphere from a lower-radius one.

We use $S(r)$ to denote a naive sphere of radius r and $S_1(r)$ to denote its 1st q-octant ($0 \leqslant x \leqslant y \leqslant z \leqslant r$). As explained in [8] and recently proved in [9], a naive sphere $S(r)$ is an irreducible 2-separating set of 3D integer points (equivalently, voxels) such that $\max_{p \in S(r)} d_\perp(p, \mathcal{S}(r))$ is minimized. Furthermore, the set $S(r)$ is unique and its closed form is as follows.

Definition 1 (Naive Sphere [8]). *A naive sphere of radius r is given by* $S(r) = \{p \in \mathbb{Z}^3 : (r^2 - \lambda \leqslant s < r^2 + \lambda) \wedge ((s \neq r^2 + \lambda - 1) \vee (\mu \neq \lambda))\}$, *where,* $p = (i, j, k), s = i^2 + j^2 + k^2, \lambda = \max\{|i|, |j|, |k|\}$, *and* $\mu = \text{med}\{|i|, |j|, |k|\}$.

By $\text{med}\{\cdot\}$, we denote the median of values. The upper bound for $S(r)$ is $r^2 + \lambda$ whereas the lower bound for $S(r + 1)$ is $(r + 1)^2 - \lambda$. Therefore, all the voxels falling in between these two values do not belong to $S(r)$ or $S(r + 1)$ or any other naive sphere. Therefore, we have the following observation.

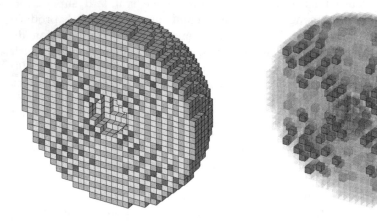

Fig. 2. Consecutive naive hemispheres from $S(4)$ to $S(15)$ shown in alternate gray shades; voxels in between them are shown in red (filler voxels) and blue (simple voxels). (Color figure online)

Observation 1. *Naive sphere is not space filling.*

Figure 2 shows consecutive naive spheres in two alternate shades of gray; voxels not belonging to any naive sphere are in red and blue. Red voxels denote the *filler voxels*, which are more than $\frac{1}{2}$ isothetic distance away from any real sphere of integer radius (as explained in Sect. 2.2). Blue ones are the *simple voxels*, which remain sandwiched between $S(r)$ and $S(r + 1)$, and lie within isothetic distance $\frac{1}{2}$ from the real sphere of radius r [8]. Characterization of simple voxels and filler voxels eventually simplifies the generation of space filling digital spheres, as we show in this paper. From the definition of simple voxels given in [8], we make the following observation.

Observation 2. *The set of simple voxels corresponding to $S(r)$ is given by $\chi(r) = \{p \in \mathbb{Z}^3 : (s = r^2 + \lambda - 1) \wedge (\mu = \lambda)\}$, where, $p = (i, j, k), s = i^2 + j^2 + k^2, \lambda = \max\{|i|, |j|, |k|\}$ and $\mu = \mathrm{med}\{|i|, |j|, |k|\}$.*

2.2   Characterization of Space Filling Digital Sphere

We define the (isothetic) $\frac{1}{2}$-offset digital sphere of radius r as the set of voxels whose isothetic distance from $\mathcal{S}(r)$ is at most $\frac{1}{2}$. For brevity, henceforth we refer this model simply as 'digital sphere'. This voxel set is effectively $S(r)$ in union with all the simple voxels corresponding to radius r, and defined precisely as follows based on Observation 2.

Definition 2 ($\frac{1}{2}$-**offset digital sphere**). *For any positive integer r, the $\frac{1}{2}$-offset digital sphere of radius r is given by $\mathbb{S}(r) = \{p \in \mathbb{Z}^3 : r^2 - \lambda \leqslant s < r^2 + \lambda\}$, where, $p = (i, j, k), s = i^2 + j^2 + k^2$, and $\lambda = \max\{|i|, |j|, |k|\}$.*

The voxels other than $\chi(r)$ which lie in between $S(r)$ and $S(r+1)$ comprise the set of filler voxels. This is the same set of voxels lying in between $\mathbb{S}(r)$ and $\mathbb{S}(r+1)$ and given as follows.

Observation 3. *The set of filler voxels lying in between $\mathbb{S}(r-1)$ and $\mathbb{S}(r)$ is given by $F(r) = \{p \in \mathbb{Z}^3 : (r-1)^2 + \lambda \leqslant s < r^2 - \lambda\}$, where, $p = (i, j, k), s = i^2 + j^2 + k^2$, and $\lambda = \max\{|i|, |j|, |k|\}$.*

We derive a few interesting properties pertaining to the set of filler voxels, which leads to an efficient algorithm for the construction of space filling spheres. We denote by $F_1(r)$ the first q-octant of $F(r)$, and require the following lemmas for this.

Lemma 1. *For any voxel in $F_1(r)$, we have $j + k \geqslant r$.*

Proof. By Observation 3, for any voxel (i, j, k) in $F_1(r)$, we have

$$(r-1)^2 + k \leqslant i^2 + j^2 + k^2 < r^2 - k$$
$$\Rightarrow r^2 - 2r + 1 + k - k^2 \leqslant i^2 + j^2 < r^2 - k - k^2. \tag{1}$$

Since $i \leqslant j$ in the first q-octant, from Eq. 1, we get $2j^2 \geqslant i^2 + j^2 \geqslant r^2 - 2r + 1 + k - k^2$. Let us assume, to the contrary of the lemma statement, that $j + k < r$, or, $j < r - k$, which yields $2(r - k)^2 \geqslant r^2 - 2r + 1 + k - k^2 \Rightarrow f(k) := 3k^2 - (4r + 1)k + (r^2 + 2r - 1) \geqslant 0$. For a constant value of r, the function $f(k) = 0$ is an upward facing parabola, and hence the maximum values can be reached only at the extremes. The domain of k satisfied by $\mathsf{F}_1(r)$ is $(\frac{r}{\sqrt{3}}, r]$, and we get the values of $f(k)$ at these extreme points as follows.

$$f(\tfrac{r}{\sqrt{3}}) : 3r^2 - 4r^2 - r + r^2 + 2r - 1 = -r - 1 < 0$$
$$f(r) : r^2 - \tfrac{4}{\sqrt{3}}r^2 - \tfrac{r}{\sqrt{3}} + r^2 + 2r - 1$$
$$= -(\tfrac{4}{\sqrt{3}} - 2)r^2 - (2 + \tfrac{1}{\sqrt{3}})r - 1 < 0$$

The above negative values indicate a contradiction, whence the proof. □

Lemma 2. *The filler voxels in the first q-octant are functional on both xy- and xz-planes.*

Proof. The lemma tells that if $(i, j, k) \in \mathsf{F}_1(r)$, then $(i, j, k + 1) \notin \mathsf{F}_1(r)$ (for being functional on xy-plane) and $(i, j + 1, k) \notin \mathsf{F}_1(r)$ (for being functional on xz-plane). Let us assume the contrary. Hence, first we take both (i, j, k) and $(i, j, k + 1)$ belong to $\mathsf{F}_1(r)$.

$$r^2 - 2r + k + 1 \leqslant i^2 + j^2 + k^2 < r^2 - k \tag{2}$$

$$r^2 - 2r + k + 2 \leqslant i^2 + j^2 + k^2 + 2k + 1 < r^2 - k - 1$$
$$\Rightarrow r^2 - 2r + k + 2 - 2k - 1 \leqslant i^2 + j^2 + k^2 < r^2 - k - 2k - 1 \tag{3}$$
$$\Rightarrow r^2 - 2r - k + 1 \leqslant i^2 + j^2 + k^2 < r^2 - 3k - 1$$

To make the two inequalities simultaneously true, we must have $r^2 - 2r + k + 1 < r^2 - 3k - 1$, which implies $2k < r - 1$. By Lemma 1, $j + k \geqslant r$ for $\mathsf{F}_1(r)$. Since $j \leqslant k$ in the 1st octant, we get $2k \geqslant r$, which contradicts our initial assumption. The other part of the theorem regarding the functionality on xz-plane can also be proved in a similar way. □

Lemma 3. *The filler voxels in the first q-octant follow circular paths.*

Proof. The voxels in $\mathsf{F}_1(r)$ satisfies $(r - 1)^2 + k \leqslant i^2 + j^2 + k^2 < r^2 - k$. For a fixed value of k, we get $r^2 - k^2 + k \leqslant i^2 + j^2 < (r + 1)^2 - k^2 - k \Rightarrow r^2 - k^2 + k \leqslant i^2 + j^2 < r^2 - k^2 + 2r - k + 1$. For a constant value of k, the set of voxels are bounded by the range $[r^2 - k^2 + k, r^2 - k^2 + 2r - k + 1)$ and has x-axis as the functional axis (as $\mathsf{F}_1(r)$ is functional on both xy- and xz-planes by Lemma 2). Hence, for each k, the filler voxels belong to a discrete circle. □

The properties of filler voxels are used to define space filling digital spheres suitable for spherical propagation. This is evident from the following theorem.

Theorem 1. $\mathsf{S}(r) := \mathbb{S}(r) \cup \mathsf{F}(r) \cup \chi(r) = \mathbb{S}(r) \cup \mathsf{F}(r)$ *is a space filling digital sphere, and hence given by* $\mathsf{S}(r) = \{p \in \mathbb{Z}^3 : (r - 1)^2 + \lambda \leqslant s < r^2 + \lambda\}$, *where,* $p = (i, j, k), s = i^2 + j^2 + k^2,$ *and* $\lambda = \max\{|i|, |j|, |k|\}$.

Proof. From Definition 2, Observation 3, and Lemma 3, we get

$$
\begin{aligned}
\mathbf{S}(r) &= \mathsf{F}(r) \cup \mathbb{S}(r) \cup \chi(r) \\
&= \mathsf{F}(r) \cup \mathbb{S}(r) \\
&= \left\{ p \in \mathbb{Z}^3 : (r-1)^2 + \lambda \leqslant s < r^2 - \lambda \right\} \cup \left\{ p \in \mathbb{Z}^3 : r^2 - \lambda \leqslant s < r^2 + \lambda \right\} \\
&= \left\{ p \in \mathbb{Z}^3 : ((r-1)^2 + \lambda \leqslant s < r^2 - \lambda) \vee (r^2 - \lambda \leqslant s < r^2 + \lambda) \right\} \\
&= \left\{ p \in \mathbb{Z}^3 : (r-1)^2 + \lambda \leqslant s < r^2 + \lambda \right\}.
\end{aligned}
$$

$\square$

3 Frontier Propagation

By 'frontier', we mean the boundary of a voxel set, which is typically a solid digital sphere or union of many such in our work. We propose an incremental algorithm for frontier propagation, which uses the voxel set of the immediate lower-radius sphere or circle to generate the higher-radius sphere or circle. For this, the difference between these two are characterized very specifically. By Theorem 1, $\mathbf{S}(r)$ is the union of $\mathsf{F}(r)$ and $\mathbb{S}(r)$. Circular characterization of $\mathsf{F}(r)$ by Lemma 3 helps to generate the same in an efficient manner. If we generate the digital spheres independently of each other over successive radii, the algorithm will not be efficient. So, we use the voxel set of $\mathbb{S}(r-1)$ along with $\mathsf{F}(r)$ to generate $\mathbf{S}(r)$ by the incremental algorithm. In this section, we first identify the challenges in this approach and then propose the algorithm by overcoming those.

Consider $\mathbb{S}_1(r-1)$. Observe that it is functional on xy-plane. Hence, for each (i,j) pair, the k value gets increased in $\mathbb{S}_1(r)$ by 1 if $(i,j,k+1)$ is not a filler voxel or by 2 otherwise. Based on this observation, we first increase the k value for each (i,j) pair by 1 to produce a new voxel set, namely, Γ, which contains all the voxels of $\mathsf{F}_1(r)$ excepting the ones that have $j = k$. Note that $(i,k,k-1) \notin \mathbb{S}_1(r-1)$ and the remaining voxels of $\mathsf{F}_1(r)$ are always 2-adjacent to the voxels of $\mathbb{S}_1(r-1)$. Each of the voxels of $\mathsf{F}_1(r)$ are also 2-adjacent to some voxels of $\mathbb{S}_1(r)$. We traverse the circles of filler voxels and add the voxel $(i,j,k+1)$ (both $(i,j,k+1)$ and (i,j,k) if $j = k$) to Γ if (i,j,k) is the filler voxel, as others are already included in the last step. Note that the voxel set of Γ now contains a set of voxels (i,j,k) for which we always have $(i,j,k-1) \in \mathbb{S}_1(r)$ or $(i,j,k-1) \in \mathsf{F}_1(r)$. Therefore, the voxels of $\mathbf{S}(r)$ having the form (i,k,k) are not added in Γ yet, as $(i,k,k-1) \notin \mathbb{S}_1(r)$ or $\mathsf{F}_1(r)$. Hence, the explained two steps (along with the other 47 symmetric voxels for each voxel added) produce almost all the voxels of $\mathbf{S}(r)$ excepting these voxels lying on the inter-octant boundaries having $\max\{|i|,|j|,|k|\} = \mathrm{med}\{|i|,|j|,|k|\}$. Clearly these voxels do not have immediate successor on their same q-octant. We have the following observation regarding these extra voxels which are to be added in Γ to make it equivalent to $\mathbb{S}_1(r)$.

Observation 4. *The set of inter-octant boundary voxels of* $\mathbb{S}(r)$ *is given by* $\mathsf{B}(r) = \left\{ p \in \mathbb{Z}^3 : r^2 - \lambda \leqslant s < r^2 + \lambda \wedge (\mu = \lambda) \right\}$, *where,* $p = (i,j,k)$, $s = i^2 + j^2 + k^2$, *and* $\lambda = \max\{|i|,|j|,|k|\}$.

Theorem 2. *First q-octant of the space filling digital sphere of radius r is given by*

$$\mathbf{S}_1(r) = \left\{ \begin{matrix} (i,j,k) \in \mathbb{Z}^3 : (i,j,k-1) \in \mathbb{S}_1(r-1) \vee (i,j,k-1) \in \mathsf{F}_1(r) \\ \vee (j = k \wedge (i,j,k) \in \mathsf{F}_1(r)) \vee (i,j,k) \in \mathsf{B}_1(r) \end{matrix} \right\} \quad (4)$$

Proof. Follows from Theorem 1, Observation 4, and our discussion before Observation 4.   □

Clearly, a solid digital sphere, which is given by $\mathbf{S}^*(r) = \{p \in \mathbb{Z}^3 : s < r^2 + \lambda\}$, is the union of the solid digital sphere of radius $r - 1$ and the space filling digital sphere of radius r, i.e., $\mathbf{S}^*(r) = \mathbf{S}^*(r-1) \cup \mathbf{S}(r)$. During spherical propagation, we increase the radius of the solid digital sphere continuously by unit increment in each iteration. This is discussed in the forthcoming section.

3.1   Spherical Propagation

The algorithm for spherical propagation uses the characterization of filler and inter-octant boundary voxels to generate the successive space filling spheres in a very efficient manner. In this section, for simplicity, we discuss the algorithm for building $\mathbf{S}^*(r)$ from $\mathbf{S}^*(r-1)$ without going into much detail. We resort to a new notion of *axis-parallel 2-neighborhood* of length l defined as follows.

$$N^{(2)}(p, l) = \{q \in \mathbb{Z}^3 : (d_x(p,q) + d_y(p,q) + d_z(p,q)) = d_\perp(p,q) \leqslant l\}$$

Here $d_x(p,q)$, $d_y(p,q)$, and $d_z(p,q)$ denote the axis-parallel distances between points p and q along x-, y-, and z-axes and $d_\perp(p,q)$ denotes the isothetic distance.

From the analysis of the space filling digital sphere, it is easy to observe that any new voxel to be added in $\mathbf{S}^*(r)$ falls in $N^{(2)}(p,2)$ of some voxel p lying on the surface of $\mathbf{S}^*(r-1)$. Also observe that the voxels lying on the surface of $\mathbf{S}^*(r-1)$ are none other than the voxels of $\mathbb{S}(r-1)$ and can easily be found by checking if any of the voxels lying in $N^{(2)}(p,1)$ for voxel p is not yet included in $\mathbf{S}^*(r)$. This neighborhood characterization readily simplifies the steps for spherical propagation, as shown in Algorithm 1.

Algorithm 1. Spherical Propagation

    Input: $\mathbf{S}^*(r-1)$, r
    Output: $\mathbf{S}^*(r)$
1 $A \leftarrow \mathbf{S}^*(r-1), B \leftarrow \phi$
2 **for** $(p \in A) \wedge (q \in N^{(2)}(p,1)) \wedge (q \notin A)$ **do**
3 | $B \leftarrow B \cup p$
4 **end**
5 **for** $(p \in B) \wedge (q \in N^{(2)}(p,2)) \wedge (q \notin A)$ **do**
6 | **if** $q \in \mathbf{S}(r)$ **then**
7 | | $A \leftarrow A \cup q$
8 | **end**
9 **end**
10 **return** A

3.2   Circular Propagation

Our characterization of space filling digital sphere can also be utilized for circular propagation on 3D digital planes. Along with checking the membership of a voxel in space filling digital sphere, we also check for its membership in the given digital plane. This makes the neighborhood to be considered slightly larger, as now we cannot say about surface voxels by just checking their 2-adjacent neighbors or even newly added voxels does not maintain axis-parallel 2-neighborhood with the existing voxels. Hence, we define the axis-parallel 0-neighborhood of a voxel p, within an isothetic distance l from it, as follows.

$$N^{(0)}(p, l) = \left\{ q \in \mathbb{Z}^3 : d_\perp(p, q) \leqslant l \right\}$$

We utilize this neighborhood for propagating circularly from a given point. Algorithm 2 takes a disc $\mathbf{C}^*(r-1)$ of radius $r-1$ on a given digital plane P and produces $\mathbf{C}^*(r)$ by adding the voxels belonging to both P and $\mathbf{S}(r)$.

Algorithm 2. Circular Propagation

   Input: $\mathbf{C}^*(r-1), r, \mathsf{P}$
   Output: $\mathbf{C}^*(r)$
1 $A \leftarrow \mathbf{C}^*(r-1), B \leftarrow \phi$
2 **for** $(p \in A) \wedge (q \in N^{(0)}(p, 1)) \wedge (q \in \mathsf{P}) \wedge (q \notin A)$ **do**
3 | $B \leftarrow B \cup p$
4 **end**
5 **for** $(p \in B) \wedge (q \in N^{(0)}(p, 2)) \wedge (q \in \mathsf{P}) \wedge (q \notin A)$ **do**
6   **if** $q \in \mathbf{S}(r)$ **then**
7 | $A \leftarrow A \cup q$
8   **end**
9 **end**
10 **return** A

---

4   Test Result

In this section, we furnish some preliminary test results on both spherical propagation in 3D space and circular propagation on 3D discrete plane. We use these propagation techniques to generate digital Voronoi diagram from a given set of seed points.

A Voronoi diagram (VD) is a partitioning of a space into regions based on distance from a specific set of seed points as input. For each seed, there is a corresponding region consisting of all points closer to that seed than to any other. These regions are called *Voronoi cells* or *Voronoi regions*. For a given distance metric d, the Voronoi region R_i corresponding to a seed p_i $(1 \leqslant i \leqslant n)$ can be defined as follows.

$$R_i = \{q : d(q, p_i) \leqslant d(q, p_j) \ \forall \ j = 1, 2, \ldots, n\} \tag{5}$$

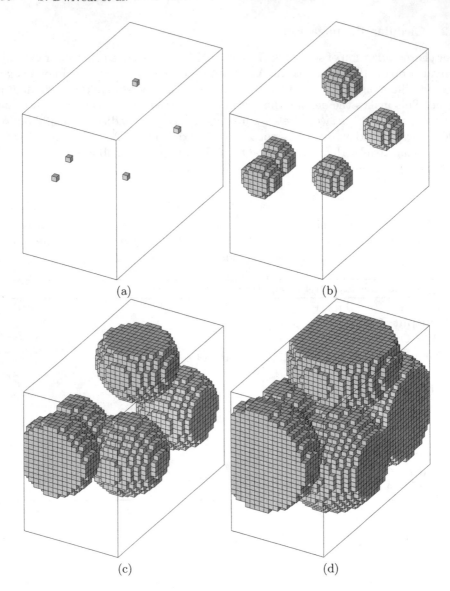

(a)   (b)

(c)   (d)

Fig. 3. Spherical propagation for Voronoi diagram in a confined 3D space. (a) Seed points, (b) After iteration 3, (c) After iteration 7, (d) After iteration 10.

As known, Voronoi diagram (VD) in 2D and in 3D real spaces is a well-researched topic in computational geometry [2,3], but there has not been any significant progress on this to date for 3D digital space. Some interesting work have been reported on characteristics and digital-geometric properties of VD in recent time on the 2D digital plane [12,19].

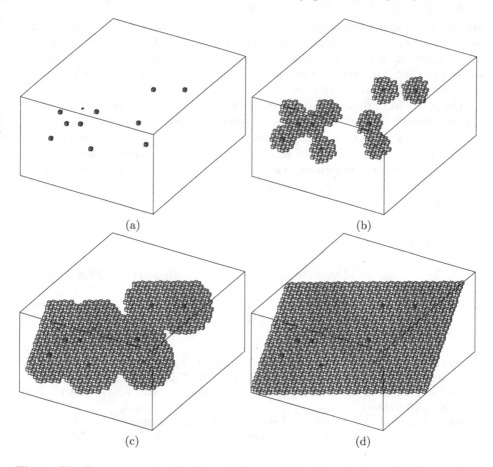

Fig. 4. Circular propagation for Voronoi diagram on a discrete plane. (a) Seed points, (b) After iteration 4, (c) After iteration 9, (d) After iteration 19.

We use Algorithms 1 and 2 to grow in parallel the Voronoi regions over all seeds. The effectiveness of the parallel and incremental region growing algorithm can be utilized for generating the boundaries of the Voronoi regions where the spheres or circles meet. Figure 3(b–d) shows the generation of Voronoi diagram in a confined 3D space by parallel spherical propagation from all the seed points as given in Fig. 3(a). Similarly, in Fig. 4(a), a set of seed points is given on a 3D discrete plane, and Fig. 4(b–d) show the circular propagation to form Voronoi regions on the discrete plane.

5   Concluding Note

We have proposed a model of space filling digital sphere that can be used for efficient propagation of spherical or circular frontier in the digital space. Necessary theoretical analysis and characterization have been provided with related proofs. As an immediate application, we have also shown how this space filling model of digital sphere can be used for construction of digital Voronoi diagrams in 3D space or along a 3D plane. Naturally, this brings in several interesting issues like digital convexity of the Voronoi regions, thus formed, which needs to be addressed in future work.

On real-world applications, a very specific use of space filling digital sphere can be related to discrete 3D terrains for solving various computational problems related to geography information system. A suitable distance measure for construction of well-defined Voronoi diagram on an arbitrary digital surface, e.g., a digital terrain whose underlying real surface is unknown, seems to be an interesting and challenging task. The work presented in this paper can be used to meet these challenges in future.

References

1. Andres, E., Jacob, M.-A.: The discrete analytical hyperspheres. IEEE TVCG **3**, 75–86 (1997)
2. Aurenhammer, F.: Voronoi diagrams–a survey of a fundamental geometric data structure. ACM Comput. Surv. **23**, 345–405 (1991)
3. Aurenhammer, F., Klein, R., Lee, D.: Voronoi Diagrams and Delaunay Triangulations. World Scientific, Singapore (2013)
4. Bera, S., Bhowmick, P., Bhattacharya, B.B.: A digital-geometric algorithm for generating a complete spherical surface in $\mathbb{Z}^3$. In: Gupta, P., Zaroliagis, C. (eds.) ICAA 2014. LNCS, vol. 8321, pp. 49–61. Springer, Cham (2014). doi:10.1007/978-3-319-04126-1_5
5. Bera, S., Bhowmick, P., Bhattacharya, B.B.: On the characterization of absentee-voxels in a spherical surface and volume of revolution in $\mathbb{Z}^3$. JMIV **56**, 535–553 (2016)
6. Bera, S., Bhowmick, P., Stelldinger, P., Bhattacharya, B.B.: On covering a digital disc with concentric circles in $\mathbb{Z}^2$. TCS **506**, 1–16 (2013)
7. Biswas, R., Bhowmick, P.: Layer the sphere. Vis. Comput. **31**, 787–797 (2015)
8. Biswas, R., Bhowmick, P.: From prima quadraginta octant to lattice sphere through primitive integer operations. TCS **624**, 56–72 (2016)
9. Biswas, R., Bhowmick, P.: On the functionality and usefulness of quadraginta octants of naive sphere. JMIV (2017). doi:10.1007/s10851-017-0718-4
10. Biswas, R., Bhowmick, P., Brimkov, V.E.: On the connectivity and smoothness of discrete spherical circles. In: Barneva, R.P., Bhattacharya, B.B., Brimkov, V.E. (eds.) IWCIA 2015. LNCS, vol. 9448, pp. 86–100. Springer, Cham (2015). doi:10.1007/978-3-319-26145-4_7
11. Brimkov, V.E.: Formulas for the number of $(n-2)$-gaps of binary objects in arbitrary dimension. DAM **157**, 452–463 (2009)
12. Cao, T., Edelsbrunner, H., Tan, T.: Triangulations from topologically correct digital Voronoi diagrams. Comput. Geom. **48**, 507–519 (2015)
13. Cohen-Or, D., Kaufman, A.: 3D line voxelization and connectivity control. IEEE Comput. Graph **17**, 80–87 (1997)
14. Gouraud, H.: Continuous shading of curved surfaces. IEEE Trans. Comput. **20**, 623–629 (1971)

15. Kaufman, A.: Efficient algorithms for 3D scan-conversion of parametric curves, surfaces, and volumes. SIGGRAPH **21**, 171–179 (1987)
16. Klette, R., Rosenfeld, A.: Digital Geometry: Geometric Methods for Digital Picture Analysis. Morgan Kaufmann, San Francisco (2004)
17. Roget, B., Sitaraman, J.: Wall distance search algorithm using voxelized marching spheres. In: ICCFD 2012, pp. 1–23 (2012)
18. Roget, B., Sitaraman, J.: Wall distance search algorithm using voxelized marching spheres. J. Comput. Phys. **241**, 76–94 (2013)
19. Rong, G., Tan, T.: Jump flooding in GPU with applications to Voronoi diagram and distance transform. In: Symposium on International 3D Graphics & Games, pp. 109–116 (2006)

Models for Discrete Geometry

Study on the Digitization Dual Combinatorics and Convex Case

Loïc Mazo$^{(\boxtimes)}$ and Étienne Baudrier

ICube-UMR 7357, 300 Bd Sébastien Brant - CS 10413,
67412 Illkirch Cedex, France
loic.mazo@unistra.fr

Abstract. The action of a translation on a continuous object before its digitization generates several digitizations. The *dual*, introduced by the authors in a previous paper, stands for these digitizations in function of the translation parameters. This paper focuses on the combinatorics of the dual by making a link between the digitization number and the boundary curve, especially through its dual representation. The convex case is then studied and a few significant examples are exhibited.

1 Introduction

For a given grid step and a given digitization method, a planar object produces several digitizations in function of its position on the grid. The object digital properties and digitally estimated characteristics depend on the obtained digitization. Thereby, this study of the digitization variability is an important issue in image analysis.

This field has been explored for some geometrical primitives. For instance, the set of straight segment digitizations in function of the segment slope and offset is known as the segment *preimage* and is used for digital straight segment recognition [3]. Several papers are also dedicated to the study of the generation and combinatorics of the disc digitization set in function of its radius and center position [4,6–10,12–14] and the combinatorics of the strictly convex sets [5]. In the general case, the digitization set can be seen as the consequence of a group action on the object. A function, so-called *dual*, linking the group action and the produced digitization is used by the authors to study the digitization set up to a translation, for function graphs in [1] and for planar object in [11]. In the latter case, the dual has been proved to be piecewise constant in function of the translation. This paper focuses on the dual combinatorics.

Two upper bounds are given for the number of digitizations of a planar object whose boundary is a Jordan curve. The first one is expressed in terms of the number of grid cells crossed by the boundary and the second one in terms of the intersection number when plotting the boundary on the torus $\mathbb{R}^2/\mathbb{Z}^2$. The latter bound is proved to be quadratic in the convex case. Some examples are provided in order to compare the two upper bounds both in the convex and the non-convex cases. A conclusion and some perspectives end the paper.

© Springer International Publishing AG 2017
W.G. Kropatsch et al. (Eds.): DGCI 2017, LNCS 10502, pp. 363–374, 2017.
DOI: 10.1007/978-3-319-66272-5_29

2   Background

Let us consider a connected compact set S in $\mathbb{R}^2$ hose boundary is a simple closed (Jordan) curve Γ. Thanks to the Jordan curve theorem, we may assume a continuous map $f \colon \mathbb{R}^2 \to \mathbb{R}$ such that Γ and S are implicitly defined by

$$\Gamma = \{f(x) = 0 \mid x \in \mathbb{R}^2\} \quad \text{and} \quad S = \{x \in \mathbb{R}^2 \mid f(x) \le 0\}.$$

We are interested in the variability of the Gauss digitization when the group of the translations acts on S, that is on the sets $(u + S) \cap \mathbb{Z}^2$, $u \in \mathbb{R}^2$. In this paper, we focus on the combinatorial aspects of this variance. Of course, the variance has to be understood "up to integer translations". This is the reason why we defined in a previous paper [11] the *dual by translation*[1] of the digitizations of S as a set-valued function φ_S defined on the torus $\mathbb{T} = \mathbb{R}^2/\mathbb{Z}^2$ which maps each point $t \in \mathbb{T}$ to the digitization, up to integer translations, of $u + S$ where the vector u is any representative of t in $\mathbb{R}^2$. Let us pick a representative in each class of $\mathbb{T}$ so as to form a connected set M called the *structuring element*. We note C the symmetric of M with respect to the origin: $C = -M$. The family of sets $p + C$, $p \in \mathbb{Z}^2$, is a tiling of $\mathbb{R}^2$. For instance, we can take $M = [0,1)^2$, $C = (-1,0]^2$. From now on, to simplify the notations, we identify $t \in T$ with its representative in M and the subsets of $\mathbb{Z}^2$ with their orbits for the action of the integer translations so we can write $\varphi_S(t) = (t + S) \cap \mathbb{Z}^2$. For any point or set X in $\mathbb{R}^2$, we denote by $\mathrm{proj}(X)$ its projection on the quotient space $\mathbb{T}$. We proved in [11] that the plot of the curve Γ on the torus $\mathbb{T}$, $\mathrm{proj}(\Gamma)$, delineates regions on which the dual function φ_S is constant.

We define the *grid boundary* $\mathcal{B}$ as the set of grid points that lie in the (morphological) dilation of the boundary Γ of S by the structuring element M:

$$\mathcal{B} = (\Gamma \oplus M) \cap \mathbb{Z}^2,$$

where $\oplus$ denotes the Minkowski sum. The set $\mathcal{B}$ contains all the points of $\mathbb{Z}^2$ whose membership to the digitization may change when the set S is shifted by a vector $u \in M$. Nevertheless, since $\mathbb{R}^2$ is connected, M is not open so there may exist some points in $\mathcal{B} \cap S$ not liable to change, namely those points p in $\mathcal{B}$ for which $p + C \subseteq S$ (see Fig. 1). That is why we have in fact to consider the *toggling boundary* $\mathfrak{B}$ as the set of grid points whose membership effectively toggle for some translation by a vector $u \in M$:

$$\mathfrak{B} = \mathcal{B} \setminus \{p \in \mathbb{Z}^2 \mid p + C \subseteq S\}.$$

The set $S \cap \mathbb{Z}^2 \setminus \mathfrak{B}$ of the grid points that are in any digitization of $u + S$, $u \in M$, is called the *digitization core*.

---

[1] The "dual" term is an analogy with the algebraic dual because our construction transforms a set of binary images on a discrete set ($\mathbb{Z}^2$) in a labeled image on a set of transformations (the translations).

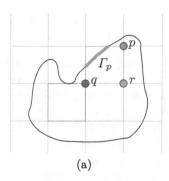 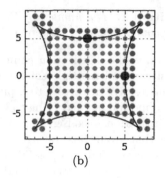

(a)   (b)

Fig. 1. (a) A Jordan curve Γ and a set of tiles $z + C$, $z \in \mathbb{Z}^2$ and $C = (-1,0]^2$. The point p is in $\mathfrak{B}$ for the square $p + C$ intersects both Γ and $\mathbb{Z}^2 \setminus S$; the point q is in $\mathcal{B} \setminus \mathfrak{B}$ for the square $q + C$ intersects the boundary Γ but is included in S; the point r is in the digitization core for the square $r + C$ is included in S and does not intersect Γ (b) Black: a Jordan curve, medium red discs: the toggling boundary, big blue discs: two points that lie in the grid boundary but that cannot change their membership, small green discs: the digitization core. (Color figure online)

Finally, for any $p \in \mathbb{Z}^2$, we set

$$\Gamma_p = -p + \big(\Gamma \cap (p + C)\big) = (-p + \Gamma) \cap C,$$

so $\Gamma_p = \emptyset$ if $p \notin \mathcal{B}$, and we denote by $\mathbb{1}_p$ the indicator function of the set Γ_p, so $\mathbb{1}_p$ is not constant iff $p \in \mathfrak{B}$. Then,

$$\varphi_S(t) = \{p \in \mathbb{Z}^2 \mid \mathbb{1}_p(-t) = 1\}.$$

In the rest of the article, we use the notation $|E|$ for the cardinal of a set E ($|E| \in \mathbb{N} \cup \{\infty\}$), $CC(A)$ for the family of the connected components of a subset A of $\mathbb{R}^2$ and $I \sqcup J$ for the disjoint union of the sets I and J.

In the following section, upper bounds are given for the number of digitizations up to a translation. The first upper bound is naive. For the second one, the idea is to link the digitization number to the number of the intersection of the translated curves Γ_p. Indeed, the dual induces a partition of the torus whose cells frontiers are arcs of the curve $\mathrm{proj}(\Gamma)$. Then, we bound from above the partition size by counting the number of curve intersections in $\mathrm{proj}(\Gamma)$.

3   Contribution

3.1   Bounding up by Counting the Crossed Tiles

Since the grid boundary $\mathcal{B}$ contains any integer point whose value may change when shifting the set S, we have a first, obvious, upper bound on the number of Gauss digitizations given by $2^{|\mathcal{B}|}$. As $|\mathcal{B}|$ is also the number of tiles $p + C$, $p \in \mathbb{Z}^2$, crossed by the frontier of S, we can state the following proposition.

Proposition 1. *The number of Gauss digitizations, up to integer translations, is upper bounded by* $2^{\mathrm{a}(\Gamma)}$ *where* $\mathrm{a}(\Gamma)$ *is the number of tiles crossed by* Γ.

Generally, the digitization enumeration provided by Proposition 1 includes false positives and multiple counts. For instance, the grid boundary of a circle with diameter 1.7 involves 4 to 8 pixels, depending on the grid position, which gives an upper bound according to Proposition 1 equals to 16, while there exists only 8 digitizations (see Fig. 2). Nevertheless, from any set, it is possible to build a new set that avoids false positives by replacing the initial boundary by a family of Hilbert curves and it is possible to extend the set to prevent multiple counts so that the theoretical upper bound $2^{\mathfrak{B}}$ is obtained (see Appendix B).

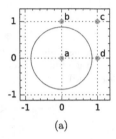
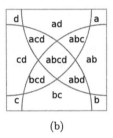

(a)   (b)

Fig. 2. (a) A circle with diameter 1.7 and the four pixels in the toggling boundary (which is equal to the grid boundary): **a**, **b**, **c**, **d**. (b) The (flatten) dual of the closed disc bounded by the circle. Among the $16 = 2^4$ potential digitizations provided by the 4-pixels **a**, **b**, **c**, **d**, three of them does not actually appear ($\emptyset$, **ac** and **bd**) and some others appear multiple times (the four singular digitizations, which are obviously congruent, and the vertical and horizontal pairs which each appear twice. Eventually, there are only 8 digitizations (and actually the (sewed) dual is divided into 8 regions).

3.2   Bounding up by Counting the Intersections

In Sect. 3.2, we assume a parametrization of Γ. It induces an order on the points of the curve Γ (for Γ is simple) that is used in the proof of the following proposition.

The dual φ_S can be regarded as the projection, on the torus $\mathbb{T}$ of a finite labeled partition of the tile C (whose cells need not be connected). For the order by refinement on the partitions, this partition is lower bounded by the infimum of the binary partitions $\mathcal{P}_p$ associated to the indicator functions $\mathbb{1}_p$, $p \in \mathfrak{B}$ (actually, $\mathrm{proj}(\mathbb{1}_p)$, $p \in \mathfrak{B}$). Indeed some cells of $\bigwedge_{p \in \mathfrak{B}} \mathcal{P}_p$ need to be merged whenever the corresponding digitizations are equivalent up to an integer translation. Then, the next proposition proposes an upper bound on the size of the partition $\bigwedge_{p \in \mathfrak{B}} \mathcal{P}_p$—and thereby an upper bound to the number of digitizations—by inductively counting the intersections between the curves Γ_p, $p \in \mathfrak{B}$. The idea is to count the partition cells created when adding a curve Γ_p. To do this we count the intersection of Γ_p with the already added curves.

Nevertheless, such an intersection can be with one or more curves and can be a singleton, a set of several points or an arc. Then, let us specify how we handle intersections in this context (the definition is illustrated in Fig. 3).

Definition 1. *Let* $\mathfrak{B} = \{b_1, \cdots, b_i, \cdots, b_n\}$, $n \geq 1$. *Let* $m \in [2, n]$. *Then,*

$$\mathrm{inter}_{\gamma, I} = \mathrm{CC}\left(\gamma \cap \left(\bigcap_{i \in I} \Gamma_{b_i} \setminus \bigcup_{j \in J} \Gamma_{b_j}\right)\right)$$

$$\textit{where } \gamma \in \mathrm{CC}(\Gamma_{b_m}) \textit{ and } I \sqcup J = [1, m - 1],$$

$$\# \mathrm{inter}_m = \sum_{\substack{\gamma \in \mathrm{CC}(\Gamma_{b_m}) \\ \emptyset \subset I \subseteq [1, m-1]}} w_I \left|\mathrm{inter}_{\gamma, I}\right| \quad \textit{where } w_I = \min(|I|, 2).$$

Note that the components of all the $\mathrm{inter}_{\gamma, I}$ are two by two disjoint and that $\#\mathrm{inter}_m$ may be infinite.

The set $\mathrm{inter}_{\gamma, I}$ stands for the intersection of the curve γ with the curves whose indexes are in a subset I, excluding any other curve Γ_{b_i} whose index is not in I. The necessity to have all the index subsets comes from the fact that intersections between multiple curves can occur.

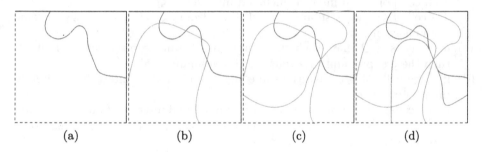

(a)   (b)   (c)   (d)

Fig. 3. (a) Blue: Γ_{b_1}. (b) Green: Γ_{b_2}; $\#\mathrm{inter}_2 = \left|\mathrm{inter}_{\Gamma_{b_2}, \{1\}}\right| = 4$. (c) Brown: Γ_{b_3}; $\#\mathrm{inter}_3 = \left|\mathrm{inter}_{\Gamma_{b_3}, \{1\}}\right| + \left|\mathrm{inter}_{\Gamma_{b_3}, \{2\}}\right| + 2\left|\mathrm{inter}_{\Gamma_{b_3}, \{1,2\}}\right| = 0 + 0 + 2 = 2$. (d) Red: Γ_{b_4}; $\#\mathrm{inter}_4 = \left|\mathrm{inter}_{\Gamma_{b_4}, \{1\}}\right| + 2\left|\mathrm{inter}_{\Gamma_{b_4}, \{1,2,3\}}\right| = 2 + 2 = 4$. (Color figure online)

We are now able to state the proposition that relates the number of digitizations and the number of intersections $\#\mathrm{inter}_m$.

Proposition 2. *The size of the partition* $\bigwedge_{p \in \mathfrak{B}} \mathcal{P}_p$ *is upper bounded by*

$$2 + \sum_{m=2}^{n} \#\mathrm{inter}_m + |\mathrm{CC}(\Gamma_{b_m})|. \tag{1}$$

Proof. The proposition is proved by induction. For $m = 1$, the result is obvious since, for any $p \in \mathfrak{B}$, $\mathcal{P}_p$ is a binary partition. Let $m > 1$.

We assume that the number of cells of the partition $\bigwedge_{i=1}^{m-1} \mathcal{P}_{b_i}$ is upper bounded by $2 + \sum_{k=2}^{m-1} \# \mathrm{inter}_k + |\mathrm{CC}(\Gamma_{b_k})|$. The cells of $\bigwedge_{i=1}^{m-1} \mathcal{P}_{b_i}$ which are included in one of the two cells of $\mathcal{P}_{b_m}$, namely the sets $(\mathbb{1}_{b_m} = 0)$ and $(\mathbb{1}_{b_m} = 1)$, stay unchanged in the partition $\bigwedge_{i=1}^{m} \mathcal{P}_{b_i}$. Conversely, the cells of $\bigwedge_{i=1}^{m-1} \mathcal{P}_{b_i}$ that are intersected by both $(\mathbb{1}_{b_m} = 0)$ and $(\mathbb{1}_{b_m} = 1)$, which obviously are cells intersected by Γ_{b_m}, are each divided in two new cells. Thereby, the number N_m of new cells is upper bounded by N_m', the number of cells in $\bigwedge_{i=1}^{m-1} \mathcal{P}_{b_i}$ intersected by Γ_{b_m}. Besides, the partition $\bigwedge_{i=1}^{m-1} \mathcal{P}_{b_i}$ of the tile C induces a partition $\mathcal{Q}$ of Γ_{b_m}, as a subset of C, with N_m' cells (these cells of Γ_{b_m} need not be connected). Then, the idea of the proof is to map each cell of $\mathcal{Q}$ to its supremum, for the order induced by the parametrization of Γ—namely to the intersection, as defined in Definition 1—its supremum belongs to, or to the empty set when no such intersection exists. Then, though this mapping is not one-to-one, a careful examination of the different cases will permit to conclude that N_m' is upper bounded by $\# \mathrm{inter}_m + |\mathrm{CC}(\Gamma_{b_m})|$.

Let γ be a connected component of Γ_{b_m} and s_∞ be the supremum of γ for the order induced by the parametrization of Γ. Let $\mathcal{Q}_\gamma$ be the restriction of $\mathcal{Q}$ to γ. We set $\mathrm{inter}_\gamma = \bigcup_{\emptyset \subset I \subseteq [1,m-1]} \mathrm{inter}_{\gamma,I}$, and we assume the following facts that will be prove further:

(a) each component in inter_γ is included in a cell of $\mathcal{Q}_\gamma$;
(b) if a cell of $\mathcal{Q}_\gamma$ has a supremum s distinct from s_∞, then s belongs to some component of inter_γ;
(c) if two cells of $\mathcal{Q}_\gamma$ have their supremums in the same component K of inter_γ, then these supremums are equal to the infimum of K.
(d) if three cells of $\mathcal{Q}_\gamma$ have the same supremum s, then inter_γ has infinitely many components;
(e) if two cells of $\mathcal{Q}_\gamma$ have the same supremum s and inter_m is finite then either $s = s_\infty$ and s_∞ belongs to some component of inter_γ, or s belongs to at least two curves Γ_{b_i} and Γ_{b_j} where $i,j < m$.

Then, with these five assumptions, we deduce the desired upper bound as follows. We map each cell of $\mathcal{Q}_\gamma$ to the component of inter_γ its supremum belongs to, if any. The other cells are map on the empty set. We denote by ψ this mapping. Hence, from Fact b,the supremum of the preimages by ψ of $\emptyset$, if any, is s_∞ and so, $s_\infty \notin \mathrm{inter}_\gamma$ if such preimages exist. If a component K of inter_γ has exactly two preimages K_1 and K_2 by the mapping ψ, then from Fact c, we derive that the two cells share the same supremum and from Fact e, and the very definition of $\# \mathrm{inter}_m$, we see that the weight of K in $\# \mathrm{inter}_m$ is 2, which corresponds to the number of its preimages, or $s = s_\infty$ and the empty set has no preimage. If the empty set has two or more preimages by ψ, then $\# \mathrm{inter}_m = \infty$ (Fact e). If a component K of inter_γ has three or more preimages by ψ, these preimages share the same supremum (Fact c) and from Fact d we derive that $\# \mathrm{inter}_m = \infty$. We readily conclude that the number of cells of $\mathcal{Q}_\gamma$ is upper bounded by $1 + \sum_{\emptyset \subset I \subseteq [1,m-1]} w_I |\mathrm{inter}_{\gamma,I}|$ where $w_I = \min(2, |I|)$. By summing on all the connected components of Γ_{b_m}, we derive that N_m', the number of cells of $\mathcal{Q}$, is upper bounded by $\# \mathrm{inter}_m + |\mathrm{CC}(\Gamma_{b_m})|$ which achieves the proof.

Let us now prove the five facts stated above.

(a) We demonstrate Fact a by contradiction. Suppose it exists K in $\text{inter}_{\gamma,I}$, $I \subseteq [1, m-1]$, and two distinct cells K_1, K_2 of $\bigwedge_{i=1}^{m-1} \mathcal{P}_{b_i}$ such that $K \cap K_1 \neq \emptyset$ and $K \cap K_2 \neq \emptyset$. Let $c_1 \in K \cap K_1$ and $c_2 \in K \cap K_2$. Since $K_1 \neq K_2$, there exists $i \in [1, m-1]$ such that $\mathbb{1}_{b_i}(c_1) \neq \mathbb{1}_{b_i}(c_2)$. Then, the segment $[c_1, c_2]$ in γ contains a point $c_3 \in \Gamma_{b_i}$ which also belongs to K for K is connected. Then, on the one hand $i \in I$ for $c_3 \in \Gamma_{b_i} \cap K$ and, on the other hand, c_1, or c_2, is in $(\mathbb{1}_{b_i} = 0)$, that is $i \notin I$. Contradiction.

(b) Suppose that the supremum s of the cell K of $\mathcal{Q}_\gamma$ is not in any component of inter_m. Then, for any $i \in [1, m-1]$, $s \notin \Gamma_{b_i}$. In other words, for any $i \in [1, m-1]$, s is in the open set $(\mathbb{1}_{b_i} = 0)$ or in the interior of $(\mathbb{1}_{b_i} = 1)$. Thus, there exists an open neighborhood U of s in the tile C which does not intersect any curve Γ_{b_i}, $1 \leq i < m$. Thereby, U is included in a cell of the partition $\bigwedge_{i=1}^{m-1} \mathcal{P}_{b_i}$. Since $s = \sup(K)$, there is no point t in $\gamma \cap U$ such that $t > s$. Thus, $s = s_\infty$.

(c) Let K_1, K_2 be two cells of $\mathcal{Q}_\gamma$ whose supremums are in $K \in \text{inter}_\gamma$. Let K_3 be the cell of $\mathcal{Q}_\gamma$ which includes this component (Fact a). If s is not the infimum of K then there exists an interval $[u, s]$, with $u < s$, in $K \subseteq K_3$ (for K is connected) and s is not the supremum of both K_1 and K_2. Contradiction.

(d) Let K_1, K_2, K_3 be three cells of $\mathcal{Q}_\gamma$ that share the same supremum s. Since s belongs to at most one of these three cells, it is a limit point for the two others. For instance, assume that s is a limit point for K_1 and K_2. Let $k \in [1, m-1]$ such that $\mathbb{1}_{b_k}$ takes two distinct values on K_1 and K_2. For instance, $K_1 \subseteq (\mathbb{1}_{b_k} = 0)$ and $K_2 \subseteq (\mathbb{1}_{b_k} = 1)$. Then, we can inductively build an infinite sequence $c_1 < c_2 < \cdots < c_i < \cdots < s$ such that $c_{2i-1} \in K_1$ and $K_{2i} \in K_2$ for any $i \geq 1$. In particular, we have $\mathbb{1}_{b_k}(c_{2i-1}) = 0$ and $\mathbb{1}_{b_k}(c_{2i}) = 1$ for any $i \geq 1$. Then, $c_{2i-1} \notin \Gamma_{b_k}$ and $[c_{2i-1}, c_{2*i}]$ intersects Γ_{b_k}. Therefore, inter_γ contains infinitely may components between c_1 and s.

(e) We assume $\#\text{inter}_m < \infty$. Let K_1, K_2 be two cells of $\mathcal{Q}_\gamma$ that share the same supremum s. This supremum cannot be a limit point both in K_1 and K_2 otherwise we could make the same reasoning as in the previous item and concludes that $\#\text{inter}_m = \infty$ which contradicts our assumption. So, s is an isolated point in one of the two cells, for instance K_1 (therefore, $s \in K_1$, $s \notin K_2$ and s is a limit point in K_2). From Fact b, there exist a subset I of $[1, m-1]$ and a component K in $\text{inter}_{\gamma,I}$ such that $s \in K$. As in the proof of Fact b, we derive that there exists an open neighborhood U of s in C which does not intersect any curve Γ_{b_i}, $i \notin I$. Suppose that I is a singleton, say $I = \{b_1\}$. Then, taking a point in $K_2 \cap U$ (recall that s is a limit point of K_2), we found that necessarily $\mathbb{1}_{b_1} = 0$ on K_2, while $\mathbb{1}_{b_1} = 1$ on K_1, and $\mathbb{1}_{b_i}$, $i \neq 1$, coincides on K_1 and K_2. If $s \neq s_\infty$, there exists a point u in $\gamma \cap U$ greater than s. This point is not in K_1 nor in K_2 for $s = \sup(K_1) = \sup(K_2)$. Thereby, $\mathbb{1}_{b_1}(u) \neq 0$ and $\mathbb{1}_{b_1}(u) \neq 1$ which is absurd. Thus, either $s_\infty = s \in K_1$ or I is not a singleton. $\square$

In the simplest case where the curve segments Γ_{b_m} all have one connected component and where the intersections are sets of points belonging to at most two curves, Formula 1 reduces to $1 + |\mathfrak{B}| + I$ where I is the number of inter-section points between the curve segments $\varGamma_{b_m}$. Furthermore, when projecting the partition $\bigwedge_{i=1}^{n} \mathcal{P}_{b_i}$ on the torus $\mathbb{T}$, the cells that touch the boundary of C are identified two by two, which decreases the number of cells in the partition. Unfortunately, it is difficult to count these cells in the general case.

3.3   The Convex Case

It seems plain that the structure of the dual should be simpler when the set S is convex compared to a set with a winding boundary. We could even hope that the digitizations coincide with the cells of the partition $\bigwedge_{p \in \mathfrak{B}} \varGamma_p$, after sewing the boundary of the tile C, and with the connected regions of the torus $\mathbb{T}$ delineated by the curve $\varGamma$. Figure 4 annihilates this hope by exhibiting a convex object and one of its digitizations whose inverse image by the dual is not connected.

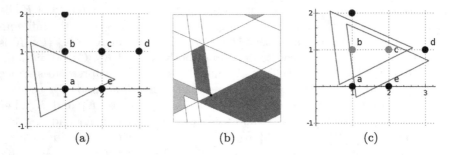

Fig. 4. (a) A triangle and its toggling boundary. (b) The dual of the (filled) triangle. The four colored regions correspond to the same digitization: the horizontal pair. Red region: cell **bc** of the partition $\bigwedge_{p \in \mathfrak{B}} \varGamma_p$. The red region is disconnected by the unique point of the cell **abcd**. Orange region: cell **cd**. Pink region: cell **ae**. (c) The configurations of the shifted triangle in the two red components. Since they are disconnected, it is not possible to continuously move to one configuration to the other without hurting a black toggling point. Note that the triangle could be slightly inflated so as to provide a strictly convex counterexample. (Color figure online)

Nevertheless, in this section we show that the structure of the sets $\text{inter}_{\gamma,I}$ is simple when the set S is strictly convex and permits to obtain a quadratic bound for the complexity of the dual in terms of the grid boundary size (to be compare to the exponential bound of Proposition 1).

Proposition 3. *We assume a convex quadrilateral structuring element M. Let $n = |\mathfrak{B}|$. The number of Gauss digitizations of a strictly convex planar object up to a translation is upper bounded by*

$$4n^2 + 4n - 6.$$

Proof. The proof invokes Lemma 3 which is stated and proved in the appendix. Let N be the digitization number. Thanks to Proposition 2, we have

$$N \leq 2 + \sum_{m=2}^{n} \# \text{inter}_m + |CC(\Gamma_{b_m})|. \tag{2}$$

From Lemma 3, we derive that, for any $m \in [1, n]$, $\gamma \in CC(\Gamma_{b_m})$, $i \in [1, m-1]$,

$$\sum_{I \ni i} |\text{inter}_{\gamma, I}| \leq 2 |CC(\Gamma_{b_i})|.$$

Then,

$$\# \text{inter}_m \leq 2 \sum_{k=1}^{m-1} |CC(\Gamma_{b_k})|.$$

Eventually, Eq. (2) turns into

$$N \leq 2 + 2 \sum_{m=2}^{n} \sum_{k=1}^{m-1} |CC(\Gamma_{b_k})| + \sum_{m=2}^{n} |CC(\Gamma_{b_m})|$$

$$\leq 2 \left(1 + \sum_{k=1}^{n} (n - k + 1) |CC(\Gamma_{b_k})| - |CC(\Gamma_{b_1})| \right).$$

It can easily be seen (from Lemma 2 for instance) that, for any convex curve Γ and any convex polygon P with c edges, the number of connected arcs in the intersection $\Gamma \cap P$ is upper bounded by c. Then, if we assume that the tile C is a convex quadrilateral, we straightforwardly obtain the desired bound.   □

The term $4n^2$ in Proposition 3 comes down to n^2 for sufficiently high resolutions because each curve Γ_{b_i} then have just one connected component instead of possibly 4 in the general case. Then, when S a disc of radius r, the result of Proposition 3 is close to the one in [7] which states that the number of digitizations of the disc is asymptotically $4\pi r^2 + O(r)$. As the ratio between the radius r and the size of the grid boundary is $\pi/4$ for the disc, our upper bound in function of the radius r of the disc is asymptotically equivalent to $(16/\pi^2)r^2$.

4   Conclusion

We present in this paper two upper bounds on the number of digitizations obtained from all the translated of a continuous object. The first one is exponential in the number of toggling object-boundary pixels and a generic example reaching this bound is given. The second one is based on the passage from the dual connected-component count to the curve intersection count which, if the curve is parameterized, comes down to count some equation solutions. In the convex case, the latter upper bound is shown to yield a quadratic digitization number in term of the grid boundary size. An example of a convex object is given where the set of translation parameter classes corresponding to a given digitization is not connected.

The perspectives of this work are first to explicit the second upper bound under assumptions less restrictive that the convexity, *e.g.* bounded curvature; then to study the combinatorics of the digitization under the rigid transformations and to propose an algorithm for the digitization generation.

Acknowledgements. We thank Renan Lauretti for his idea to link the dual regions to the dual region border crossings in his study of the function graph dual.

Appendix

A   Convex Sets

The proof of Lemma 3 relies on the two following lemmas about convex sets that seem obvious at the first glance. Nevertheless, since we did not find any result related to these lemmas in the literature, we provide our own justifications of the two statements.

Lemma 1 (Chords of convex sets). *Let $[a, b]$ be a chord of the boundary Γ of a closed convex set S. If $[a, b] \not\subseteq \Gamma$, then $(ab) \cap S = [a, b]$ and $(ab) \cap \Gamma = \{a, b\}$.*

Proof. Since $[a, b] \not\subseteq \Gamma$, the line (ab) does not support S at any point (the notion of supporting line of a convex set is exposed for instance in [2]). So, there exists two supporting lines of S at a and b that cross the line (ab). Then, $(ab) \cap S$ is included in $[a, b]$. Let $c \in [a, b] \cap \Gamma$, $c \neq a$. Applying the first part of the proof to the chord $[a, c]$, we derive that $(ac) \cap S \subseteq [a, c]$ and, since $b \in (ac) \cap S$, we conclude that $b = c$. $\square$

Lemma 2 (Cuts of convex sets). *Let a, b be two points of the boundary Γ of a closed convex set S. If $[a, b] \not\subseteq \Gamma$, then the open curve segments of Γ, σ_1, σ_2, whose extremities are a and b are included in distinct open half-planes bounded by the line (ab).*

Proof. Let H^- and H^+ be the two open half-planes bounded by (ab). Since $[a, b] \not\subseteq \Gamma$, from Lemma 1, $(ab) \cap \Gamma = \{a, b\}$. Thereby, by connectivity, either $\sigma_1 \subset H^-$ or $\sigma_1 \subset H^+$ and $\sigma_2 \subset H^-$ or $\sigma_2 \subset H^+$. Suppose for instance that $\sigma_1 \subset H^-$ and $\sigma_2 \subset H^-$. Then, S, which is the connected subset of $\mathbb{R}^2$ bounded by $\sigma_1 \cup \sigma_2 \cup \{a, b\}$ is included in $H^- \cup (ab)$ and, since $(ab) \cap S = [a, b]$ from Lemma 1, $[a, b] \subset \Gamma$: contradiction. $\square$

Lemma 3 (Intersection of two segments of a convex curve). *Let Γ be a Jordan curve whose interior is convex. Let Γ_1 and Γ_2 two disjoint closed segments of the curve Γ and τ a vector of $\mathbb{R}^2$. Then the intersection of Γ_1 and $\tau + \Gamma_2$ is composed of none, one, two points or a line segment.*

Proof. Let p, q be two distinct points in $\Gamma_1 \cap (\tau + \Gamma_2)$ if such a pair exists. We denote by Σ_1 the open segment of Γ_1 between p and q. Alike, Σ_2 is the open segment of Γ_2 between $-\tau + p$ and $-\tau + q$. We set $\overline{\Sigma}_1 = \Sigma_1 \cup \{p, q\}$ and $\overline{\Sigma}_2 = \Sigma_2 \cup \{p, q\}$. Firstly, we prove that $\Sigma_1 \cup (\tau + \Sigma_2)$ is a straight line segment whenever it contains more than two points. First case: $\Sigma_1 \cup (\tau + \Sigma_2) \subseteq (pq)$. Then, since $\overline{\Sigma}_1$ and $\overline{\Sigma}_2$ are connected and Γ is simple, $\overline{\Sigma}_1 = \overline{\Sigma}_2 = [p, q]$. Second case: $\exists x \in \left(\Sigma_1 \cup (\tau + \Sigma_2) \right) \setminus (pq)$. For instance, we assume $x \in \Sigma_1 \setminus (pq)$. By Lemma 2, $\Sigma_1 = \Gamma \cap H_1$, where H_1 is the open half-plane bounded by the line (pq) and containing x, and $-\tau + p$ is in $\mathbb{R}^2 \setminus H_1$. Then, it can easily be seen that $-\tau + H_1$ is the open half-plane bounded by the line joining $-\tau + p$ and $-\tau + q$ and including p. Thanks to Lemma 2, we derive that Σ_2 does not intersect $-\tau + H_1$. Thus, $\tau + \Sigma_2$ does not intersect H_1. In particular, $(\tau + \Sigma_2) \cap \Sigma_1 = \emptyset$. This achieves the first part of the proof. Now, let r be a point in $\Gamma_1 \cap (\tau + \Gamma_2)$ which is not in Σ_1 (if such a point exists). For instance, p belongs to the segment of Γ_1 between q and r. Then, the first part of the proof, applied to the points q and r, implies that $\Gamma_1 \cap (\tau + \Gamma_2)$ includes the segments $[q, r]$. We straightforwardly concludes that either the intersection of Γ_1 and $\tau + \Gamma_2$ is composed of at most two points or it is a line segment. $\qquad\square$

B   Examples and Counterexamples

B.1   Building Examples Without Proper Congruent Digitizations in the Image of the Dual

Let u and v be two vectors in $[0, 1)^2$ such that the sets $u+S$ and $v+S$ have distinct but congruent digitizations. Then, there exists an integer vector w, $w \neq 0$, such that $(u + S) \cap \mathbb{Z}^2 = w + ((v + S) \cap \mathbb{Z}^2) = (w + v + S) \cap \mathbb{Z}^2$. Let p be a point in the digitization core. Then, $p \in (u + S) \cap \mathbb{Z}^2$ and $p \in (v + S) \cap \mathbb{Z}^2$. Therefore, $w + p \in (u + S) \cap \mathbb{Z}^2$ and $-w + p \in (v + S) \cap \mathbb{Z}^2$, which can be rewritten as $p \in ((-w + u) + S) \cap \mathbb{Z}^2$ and $p \in ((w + v) + S) \cap \mathbb{Z}^2$. Then, at least one of the vectors $w + v$ or $-w + u$ has one of its coordinates which is negative. We derive that if there exists a point in the digitization core which is maximal in S for both coordinates then there is no proper congruent digitizations in the dual.

B.2   Building Toric Partitions in One-to-one Correspondence with the Power Set of the Toggling Boundary

In this section, we exhibit a way to modify the boundary of the set S in order to ensure that any subset of the toggling boundary is represented in the dual. To do so, we move along $\mathcal{B}$, ordered in the same way as in Definition 1. Then, a new boundary is built thanks to the approximations of the Hilbert filling curve: the segment of Γ intersecting the n-th cell of $\mathcal{B}$ is replaced by a n-th approximation of the Hilbert filling curve H_n (extended at its extremities to ensure the continuity of the boundary). We consider the family of binary partitions $\mathcal{P}_n$ of the unit square that comes with the curves H_n. We claim that each curve H_{n+1} crosses each cell of the partition $\bigwedge_{i=1}^{n} \mathcal{P}_i$ so that the size of the final torus partition is 2^N where N is the cardinal of the toggling boundary. To justify our claim, we divide the unit square in a family of $2^n \times 2^n$ small squares $(K_{i,j}^n)_{1 \leq i,j \leq 2^n}$ $(n \geq 0)$ whose sizes are $\frac{1}{2^n} \times \frac{1}{2^n}$. It can be seen that on the one hand, the Hilbert curve H_n passes through the center of each of the squares $K_{i,j}^n$ and, on the other hand, does not intersect any of the interior of the squares $K_{i,j}^{n+1}$ (H_0 is just the center of the unit square). Thereby, the partition $\bigwedge_{i=1}^{n} \mathcal{P}_i$ is coarser than the partition $\{K_{i,j}^{n+1} \mid 1 \leq i,j < n\}$ (the boundaries of the squares $K_{i,j}^k$ are assigned to the cells so as to coincide with $\mathcal{P}_n$). Since H_{n+1} passes through the center of each of the squares $K_{i,j}^{n+1}$, it passes in each cell of $\bigwedge_{i=1}^{n} \mathcal{P}_i$ which gives the claim.

References

1. Baudrier, É., Mazo, L.: Curve digitization variability. In: Normand, N., Guédon, J., Autrusseau, F. (eds.) DGCI 2016. LNCS, vol. 9647, pp. 59–70. Springer, Cham (2016). doi:10.1007/978-3-319-32360-2_5
2. Berger, M.: Geometry, vol. 1 (1987)
3. Dorst, L., Smeulders, A.W.M.: Discrete Representation of Straight Lines. IEEE Trans. Pattern Anal. Mach. Intell. **6**(4), 450–463 (1984). http://ieeexplore.ieee.org/lpdocs/epic03/wrapper.htm?arnumber=4767550
4. Heath-Brown, D.: Lattice points in the sphere. Number Theory Prog. **2**, 883–892 (1997)
5. Huxley, M.N.: The number of configurations in lattice point counting I. Forum Mathematicum **22**(1), 127–152 (2010)
6. Huxley, M.N., Žunić, J.: On the number of digitizations of a disc depending on its position. In: Klette, R., Žunić, J. (eds.) IWCIA 2004. LNCS, vol. 3322, pp. 219–231. Springer, Heidelberg (2004). doi:10.1007/978-3-540-30503-3_17
7. Huxley, M.N., Žunić, J.D.: Different digitisations of displaced discs. Found Comput. Math. **6**(2), 255–268 (2006). doi:10.1007/s10208-005-0177-y
8. Huxley, M.N., Zunić, J.D.: The number of N-point digital discs. IEEE Trans. Pattern Anal. Mach. Intell. **29**(1), 159–161 (2007). doi:10.1109/TPAMI.2007.250606
9. Huxley, M.N., Zunić, J.D.: The number of different digital N-discs. J. Math. Imaging Vis. **56**(3), 403–408 (2016). doi:10.1007/s10851-016-0643-y
10. Mazo, J.E., Odlyzko, A.M.: Lattice points in high-dimensional spheres. Monatshefte für Mathematik **110**(1), 47–61 (1990)
11. Mazo, L., Baudrier, É.: Object digitization up to a translation, September 2016. preprint https://hal.archives-ouvertes.fr/hal-01384377
12. Nagy, B.: An algorithm to find the number of the digitizations of discs with a fixed radius. Electron. Notes Discr. Math. **20**, 607–622 (2005). doi:10.1016/j.endm.2005.04.006
13. Žunić, J.D.: On the number of digital discs. J. Math. Imaging Vis. **21**(3), 199–204 (2004). doi:10.1023/B:JMIV.0000043736.15525.ed
14. Žunić, J.D.: On the number of ways to occupy n lattice points by balls in d-dimensional space. J. Number Theor. **110**(2), 396–402 (2004)

Algorithmic Construction of Acyclic Partial Matchings for Multidimensional Persistence

Madjid Allili[1]([✉]), Tomasz Kaczynski[2], Claudia Landi[3], and Filippo Masoni[3]

[1] Department of Computer Science, Bishop's University,
Sherbrooke, QC J1M1Z7, Canada
mallili@ubishops.ca
[2] Department of Mathematics, Université de Sherbrooke,
Sherbrooke, QC J1K2R1, Canada
t.kaczynski@usherbrooke.ca
[3] Dipartimento di Scienze e Metodi dell'Ingegneria,
Università di Modena e Reggio Emilia, Reggio Emilia, Italy
claudia.landi@unimore.it

Abstract. Given a simplicial complex and a vector-valued function on its vertices, we present an algorithmic construction of an acyclic partial matching on the cells of the complex. This construction is used to build a reduced filtered complex with the same multidimensional persistent homology as of the original one filtered by the sublevel sets of the function. A number of numerical experiments show a substantial rate of reduction in the number of cells achieved by the algorithm.

Keywords: Multidimensional persistent homology · Discrete Morse theory · Acyclic partial matchings · Matching algorithm · Reduced complex

1 Introduction

Persistent homology has been established as an important tool for the topological analysis of discrete data. However, its effective computation remains a challenge due to the huge size of complexes built from data. Some recent works focussed on algorithms that reduce the original complexes generated from data to much smaller cellular complexes, homotopically equivalent to the initial ones by means of *acyclic partial matchings* of discrete Morse theory.

Although algorithms computing acyclic partial matchings have primarily been used for persistence of one-dimensional filtrations, see e.g. [11,13,16], there is currently a strong interest in combining persistence information coming from multiple functions in multiscale problems, e.g. in biological applications [20], which motivates extensions to generalized types of persistence. The extension

T. Kaczynski—This work was partially supported by NSERC Canada Discovery Grant and IMA Minnesota.

C. Landi—Work performed under the auspices of INdAM-GNSAGA.

W.G. Kropatsch et al. (Eds.): DGCI 2017, LNCS 10502, pp. 375–387, 2017.
DOI: 10.1007/978-3-319-66272-5_30

of persistent homology to multifiltrations is studied in [5]. Other related directions are explored e.g. by the authors of [19] who do statistics on a set of one-dimensional persistence diagrams varied as coordinate system rotates, and in [8], where persistence modules on quiver complexes are studied.

Our attempt parallel to [8] is [2], where an algorithm given by King et al. in [11] is extended to multifiltrations. The algorithm produces a partition of the initial complex into three sets (A, B, C) and an acyclic partial matching $m : A \rightarrow B$. Any simplex which is not matched is added to C and defined as critical. The matching algorithm of [2] is used for reducing a simplicial complex to a smaller one by elimination of matched simplices. First experiments with filtrations of triangular meshes show that there is a considerable amount of cells identified by the algorithm as critical but which seem to be spurious, in the sense that they appear in clusters of adjacent critical faces which do not seem to carry significant topological information.

The aim of this paper is to improve our previous matching method for optimality, in the sense of reducing the number of spurious critical cells. Our new matching algorithm extends the one given in [16] for cubical complexes, which processes lower stars rather than lower links, and improves the result of [2] for optimality. Next, the new matching algorithm presented here emerges from the observation that, in the multidimensional setting, it is not enough to look at lower stars of vertices: one should take into consideration the lower stars of simplices of all dimensions, as there may be vertices of a simplex which are not comparable in the partial order of the multifiltration. The vector-valued function initially given on vertices of a complex is first extended to simplices of all dimensions. Then the algorithm processes the lower stars of all simplices, not only the vertices. The resulting acyclic partial matching is used as in [2] to construct a reduced filtered Lefschetz complex with the same multidimensional persistent homology as the original simplicial complex filtered by the sublevel sets of the function. Our reduction is derived from the works of [10,13,14]. A recent related work on reduction techniques can be found among others in [6].

The paper is organized as follows. In Sect. 2, the preliminaries are introduced. In Sect. 3, the main Algorithm 2 is presented and its correctness is discussed. At the section end, the complementing reduction method is recalled from [2]. In Sect. 4 experiments on synthetic and real 3D data are presented. In Sect. 5, we comment on open questions and prospects for future work.

2   Preliminaries

Let $\mathcal{K}$ be a finite geometric simplicial complex, that is a finite set composed of vertices, edges, triangles, and their q-dimensional counterparts, called simplices. A q-dimensional simplex is the convex hull of affinely independent vertices $v_0, \ldots v_q \in \mathbb{R}^n$ and is denoted by $\sigma = [v_0, \ldots v_q]$. The set of q-simplices of $\mathcal{K}$ is denoted by $\mathcal{K}_q$. A *face* of a q-simplex $\sigma \in \mathcal{K}$ is a simplex τ whose vertices constitute a subset of $\{v_0, v_1, \ldots, v_q\}$. If $\dim \tau = q - 1$, it is called a *facet* of σ. In this case, σ is called a *cofacet* of τ, and we write $\tau < \sigma$.

A *partial matching* (A, B, C, m) on $\mathcal{K}$ is a partition of $\mathcal{K}$ into three sets A, B, C together with a bijective map $m : A \to B$, also called *discrete vector field*, such that, for each $\tau \in A$, $m(\tau)$ is a cofacet of τ. The intuition behind is that projection from τ to the complementing part of the boundary of $m(\tau)$ induces a homotopy equivalence between $\mathcal{K}$ and a smaller complex. An *m-path* is a sequence $\sigma_0, \tau_0, \sigma_1, \tau_1, \ldots, \sigma_p, \tau_p, \sigma_{p+1}$ such that, for each $i = 0, \ldots, p$, $\sigma_{i+1} \neq \sigma_i$, $\tau_i = m(\sigma_i)$, and τ_i is a cofacet of σ_{i+1}.

A partial matching (A, B, C, m) on $\mathcal{K}$ is called *acyclic* if there does not exist a closed m-path, that is an m-path such that, $\sigma_{p+1} = \sigma_0$.

The main goal of this paper is to produce an acyclic partial matching which preserves the filtration of a simplicial complex $\mathcal{K}$ by sublevel sets of a vector-valued function $f : \mathcal{K}_0 \to \mathbb{R}^k$ given on the set of vertices of $\mathcal{K}$. We assume that $f : \mathcal{K}_0 \to \mathbb{R}^k$ is a function which is *component-wise injective*, that is, whose components f_i are injective. This assumption is used in [3] for proving correctness of the algorithm. Given any function $\tilde{f} : \mathcal{K}_0 \to \mathbb{R}^k$, we can obtain a component-wise injective function f which is arbitrarily close to $\tilde{f}$ via the following procedure. Let n denote the cardinality of $\mathcal{K}_0$. For $i = 1, \ldots, k$, let us set $\eta_i = \min\{|\tilde{f}_i(v) - \tilde{f}_i(w)| : v, w \in \mathcal{K}_0 \wedge \tilde{f}_i(v) \neq \tilde{f}_i(w)\}$. For each i with $1 \leq i \leq k$, we can assume that the n vertices in $\mathcal{K}_0$ are indexed by an integer index j, with $1 \leq j \leq n$, increasing with $\tilde{f}_i$. Thus, the function $f_i : \mathcal{K}_0 \to \mathbb{R}$ can be defined by setting $f_i(v_j) = \tilde{f}_i(v_j) + j\eta_i/n^s$, with $s \geq 1$ (the larger s, the closer f to $\tilde{f}$). Finally, it is sufficient to set $f = (f_1, f_2, \ldots, f_k)$. We extend f to a function $f : \mathcal{K} \to \mathbb{R}^k$ as follows:

$$f(\sigma) = (f_1(\sigma), \ldots, f_k(\sigma)) \qquad \text{with} \qquad f_i(\sigma) = \max_{v \in \mathcal{K}_0(\sigma)} f_i(v). \qquad (1)$$

Any function $f : \mathcal{K} \to \mathbb{R}^k$ that is an extension of a component-wise injective function $f : \mathcal{K}_0 \to \mathbb{R}^k$ defined on the vertices of the complex $\mathcal{K}$ in such a way that f satisfies Eq. (1) will be called *admissible*. In $\mathbb{R}^k$ we consider the following partial order. Given two values $a = (a_i), b = (b_i) \in \mathbb{R}^k$ we set $a \preceq b$ if and only if $a_i \leq b_i$ for every i with $1 \leq i \leq k$. Moreover we write $a \npreceq b$ whenever $a \preceq b$ and $a \neq b$. The *sublevel set filtration* of $\mathcal{K}$ induced by an admissible function f is the family $\{\mathcal{K}^a\}_{a \in \mathbb{R}^k}$ of subsets of $\mathcal{K}$ defined as follows:

$$\mathcal{K}^a = \{\sigma = [v_0, v_1, \ldots, v_q] \in \mathcal{K} \mid f(v_i) \preceq a, \ i = 0, \ldots, q\}.$$

It is clear that, for any parameter value $a \in \mathbb{R}^k$ and any simplex $\sigma \in \mathcal{K}^a$, all faces of σ are also in $\mathcal{K}^a$. Thus $\mathcal{K}^a$ is a simplicial subcomplex of $\mathcal{K}$ for each a. The changes of topology of $\mathcal{K}^a$ as we change the multiparameter a permit recognizing some features of the shape of $|\mathcal{K}|$ if f is appropriately chosen. For this reason, the function f is called in the literature a *measuring function* or, more specifically, a *multidimensional measuring function* [4]. The *lower star* of a simplex is the set

$$L(\sigma) = \{\alpha \in \mathcal{K} \mid \sigma \subseteq \alpha \quad \text{and} \quad f(\alpha) \preceq f(\sigma)\},$$

and the *reduced lower stars* is the set $L_*(\sigma) = L(\sigma) \setminus \{\sigma\}$.

2.1   Indexing Map

An *indexing map* on the simplices of the complex $\mathcal{K}$ of cardinality N, compatible with an admissible function f, is a bijective map $I : \mathcal{K} \to \{1, 2, \ldots, N\}$ such that, for each $\sigma, \tau \in \mathcal{K}$ with $\sigma \neq \tau$, if $\sigma \subseteq \tau$ or $f(\sigma) \not\supseteq f(\tau)$ then $I(\sigma) < I(\tau)$.

To build an indexing map I on the simplices of the complex $\mathcal{K}$, we will revisit the algorithm introduced in [2] that uses the topological sorting of a Directed Acyclic Graph (DAG) to build an indexing for vertices of a complex that is compatible with the ordering of values of a given function defined on the vertices. We will extend the algorithm to build an indexing for all cells of a complex that is compatible with both the ordering of values of a given admissible function defined on the cells and the ordering of the dimensions of the cells. We recall that a topological sorting of a directed graph is a linear ordering of its nodes such that for every directed edge (u, v) from node u to node v, u precedes v in the ordering. This ordering is possible if and only if the graph has no directed cycles, that is, if it is a DAG. A simple well known algorithm (see [17]) for this task consists of successively finding nodes of the DAG that have no incoming edges and placing them in a list for the final sorting. Note that at least one such node must exist in a DAG, otherwise the graph must have at least one directed cycle.

Algorithm 1. Topological sorting

1: **Input:** A DAG whose list of nodes with no incoming edges is I
2: **Output:** The list L containing the sorted nodes
3: **while** there are nodes remaining in I **do**
4:     remove a node u from I
5:     add u to L
6:     **for** each node v with an edge e from u to v **do**
7:         remove edge e from the DAG
8:         **if** v has no other incoming edges **then**
9:             insert v into I
10:        **end if**
11:    **end for**
12: **end while**

When the graph is a DAG, there exists at least one solution for the sorting problem, which is not necessarily unique. We can easily see that each node and each edge of the DAG is visited once by the algorithm, therefore its running time is linear in the number of nodes plus the number of edges in the DAG. The following lemma whose proof is based on Algorithm 1 is proved in [3].

Lemma 1. *Let* $f : \mathcal{K} \to \mathbb{R}^k$ *be an admissible function. There exists an injective function* $I : \mathcal{K} \to \mathbb{N}$ *such that, for each* $\sigma, \tau \in \mathcal{K}$ *with* $\sigma \neq \tau$, *if* $\sigma \subseteq \tau$ *or* $f(\sigma) \not\supseteq f(\tau)$ *then* $I(\sigma) < I(\tau)$.

3   Matching Algorithm

The main contribution of this paper is the Matching Algorithm 2. It uses as input a finite simplicial complex $\mathcal{K}$ of cardinality N, an admissible function $f : \mathcal{K} \to \mathbb{R}^k$ built from a component-wise injective function $f : \mathcal{K}_0 \to \mathbb{R}^k$ using the extension formula given in Eq. (1), and an indexing map I compatible with f. It can be precomputed using the topological sorting Algorithm 1. Given a simplex σ, we use unclass_facets$_\sigma(\alpha)$ to denote the set of facets of a simplex α that are in $L(\sigma)$ and have not been classified yet, that is, not inserted in either A, B, or C, and num_unclass_facets$_\sigma(\alpha)$ to denote the cardinality of unclass_facets$_\sigma(\alpha)$. We initialize classified$(\sigma) = $ **false** for every $\sigma \in \mathcal{K}$. We use priority queues PQzero and PQone which store candidates for pairings with zero and one unclassified facets respectively in the order given by I. We initialize both as empty sets. The algorithm processes cells in the increasing order of their indexes. Each cell σ can be set to the states of classified$(\sigma) = $ **true** or classified$(\sigma) = $ **false** so that if it is processed as part of a lower star of another cell it is not processed again by the algorithm. The algorithm makes use of extra routines to calculate the cells in the lower star $L(\sigma)$ and the set of unclassified facets unclass_facets$_\sigma(\alpha)$ of α in $L_*(\sigma)$ for each cell $\sigma \in \mathcal{K}$ and each cell $\alpha \in L_*(\sigma)$.

The goal of the process is to build a partition of $\mathcal{K}$ into three lists A, B, and C where C is the list of critical cells and in which each cell in A is paired in a one-to-one manner with a cell in B which defines a bijective map $m : A \to B$. When a cell σ is considered, each cell in its lower star $L(\sigma)$ is processed exactly once as shown in [3]. The cell σ is inserted into the list of critical cells C if $L_*(\sigma) = \emptyset$. Otherwise, σ is paired with the cofacet $\delta \in L_*(\sigma)$ that has minimal index value $I(\delta)$. The algorithm makes additional pairings which can be interpreted topologically as the process of constructing $L_*(\sigma)$ with simple homotopy expansions or the process of reducing $L_*(\sigma)$ with simple homotopy contractions. When no pairing is possible a cell is classified as critical and the process is continued from that cell. A cell α is candidate for a pairing when unclass_facets$_\sigma(\alpha)$ contains exactly one element λ that belongs to PQzero. For this purpose, the priority queues PQzero and PQone which store cells with zero and one available unclassified faces respectively are created. As long as PQone is not empty, its front is popped and either inserted into PQzero or paired with its single available unclassified face. When PQone becomes empty, the front cell of PQzero is declared as critical and inserted in C.

We illustrate the algorithm by a simple example. We use the simplicial complex S from our first paper [2, Figure 2] to compare the outputs of the previous matching algorithm and the new one. Figure 1(a) displays S and the output of [2, Algorithm 6]. The coordinates of vertices are the values of the function considered in [2]. Since that function is not component-wise injective, we denote it by $\tilde{f}$ and we start from constructing a component-wise injective approximation f discussed at the beginning of Sect. 2. If we interpret the passage from $\tilde{f}$ to f as a displacement of the coordinates of vertices, the new complex $\mathcal{K}$ is illustrated by Fig. 1(b). The partial order relation is preserved when passing from $\tilde{f}$ to f, and the indexing of vertices in [2, Figure 2] may be kept for f. Hence, it is easy to see that [2, Algorithm 6] applied to $\mathcal{K}$ gives the same result as

---

Algorithm 2. Matching

1: **Input:** A finite simplicial complex $\mathcal{K}$ with an admissible function $f : \mathcal{K} \to \mathbb{R}^k$ and an indexing map $I : \mathcal{K} \to \{1, 2, \dots, N\}$ on its simplices compatible with f.

2: **Output:** Three lists A, B, C of simplices of $\mathcal{K}$, and a function m : A $\to$ B.

3: **for** $i = 1$ to N **do**

4:     $\sigma := I^{-1}(i)$

5:     **if** classified(σ)=false **then**

6:         **if** $L_*(\sigma)$ contains no cells **then**

7:             add σ to C, classified(σ)=true

8:         **else**

9:             $\delta :=$ the cofacet in $L_*(\sigma)$ of minimal index $I(\delta)$

10:            add σ to A and δ to B and define m(σ) = δ, classified(σ)=true, classified(δ)=true

11:            add all $\alpha \in L_*(\sigma) - \{\delta\}$ with num_unclass_facets$_\sigma(\alpha) = 0$ to PQzero

12:            add all $\alpha \in L_*(\sigma)$ with num_unclass_facets$_\sigma(\alpha) = 1$ and $\alpha > \delta$ to PQone

13:            **while** PQone $\neq \emptyset$ or PQzero $\neq \emptyset$ **do**

14:                **while** PQone $\neq \emptyset$ **do**

15:                    $\alpha :=$ PQone.pop_front

16:                    **if** num_unclass_facets$_\sigma(\alpha) = 0$ **then**

17:                        add α to PQzero

18:                    **else**

19:                        add $\lambda \in$ unclass_facets$_\sigma(\alpha)$ to A, add α to B and define m(λ) = α, classified(α)=true, classified(λ)=true

20:                        remove λ from PQzero

21:                        add all $\beta \in L_*(\sigma)$ with num_unclass_facets$_\sigma(\beta) = 1$ and either $\beta > \alpha$ or $\beta > \lambda$ to PQone

22:                    **end if**

23:                **end while**

24:                **if** PQzero $\neq \emptyset$ **then**

25:                    $\gamma :=$ PQzero.pop_front

26:                    add γ to C, classified(γ)=true

27:                    add all $\tau \in L_*(\sigma)$ with num_unclass_facets$_\sigma(\tau) = 1$ and $\tau > \gamma$ to PQone

28:                **end if**

29:            **end while**

30:         **end if**

31:     **end if**

32: **end for**

---

that displayed in Fig. 1(a). In order to apply our new Algorithm 2, we need to index all 14 simplices of $\mathcal{K}$. For convenience of presentation, we label the vertices w_i, edges e_i, and triangles t_i by the index values $i = 1, 2, \dots, 14$. The result is displayed in Fig. 1(b). The sequence of vertices $(v_0, v_1, v_2, v_3, v_4)$ is replaced by $(w_1, w_2, w_4, w_8, w_{12})$. Here are the main steps of the algorithm:

$i = 1$	$L_*(w_1) = \emptyset$, $w_1 \in \mathbf{C}$
$i = 2$	$L_*(w_2) = \{e_3\}$, $\mathbf{m}(w_2) = e_3$
$i = 3$	e_3 classified
$i = 4$	$L_*(w_4) = \{e_5, e_6, t_7\}$,
	$\mathbf{m}(w_4) = e_5$,
	$e_6 \in$ PQzero, $t_7 \in$ PQone,
	line 15, $\alpha = t_7 \notin$ PQone,
	line 19, $\lambda = e_6$, $\mathbf{m}(e_6) = t_7$,
	$e_6 \notin$ PQzero.
$5, 6, 7$	e_5, e_6, t_7 classified

$i = 8$	$L_*(w_8) = \{e_9, e_{10}, t_{11}\}$,
	$\mathbf{m}(w_8) = e_9$,
	$e_{10} \in$ PQzero, $e_{11} \in$ PQone,
	line 15, $\alpha = t_{11} \notin$ PQone,
	line 19, $\lambda = e_{10}$, $\mathbf{m}(e_{10}) = t_{11}$,
	$e_{10} \notin$ PQzero.
$9, 10, 11$	e_9, e_{10}, t_{11} classified
$i = 12$	$L_*(w_{12}) = \{e_{13}, e_{14}\}$,
	$\mathbf{m}(w_{12}) = e_{13}$,
	$e_{14} \in$ PQzero, PQone $= \emptyset$,
	Line 25, $\gamma = e_{14} \in \mathbf{C}$.
$13, 14$	e_{13}, e_{14} classified

The output is displayed in Fig. 1(b).

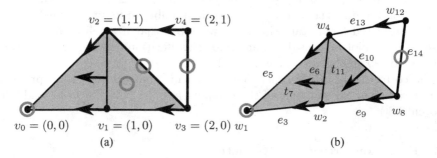

$v_2 = (1, 1)$ $v_4 = (2, 1)$

$v_0 = (0,0)$ $v_1 = (1,0)$ $v_3 = (2,0)\ w_1$

(a)

w_{12}

w_4 e_{13}

e_{10} e_{14}

e_5 $e_6\ \ t_{11}$

t_7

e_3 w_2 e_9 w_8

(b)

Fig. 1. In (a), the complex and output of Algorithm 6 of [2] are displayed. Gray-shaded triangles are those which are present in the simplicial complex. Critical simplices are marked by red circles and the matched simplices are marked by arrows. In (b), the complex is modified so to satisfy the coordinate-wise injectivity assumption. Labeling of all simplices by the indexing function and the output of Algorithm 2 are displayed. (Color figure online)

3.1   Correctness

The full and detailed proof of correctness of Algorithm 2 is given in [3]. We present here the main steps of the proof. Recall that $f = (f_1, \ldots, f_k) : \mathcal{K} \to \mathbb{R}^k$ is an admissible function. We start with some preparatory results that allow to make a proof by induction that every cell of the complex is classified exactly once by the algorithm and the algorithm produces effectively an acyclic partial matching of the complex.

- First, we note that lower stars of distinct simplices are not necessarily disjoint. However when two lower stars meet, they get automatically classified at the same time by the algorithm (see [3, Lemma 3.1]).

- If a cell σ is unclassified when Algorithm 2 reaches line 5, then either $L_*(\sigma)$ is empty for which case σ is classified as critical. Alternatively, $L_*(\sigma)$ contains at least one cofacet of σ that has σ as a unique facet in $L(\sigma)$. Then σ is paired with the cofacet with minimal index and the remainder of its cofacets in $L_*(\sigma)$, have no unclassified faces in $L_*(\sigma)$ and hence they must enter PQzero at line 11 of Algorithm 2. Moreover, if for every cell α with $I(\alpha) < I(\sigma)$, $L(\alpha)$ consists only of classified cells, then all cells in $L(\sigma)$ are also unclassified (see [3, Lemmas 3.2, 3.4]).
- Finally, we prove by induction on the index of the cells that every cell is classified in a unique fashion by the algorithm. The proof is simple when the index takes values 1 and 2 since the cells can be only vertices or edges (see [3, Lemma 3.3]). For the general index, we first prove that for any cell popped from PQone, its unique unclassified facet belongs to PQzero and the two cells get paired at line 19 of the algorithm (see [3, Lemma 3.5]). Afterwards, we state that each cell that is still unclassified after line 10 of the algorithm ultimately enters PQone or PQzero and gets classified. This requires an argument based on the dimension of the cell and the number of its unclassified faces in the considered lower star. Moreover, we prove that if a cell is already classified, it cannot be considered again for classification or enter the structures PQone or PQzero (see [3, Lemma 3.6]).
- The correctness proof is concluded by proving that the algorithm produces an acyclic partial matching of the complex (see [3, Proposition 3.7 and Theorem 3.8]).

3.2 Filtration Preserving Reductions

Lefchetz complexes introduced by Lefschetz in [12] are developed further in [14] under the name S-complex. In our context, these complexes are produced by applying the reduction method [10,13,14] to an initial simplicial complex $\mathcal{K}$, with the use of the matchings produced by our main Algorithm 2. Both concepts of partial matchings and *sublevel set filtration* of $\mathcal{K}$ induced by $f : \mathcal{K} \to \mathbb{R}^k$ introduced in Sect. 2 naturally extend to Lefschetz complexes as proved in [2]. Persistence is based on analyzing the homological changes occurring along the filtration as the multiparameter $a \in \mathbb{R}^k$ varies. This analysis is carried out by considering, for $a \preceq b$, the homomorphism $H_*(j^{(a,b)}) : H_*(\mathsf{S}^a) \to H_*(\mathsf{S}^b)$ induced by the inclusion map $j^{(a,b)} : \mathsf{S}^a \hookrightarrow \mathsf{S}^b$. The image of the map $H_q(j^{(a,b)})$ is known as the *q-th multidimensional persistent homology group* of the filtration at (a, b) and we denote it by $H_q^{a,b}(\mathsf{S})$. It contains the homology classes of order q born not later than a and still alive at b.

   If we assume that $(\mathsf{A}, \mathsf{B}, \mathsf{C}, \mathsf{m})$ is an acyclic matching on a filtered Lefschetz complex S obtained from the original simplicial complex $\mathcal{K}$ by reduction, the following result holds which asserts that the multidimensional persistent homology of the reduced complex is the same as of the initial complex (see [2]).

Corollary 2. *For every* $a \preceq b \in \mathbb{R}^k$, $H_*^{a,b}(\mathcal{C}) \cong H_*^{a,b}(\mathcal{K})$.

3.3   Complexity Analysis

Given a simplex $\sigma \in \mathcal{K}$, the coboundary cells of σ are given by $\mathbf{cb}(\sigma) := \{\tau \in \mathcal{K} \mid \sigma$ is a face of $\tau\}$. It is immediate from the definitions that $L_*(\sigma) \subset \mathbf{cb}(\sigma)$. We define the coboundary mass γ of $\mathcal{K}$ as $\gamma = \max_{\sigma \in \mathcal{K}}$ card $\mathbf{cb}(\sigma)$, where card denotes cardinality. While γ is trivially bounded by N, the number of cells in $\mathcal{K}$, this upper bound is a gross estimate of γ for many complexes of manifolds and approximating surface boundaries of objects. For the simplicial complex $\mathcal{K}$, we assume that the boundary and coboundary cells of each simplex are computed offline and stored in such a way that access to every cell is done in constant time. Given an admissible function $f : \mathcal{K} \to \mathbb{R}^k$, the values by f of simplices $\sigma \in \mathcal{K}$ are stored in the structure that stores the complex $\mathcal{K}$ in such a way that they are accessed in constant time. We assume that adding cells to the lists A, B, and C is done in constant time. Algorithm 2 processes every cell σ of the simplicial complex $\mathcal{K}$ and checks whether it is classified or not. In the latter case, the algorithm requires a function that returns the cells in the reduced lower star $L_*(\sigma)$ which is read directly from the structure storing the complex. In the best case, $L_*(\sigma)$ is empty and the cell is declared critical. Since $L_*(\sigma) \subset \mathbf{cb}(\sigma)$, it follows that card $L_*(\sigma) \leq \gamma$. From Algorithm 2, we can see that every cell in $L_*(\sigma)$ enters at most once in PQzero and PQone. It follows that the while loops in the algorithm are executed all together in at most 2γ steps. We may consider the operations such as finding the number of unclassified faces of a cell to have constant time except for the priority queue operations which are logarithmic in the size of the priority queue when implemented using heaps. Since the sizes of PQzero and PQone are clearly bounded by γ, it follows that $L_*(\sigma)$ is processed in at most $O(\gamma \log \gamma)$ steps. Therefore processing the whole complex incurs a worst case cost of $O(N \cdot \gamma \log \gamma)$.

4   Experimental Results

We have successfully applied the algorithms from Sect. 3 to different sets of triangle meshes. In each case the input data is a 2-dimensional simplicial complex $\mathcal{K}$ and a function f defined on the vertices of $\mathcal{K}$ with values in $\mathbb{R}^2$. The first step is to slightly perturb f in order to achieve injectivity on each component as described in Sect. 2. The second step is to construct an index function defined on all the simplices of the complex and satisfying the properties of Lemma 1. Then we build the acyclic matching m and the partition (A, B, C) in the simplices of the complex using Algorithm 2. In particular, the number of simplices in C out of the total number of simplices of $\mathcal{K}$ is relevant, because it determines the amount of reduction obtained by our algorithm to speed up the computation of multidimensional persistent homology.

4.1   Examples on Synthetic Data

We consider two well known 2-dimensional manifolds — the sphere and the torus. For the sphere, we consider its triangulations of five different sizes and we

Table 1. Reduction performance on different triangulations of a sphere and torus.

	sphe_1	sphe_2	sphe_3	sphe_4	sphe_5	tor_96	tor_4608	tor_7200
#$\mathcal{K}$	38	242	962	1538	2882	96	4608	7200
#C	4	20	98	178	278	8	128	156
%	10.53	8.26	10.19	11.57	9.65	8.33	2.78	2.17

take $f(x, y, z) = (x, y)$. The triangulated sphere with the maximum number of simplices is shown in Fig. 2(left), where cells found critical by the algorithm are colored. The comparison with other triangulations of the sphere is shown in the Table 1: the first row shows the number of simplices in each considered mesh $\mathcal{K}$; the middle row shows the number of critical cells obtained by using our matching algorithm to reduce $\mathcal{K}$; the bottom row shows the ratio between the second and the first lines, expressing them in percentage points. In the case of the torus, we again consider triangulations of different sizes and we take $f(x, y, z) = (x, y)$. The numerical results are shown in the same table and also displayed in Fig. 2(right).

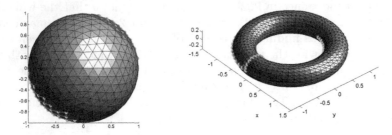

Fig. 2. A triangulation a sphere with 2882 simplices and one of the torus with 7200 simplices, with respect to a component-wise injective perturbation of the function defined of its vertices by $f(x, y, z) = (x, y)$. Critical vertices are in yellow, critical edges in blue, and critical triangles in red. (Color figure online)

In conclusion, our experiments on synthetic data confirm that the current simplex-based matching algorithm scales well with the size of the complex.

4.2 Examples on Real Data

We consider four triangle meshes (available at [1]). For each mesh the input 2-dimensional measuring function f takes each vertex v of coordinates (x, y, z) to the pair $f(v) = (|x|, |y|)$. In Table 2, the first row shows on the top line the number of vertices in each considered mesh, and in the middle line the reduction ratio achieved by our algorithm on those cells, expressing them in percentage points. Finally, it also displays in the bottom line the analogous ratio achieved by our previous algorithm [2]. The second and the third rows show

Table 2. Percentage of reduction achieved by Algorithm 2 on some natural triangle meshes compared to that of [2].

Dataset	tie	space_shuttle	x_wing	space_station
$\#\mathcal{K}_0$	2014	2376	3099	5749
%	27.5	9.5	19.8	30.8
% [2]	11.3	5.1	5.6	32.7
$\#\mathcal{K}_1$	5944	6330	9190	15949
%	20.1	3.8	13.4	16.0
%[2]	56.2	58.4	39.2	70.0
$\#\mathcal{K}_2$	3827	3952	6076	10237
%	14.1	0.4	9.9	8.0
%[2]	78.7	90.5	56.2	56.2
$\#\mathcal{K}$	11785	12658	18365	31935
%	19.4	3.8	13.3	16.1
% [2]	55.9	58.4	39.2	70.0

similar information for the edges and the faces. Finally, the fourth row show the same information for the total number of cells of each considered mesh $\mathcal{K}$.

Our experiments confirm that the current simplex-based matching algorithms produce a fair rate of reduction for simplices of any dimension also on real data. In particular, it shows a clear improvement with respect to the analogous result presented in [2] and obtained using a vertex-based and recursive matching algorithm.

4.3   Discussion

The experiments on synthetic data confirmed two aspects: (1) The discrete case seems to behave much as the differentiable case for two functions [18] because critical cells are still localized along curves; (2) The number of critical cells scales well with the total number of cells, indicating that we are not detecting too many spurious critical cells.

We should point here a fundamental difference between Morse theory for one function whose critical points are isolated and extensions of Morse theory to vector-valued functions where, even in the generic case, critical points form stratified submanifolds. For example, for two functions on a surface, they form curves. Hence the topological complexity depends not on the number of critical points but on the number of such curves. As a consequence, the finer the triangulation, the finer the discretization of such curves and the larger number of critical cells we get.

On the other hand, experiments on real data show the improvement with respect to our previous algorithm [2] already observed in the toy example of

Fig. 1. We think that the new algorithm performs better because it is simplex-based rather than vertex-based. So, the presence of many non-comparable vertices has a limited impact on it.

5   Conclusion

The point of this paper is the presentation of Algorithm 2 to construct an acyclic partial matching from which a gradient compatible with multiple functions can be obtained. As such, it can be useful for specific purposes such as multidimensional persistence computation, whereas it is not meant to be a competitive algorithm to construct an acyclic partial matching for general purposes.

Some questions remain open. First, since indexing map is not unique, one may ask what is the effect of its choice on the output. We believe that the size of the resulting complex should be independent. This is a subject for future work.

A deeper open problem arises from the fact that the optimality of reductions is not yet well defined in the multidimensional setting, although the improvement is observed in practice. As commented earlier, even in the classical smooth case, the singularities of vector-valued functions on manifolds are not isolated. An appropriate application-driven extension of the Morse theory to multidimensional functions is not much investigated yet. Some related work is that of [15] on Jacobi sets and of [7] on preimages of maps between manifolds. However there are essential differences between those concepts and our sublevel sets with respect to the partial order relation.

The experiments presented in this paper show an improvement with respect to the algorithm in [2] that, to the best of our knowledge, is still the only other algorithm available for this task. Such experiments were obtained with a non-optimized implementation. For an optimized implementation of it, we defer the reader to [9], where experiments on larger data sets can be found.

References

1. The GTS Library. http://gts.sourceforge.net/samples.html
2. Allili, M., Kaczynski, T., Landi, C.: Reducing complexes in multidimensional persistent homology theory. J. Symbolic Comput. **78**, 61–75 (2017)
3. Allili, M., Kaczynski, T., Landi, C., Masoni, F.: A new matching algorithm for multidimensional persistence. arXiv:1511.05427v3
4. Biasotti, S., Cerri, A., Frosini, P., Giorgi, D., Landi, C.: Multidimensional size functions for shape comparison. J. Math. Imaging Vis. **32**(2), 161–179 (2008)
5. Carlsson, G., Zomorodian, A.: The theory of multidimensional persistence. In: Proceedings of the 23rd Annual Symposium on Computational Geometry (SCG 2007), pp. 184–193, New York, NY, USA. ACM (2007)
6. Dłotko, P., Wagner, H.: Simplification of complexes for persistent homology computations. Homology Homotopy Appl. **16**(1), 49–63 (2014)
7. Edelsbrunner, H., Harer, J.: Jacobi sets of multiple Morse functions. In: Foundations of Computational Mathematics (FoCM 2002), Minneapolis, pp. 37–57. Cambridge Univ. Press, Cambridge (2002)

8. Escolar, E.G., Hiraoka, Y.: Computing persistence modules on commutative ladders of finite type. In: Hong, H., Yap, C. (eds.) ICMS 2014. LNCS, vol. 8592, pp. 144–151. Springer, Heidelberg (2014). doi:10.1007/978-3-662-44199-2_25

9. Iuricich, F., Scaramuccia, S., Landi, C., De Floriani, L.: A discrete morse-based approach to multivariate data analysis. In: SIGGRAPH ASIA 2016 Symposium on Visualization (SA 2016), New York, NY, USA, pp. 5:1–5:8. ACM (2016)

10. Kaczynski, T., Mrozek, M., Slusarek, M.: Homology computation by reduction of chain complexes. Comput. Math. Appl. **35**(4), 59–70 (1998)

11. King, H., Knudson, K., Mramor, N.: Generating discrete Morse functions from point data. Exp. Math. **14**(4), 435–444 (2005)

12. Lefschetz, S.: Algebraic Topology, vol. 27. American Mathematical Society, Providence (1942). Colloquium Publications

13. Mischaikow, K., Nanda, V.: Morse theory for filtrations and efficient computation of persistent homology. Discrete Comput. Geometry **50**(2), 330–353 (2013)

14. Mrozek, M., Batko, B.: Coreduction homology algorithm. Discrete Comput. Geometry **41**, 96–118 (2009)

15. Patel, A.: Reeb spaces and the robustness of preimages. Ph.D. thesis, Duke University (2010)

16. Robins, V., Wood, P.J., Sheppard, A.P.: Theory and algorithms for constructing discrete Morse complexes from grayscale digital images. IEEE Trans. Pattern Anal. Mach. Intell. **33**(8), 1646–1658 (2011)

17. Sedgewick, R., Wayne, K.: Algorithms. Addison-Wesley, Boston (2011)

18. Smale, S.: Global analysis and economics. I. Pareto optimum and a generalization of Morse theory. In: Dynamical systems (Proceedings of Symposium, Univ. Bahia, Salvador, 1971), pp. 531–544. Academic Press, New York (1973)

19. Turner, K., Murkherjee, A., Boyer, D.M.: Persistent homology transform for modeling shapes ans surfaces. Inf. Inference **3**(4), 310–344 (2014)

20. Xia, K., Wei, G.-W.: Multidimensional persistence in biomolecular data. J. Comput. Chem. **36**(20), 1502–1520 (2015)

Digital Primitives Defined by Weighted Focal Set

Eric Andres[1]([✉]), Ranita Biswas[2], and Partha Bhowmick[3]

[1] Université de Poitiers, Laboratoire XLIM, ASALI, UMR CNRS 7252,
Futuroscope, BP 30179, 86962 Chasseneuil, France
`eric.andres@univ-poitiers.fr`
[2] Department of Computer Science and Engineering,
Indian Institute of Technology, Roorkee, India
`biswas.ranita@gmail.com`
[3] Department of Computer Science and Engineering,
Indian Institute of Technology, Kharagpur, India
`bhowmick@gmail.com`

Abstract. This papers introduces a definition of digital primitives based on focal points and weighted distances (with positive weights). The proposed definition is applicable to general dimensions and covers in its gamut various regular curves and surfaces like circles, ellipses, digital spheres and hyperspheres, ellipsoids and k-ellipsoids, Cartesian k-ovals, etc. Several interesting properties are presented for this class of digital primitives such as space partitioning, topological separation, and connectivity properties. To demonstrate further the potential of this new way of defining digital primitives, we propose, as extension, another class of digital conics defined by focus-directrix combination.

Keywords: Digital primitive · Focus · Hypersphere · Ellipse · Ellipsoid · k-ellipse · Cartesian oval · Conic

1 Introduction

In this paper we introduce digital primitives defined by a weighted focal set. Continuous geometric objects defined by foci have been well studied but nothing, to our knowledge, has been proposed so far in digital geometry.

The word *focus* is the Latin word for *fireplace*. This comes from classical experiment which consists in converging the sunlight, in the focal point of a lens, on a piece of paper to ignite it. Focal points play a fundamental role in the geometry of lenses and study of lenses played an important role in the early development of mathematical physics. The historic importance of the research in optics can even be traced in our common language with expressions such as *"staying focused"*.

Classically a foci based continuous geometric object is defined as all the points such that the sum of the distances to the foci is a constant. The distances may have different weights in an even more general definition [9]. In this paper

© Springer International Publishing AG 2017
W.G. Kropatsch et al. (Eds.): DGCI 2017, LNCS 10502, pp. 388–398, 2017.
DOI: 10.1007/978-3-319-66272-5_31

we introduce foci defined digital geometric primitives. The definition we propose covers digital objects with multiple foci, in arbitrary dimension. Our definition includes weighted distances with positive weights. Contrary to the continuous definition, in the definition that we propose, the weighted sum of the distances to the foci is not a constant but lies in an interval. This definition generalizes the Andres digital hyperspheres [1,2] that has only one focal point. We show that with a well defined interval, we can prove some important topological properties for our digital objects such as $(n-1)$-separation and $(n-2)$-connectivity properties. With an appropriate sequence of such intervals, it is easy to see that we can provide a space partition by such foci based digital primitives.

In a second part of the paper, we propose two extensions. Firstly, we propose an immediate extension of the definition that allows to define m-separating digital foci based primitives. As a second extension, we propose a new type of digital conics whose definition is based on a focal point and a directrix. Again, it is possible to show that we have topological separation properties. This illustrates that the exploration on the possibilities provided by this new approach offers many opportunities for further research.

After this introduction and basic notions and notations, we introduce in Sect. 2.1, our definition of foci based digital objects and propose several fundamental properties of such objects. In Sect. 3, we propose two types of extensions: foci and directrix defined digital objects with properties, and boundary foci based digital objects with properties. We conclude and discuss perspectives in Sect. 4.

Basic Notions and Notations

Let $\{e_1, \ldots, e_n\}$ denote the canonical basis of the n-dimensional Euclidean vector space. Let $\mathbb{Z}^n$ be the subset of $\mathbb{R}^n$ that consists of all the integer points. A *digital (resp. Euclidean) point* is an element of $\mathbb{Z}^n$ (resp. $\mathbb{R}^n$). We denote by x_i the i-th coordinate of a point or a vector x, that is its coordinate associated to e_i. We denote by c_i the i-th element of a list or sequence C. A *digital (resp. Euclidean) object* is a set of digital (resp. Euclidean) points. When not otherwise stated, the distance we are considering in this paper is the Euclidean distance $d(\cdot)$ with $d(p, q) = \sqrt{\sum_{i=1}^{n}(p_i - q_i)^2}$ for $p, q \in \mathbb{R}^n$.

For all $k \in \{0, \ldots, n-1\}$, two integer points p and q are said to be *k-adjacent* or *k-neighbors*, if for all $i \in \{1, \ldots, n\}$, $|p_i - q_i| \leq 1$ and $\sum_{j=1}^{n} |p_j - q_j| \leq n-k$. In the 2-dimensional plane, the 0- and 1-neighborhood notations correspond respectively to the classical 8- and 4-neighborhood notations. In the 3-dimensional space, the 0-, 1- and 2-neighborhood notations correspond respectively to the classical 26-, 18- and 6-neighborhood notations [8].

A *k-path* is a sequence of integer points such that every two consecutive points in the sequence are k-adjacent. A digital object C is *k-connected* if there exists a k-path in C between any two points of C. Let us suppose that the complement of a digital object E, $\mathbb{Z}^n \backslash E$ admits a set of k-connected components C, or in other words that there exists no k-path joining integer points of any two connected components of the set C then E is said to be *k-separating*, or *k-tunnel free*, in $\mathbb{Z}^n$.

If there is no path from any two connected components of the set C then E is said to be 0-separating or simply *separating*.

2   Foci Based Digital Primitives

2.1   Definition

The classical continuous primitives that are defined by a set of focal points can be summarized by the following definition.

Definition 1. *A foci based continuous nD primitive is defined as all the continuous points:*

$$\left\{ x \in \mathbb{R}^n : \sum_{i=1}^{k} \alpha_i d\left(x, f_i\right) = r \right\}$$

where f is the list of foci $f = (f_1, \ldots, f_k)$ with $f_i \in \mathbb{R}^n$, $\alpha_i \in \mathbb{R}^{*+}$ the weight of the distance to focus f_i, and $r \in \mathbb{R}$ the generalized radius.

This definition covers hyperspheres (with one only focus traditionally called the center of the hypersphere), ellipsoids and k-ellipsoids (with respectively two and k foci, and all weights equal to unity), Cartesian ovals and k-Cartesian ovals (with respectively two and k focal points and arbitrary non-zero weights).

We are now going to introduce a digital version of this definition with some restrictions. Firstly, let us note that a digital primitive is generally not defined mathematically by an implicit equation $\{p \in \mathbb{Z}^n : g(p) = 0\}$ because there is usually no particular reason for an integer point to lie on the continuous curve $g(x) = 0$. Instead, J.-P. Reveillés [3,7] proposed to define a digital line as the digital points in a band defined by a thickness interval $\{p \in \mathbb{Z}^2 : 0 \le ap_1 + bp_2 + c < \omega\}$. This captures the general idea that digital primitives are based on grid points where neighborhoods are defined by points that are at a certain, non-zero, distance from each other. To define topologically sound objects, this distance between neighboring points has to be part of the digital definition. This idea of defining primitives as points with an interval has been extended by E. Andres to circles and hyperspheres [1,2] with definitions based on annulus in 2D, concentric hyperspheres in nD. We now propose a new extension for foci based primitives.

Definition 2. *A foci based digital nD primitive $\mathcal{F}_k^n(f, \alpha, r)$ is defined as all the integer points verifying*

$$\mathcal{F}_k^n(f, \alpha, r) = \left\{ p \in \mathbb{Z}^n : \left(\sum_{i=1}^{k} \alpha_i\right)\left(r - \frac{1}{2}\right) \le \sum_{i=1}^{k} \alpha_i d\left(p, f_i\right) < \left(\sum_{i=1}^{k} \alpha_i\right)\left(r + \frac{1}{2}\right) \right\}$$

where, f is the list of foci $f = (f_1, \ldots, f_k)$ with $f_i \in \mathbb{R}^n$, $\alpha_i \in \mathbb{R}^{*+}$ the weight of the distance to focus f_i, and $r \in \mathbb{R}$ the generalized radius.

Let us note that Definition 2 is slightly less general than Definition 1, since we are considering only strictly positive weights for the digital primitives. This restriction comes from the fact that we are seeking digital primitives with some specific topological properties (see Sect. 2.2) that do not stand with this definition if some weights are taken negative. Let us note that if the general radius is too small then the digital object might be empty. For instance, if $\sum_{i=1}^{k} \left(r + \frac{1}{2} \right) < \min_{i=1}^{k} \left(\sum_{j=1}^{k} d\left(f_i, f_j \right) \right)$, then the digital primitive will be empty.

The Andres digital hypersphere [1,2] of center c and radius r has been defined as the set $H(c, r)$:

$$H(c, r) = \left\{ p \in \mathbb{Z}^n : \left(r - \frac{1}{2} \right)^2 \leq \sum_{i=1}^{n} (p_i - c_i)^2 < \left(r + \frac{1}{2} \right)^2 \right\}$$

It is easy to see that $H(c, r) = \mathcal{F}_1^n\left((c), (1), r \right)$ (see Fig. 1). A Andres digital hypersphere is a foci based digital primitive with one focal point, classically called the center of the hypersphere.

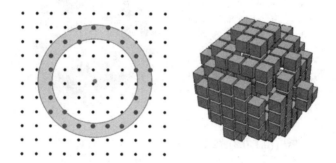

Fig. 1. Andres circle $\mathcal{F}_1^2\left(((0.1, 0.2)), (1), 3.5 \right)$ and Andres sphere $\mathcal{F}_1^3(((0.1, 0.2, 0.3)), (1), 3.5)$

Definition 2 defines a new type of 2D digital ellipse, the foci based digital ellipse $\mathcal{F}_2^2((f_1, f_2), (1, 1), r)$ with two focal points and equal weights of 1 (or any strictly positive equal weights) and a new type of foci based digital ellipsoid $\mathcal{E}_2^n((f_1, f_2), (1, 1), r)$ (see Fig. 2). Contrary to some classical or more recent digital ellipse definitions [4–6], the digital ellipses are not limited to axis aligned ellipses. Another major property of this definition is that it is dimension independent.

With Definition 2, we propose the first definition of a digital k-ellipse and k-ellipsoid: $\mathcal{F}_k^n\left((f_1, \ldots, f_k), (1, \ldots, 1), r \right)$ (see Fig. 3). And lastly, it allows to define foci based digital *Cartesian ovals* with weighted distances $\mathcal{F}_k^2(F, \alpha, r)$ (See Fig. 4). As we can see, this simple definition allows to define a wide range of new types of digital objects.

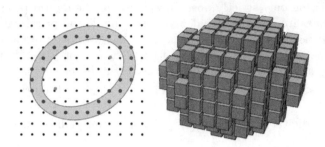

Fig. 2. Foci based digital ellipse $\mathcal{F}_2^2\left(\left(\left(0.1, 0.2\right),\left(5.1, 3.1\right)\right),\left(1, 1\right), 4.5\right)$ and ellipsoid $\mathcal{F}_2^3\left(\left(\left(0.1, 0.2, 0.3\right),\left(5.1, 3.1, 1.1\right)\right),\left(1, 1\right), 4.5\right)$

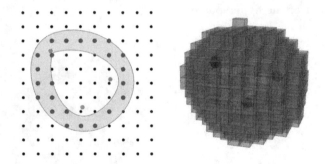

Fig. 3. 3-ellipse $\mathcal{F}_3^2\left(\left(\left(0.1, 0.3\right),\left(2.1, 2.3\right),\left(-2.1, 4.3\right)\right),\left(1, 1, 1\right), 3.5\right)$ and 3-ellipsoid $\mathcal{F}_3^3\left(\left(\left(0.1, 0.3, 0.5\right),\left(2.1, 2.3, -2.5\right),\left(-2.1, 4.3, -5.5\right)\right),\left(1, 1, 1\right), 5.5\right)$

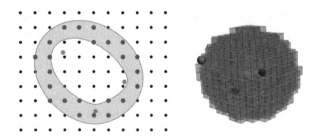

Fig. 4. Cartesian oval $\mathcal{F}_3^2\left(\left(\left(0.1, 0.3\right),\left(2.1, 2.3\right),\left(-2.1, 4.3\right)\right),\left(2, 0.5, 0.5\right), 3.5\right)$ and $\mathcal{F}_3^3\left(\left(\left(0.1, 0.3, 0.5\right),\left(2.1, 2.3, -2.5\right),\left(-2.1, 4.3, -5.5\right)\right),\left(2, 0.5, 0.5\right), 5.5\right)$

2.2   Properties

Let us have a look now at the properties of such digital objects. As we will see in what follows, the foci based digital primitives have interesting structural properties:

Theorem 1. *A foci based digital nD primitive $\mathcal{F}_k^n(f, \alpha, r)$ is $(n-1)$-separating in $\mathbb{Z}^n$.*

Let us, for the sake of simplicity of language, call the Euclidean region(s) defined by $\sum_{i=1}^k \alpha_i d(x, f_i) < \left(\sum_{i=1}^k \alpha_i\right)(r - \frac{1}{2})$ the interior of the digital object and the region(s) defined by $\sum_{i=1}^k \alpha_i d(x, f_i) \geq \left(\sum_{i=1}^k \alpha_i\right)(r + \frac{1}{2})$ the outside. Let us note that nothing in Definition 2 requires the inside or outside regions to be composed of a singular connected component. A good example is given by hyperbolic type curves that divide space into three regions and not two.

Proof. Let us consider two digital points a and b and $E = \mathcal{F}_k^n(f, \alpha, r)$ a foci based digital nD object such that a is inside E and b outside E: $\sum_{i=1}^k \alpha_i d(a, f_i) < \left(\sum_{i=1}^k \alpha_i\right)(r - \frac{1}{2})$ and $\sum_{i=1}^k \alpha_i d(b, f_i) \geq \left(\sum_{i=1}^k \alpha_i\right)(r + \frac{1}{2})$.
Here we suppose that the inside of E contains at least one digital point.
Since $\sum_{i=1}^k \alpha_i d(a, f_i) < \left(\sum_{i=1}^k \alpha_i\right)(r - \frac{1}{2})$, we have $-\sum_{i=1}^k \alpha_i d(a, f_i) \geq -\left(\sum_{i=1}^k \alpha_i\right)(r - \frac{1}{2}) + \epsilon$ with ϵ a strictly positive real value.
This means that $\sum_{i=1}^k \alpha_i (d(b, f_i) - d(a, f_i)) > \left(\sum_{i=1}^k \alpha_i\right)$.
Since d is a distance, it verifies the triangular inequality $d(f_i, a) + d(a, b) \geq d(f_i, b)$. With $\alpha_i > 0$, this means that $\alpha_i d(a, b) \geq \alpha_i (d(f_i, b) - d(f_i, a))$.
Therefore $\left(\sum_{i=1}^k \alpha_i\right) d(a, b) \geq \sum_{i=1}^k (\alpha_i (d(f_i, b) - d(f_i, a))) > \left(\sum_{i=1}^k \alpha_i\right)$, and thus $d(a, b) > 1$. Now, it is easy to see that if there exist a $(n-1)$ path linking a to b without intersecting the object then there has to be a point on the path inside that is $(n-1)$-neighbor to a point outside. The distance between two such points is at least 1 which proves that E is $(n-1)$-separating in $\mathbb{Z}^n$. $\square$

Let us note some important points here. Nowhere in this proof (or more generally in the definition) appears the type of distance. So far in the images we have considered the Euclidean distance but that any distance. It is the triangular inequality property of the distance that is used in the proof.

The next proposition concerns partitioning properties similar to those already seen for the Andres circles and hyperspheres [1,2]. For the sake of simplicity, we are going to consider, in what follows, foci based digital primitives with consecutive integer general radii. In all generality, it can be any set of consecutive sequence of radii as long as the difference between two consecutive radii is one.

Proposition 1. *A set of foci based digital nD objects with consecutive general radii is partitioning space:*
For $r_1, r_2 \in \mathbb{Z}, r_1 \neq r_2$, we have $\mathcal{F}_k^n(f, \alpha, r_1) \cap \mathcal{F}_k^n(f, \alpha, r_2) = \varnothing$,
and for $r \in \mathbb{N}$, $\uplus_{r=-\infty}^{\infty} \mathcal{F}_k^n(f, \alpha, r) = \mathbb{Z}^n$.

Proof. The property is a direct consequence from the definition: the intervals $\left[r_1 - \frac{1}{2}, r_1 + \frac{1}{2}\right[$ and $\left[r_2 - \frac{1}{2}, r_2 + \frac{1}{2}\right[$ are disjoint for $r_1, r_2 \in \mathbb{Z}$, $r_1 \neq r_2$. And the intervals of consecutive integer general radii r partition the natural number set $\biguplus_{r=-\infty}^{\infty} \left[r - \frac{1}{2}, r + \frac{1}{2}\right[= \mathbb{Z}$. □

This property is a direct extension of the space partitioning property already seen for the Andres hyperspheres [2]. It is interesting here to note that this property comes directly in contradiction with another property that is often sought which is minimal thickness. It is easy to see that it is not possible to partition space with, for example digital ellipses, without having local non-uniform thickness. Even for the most regular of all those type of figure, circles, this is not possible. Another point to be made about the local thickness of this digital curves and surfaces comes from the proof of Theorem 1. In the proof, the radius disappears when we consider the distance between a point inside and outside the digital object. What that means is that the proposed bounds $\left(\sum_{i=1}^{k} \alpha_i\right)\left(r \pm \frac{1}{2}\right)$ are the bounds that ensure that all the curves and surfaces that partition space (independently of r) are $(n-1)$-separating. From a general perspective, this can be understood quite easily. For a very large generalized radius r, all the focus points become basically one focus point and the shape of the focus based primitives becomes a hypersphere. It is not very difficult to see that the minimal thickness to ensure the $(n-1)$-separation property is $r + \frac{1}{2} - (r - \frac{1}{2}) = 1$ (one can always divide the formula of Definition 2 by $\sum_{i=1}^{k} \alpha_i$).

Fig. 5. Partitioning ellipses and partitioning ellipsoids.

Figure 5 illustrates the partitioning property of foci based digital primitives. Note that in the 2D case presented in the figure with focal points $(0.1, 0.3)$ and $(5.1, 3.3)$, the ellipses of radii $0, 1$ and 2 are of course empty since the distance between both foci is around 5.83 and with two foci and weights of 1, the definition is given by $2r - 1 \leq d(p, f_1) + d(p, f_2) < 2r + 1$.

3   Extensions

Let us know look at some extensions of proposed Definition 2 of digital foci based primitives. The first extension is an immediate extension of the definition to more general thicknesses which allows more general separation properties. The second extension corresponds to the classical approach where a digital conic is defined by a focal point and a directrix.

3.1 m-Separating Digital Foci Based Primitives

At first, let us expand Definition 2 to include foci based nD primitives that are m-separating, for $0 \leq m < n - 1$ rather than only $(n - 1)$.-separating (Fig. 6).

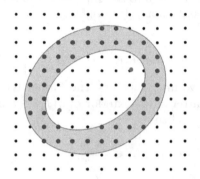

Fig. 6. Foci based digital 0-separating ellipse $\mathcal{F}_2^{2,0}\left(((0.1, 0.2), (5.1, 3.1)), (1, 1), 4.5\right)$

Definition 3. *The m-separating foci based digital nD primitive is defined by:*
$$\mathcal{F}_k^{n,m}(f, \alpha, r) =$$
$$\left\{ p \in \mathbb{Z}^n : \left(\sum_{i=1}^k |\alpha_i| \right) \left(r - \frac{\sqrt{n-m}}{2} \right) \leq \sum_{i=1}^k \alpha_i d(p, f_i) < \left(\sum_{i=1}^k |\alpha_i| \right) \left(r + \frac{\sqrt{n-m}}{2} \right) \right\}$$

Proposition 2. *The digital primitive $\mathcal{F}_k^{n,m}(f, \alpha, r)$ is m-separating in $\mathbb{Z}^n$.*

Proof. The proof for the separation property is similar to the one of Proposition 1 with simply a different constant. It results in $d(a, b) > \sqrt{n - m}$ which proves that the digital object is m-separating. □

3.2   Primitives with One Focal Point and a Directrix

There are many different ways of defining 2D conics. One way is to define a conic with a focal point, a directrix (a straight line) and a constant e called the eccentricity. We are going to propose now a digital definition of conics based on such parameters:

Definition 4. *A digital conic $\mathcal{C}(f, L, e)$ in 2D is given by*
$$\mathcal{C}(f, L, e) = \left\{ p \in \mathbb{Z}^2 : -\frac{e+1}{2} \leq d(p, f) - e \cdot d(p, L) < \frac{e+1}{2} \right\} \tag{1}$$

where $f \in \mathbb{R}^2, L \subset \mathbb{R}^2, e > 0$ denote the respective focal point, directrix, and eccentricity of the corresponding real conic.

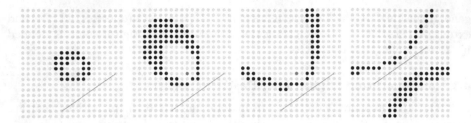

Fig. 7. Conics $\mathcal{C}((-2, 2), L, e)$, for directrix L passing through $(-5, -5)$ and $(5, 2)$ and eccentricity $e = 0.5, 0.7, 1.0, 2.0$ from left to right.

Theorem 2. *A digital conic* $\mathcal{C}(f, L, e)$ *given by Eq. 1 is 1-separating in* $\mathbb{Z}^2$.

Proof. Let a and b be two integer points, the former lying in the interior and the latter in the exterior of $\mathcal{C}(f, L, e)$, as shown in the inset figure. Then,

$$d(a, f) - e \cdot d(a, L) < -\frac{e+1}{2}$$

$$\implies -d(a, f) + e \cdot d(a, L) > \frac{e+1}{2} \tag{2}$$

$$\text{and} \quad d(b, f) - e \cdot d(b, L) \geq \frac{e+1}{2}. \tag{3}$$

Adding Eqs. 2 and 3, we get

$$d(b, f) - d(a, f) + e(d(a, L) - d(b, L)) > e + 1. \tag{4}$$

We have two possible cases:

(i) $d(b, f) - d(a, f) > 1.$ by triangle inequality, $d(a, b) > 1$.
(ii) $d(b, f) - d(a, f) \leq 1$. By Eq. 4, $e(d(a, L) - d(b, L)) > e$, or, $d(a, L) - d(b, L) > 1$, which implies by Pythagorean theorem, $d(a, b) > 1$.

As $d(a, b) > 1$ for either case, $\mathcal{C}(f, L, e)$ is 1-separating.

4   Conclusion and Perspectives

In this paper we are proposing a new class of digital primitives with definitions based on focal points. The definition allows any number of foci and weighted distances. The proposed definition generalizes Andres hyperspheres [1,2]. These primitives are defined in dimension n, have a space partitioning property, and their thickness can be controlled so that they are guaranteed to be m-separating in space. We propose an extension based on a similar principle, where we define a new class of digital conics defined by a directrix (a straight line), a focal point, and a parameter e called eccentricity. What we would like to highlight with this way of defining digital primitives is that it allows power and flexibility to the design of digital primitives.

This work has opened many possibilities for future work and further extensions. Instead of focal points, one can imagine considering distances to objects which could make an interesting link with distance transforms and skeletonization. Can we keep topological m-separation properties? As we can see in Fig. 7 for instance, the foci based primitives do not, in general, have a constant thickness. One could imagine primitives that are defined as the outer or the inner k-connected boundary of the foci based primitives as we have defined them. As mentioned, the proposed formula ensures that, for a set of foci and weights, the digital primitives separate space regardless of the generalized radius. Now, what would we have to change in order to ensure separation for a primitive of a given generalized radius? It would be interesting to compare such primitives to more classically defined digital primitives. For that matter, do the classically defined ellipses, parabola, hyperbola, etc. respect distance sum properties to some focal point?

In the proof of Theorem 1, the only thing that appears is a notion of distance and the minimal distance to ensure a separation property. As one can see in

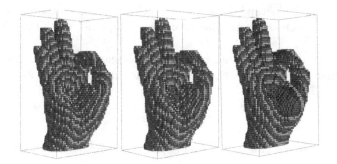

Fig. 8. Foci based digital primitives on arbitrary digital surfaces

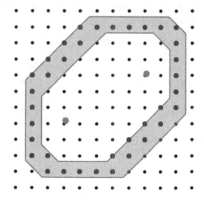

Fig. 9. A digital ellipse based on the Chebychev distance

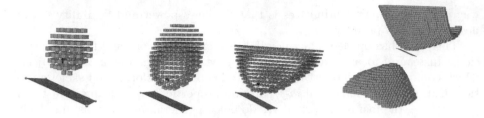

Fig. 10. Conics $\mathcal{C}((0,0,5), L, e)$, for directrix plane L of equation $ax + by + cz + d = 0$ with $(a, b, c, d) = (18.02, -33.10, 92.62, 0)$ and eccentricity $e = 0.5, 0.7, 1, 1.4$ from left to right.

Fig. 8, one can define such focus based digital objects on arbitrary digital surfaces. It would be interesting to extend such notions to graphs and triangular meshes (as long as the triangles in the mesh are somewhat regular). Nothing in the definition limits us to the Euclidean distance. Experimentation with different distances could be very interesting as well (See Fig. 9 as an example).

It is interesting to notice that nothing in Definition 4 or in the proof of Theorem 2 limits our definition to dimension two. As one can see in Fig. 10, with a 3D plane as directrix for instance, that one can create digital ellipsoids, paraboloids, and hyperboloids. One can imagine replacing the plane in 3D by a 3D straight line.

Lastly, general questions can be raised: how can such primitives be recognized? At what more precise conditions are such primitives empty?

References

1. Andres, E.: Discrete circles, rings and spheres. Comput. Graph. **18**(5), 695–706 (1994)
2. Andres, E., Jacob, M.-A.: The discrete analytical hyperspheres. IEEE Trans. Vis. Comput. Graph. **3**(1), 75–86 (1997)
3. Brimkov, V.E., Coeurjolly, D., Klette, R.: Digital planarity - a review. Discrete Appl. Math. **155**(4), 468–495 (2007)
4. Fellner, D.W., Helmberg, C.: Robust rendering of general ellipses and elliptical arcs. ACM Trans. Graph. **12**(3), 251–276 (1993)
5. Mahato, P., Bhowmick, P.: Construction of digital ellipse by recursive integer intervals. In: Normand, N., Guédon, J., Autrusseau, F. (eds.) DGCI 2016. LNCS, vol. 9647, pp. 295–308. Springer, Cham (2016). doi:10.1007/978-3-319-32360-2_23
6. McIlroy, M.D.: Getting raster ellipses right. ACM Trans. Graph. **11**(3), 259–275 (1992)
7. Reveillès, J.-P.: Calcul en Nombres Entiers et Algorithmique. Ph.D. thesis, Université Louis Pasteur, Strasbourg, France (1991)
8. Rosenfeld, A.: Adjacency in digital pictures. Inf. Control **26**(1), 24–33 (1974)
9. Sz-Nagy, G.: Tschirnhaus'sche eiflachen und eikurven. Acta Mathematica Academiae Scientiarum Hungarica **1**(2), 36–45 (1950)

Author Index

Printed in the United States
By Bookmasters